Sugawara Operators for Classical Lie Algebras

Mathematical
Surveys
and
Monographs

Volume 229

Sugawara Operators for Classical Lie Algebras

Alexander Molev

AMERICAN MATHEMATICAL SOCIETY
Providence, Rhode Island

2010 *Mathematics Subject Classification.* Primary 17B35, 17B63, 17B67, 17B69, 16S30.

For additional information and updates on this book, visit
www.ams.org/bookpages/surv-229

Library of Congress Cataloging-in-Publication Data

Names: Molev, Alexander, 1961- author.
Title: Sugawara operators for classical Lie algebras / Alexander Molev.
Description: Providence, Rhode Island : American Mathematical Society, [2018] | Series: Mathematical surveys and monographs ; volume 229 | Includes bibliographical references and index.
Identifiers: LCCN 2017041529 | ISBN 9781470436599 (alk. paper)
Subjects: LCSH: Lie algebras. | Affine algebraic groups. | Kac-Moody algebras. | AMS: Nonassociative rings and algebras – Lie algebras and Lie superalgebras – Universal enveloping (super)algebras. msc | Nonassociative rings and algebras – Lie algebras and Lie superalgebras – Poisson algebras. msc | Nonassociative rings and algebras – Lie algebras and Lie superalgebras – Kac-Moody (super)algebras; extended affine Lie algebras; toroidal Lie algebras. msc | Nonassociative rings and algebras – Lie algebras and Lie superalgebras – Vertex operators; vertex operator algebras and related structures. msc | Associative rings and algebras – Rings and algebras arising under various constructions – Universal enveloping algebras of Lie algebras. msc
Classification: LCC QA252.3 .M6495 2018 | DDC 512/.482–dc23
LC record available at https://lccn.loc.gov/2017041529

∞ The paper used in this book is acid-free and falls within the guidelines
established to ensure permanence and durability.
Visit the AMS home page at `http://www.ams.org/`

10 9 8 7 6 5 4 3 2 1   23 22 21 20 19 18

Оруженосцу Кашке

Contents

Preface

In representation theory of Lie algebras, *Casimir operators* are commonly understood as certain expressions constructed from generators of a Lie algebra which commute with its action. Their spectra are useful for understanding the representation. In particular, finite-dimensional irreducible representations of a simple Lie algebra $\mathfrak{g}$ over the field of complex numbers are characterized by the eigenvalues of the Casimir operators. This fact is based on a theorem of Harish-Chandra describing the center $\mathrm{Z}(\mathfrak{g})$ of the associated universal enveloping algebra $\mathrm{U}(\mathfrak{g})$. The center is isomorphic to an algebra of polynomials via the *Harish-Chandra isomorphism*

$$\mathrm{Z}(\mathfrak{g}) \cong \mathbb{C}[L_1, \dots, L_n]. \tag{0.1}$$

Here n is the rank of $\mathfrak{g}$ and $L_1, \dots, L_n$ are polynomial functions in the highest weights of the representations, each L_i is invariant under a certain action of the Weyl group of $\mathfrak{g}$. The isomorphism (0.1) relies on a theorem of Chevalley which can also be recovered as a 'classical limit' of (0.1). Namely, the symmetric algebra $\mathrm{S}(\mathfrak{g})$ is isomorphic to the graded algebra gr $\mathrm{U}(\mathfrak{g})$, and the subalgebra of $\mathfrak{g}$-invariants in $\mathrm{S}(\mathfrak{g})$ is isomorphic to gr $\mathrm{Z}(\mathfrak{g})$. Taking the symbols M_i of the polynomials L_i, we get the *Chevalley isomorphism*

$$\mathrm{S}(\mathfrak{g})^{\mathfrak{g}} \cong \mathbb{C}[M_1, \dots, M_n]. \tag{0.2}$$

The respective degrees $d_1, \dots, d_n$ of the Weyl group invariants $M_1, \dots, M_n$ coincide with the exponents of $\mathfrak{g}$ increased by 1.

A vast amount of literature both in mathematical physics and representation theory has been devoted to understanding the correspondence in (0.1) in terms of concrete generators on both sides, especially for the Lie algebras $\mathfrak{g}$ of classical types A, B, C and D. Various families of generators of the center $\mathrm{Z}(\mathfrak{g})$ were discovered together with their Harish-Chandra images.

The simple Lie algebras $\mathfrak{g}$ can be regarded as a part of the family of *Kac–Moody algebras* parameterized by generalized Cartan matrices. Of particular importance is the class of *affine Kac–Moody algebras* which admits a simple presentation. The (untwisted) affine Kac–Moody algebra $\widehat{\mathfrak{g}}$ is the central extension $\mathfrak{g}[t, t^{-1}] \oplus \mathbb{C}K$ of the Lie algebra of Laurent polynomials with coefficients in $\mathfrak{g}$. Basic results of representation theory of these Lie algebras together with applications to conformal field theory, modular forms and soliton equations can be found in the book by V. Kac [**86**]. Motivated by the significance of the Lie algebras $\widehat{\mathfrak{g}}$, one comes to wonder what the center of $\mathrm{U}(\widehat{\mathfrak{g}})$ looks like. However, this straight question turns out to be too naive to have a meaningful answer. First of all, the enveloping algebra is 'too small' to contain central elements beyond polynomials in K. The canonical quadratic Casimir element is already a formal series of elements of the algebra $\mathrm{U}(\widehat{\mathfrak{g}})$, so it is necessary to consider its completion. As a natural choice, one requires that the action of elements for such a completion is well-defined on certain *smooth modules* over $\widehat{\mathfrak{g}}$. Secondly, the central element K must be given a unique constant value known as the *critical level.* With a standard choice of the invariant bilinear form on $\mathfrak{g}$, this value is the negative of the dual Coxeter number, $K = -h^\vee$. The suitably completed universal enveloping algebra $\widetilde{\mathrm{U}}_{-h^\vee}(\widehat{\mathfrak{g}})$ at the critical level does contain a large center $\mathrm{Z}(\widehat{\mathfrak{g}})$, and the qualified question has a remarkably comprehensive answer which is explained in detail in the book by E. Frenkel [**46**]. Namely, similar to (0.1), the center $\mathrm{Z}(\widehat{\mathfrak{g}})$ is a completion of the algebra of polynomials

$$\mathbb{C}\big[S_{1\,[r]},\dots,S_{n\,[r]}\mid r\in\mathbb{Z}\big]$$

in infinitely many variables. Moreover, the elements $S_{i\,[r]}$ which are known as *Sugawara operators*, can be produced from a family of generators $S_1,\dots,S_n$ of a commutative differential algebra $\mathfrak{z}(\widehat{\mathfrak{g}})$ by employing instruments of the *vertex algebra theory*: the vacuum module at the critical level over $\widehat{\mathfrak{g}}$ is equipped with a vertex algebra structure, and $\mathfrak{z}(\widehat{\mathfrak{g}})$ is the *center* of this vertex algebra.

Thus the key to understanding the center $\mathrm{Z}(\widehat{\mathfrak{g}})$ lies within the smaller object $\mathfrak{z}(\widehat{\mathfrak{g}})$. Its structure was described by a theorem of B. Feigin and E. Frenkel [**39**] and hence is known as the *Feigin–Frenkel center.* The theorem states that $\mathfrak{z}(\widehat{\mathfrak{g}})$ is an algebra of polynomials

$$\mathfrak{z}(\widehat{\mathfrak{g}})=\mathbb{C}\big[T^rS_1,\dots,T^rS_n\mid r=0,1,\dots\big],$$

where T is a derivation defined as the *translation operator* of the vertex algebra. For type A this theorem can be derived from a previous work of R. Goodman and N. Wallach [**58**], and for types A, B, C from an independent work of T. Hayashi [**65**]. Both papers were concerned with a derivation of the character formula for the irreducible quotient $L(\lambda)$ of the Verma module $M(\lambda)$ at the critical level over $\widehat{\mathfrak{g}}$. The Sugawara operators form a commuting family of $\widehat{\mathfrak{g}}$-endomorphisms of $M(\lambda)$ which leads to a computation of the character and thus proves the Kac–Kazhdan conjecture [**89**]. Our choice for the title of the book was motivated by the terminology used in both pioneering papers [**58**] and [**65**], although the term *Segal–Sugawara operators* is also common in the literature. The origins of the terminology go back to the paper by H. Sugawara [**144**] and an unpublished work of Graeme Segal; see e.g. I. Frenkel [**52**]. We chose to reserve the longer name, *Segal–Sugawara vectors*, for elements of $\mathfrak{z}(\widehat{\mathfrak{g}})$ to make a clearer distinction between the vectors and operators.

More recently, new families of Segal–Sugawara vectors were constructed by A. Chervov and D. Talalaev [**24**] for type A, by the author [**110**] in types B, C and D, and in joint work with E. Ragoucy and N. Rozhkovskaya [**116**] in type G_2. Furthermore, these constructions lead to a direct proof of the Feigin–Frenkel theorem in those cases relying on an affine analogue of the Chevalley isomorphism (0.2). This analogue provides an isomorphism

$$\mathrm{S}\left(t^{-1}\mathfrak{g}[t^{-1}]\right)^{\mathfrak{g}[t]}\cong\mathbb{C}\big[T^rM_1,\dots,T^rM_n\mid r=0,1,\dots\big],$$

for the 'classical limit' of $\mathfrak{z}(\widehat{\mathfrak{g}})$, and is due to M. Raïs and P. Tauvel [**134**] and to A. Beilinson and V. Drinfeld; see [**46**, Theorem 3.4.2].

Our goal in the book is to review these constructions of Segal–Sugawara vectors and to give an introduction to the subject. We hope that together with the general results explained in the book [**46**], they would bring more content to make the beautiful theory more accessible via concrete examples. The explicit Segal–Sugawara vectors will also be used in the applications of the theory as envisaged by the seminal work of B. Feigin, E. Frenkel and N. Reshetikhin [**40**]. Elements $S \in \mathfrak{z}(\widehat{\mathfrak{g}})$ give rise to Hamiltonians of the Gaudin model describing quantum spin chain. Their eigenvalues on the Bethe vectors can be calculated by using an affine version of the Harish-Chandra isomorphism for the algebra $\mathfrak{z}(\widehat{\mathfrak{g}})$. The role of the invariant polynomials occurring in (0.1) will now be played by elements of the *classical $\mathcal{W}$-algebra* $\mathcal{W}({}^L\mathfrak{g})$ associated with the Langlands dual Lie algebra ${}^L\mathfrak{g}$. In parallel to the finite-dimensional theory, the affine Harish-Chandra isomorphism can be understood via the action of elements of the center $Z(\widehat{\mathfrak{g}})$ in the *Wakimoto modules* over $\widehat{\mathfrak{g}}$: central elements act by scalar multiplication with the scalars interpreted as the Harish-Chandra images.

As another application of the constructions of Segal–Sugawara vectors, an explicit solution of E. Vinberg's quantization problem [**149**] will be given. It is based on the general results of L. Rybnikov [**139**] and B. Feigin, E. Frenkel and V. Toledano Laredo [**42**] which provide algebraically independent families of generators of commutative subalgebras of $U(\mathfrak{g})$ from generators of the algebra $\mathfrak{z}(\widehat{\mathfrak{g}})$.

All constructions of the Segal–Sugawara vectors which we discuss in the book can be explained in a uniform way with the use of the *fusion procedure* allowing one to represent primitive idempotents for the centralizer algebras associated with representations of $\mathfrak{g}$, as products of rational *R-matrices*. This approach is therefore applicable, in principle, to all simple Lie algebras, depending on the availability of such a procedure. Its development for the exceptional types would give a uniform description of the Feigin–Frenkel center.

An R-matrix is a solution of the *Yang–Baxter equation*. Given such a solution, one can define the corresponding algebra by an *RTT-relation*, where the generators of the algebra are combined into a matrix. This general approach originated in the work of L. Faddeev and the St. Petersburg (Leningrad) school on the *quantum inverse scattering method* in the early 1980s. Motivated by this work, V. Drinfeld [**30**] and M. Jimbo [**81**] came to the discovery of *quantum groups*. Deformations of universal enveloping algebras in the class of Hopf algebras form one of the most important families of quantum groups. The presentations of these Hopf algebras involving R-matrices give rise to special algebraic methods often referred to as the *R-matrix techniques*, to investigate their structure and representations; see e.g. [**32**], [**96**], [**136**] and references therein for more details on the origins of the methods. Moreover, these techniques can also be used to study the underlying Lie algebras themselves to bring new insights into their properties. It is these techniques which will underpin our approach. As a starting point, we will consider their applications to the simple Lie algebras $\mathfrak{g}$ of classical types. Then we apply the R-matrix techniques to the corresponding affine Kac–Moody algebras $\widehat{\mathfrak{g}}$ and a class of quantum groups $Y(\mathfrak{g})$ known as *Yangians*. In both cases, the defining relations of the algebras will be written in terms of certain generator matrices which can be understood as 'operators' on the space of tensors $(\mathbb{C}^N)^{\otimes m}$ with coefficients in the respective algebras. Therefore, as essential role will be played by the *Schur–Weyl duality* involving natural actions of the classical Lie algebras on the space $(\mathbb{C}^N)^{\otimes m}$ and the commuting actions of the symmetric group in type A or the Brauer algebra in types B, C and D.

We will begin by reviewing constructions of primitive idempotents for the symmetric group and the Brauer algebra based on the respective fusion procedures which provide multiplicative R-matrix formulas for these idempotents (Chapter 1). We apply them to construct invariants in symmetric algebras $\mathrm{S}(\mathfrak{g})$ in Chapter 2. Then we use the R-matrix techniques to derive some basic algebraic properties of *Manin matrices* (Chapter 3). They will be applied for constructions of Casimir elements for the general linear Lie algebras (Chapter 4). Similar constructions based on symmetrizers and anti-symmetrizers for the Brauer algebra will be used for the orthogonal and symplectic Lie algebras (Chapter 5).

In Chapter 6 we introduce the center of the affine vertex algebra at the critical level associated with the affine Kac–Moody algebra $\widehat{\mathfrak{g}}$. We will produce explicit generators of the center in the classical types in Chapters 7 and 8 and show how this leads to a proof of the Feigin–Frenkel theorem. In Chapter 9 the generators are used to construct commutative subalgebras of the classical universal enveloping algebras which 'quantize' the shift of argument subalgebras of the symmetric algebras.

Our calculation of the Harish-Chandra images of the Segal–Sugawara vectors will be based on explicit formulas for the characters of some finite-dimensional representations of the Yangian $\mathrm{Y}(\mathfrak{g})$. The R-matrix techniques will play a key role in the derivation of the character formulas which we review in Chapters 10 and 11. In Chapter 12 we discuss the classical $\mathcal{W}$-algebras and construct their generators. The images of the Segal–Sugawara vectors with respect to an affine version of the Harish-Chandra isomorphism will be calculated in Chapter 13. This will produce special families of generators of the classical $\mathcal{W}$-algebra $\mathcal{W}({}^L\mathfrak{g})$ associated with the Langlands dual Lie algebra ${}^L\mathfrak{g}$. Applications to the Gaudin model will be discussed in Chapter 14. In the final Chapter 15 we will give a construction of the Wakimoto modules over $\widehat{\mathfrak{g}}$ for all classical types and calculate the eigenvalues of the Sugawara operators in these modules.

Bibliographical notes at the end of each chapter contain some comments on the origins of the results and references.

An initial version of the exposition was based on the lecture courses delivered by the author at the *Second Sino–US Summer School on Representation Theory* at the South China University of Technology in 2011, organized by Loek Helminck and Naihuan Jing, and the *International Workshop on Tropical and Quantum Geometries* at the Research Institute for Mathematical Sciences, Kyoto, in 2012, organized by Anatol Kirillov and Shigefumi Mori. I am grateful to the organizers of both events for the invitation to speak. My warm thanks extend to Alexander Chervov, Vyacheslav Futorny, Alexey Isaev, Evgeny Mukhin and Eric Ragoucy for collaboration on the projects which have formed the backbone of the book.

Alexander Molev

Sydney, July 2017

CHAPTER 1

Idempotents and traces

We begin by reviewing some basic facts on representations of the symmetric group and the Brauer algebra. In addition to standard material which can be found in the books by Goodman and Wallach [**59**], James and Kerber [**79**] and Sagan [**141**], we discuss fusion procedures providing multiplicative formulas for primitive idempotents in both cases. The idempotents associated with one-dimensional representations will play a key role in our constructions of generators of the Feigin–Frenkel center in Chapters 7 and 8. We also introduce trace maps on the Brauer algebra and connect them with the traces of linear operators via natural actions of the symmetric group and the Brauer algebra in tensors.

1.1. Primitive idempotents for the symmetric group

We let $\mathfrak{S}_m$ denote the symmetric group whose elements are permutations of the set $\{1,\dots,m\}$. We will identify the group $\mathfrak{S}_{m-1}$ with the subgroup of $\mathfrak{S}_m$ which consists of the permutations s such that $s(m)=m$. For $1\leqslant a<b\leqslant m$ we often denote by $s_{ab}\in\mathfrak{S}_m$ the transposition $(a\,b)$ of the elements a and b.

A *partition* λ is a weakly decreasing sequence $\lambda=(\lambda_1,\dots,\lambda_n)$ of integers such that $\lambda_1\geqslant\dots\geqslant\lambda_n\geqslant 0$. We will identify a partition λ with its (*Young*) *diagram*, which is a left-justified array of rows of unit boxes such that the top row contains λ_1 boxes, the next row contains λ_2 boxes, etc. The number of nonempty rows in the diagram is called the *length* of λ and denoted by $\ell(\lambda)$. The boxes of a diagram will be identified by their row and column numbers so that the box $\alpha=(i,j)$ is found at the intersection of row i and column j. The *content* of this box is the number $c(\alpha)=j-i$. We set $|\lambda|=\lambda_1+\dots+\lambda_n$. If $|\lambda|=m$, then λ is a *partition of* m, written $\lambda\vdash m$. The following figure illustrates the diagram of the partition $(5,4,4,2)$ of 15, its length is 4:

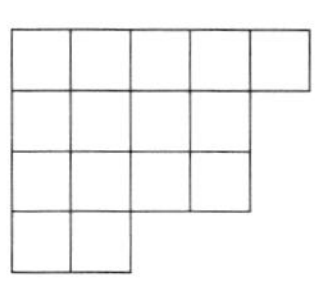

A box of λ is called *removable* if its removal leaves a diagram. Similarly, a box outside λ is *addable* to λ if the union of λ and the box is a diagram. In the example above, the boxes $(1,5)$, $(3,4)$ and $(4,2)$ are removable, while the boxes $(1,6)$, $(2,5)$, $(4,3)$ and $(5,1)$ are addable.

A *tableau* $\mathcal{U}$ *of shape* $\lambda\vdash m$ (or a λ-*tableau* $\mathcal{U}$) is obtained by filling in the boxes of the diagram with the numbers in a given set $\{1,\dots,N\}$. The tableau is called *semistandard* if the entries weakly increase along each row from left to right and strictly increase in each column from top to bottom. We write $\mathrm{sh}(\mathcal{U})=\lambda$ if the shape of $\mathcal{U}$ is λ.

A tableau $\mathcal{U}$ with entries in $\{1,\dots,m\}$ which are filled in the boxes bijectively is called *standard* if its entries strictly increase along the rows and down the columns. The following is a standard tableau of shape $(4,4,1)$:

1	3	4	5
2	6	7	9
8			

The irreducible representations of $\mathfrak{S}_m$ over $\mathbb{C}$ are parameterized by partitions of m. Given $\lambda \vdash m$ denote the corresponding irreducible representation of $\mathfrak{S}_m$ by V_λ. The vector space V_λ is equipped with an $\mathfrak{S}_m$-invariant inner product $(\ ,\ )$. The orthonormal *Young basis* $\{v_{\mathcal{U}}\}$ of V_λ is parameterized by the set of standard λ-tableaux $\mathcal{U}$. The action of the generators $s_a = s_{a\,a+1}$ of $\mathfrak{S}_m$ in the Young basis is described as follows. We denote by $c_b = c_b(\mathcal{U})$ the content of the box occupied by the number b in a standard λ-tableau $\mathcal{U}$. Then for any $a \in \{1,\dots,m-1\}$ we have

$$s_a \cdot v_{\mathcal{U}} = d\, v_{\mathcal{U}} + \sqrt{1-d^2}\, v_{s_a\mathcal{U}}, \qquad d = (c_{a+1}-c_a)^{-1}, \tag{1.1}$$

where the tableau $s_a\mathcal{U}$ is obtained from $\mathcal{U}$ by swapping the entries a and $a+1$, and we assume $v_{s_a\mathcal{U}} = 0$ if the tableau $s_a\mathcal{U}$ is not standard.

The group algebra $\mathbb{C}[\mathfrak{S}_m]$ is isomorphic to the direct sum of matrix algebras

$$\mathbb{C}[\mathfrak{S}_m] \cong \bigoplus_{\lambda\vdash m} \mathrm{Mat}_{f_\lambda}(\mathbb{C}), \tag{1.2}$$

where $f_\lambda = \dim V_\lambda$ is the number of standard tableaux of shape λ. This number can be found by the *hook length formula*

$$f_\lambda = \frac{m!}{h(\lambda)}, \qquad h(\lambda) = \prod_{(i,j)\in\lambda} (\lambda_i + \lambda'_j - i - j + 1), \tag{1.3}$$

where λ'_j denotes the number of boxes in the column j of the diagram λ.

The matrix units $e_{\mathcal{U}\mathcal{U}'} \in \mathrm{Mat}_{f_\lambda}(\mathbb{C})$ are parameterized by pairs of standard λ-tableaux $(\mathcal{U},\mathcal{U}')$. We will identify $\mathbb{C}[\mathfrak{S}_m]$ with the direct sum of matrix algebras by the formulas

$$e_{\mathcal{U}\mathcal{U}'} = \frac{f_\lambda}{m!}\, \phi_{\mathcal{U}\mathcal{U}'}, \tag{1.4}$$

where $\phi_{\mathcal{U}\mathcal{U}'}$ is the matrix element corresponding to the basis vectors $v_{\mathcal{U}}$ and $v_{\mathcal{U}'}$ of the representation V_λ,

$$\phi_{\mathcal{U}\mathcal{U}'} = \sum_{s\in\mathfrak{S}_m} (s\cdot v_{\mathcal{U}}, v_{\mathcal{U}'})\, s^{-1} \in \mathbb{C}[\mathfrak{S}_m]. \tag{1.5}$$

We will write $\phi_{\mathcal{U}} = \phi_{\mathcal{U}\mathcal{U}}$ for brevity. The diagonal matrix units $e_{\mathcal{U}} = e_{\mathcal{U}\mathcal{U}}$ are *primitive idempotents* of $\mathbb{C}[\mathfrak{S}_m]$. They are pairwise orthogonal,

$$e_{\mathcal{U}}\, e_{\mathcal{V}} = 0 \qquad \text{if} \quad \mathcal{U} \neq \mathcal{V},$$

with $e_{\mathcal{U}}^2 = e_{\mathcal{U}}$ and yield a decomposition of the identity element in $\mathbb{C}[\mathfrak{S}_m]$,

$$1 = \sum_{\lambda\vdash m}\ \sum_{\mathrm{sh}(\mathcal{U})=\lambda} e_{\mathcal{U}}.$$

The following simple properties of the matrix units will be needed later on.

LEMMA 1.1 1. *Let $\mathcal{U}$ be a standard tableau of shape λ and let $a \in \{1, \dots, m-1\}$. Then*

$$e_{\mathcal{U}}\,(s_a - d) = e_{\mathcal{U}}\, s_a\, e_{s_a\mathcal{U}} \tag{1.6}$$

and

$$e_{s_a\mathcal{U}}\, s_a\, e_{\mathcal{U}}\, s_a\, e_{s_a\mathcal{U}} = (1 - d^2)\, e_{s_a\mathcal{U}}, \tag{1.7}$$

where d is defined in (1.1) *and we suppose that $e_{s_a\mathcal{U}} = 0$ if the tableau $s_a\mathcal{U}$ is not standard.*

PROOF. We have

$$\phi_{\mathcal{U}}\, s_a = \sum_{s\in\mathfrak{S}_m} (s\cdot v_{\mathcal{U}}, v_{\mathcal{U}})\,(s_a s)^{-1} = \sum_{t\in\mathfrak{S}_m} (s_a t\cdot v_{\mathcal{U}}, v_{\mathcal{U}})\, t^{-1} = \sum_{t\in\mathfrak{S}_m} (t\cdot v_{\mathcal{U}}, s_a\cdot v_{\mathcal{U}})\, t^{-1}.$$

Hence, applying (1.1) we find

$$\phi_{\mathcal{U}}\, s_a = d\,\phi_{\mathcal{U}} + \sqrt{1-d^2}\,\phi_{\mathcal{U}\, s_a\mathcal{U}}$$

and so, by (1.4),

$$e_{\mathcal{U}}\,(s_a - d) = \sqrt{1-d^2}\, e_{\mathcal{U}\, s_a\mathcal{U}}. \tag{1.8}$$

This element remains unchanged when multiplied by the matrix unit $e_{s_a\mathcal{U}}$ from the right. So (1.6) follows since $e_{\mathcal{U}}\, e_{s_a\mathcal{U}} = 0$. Furthermore, assuming that $e_{s_a\mathcal{U}} \neq 0$ and replacing $\mathcal{U}$ by $s_a\mathcal{U}$ in (1.8) we get

$$e_{s_a\mathcal{U}}\,(s_a + d) = \sqrt{1-d^2}\, e_{s_a\mathcal{U}\,\mathcal{U}}.$$

Together with (1.8) this yields

$$e_{s_a\mathcal{U}}\, s_a\, e_{\mathcal{U}}\, s_a\, e_{s_a\mathcal{U}} = e_{s_a\mathcal{U}}\,(s_a + d)\, e_{\mathcal{U}}\,(s_a - d)\, e_{s_a\mathcal{U}} = (1-d^2)\, e_{s_a\mathcal{U}}$$

proving (1.7). □

We will regard the *character* of the representation V_λ as an element of the group algebra

$$\chi_\lambda = \sum_{s\in\mathfrak{S}_m} \chi_\lambda(s)\, s^{-1} \in \mathbb{C}[\mathfrak{S}_m]$$

so that

$$\chi_\lambda = \sum_{\mathrm{sh}(\mathcal{U})=\lambda} \phi_{\mathcal{U}} = h(\lambda) \sum_{\mathrm{sh}(\mathcal{U})=\lambda} e_{\mathcal{U}}.$$

The following identity will also be useful: for any standard λ-tableau $\mathcal{U}$ we have

$$\chi_\lambda = \sum_{s\in\mathfrak{S}_m} s\, e_{\mathcal{U}}\, s^{-1}. \tag{1.9}$$

The primitive idempotents $e_{\mathcal{U}}$ can be expressed explicitly in terms of the *Jucys–Murphy elements* $x_1, \dots, x_m$ of the group algebra $\mathbb{C}[\mathfrak{S}_m]$ defined by

$$x_1 = 0 \qquad \text{and} \qquad x_a = s_{1\,a} + \dots + s_{a-1\,a} \qquad \text{for} \quad a = 2, \dots, m. \tag{1.10}$$

Note that x_m commutes with any element of the subgroup $\mathfrak{S}_{m-1}$. Therefore the Jucys–Murphy elements generate a commutative subalgebra of $\mathbb{C}[\mathfrak{S}_m]$. Furthermore, the vectors of the Young basis are eigenvectors for the action of x_a on V_λ; for any standard λ-tableau $\mathcal{U}$ we have

$$x_a\cdot v_{\mathcal{U}} = c_a(\mathcal{U})\, v_{\mathcal{U}}, \qquad a = 1, \dots, m.$$

These relations imply

$$x_a\, e_{\mathcal{U}} = e_{\mathcal{U}}\, x_a = c_a(\mathcal{U})\, e_{\mathcal{U}}, \qquad a = 1, \dots, m. \tag{1.11}$$

In particular, we have the identity in $\mathbb{C}[\mathfrak{S}_m]$,

$$x_m = \sum_{\lambda \vdash m} \sum_{\mathrm{sh}(\mathcal{U})=\lambda} c_m(\mathcal{U})\, e_{\mathcal{U}}, \tag{1.12}$$

so that under the identification (1.2), x_m can be viewed as a diagonal matrix.

Now let $m \geqslant 2$ and let λ be a partition of m. Fix a standard λ-tableau $\mathcal{U}$ and denote by $\mathcal{V}$ the standard tableau obtained from $\mathcal{U}$ by removing the box α occupied by m. Then the shape of $\mathcal{V}$ is a diagram which we denote by μ. We let c denote the content of the box α. Let u be a complex variable. Due to (1.12), the expression

$$e_{\mathcal{V}}\, \frac{u - c}{u - x_m} \tag{1.13}$$

is a rational function in u with values in $\mathbb{C}[\mathfrak{S}_m]$. Since $e_{\mathcal{U}_0} = 1$ for the (1)-tableau $\mathcal{U}_0$ with the entry 1, the following recurrence relations allow one to write any primitive idempotent $e_{\mathcal{U}}$ in terms of the elements x_a.

PROPOSITION 1.1.2. *We have the relation in* $\mathbb{C}[\mathfrak{S}_m]$,

$$e_{\mathcal{U}} = e_{\mathcal{V}}\, \frac{(x_m - a_1) \dots (x_m - a_l)}{(c - a_1) \dots (c - a_l)}, \tag{1.14}$$

where $a_1, \dots, a_l$ *are the contents of all addable boxes of* μ *except for* α, *while* c *is the content of the latter. Moreover, the rational function* (1.13) *is regular at* $u = c$, *and we have*

$$e_{\mathcal{U}} = e_{\mathcal{V}}\, \frac{u - c}{u - x_m}\Big|_{u=c}. \tag{1.15}$$

PROOF. The branching properties of the Young basis imply that

$$e_{\mathcal{V}} = \sum_{\mathcal{U}'} e_{\mathcal{U}'}, \tag{1.16}$$

summed over the standard tableaux $\mathcal{U}'$ obtained from $\mathcal{V}$ by adding one box with the entry m. The right hand side of (1.14) now reduces to $e_{\mathcal{U}}$ since by (1.11) the product

$$e_{\mathcal{U}'}\, \frac{(x_m - a_1) \dots (x_m - a_l)}{(c - a_1) \dots (c - a_l)}$$

is zero for all $\mathcal{U}' \neq \mathcal{U}$ and it equals 1 for $\mathcal{U}' = \mathcal{U}$. Similarly, by (1.11) and (1.16) we have

$$e_{\mathcal{V}}\, \frac{u - c}{u - x_m} = \sum_{\mathcal{U}'} e_{\mathcal{U}'}\, \frac{u - c}{u - c_m(\mathcal{U}')} = e_{\mathcal{U}} + \sum_{\mathcal{U}' \neq \mathcal{U}} e_{\mathcal{U}'}\, \frac{u - c}{u - c_m(\mathcal{U}')}.$$

Since $c_m(\mathcal{U}') \neq c$ for all standard tableaux $\mathcal{U}'$ distinct from $\mathcal{U}$, the value of this rational function at $u = c$ is $e_{\mathcal{U}}$. □

EXAMPLE 1.1.3. The trivial one-dimensional representation of $\mathfrak{S}_m$ corresponds to the partition (m); its diagram is a row with m boxes. By (1.4) and (1.5), the idempotent $e_{\mathcal{U}}$ associated with the unique standard tableau $\mathcal{U}$ of shape (m) coincides with the *symmetrizer*

$$h^{(m)} = \frac{1}{m!} \sum_{s \in \mathfrak{S}_m} s. \tag{1.17}$$

Proposition 1.1.2 yields the product formula

$$h^{(m)} = \frac{(1+x_2)(1+x_3)\dots(1+x_m)}{m!}. \tag{1.18}$$

□

EXAMPLE 1.1.4. The sign representation of $\mathfrak{S}_m$ corresponds to the partition (1^m); its diagram is a column with m boxes. By (1.4) and (1.5), the idempotent $e_{\mathcal{U}}$ associated with the unique standard tableau $\mathcal{U}$ of shape (1^m) coincides with the *anti-symmetrizer*

$$a^{(m)} = \frac{1}{m!} \sum_{s \in \mathfrak{S}_m} \operatorname{sgn} s \cdot s. \tag{1.19}$$

Proposition 1.1.2 yields the product formula

$$a^{(m)} = \frac{(1-x_2)(1-x_3)\dots(1-x_m)}{m!}. \tag{1.20}$$

□

EXAMPLE 1.1.5. For $m = 2$, the two primitive idempotents are the symmetrizer $h^{(2)}$ and anti-symmetrizer $a^{(2)}$. For $m = 3$, in addition to the symmetrizer $h^{(3)}$ and anti-symmetrizer $a^{(3)}$ there are two more primitive idempotents associated with the standard tableaux

$$\mathcal{U} = \begin{array}{|c|c|}\hline 1 & 2 \\ \hline 3 \\ \cline{1-1}\end{array} \qquad \text{and} \qquad \mathcal{V} = \begin{array}{|c|c|}\hline 1 & 3 \\ \hline 2 \\ \cline{1-1}\end{array}$$

given by

$$e_{\mathcal{U}} = \frac{1}{6}\big(1 + (1\,2)\big)\big(2 - (1\,3) - (2\,3)\big) \quad \text{and} \quad e_{\mathcal{V}} = \frac{1}{6}\big(1 - (1\,2)\big)\big(2 + (1\,3) + (2\,3)\big).$$

□

We record a property of the Jucys–Murphy elements to be used for calculation of traces in Section 1.3.

LEMMA 1.1.6. *We have the identity for rational functions in u with values in the group algebra $\mathbb{C}[\mathfrak{S}_m]$,*

$$\frac{1}{u - x_m} = s_{m-1}\,\frac{1}{u - x_{m-1}}\,s_{m-1} + \frac{1}{u - x_{m-1}}\Big(s_{m-1} + \frac{1}{u - x_m}\Big)\frac{1}{u - x_{m-1}}. \tag{1.21}$$

PROOF. We have the relation $s_{m-1}x_m = x_{m-1}s_{m-1} + 1$ in the group algebra. It implies

$$s_{m-1} + \frac{1}{u - x_m} = (u - x_{m-1})\,s_{m-1}\,\frac{1}{u - x_m}. \tag{1.22}$$

Hence, multiplying from the left by the inverse of $(u - x_{m-1})\,s_{m-1}$ we get

$$\begin{aligned}\frac{1}{u - x_m} &= s_{m-1}\,\frac{1}{u - x_{m-1}}\Big(s_{m-1} + \frac{1}{u - x_m}\Big)\\ &= s_{m-1}\,\frac{1}{u - x_{m-1}}\,s_{m-1} + s_{m-1}\,\frac{1}{(u - x_{m-1})(u - x_m)}.\end{aligned}$$

Therefore, permuting the two factors in the denominator of the last fraction and applying (1.22) once again, we come to (1.21). □

An alternative way to express the primitive idempotents $e_{\mathcal{U}}$ is provided by the *fusion procedure* originated in the work of Jucys [**84**], and its various versions have since been developed by many authors. Take m complex variables $u_1, \dots, u_m$ and consider the rational function with values in $\mathbb{C}[\mathfrak{S}_m]$ defined by

$$\phi(u_1, \dots, u_m) = \prod_{1 \leqslant a < b \leqslant m} \Big(1 - \frac{s_{ab}}{u_a - u_b}\Big), \tag{1.23}$$

where the product is taken in the lexicographical order on the set of pairs (a, b). Suppose that $\lambda \vdash m$ and let $\mathcal{U}$ be a standard λ-tableau. As before, set $c_a = c_a(\mathcal{U})$ for $a = 1, \dots, m$ so that $c_a = j - i$ if a occupies the box (i, j) in $\mathcal{U}$.

PROPOSITION 1.1.7. *The following consecutive evaluations of the rational function $\phi(u_1, \dots, u_m)$ are well-defined and the value coincides with the primitive idempotent $e_{\mathcal{U}}$ multiplied by the product of hook lengths $h(\lambda)$,*

$$\phi(u_1, \dots, u_m)\big|_{u_1 = c_1}\big|_{u_2 = c_2} \cdots \big|_{u_m = c_m} = h(\lambda)\, e_{\mathcal{U}}.$$

PROOF. The group algebra $\mathbb{C}[\mathfrak{S}_m]$ can be regarded as the quotient of the Brauer algebra $\mathcal{B}_m(\omega)$ by the ideal generated by all elements ϵ_{ab}; see Section 1.2. Therefore, the statement is a consequence of Proposition 1.2.3 below. Alternatively, a direct argument is obtained by some obvious simplifications of the proof of that proposition. □

EXAMPLE 1.1.8. The contents of the unique one-row standard tableau with m boxes are found by $c_a = a - 1$ for $a = 1, \dots, m$. Hence the symmetrizer (1.17) is given by the multiplicative formula

$$h^{(m)} = \frac{1}{m!} \prod_{1 \leqslant a < b \leqslant m} \Big(1 + \frac{s_{ab}}{b - a}\Big)$$

with the products taken in the lexicographical order on the set of pairs (a, b). □

EXAMPLE 1.1.9. The contents of the unique one-column standard tableau with m boxes are found by $c_a = -a + 1$ for $a = 1, \dots, m$. Hence the anti-symmetrizer (1.19) is given by the multiplicative formula

$$a^{(m)} = \frac{1}{m!} \prod_{1 \leqslant a < b \leqslant m} \Big(1 - \frac{s_{ab}}{b - a}\Big) \tag{1.24}$$

with the products taken in the lexicographical order on the set of pairs (a, b). □

EXAMPLE 1.1.10. By Proposition 1.1.7, the primitive idempotents $e_{\mathcal{U}}$ and $e_{\mathcal{V}}$ of Example 1.1.5 are also given by the formulas

$$e_{\mathcal{U}} = \frac{1}{3}\big(1 + (1\,2)\big)\big(1 - (1\,3)\big)\Big(1 - \frac{(2\,3)}{2}\Big),$$
$$e_{\mathcal{V}} = \frac{1}{3}\big(1 - (1\,2)\big)\big(1 + (1\,3)\big)\Big(1 + \frac{(2\,3)}{2}\Big).$$

1.2. Primitive idempotents for the Brauer algebra

Let ω be an indeterminate. An *m-diagram* d is a collection of $2m$ dots arranged into two rows with m dots in each row connected by m edges such that any dot belongs to only one edge. The product dd' of two diagrams d and d' is determined by placing d under d' and identifying the vertices of the bottom row of d' with the corresponding vertices in the top row of d. Let s be the number of closed loops obtained in this placement. The product dd' is given by ω^s times the resulting diagram without loops. For example, the product of the 8-diagrams

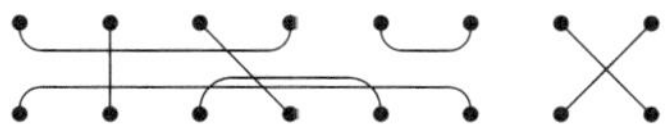 and 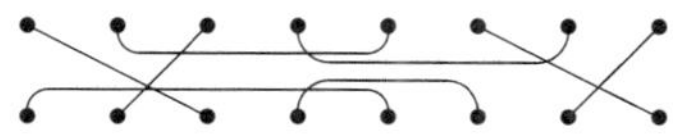

is found by

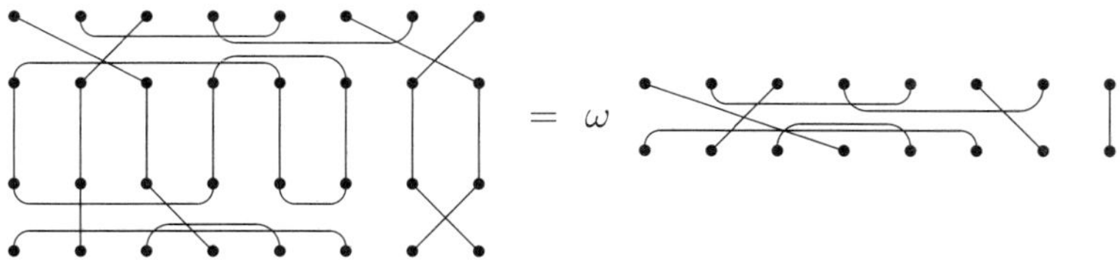

The *Brauer algebra* $\mathcal{B}_m(\omega)$ is defined as the $\mathbb{C}(\omega)$-linear span of the m-diagrams with the multiplication defined above. The dimension of the algebra equals the number of the m-diagrams which is $1 \cdot 3 \cdots (2m-1)$. For $1 \leqslant a < b \leqslant m$ denote by $s_{a\,b}$ and $\epsilon_{a\,b}$ the respective diagrams of the form

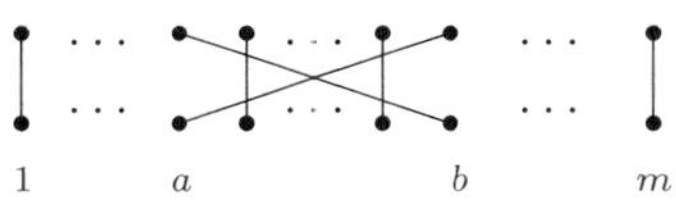 and 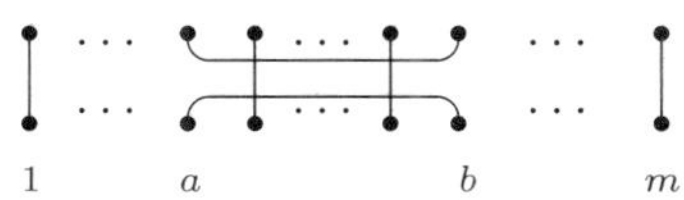

The subalgebra of $\mathcal{B}_m(\omega)$ generated over $\mathbb{C}$ by the elements $s_{a\,b}$ is isomorphic to the group algebra of the symmetric group $\mathbb{C}[\mathfrak{S}_m]$ so that the m-diagram $s_{a\,b}$ is identified with the transposition $(a\,b)$. Sometimes we also use this notation with swapped subscripts, $s_{b\,a} = s_{a\,b}$ and $\epsilon_{b\,a} = \epsilon_{a\,b}$. The Brauer algebra $\mathcal{B}_{m-1}(\omega)$ will be regarded as a natural subalgebra of $\mathcal{B}_m(\omega)$. This subalgebra is spanned by all m-diagrams in which the m-th dots in the top and bottom rows are connected by an edge. The algebra $\mathcal{B}_m(\omega)$ is generated by the elements

$$s_a = s_{a\,a+1}, \qquad \epsilon_a = \epsilon_{a\,a+1}, \qquad a = 1, \dots, m-1, \tag{1.25}$$

subject only to the following set of relations:

$$s_a^2 = 1, \qquad \epsilon_a^2 = \omega\,\epsilon_a, \qquad s_a\epsilon_a = \epsilon_a s_a = \epsilon_a, \qquad a = 1, \dots, m-1,$$

$$s_a s_b = s_b s_a, \qquad \epsilon_a\epsilon_b = \epsilon_b\epsilon_a, \qquad s_a\epsilon_b = \epsilon_b s_a, \qquad |a-b| > 1,$$

$$s_a s_{a+1} s_a = s_{a+1} s_a s_{a+1}, \qquad \epsilon_a\epsilon_{a+1}\epsilon_a = \epsilon_a, \qquad \epsilon_{a+1}\epsilon_a\epsilon_{a+1} = \epsilon_{a+1},$$

$$s_a\epsilon_{a+1}\epsilon_a = s_{a+1}\epsilon_a, \qquad \epsilon_{a+1}\epsilon_a s_{a+1} = \epsilon_{a+1}s_a, \qquad a = 1, \dots, m-2.$$

The *Jucys–Murphy elements* $y_1, \dots, y_m$ for the Brauer algebra $\mathcal{B}_m(\omega)$ are given by the formulas

$$y_b = \frac{\omega - 1}{2} + \sum_{a=1}^{b-1} (s_{a\,b} - \epsilon_{a\,b}), \qquad b = 1, \dots, m. \tag{1.26}$$

It is easily verified that y_m commutes with all elements of the subalgebra $\mathcal{B}_{m-1}(\omega)$. This implies that the elements $y_1, \dots, y_m$ of $\mathcal{B}_m(\omega)$ pairwise commute.

By Wenzl's theorem [**152**, Theorem 3.2], the algebra $\mathcal{B}_m(\omega)$ over $\mathbb{C}(\omega)$ is semisimple. It is isomorphic to a direct sum of matrix algebras with the sizes of matrices given by the dimensions of irreducible representations. Such representations are parameterized by the set of partitions λ of the numbers $m - 2f$ with $f = 0, 1, \dots, \lfloor m/2 \rfloor$. An *updown λ-tableau* is a sequence $\mathcal{U} = (\Lambda_1, \dots, \Lambda_m)$ of m diagrams such that for each $r = 1, \dots, m$ the diagram Λ_r is obtained from Λ_{r-1} by adding or removing one box, where $\Lambda_0 = \varnothing$ is the empty diagram and $\Lambda_m = \lambda$. The irreducible representation of $\mathcal{B}_m(\omega)$ corresponding to λ admits a Young-type basis labeled by all updown λ-tableaux. To each updown tableau $\mathcal{U}$ we associate the sequence of *contents* $(c_1, \dots, c_m)$, $c_r = c_r(\mathcal{U})$, where

$$c_r = \frac{\omega - 1}{2} + j - i \qquad \text{or} \qquad c_r = -\Big(\frac{\omega - 1}{2} + j - i\Big),$$

if Λ_r is obtained by adding the box (i, j) to Λ_{r-1} or by removing this box from Λ_{r-1}, respectively.

As with the isomorphism (1.2), the primitive idempotents $e_{\mathcal{U}} = e^{\lambda}_{\mathcal{U}} \in \mathcal{B}_m(\omega)$ associated with updown tableaux $\mathcal{U}$ correspond to diagonal matrix units (we omit the superscripts indicating the diagrams since they are determined by the updown tableaux). When λ runs over all partitions of $m, m-2, \dots$ and $\mathcal{U}$ runs over all updown λ-tableaux, the elements $e_{\mathcal{U}}$ yield a complete set of pairwise orthogonal primitive idempotents for $\mathcal{B}_m(\omega)$. The following analogues of relations (1.11) hold:

$$y_r\, e_{\mathcal{U}} = e_{\mathcal{U}}\, y_r = c_r(\mathcal{U})\, e_{\mathcal{U}}, \qquad r = 1, \dots, m. \tag{1.27}$$

The primitive idempotents can be found by the following recurrence formulas analogous to (1.14). Given an updown λ-tableau $\mathcal{U} = (\Lambda_1, \dots, \Lambda_m)$, set $\mu = \Lambda_{m-1}$ and consider the updown μ-tableau $\mathcal{V} = (\Lambda_1, \dots, \Lambda_{m-1})$. Let α be the box which is added to or removed from μ to get λ, and let c be the content of α.

PROPOSITION 1.2.1. *We have the relation in $\mathcal{B}_m(\omega)$,*

$$e_{\mathcal{U}} = e_{\mathcal{V}}\, \frac{(y_m - a_1) \dots (y_m - a_k)}{(c - a_1) \dots (c - a_k)}, \tag{1.28}$$

where $a_1, \dots, a_k$ are the contents of all boxes excluding α, which can be removed from or added to μ to get a diagram. Moreover, the following rational function is regular at $u = c$, and for its value we have

$$e_{\mathcal{U}} = e_{\mathcal{V}}\, \frac{u - c}{u - y_m}\Big|_{u=c}. \tag{1.29}$$

PROOF. Both parts follow by the same arguments as for the proof of Proposition 1.1.2 with the use of the branching properties of the irreducible representations of the Brauer algebra. Namely, given an updown tableau $\mathcal{V} = (\Lambda_1, \dots, \Lambda_{m-1})$, we have the relation

$$e_{\mathcal{V}} = \sum_{\mathcal{U}'} e_{\mathcal{U}'}, \tag{1.30}$$

summed over all updown tableaux of the form $\mathcal{U}' = (\Lambda_1, \dots, \Lambda_{m-1}, \Lambda'_m)$. The proof is completed by applying the properties (1.27). □

EXAMPLE 1.2.2. The Brauer algebra $\mathcal{B}_2(\omega)$ has three one-dimensional representations. The corresponding updown tableaux associated with the diagrams (2), (1^2) and $\varnothing$ are

$$\mathcal{U}_1 = (\square,\ \square\square), \qquad \mathcal{U}_2 = \Big(\square,\ \begin{smallmatrix}\square\\ \square\end{smallmatrix}\Big), \qquad \mathcal{U}_3 = (\square,\ \varnothing).$$

The respective contents (c_1, c_2) are given by

$$\Big(\frac{\omega-1}{2}, \frac{\omega+1}{2}\Big), \qquad \Big(\frac{\omega-1}{2}, \frac{\omega-3}{2}\Big), \qquad \Big(\frac{\omega-1}{2}, -\frac{\omega-1}{2}\Big).$$

The corresponding primitive idempotents are

$$e_{\mathcal{U}_1} = \frac{1+s_1}{2} - \frac{\epsilon_1}{\omega}, \qquad e_{\mathcal{U}_2} = \frac{1-s_1}{2}, \qquad e_{\mathcal{U}_3} = \frac{\epsilon_1}{\omega}. \qquad \square$$

The trivial one-dimensional representation of the Brauer algebra $\mathcal{B}_m(\omega)$ is associated with the partition (m). There is a unique updown (m)-tableau which can also be regarded as the standard tableau obtained by writing the numbers $1, \dots, m$ into the boxes of the row-diagram with m boxes from left to right. The corresponding primitive idempotent is the *symmetrizer* $s^{(m)} \in \mathcal{B}_m(\omega)$ which is a unique idempotent determined by the properties

$$s_{cb}\, s^{(m)} = s^{(m)}\, s_{ab} = s^{(m)} \qquad \text{and} \qquad \epsilon_{ab}\, s^{(m)} = s^{(m)}\, \epsilon_{ab} = 0 \tag{1.31}$$

for all $1 \leqslant a < b \leqslant m$. By (1.28), it is given in terms of the Jucys–Murphy elements as

$$s^{(m)} = \frac{1}{m!} \prod_{r=2}^{m} \frac{(y_r - c_1 + 1)(y_r + c_1 + r - 2)}{\omega + 2r - 4}, \qquad c_1 = \frac{\omega - 1}{2}. \tag{1.32}$$

There are two different versions of the fusion procedure for the Brauer algebra $\mathcal{B}_m(\omega)$. They provide expressions for each primitive idempotent $e_{\mathcal{U}}$ in terms of universal rational functions with values in $\mathcal{B}_m(\omega)$, analogous to Proposition 1.1.7. In what follows we will only use the particular cases of these versions, where λ is a partition of m. To state the results in those cases, note that an updown tableau $\mathcal{U} = (\Lambda_1, \dots, \Lambda_m)$ of shape λ can be regarded as a standard λ-tableau. It is obtained by writing each element r of the set $\{1, \dots, m\}$ into the box of the diagram λ which is added to the diagram Λ_{r-1} to get Λ_r. As before, $(c_1, \dots, c_m)$ will denote the sequence of contents $c_a = c_a(\mathcal{U})$ associated with $\mathcal{U}$.

Introduce the rational function depending on m variables $u_1, \dots, u_m$ by

$$\psi(u_1, \dots, u_m) = \prod_{1 \leqslant a < b \leqslant m} \Big(1 - \frac{\epsilon_{ab}}{u_a + u_b}\Big) \prod_{1 \leqslant a < b \leqslant m} \Big(1 - \frac{s_{ab}}{u_a - u_b}\Big) \tag{1.33}$$

with both products taken in the lexicographic order on the pairs (a, b).

PROPOSITION 1.2.3. *The following consecutive evaluations of the rational function $\psi(u_1, \dots, u_m)$ are well-defined and we have*

$$\psi(u_1, \dots, u_m)\big|_{u_1 = c_1}\big|_{u_2 = c_2} \cdots \big|_{u_m = c_m} = h(\lambda)\, e_{\mathcal{U}}, \tag{1.34}$$

where $h(\lambda)$ is the product of hooks of λ defined in (1.3).

PROOF. As a first step, we show that the following inductive formula for the function $\psi(u_1,\dots,u_m)$ holds:

$$\psi(u_1,\dots,u_m)=\psi(u_1,\dots,u_{m-1})\Big(1-\frac{\epsilon_{m-1\,m}}{u_{m-1}+u_m}\Big)\dots\Big(1-\frac{\epsilon_{1\,m}}{u_1+u_m}\Big)\times\Big(1-\frac{s_{1\,m}}{u_1-u_m}\Big)\dots\Big(1-\frac{s_{m-1\,m}}{u_{m-1}-u_m}\Big). \tag{1.35}$$

This follows by using the easily verified identities for the rational functions in u and v with values in $\mathcal{B}_m(\omega)$: if $a<b<c$ then

$$\Big(1-\frac{\epsilon_{a\,c}}{u}\Big)\Big(1-\frac{\epsilon_{b\,c}}{v}\Big)\Big(1-\frac{s_{a\,b}}{u-v}\Big)=\Big(1-\frac{s_{a\,b}}{u-v}\Big)\Big(1-\frac{\epsilon_{b\,c}}{v}\Big)\Big(1-\frac{\epsilon_{a\,c}}{u}\Big). \tag{1.36}$$

If indices a,b,c,d are distinct, then the elements $\epsilon_{a\,b}$ and $\epsilon_{c\,d}$ of $\mathcal{B}_m(\omega)$ commute. Therefore, we can represent the first product occurring in (1.33) as

$$\prod_{1\leqslant a<b\leqslant m}\Big(1-\frac{\epsilon_{a\,b}}{u_a+u_b}\Big)=\prod_{1\leqslant a<b\leqslant m-1}\Big(1-\frac{\epsilon_{a\,b}}{u_a+u_b}\Big)\times\Big(1-\frac{\epsilon_{1\,m}}{u_1+u_m}\Big)\dots\Big(1-\frac{\epsilon_{m-1\,m}}{u_{m-1}+u_m}\Big).$$

Now, using the identities (1.36) repeatedly, we get

$$\Big(1-\frac{\epsilon_{1\,m}}{u_1+u_m}\Big)\dots\Big(1-\frac{\epsilon_{m-1\,m}}{u_{m-1}+u_m}\Big)\prod_{1\leqslant a<b\leqslant m-1}\Big(1-\frac{s_{a\,b}}{u_a-u_b}\Big)$$
$$=\prod_{1\leqslant a<b\leqslant m-1}\Big(1-\frac{s_{a\,b}}{u_a-u_b}\Big)\Big(1-\frac{\epsilon_{m-1\,m}}{u_{m-1}+u_m}\Big)\dots\Big(1-\frac{\epsilon_{1\,m}}{u_1+u_m}\Big)$$

and so (1.35) follows. Now we use induction on m to prove the proposition. By the induction hypothesis, setting $u=u_m$ we get

$$\psi(u_1,\dots,u_m)\big|_{u_1=c_1}\big|_{u_2=c_2}\cdots\big|_{u_{m-1}=c_{m-1}}$$
$$=h(\mu)\,e_{\mathcal{V}}\Big(1-\frac{\epsilon_{m-1\,m}}{c_{m-1}+u}\Big)\dots\Big(1-\frac{\epsilon_{1\,m}}{c_1+u}\Big)\Big(1-\frac{s_{1\,m}}{c_1-u}\Big)\dots\Big(1-\frac{s_{m-1\,m}}{c_{m-1}-u}\Big),$$

where $\mathcal{V}$ is the standard tableau obtained from $\mathcal{U}$ by removing the box occupied by m, and μ is the shape of $\mathcal{V}$. Next, we verify the identity

$$e_{\mathcal{V}}\Big(1-\frac{\epsilon_{m-1\,m}}{c_{m-1}+u}\Big)\dots\Big(1-\frac{\epsilon_{1\,m}}{c_1+u}\Big)\Big(1-\frac{s_{1\,m}}{c_1-u}\Big)\dots\Big(1-\frac{s_{m-1\,m}}{c_{m-1}-u}\Big)=\frac{u-c_1}{u-c_m}\prod_{r=1}^{m-1}\Big(1-\frac{1}{(u-c_r)^2}\Big)\,e_{\mathcal{V}}\,\frac{u-c_m}{u-y_m}. \tag{1.37}$$

Note that the Jucys–Murphy element y_m commutes with $e_{\mathcal{V}}$, and the inverses of the expressions occurring in the product are found by

$$\Big(1-\frac{s_{r\,m}}{c_r-u}\Big)^{-1}\Big(1-\frac{1}{(u-c_r)^2}\Big)=1+\frac{s_{r\,m}}{c_r-u}$$

and

$$\Big(1-\frac{\epsilon_{r\,m}}{c_r+u}\Big)^{-1}=1+\frac{\epsilon_{r\,m}}{c_r+u-\omega},$$

where we have used the relations $s_{r\,m}^2 = 1$ and $\epsilon_{r\,m}^2 = \omega\,\epsilon_{r\,m}$. Hence, relation (1.37) is equivalent to

$$e_{\mathcal{V}}\Big(1+\frac{s_{m-1\,m}}{c_{m-1}-u}\Big)\dots\Big(1+\frac{s_{1\,m}}{c_1-u}\Big) \times\Big(1+\frac{\epsilon_{1\,m}}{c_1+u-\omega}\Big)\dots\Big(1+\frac{\epsilon_{m-1\,m}}{c_{m-1}+u-\omega}\Big)=e_{\mathcal{V}}\,\frac{u-y_m}{u-c_1}. \tag{1.38}$$

Fix an integer n such that $n \geqslant m$ and regard $\mathcal{B}_m(\omega)$ as a subalgebra of $\mathcal{B}_n(\omega)$. We will verify by induction on m the following more general identity

$$e_{\mathcal{V}}\Big(1+\frac{s_{m-1\,n}}{c_{m-1}-u}\Big)\dots\Big(1+\frac{s_{1\,n}}{c_1-u}\Big) \times\Big(1+\frac{\epsilon_{1\,n}}{c_1+u-\omega}\Big)\dots\Big(1+\frac{\epsilon_{m-1\,n}}{c_{m-1}+u-\omega}\Big)=e_{\mathcal{V}}\,\frac{u-y_m^{(n)}}{u-c_1}, \tag{1.39}$$

where

$$y_m^{(n)}=\frac{\omega-1}{2}+\sum_{a=1}^{m-1}(s_{a\,n}-\epsilon_{a\,n}).$$

It holds trivially for $m=1$ so assume that $2\leqslant m\leqslant n$. By (1.16) we have $e_{\mathcal{V}}=e_{\mathcal{V}}\,e_{\mathcal{W}}$, where $\mathcal{W}$ is the standard tableau obtained from $\mathcal{V}$ by removing the box occupied by $m-1$. Hence, using the induction hypothesis we can write the left hand side of (1.39) as

$$e_{\mathcal{V}}\Big(1+\frac{s_{m-1\,n}}{c_{m-1}-u}\Big)\,e_{\mathcal{W}}\,\frac{u-y_{m-1}^{(n)}}{u-c_1}\,\Big(1+\frac{\epsilon_{m-1\,n}}{c_{m-1}+u-\omega}\Big)$$

which equals

$$\frac{1}{u-c_1}\,e_{\mathcal{V}}\Big(u-y_{m-1}^{(n)}+\frac{s_{m-1\,n}(u-y_{m-1}^{(n)})}{c_{m-1}-u} +\frac{(u-y_{m-1}^{(n)})\,\epsilon_{m-1\,n}}{c_{m-1}+u-\omega}+\frac{s_{m-1\,n}(u-y_{m-1}^{(n)})\,\epsilon_{m-1\,n}}{(c_{m-1}-u)(c_{m-1}+u-\omega)}\Big).$$

Now use the following relations in $\mathcal{B}_n(\omega)$ which hold for $1\leqslant r<m-1$:

$$s_{m-1\,n}\,s_{r\,n}=s_{r\,m-1}\,s_{m-1\,n},\qquad s_{m-1\,n}\,\epsilon_{r\,n}=\epsilon_{r\,m-1}\,s_{m-1\,n}$$

and

$$s_{r\,n}\,\epsilon_{m-1\,n}=\epsilon_{r\,m-1}\,\epsilon_{m-1\,n},\qquad \epsilon_{r\,n}\,\epsilon_{m-1\,n}=s_{r\,m-1}\,\epsilon_{m-1\,n}.$$

They imply that

$$s_{m-1\,n}\,y_{m-1}^{(n)}=y_{m-1}\,s_{m-1\,n}$$

and

$$y_{m-1}^{(n)}\,\epsilon_{m-1\,n}=\big(\omega-1-y_{m-1}\big)\,\epsilon_{m-1\,n}.$$

Together with the relation $e_{\mathcal{V}}\,y_{m-1}=c_{m-1}\,e_{\mathcal{V}}$ implied by (1.27), this allows us to bring the left hand side of (1.39) to the form

$$\frac{1}{u-c_1}\,e_{\mathcal{V}}\Big(u-y_{m-1}^{(n)}-s_{m-1\,n}+\epsilon_{m-1\,n}\Big)=e_{\mathcal{V}}\,\frac{u-y_m^{(n)}}{u-c_1},$$

as required. This proves (1.39) and hence (1.37).

Finally, the product

$$\frac{u-c_1}{u-c_m}\prod_{r=1}^{m-1}\Big(1-\frac{1}{(u-c_r)^2}\Big)$$

depends only on the shape μ of $\mathcal{V}$ so we may choose a particular tableau $\mathcal{V}$ for its evaluation. We take $\mathcal{V}$ to be the row tableau obtained by filling in the boxes of μ with the numbers $1,\dots,m$ by consecutive rows from left to right in each row. A short calculation shows that this product is regular at $u=c_m$ and the value equals the ratio $h(\lambda)/h(\mu)$. Thus, the proof is completed by applying (1.29). □

EXAMPLE 1.2.4. The expression for the symmetrizer (1.32) provided by (1.34) takes the form

$$s^{(m)}=\prod_{1\leqslant a<b\leqslant m}\Big(1-\frac{\epsilon_{ab}}{\omega+a+b-3}\Big)\,h^{(m)}, \tag{1.40}$$

where $h^{(m)}$ is the symmetrizer for $\mathfrak{S}_m$; see Example 1.1.8. □

PROPOSITION 1.2.5. *The expression* (1.40) *can also be written as*

$$s^{(m)}=h^{(m)}\sum_{r=0}^{\lfloor m/2\rfloor}\frac{(-1)^r}{2^r\,r!}\binom{\omega/2+m-2}{r}^{-1}\sum_{a_i<b_i}\epsilon_{a_1b_1}\epsilon_{a_2b_2}\dots\epsilon_{a_rb_r} \tag{1.41}$$

with the second sum taken over the sets of disjoint pairs $\{(a_1,b_1),\dots,(a_r,b_r)\}$ *of indices from* $\{1,\dots,m\}$.

PROOF. Note that for each r the second sum commutes with any element of $\mathfrak{S}_m$ and hence commutes with $h^{(m)}$. Now we apply (1.40) and write $h^{(m)}=h^{(m-1)}\,h^{(m)}$. Since $\epsilon_{a\,m}\,\epsilon_{b\,m}\,h^{(m-1)}=\epsilon_{a\,m}\,s_{a\,b}\,h^{(m-1)}=\epsilon_{a\,m}\,h^{(m-1)}$ for distinct a and b, we get

$$\Big(1-\frac{\epsilon_{1m}}{\omega+m-2}\Big)\dots\Big(1-\frac{\epsilon_{m-1\,m}}{\omega+2m-4}\Big)\,h^{(m-1)}$$
$$=\Big(1-\frac{\epsilon_{1m}+\dots+\epsilon_{m-1\,m}}{\omega+2m-4}\Big)\,h^{(m-1)}$$

which is a version of (1.38) and is verified by a similar calculation. Using induction on m we come to verifying that if (1.41) holds for m replaced by $m-1$, then

$$s^{(m-1)}\Big(1-\frac{\epsilon_{1m}+\dots+\epsilon_{m-1\,m}}{\omega+2m-4}\Big)\,h^{(m)}$$

coincides with the right hand side of (1.41). If none of the pairs (a_i,b_i) contains an index $c\in\{1,\dots,m-1\}$, then the product $\epsilon_{a_1b_1}\epsilon_{a_2b_2}\dots\epsilon_{a_rb_r}\,\epsilon_{c\,m}$ of $r+1$ factors will contribute to the respective sum on the right hand side of (1.41). Otherwise, if $c=a_i$ or $c=b_i$ for some i, then

$$\epsilon_{a_1b_1}\epsilon_{a_2b_2}\dots\epsilon_{a_rb_r}\,\epsilon_{c\,m}\,h^{(m)}=\epsilon_{a_1b_1}\epsilon_{a_2b_2}\dots\epsilon_{a_rb_r}\,h^{(m)}$$

by the relations

$$\epsilon_{c\,b_i}\epsilon_{c\,m}\,h^{(m)}=\epsilon_{c\,b_i}s_{b_i\,m}\,h^{(m)}=\epsilon_{c\,b_i}\,h^{(m)} \tag{1.42}$$

and similar in the $c=b_i$ case. The result follows by combining similar terms. □

COROLLARY 1.2.6. *We have the expression for $s^{(m)}$ in terms of the Brauer diagrams,*

$$s^{(m)} = \frac{1}{m!} \sum_{r=0}^{\lfloor m/2 \rfloor} (-1)^r \binom{\omega/2+m-2}{r}^{-1} \sum_{d \in \mathcal{D}^{(r)}} d, \tag{1.43}$$

where $\mathcal{D}^{(r)} \subset \mathcal{B}_m(\omega)$ denotes the set of diagrams which have exactly r horizontal edges in the top (and hence in the bottom) row.

PROOF. The formula is essentially equivalent to (1.41). Indeed, the equivalence is implied by the relation

$$\sum_{s \in \mathfrak{S}_m} s \sum_{a_i < b_i} \epsilon_{a_1 b_1} \epsilon_{a_2 b_2} \dots \epsilon_{a_r b_r} = 2^r \, r! \sum_{d \in \mathcal{D}^{(r)}} d$$

which holds for any $r = 0, 1, \dots, \lfloor m/2 \rfloor$. Fix a diagram $d \in \mathcal{D}^{(r)}$ and suppose that the horizontal edges in the top row of d connect the r pairs of dots (a_i, b_i) associated with a summand on the left hand side, while the edges in the bottom row connect certain r pairs of dots $\{(g_1, h_1), \dots, (g_r, h_r)\}$. The number of permutations $s \in \mathfrak{S}_m$ with the property $s\, \epsilon_{a_1 b_1} \epsilon_{a_2 b_2} \dots \epsilon_{a_r b_r} = d$ is $2^r \, r!$. To see that, represent s as a diagram without horizontal edges. To satisfy the property, the dots g_1 and h_1 in the bottom row of s should be respectively connected to the dots a_i and b_i in the top row for a certain $i \in \{1, \dots, r\}$. This yields $2r$ choices for the edges with the ends g_1 and h_1 so the claim follows by applying the same argument to the remaining dots in the bottom row. □

COROLLARY 1.2.7. *We have the formula*

$$s^{(m)} = h^{(m)} \sum_{r=0}^{\lfloor m/2 \rfloor} \frac{(-1)^r}{r!\,(\omega+2m-4)(\omega+2m-6)\dots(\omega+2m-2r-2)} \tag{1.44}$$
$$\times \sum_{a<b} \epsilon_{ab} \Big(\sum_{a<b} \epsilon_{ab} - (\omega+2m-4) \Big) \dots \Big(\sum_{a<b} \epsilon_{ab} - (r-1)(\omega+2m-2r) \Big).$$

PROOF. This is yet another equivalent form of (1.41) which we verify by multiplying out the factors on the right hand side. Arguing by induction we come to checking the relations

$$h^{(m)} \sum_{a_i<b_i} \epsilon_{a_1 b_1} \epsilon_{a_2 b_2} \dots \epsilon_{a_{r-1} b_{r-1}} \Big(\sum_{c<d} \epsilon_{cd} - (r-1)(\omega+2m-2r) \Big) \tag{1.45}$$
$$= r\, h^{(m)} \sum_{a_i<b_i} \epsilon_{a_1 b_1} \epsilon_{a_2 b_2} \dots \epsilon_{a_r b_r}$$

for $r \leqslant \lfloor m/2 \rfloor$, with the notation as in (1.41). The terms on the left hand side corresponding to the pairs (c,d) which are disjoint to all pairs (a_i, b_i) will give the right hand side. So we need to check that all remaining terms cancel. If only one of c and d coincides with certain a_i or b_i, then similar to (1.42) we have $h^{(m)} \epsilon_{a_i b_i} \epsilon_{cd} = h^{(m)} \epsilon_{cd}$. Furthermore, if $c = a_i$ and $d = b_i$ for some i then the product of generators is calculated by $\epsilon_{cd}^2 = \omega\, \epsilon_{cd}$. Finally, we may have $c = a_i$ or $c = b_i$ while $d = a_j$ or $d = b_j$ for some $i \neq j$. In this case, taking, for instance, $c = a_i$ and $d = a_j$, we simplify the product by using

$$h^{(m)} \epsilon_{a_i b_i} \epsilon_{c_j b_j} \epsilon_{cd} = h^{(m)} \epsilon_{a_i b_i} s_{c b_j} \epsilon_{cd} = h^{(m)} s_{c b_j} \epsilon_{b_i b_j} \epsilon_{cd} = h^{(m)} \epsilon_{b_i b_j} \epsilon_{cd}.$$

Thus, counting the resulting coefficient of the product $h^{(m)}\epsilon_{a_1b_1}\epsilon_{a_2b_2}\dots\epsilon_{a_{r-1}b_{r-1}}$ on the left hand side of (1.45) we find

$$(r-1)\omega+2(r-1)(m-2r+2)+2(r-1)(r-2)-(r-1)(\omega+2m-2r)=0,$$

as desired. □

We point out one more multiplicative formula for the symmetrizer.

PROPOSITION 1.2.8. *We have*

$$s^{(m)}=\frac{1}{m!}\prod_{1\leqslant a<b\leqslant m}\Big(1+\frac{s_{ab}}{b-a}-\frac{\epsilon_{ab}}{\omega/2+b-a-1}\Big), \tag{1.46}$$

where the product is taken in the lexicographic order on the pairs (a,b).

PROOF. Set

$$\rho_{ab}(u)=1-\frac{s_{ab}}{u}-\frac{\epsilon_{ab}}{\omega/2-u-1}.$$

It is well-known by [**155**] and easily verified that the rational functions $\rho_{ab}(u)$ satisfy the *Yang–Baxter equation*

$$\rho_{ab}(u)\,\rho_{ac}(u+v)\,\rho_{bc}(v)=\rho_{bc}(v)\,\rho_{ac}(u+v)\,\rho_{ab}(u), \tag{1.47}$$

with any distinct indices a,b,c, where we set $\rho_{ba}(u)=\rho_{ab}(u)$ for $a<b$. So, (1.46) takes the form

$$s^{(m)}=\frac{1}{m!}\prod_{1\leqslant a<b\leqslant m}\rho_{ab}(a-b).$$

Given any index $c\in\{1,\dots,m-1\}$, we can use (1.47) to reorder the factors in the product in such a way that it would begin (or end) with the factor $\rho_{c\,c+1}(-1)$. Since

$$s_c\,\rho_{c\,c+1}(-1)=\rho_{c\,c+1}(-1)\,s_c=\rho_{c\,c+1}(-1)\qquad\text{and}$$
$$\epsilon_c\,\rho_{c\,c+1}(-1)=\rho_{c\,c+1}(-1)\,\epsilon_c=0,$$

we derive that the product on the right hand side of (1.46) is an idempotent in $\mathcal{B}_m(\omega)$ which satisfies (1.31) and so must coincide with $s^{(m)}$. □

EXAMPLE 1.2.9. If $\lambda=(1^m)$ is the single column-diagram with m boxes then the unique updown λ-tableau can be regarded as the standard tableau obtained by writing the numbers $1,\dots,m$ into the boxes of λ from top to bottom. The corresponding primitive idempotent is the *anti-symmetrizer* $a^{(m)}\in\mathcal{B}_m(\omega)$ which coincides with the anti-symmetrizer in the group algebra for the symmetric group $\mathfrak{S}_m$ in (1.19) and (1.24) and has the properties

$$s_{ab}\,a^{(m)}=a^{(m)}\,s_{ab}=-a^{(m)}\qquad\text{and}\qquad\epsilon_{ab}\,a^{(m)}=a^{(m)}\,\epsilon_{ab}=0 \tag{1.48}$$

for all $1\leqslant a<b\leqslant m$. □

1.3. Traces on the Brauer algebra

Given a finite set of positive integers $\{c_1,\dots,c_m\}$ denote by $\mathcal{B}_{\{c_1,\dots,c_m\}}(\omega)$ the $\mathbb{C}(\omega)$-linear span of the Brauer diagrams, where the top and bottom dots are labeled by $c_1,\dots,c_m$. We define the product of two such diagrams in the usual way as in Section 1.2 so that the algebra $\mathcal{B}_{\{c_1,\dots,c_m\}}(\omega)$ is isomorphic to $\mathcal{B}_m(\omega)$.

For any $a\in\{1,\dots,m\}$ define the *partial trace* tr_a (or the *a-th closure map*) as a linear map

$$\operatorname{tr}_a:\mathcal{B}_m(\omega)\to\mathcal{B}_{\{1,\dots,\widehat{a},\dots,m\}}(\omega), \tag{1.49}$$

where $\widehat{a}$ indicates that the index a should be skipped. Take a diagram $d \in \mathcal{B}_m(\omega)$ and add the edge connecting the top dot a with the bottom dot a. Let s be the number of closed loops obtained in this operation (so that $s = 0$ or $s = 1$). Then $\mathrm{tr}_a(d)$ equals ω^s times the resulting diagram without loops and without the dots a in the top and bottom rows. More generally, the partial trace tr_a on the algebra $\mathcal{B}_{\{c_1,\dots,c_m\}}(\omega)$ with $a \in \{c_1, \dots, c_m\}$ is defined in the same way.

Given any subset $\{a_1, \dots, a_k\}$ of $\{1, \dots, m\}$, define the corresponding partial trace $\mathrm{tr}_{a_1,\dots,a_k}$ on $\mathcal{B}_m(\omega)$ as the composition $\mathrm{tr}_{a_1} \circ \dots \circ \mathrm{tr}_{a_k}$. It is straightforward to verify that $\mathrm{tr}_a \circ \mathrm{tr}_b = \mathrm{tr}_b \circ \mathrm{tr}_a$ for any $a < b$ as maps on $\mathcal{B}_m(\omega)$. Therefore, for any diagram $d \in \mathcal{B}_m(\omega)$ its partial trace $\mathrm{tr}_{a_1,\dots,a_k}(d)$ can be found by extending the definition (1.49). Namely, for each $i = 1, \dots, k$ add the edge connecting the top dot a_i with the bottom dot a_i. Let s be the number of closed loops obtained in this operation. Then $\mathrm{tr}_{a_1,\dots,a_k}(d)$ equals ω^s times the resulting diagram without loops and without the dots a_i in the top and bottom rows.

In particular, the *full trace* is the map

$$\mathrm{tr}_{1,\dots,m} : \mathcal{B}_m(\omega) \to \mathbb{C}(\omega).$$

It is clear from the definition that it possesses the *cyclic property*

$$\mathrm{tr}_{1,\dots,m}(dd') = \mathrm{tr}_{1,\dots,m}(d'd) \tag{1.50}$$

for any two diagrams d and d'.

EXAMPLE 1.3.1. The trace of the following diagram $d \in \mathcal{B}_8(\omega)$

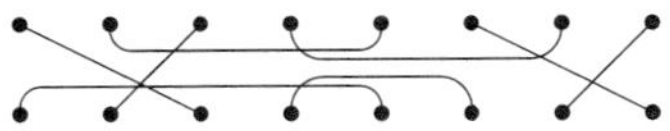

equals ω^2. □

LEMMA 1.3.2. *For the partial trace of the symmetrizer $s^{(m)} \in \mathcal{B}_m(\omega)$ we have*

$$\mathrm{tr}_m\, s^{(m)} = \frac{(\omega + m - 3)(\omega + 2m - 2)}{m\,(\omega + 2m - 4)}\, s^{(m-1)}. \tag{1.51}$$

Hence, the full trace is found by

$$\mathrm{tr}_{1,\dots,m}\, s^{(m)} = \frac{\omega + 2m - 2}{\omega + m - 2}\binom{\omega + m - 2}{m}. \tag{1.52}$$

PROOF. Expression (1.32) for the symmetrizer gives the recurrence relation

$$s^{(m)} = \frac{1}{m\,(\omega + 2m - 4)}\Big(1 + \sum_{a=1}^{m-1}(s_{a\,m} - \epsilon_{a\,m})\Big) \times \Big(\omega + m - 3 + \sum_{a=1}^{m-1}(s_{a\,m} - \epsilon_{a\,m})\Big)\, s^{(m-1)} \tag{1.53}$$

in the Brauer algebra $\mathcal{B}_m(\omega)$. We have $\mathrm{tr}_m s_{a\,m} = \mathrm{tr}_m \epsilon_{a\,m} = 1$, while

$$\mathrm{tr}_m\, s_{a\,m}\, s_{b\,m} = \mathrm{tr}_m\, \epsilon_{a\,m}\, \epsilon_{b\,m} = s_{ab} \qquad \text{and} \qquad \mathrm{tr}_m\, s_{a\,m}\, \epsilon_{b\,m} = \mathrm{tr}_m\, \epsilon_{a\,m}\, s_{b\,m} = \epsilon_{ab}$$

for $a \neq b$. Together with (1.31) this implies (1.51). □

LEMMA 1.3.3. *For the partial trace of the anti-symmetrizer $a^{(m)} \in \mathcal{B}_m(\omega)$ we have*

$$\mathrm{tr}_m\, a^{(m)} = \frac{\omega - m + 1}{m}\, a^{(m-1)}. \tag{1.54}$$

Hence, the full trace is found by

$$\operatorname{tr}_{1,\dots,m} a^{(m)} = \binom{\omega}{m}.$$

PROOF. As we pointed out in Example 1.2.9, $a^{(m)}$ coincides with the anti-symmetrizer in the group algebra for the symmetric group and is given by (1.20). Therefore, we have the recurrence relation

$$a^{(m)} = \frac{1}{m}\Big(1 - \sum_{a=1}^{m-1} s_{a\,m}\Big)\, a^{(m-1)}.$$

Since $\operatorname{tr}_m s_{a\,m} = 1$, relation (1.54) follows. □

LEMMA 1.3.4. *For the partial trace of the element $h^{(m)} \in \mathcal{B}_m(\omega)$ we have*

$$\operatorname{tr}_m h^{(m)} = \frac{\omega + m - 1}{m}\, h^{(m-1)}. \tag{1.55}$$

Hence, the full trace is found by

$$\operatorname{tr}_{1,\dots,m} h^{(m)} = \binom{\omega + m - 1}{m}.$$

PROOF. By (1.18) we have the recurrence relation

$$h^{(m)} = \frac{1}{m}\Big(1 + \sum_{a=1}^{m-1} s_{a\,m}\Big)\, h^{(m-1)}$$

which implies (1.55). □

Now we will generalize the last two lemmas and calculate the traces of primitive idempotents associated with standard tableaux. They will imply hook dimension formulas for representations of GL_N via the Schur–Weyl duality; see Section 4.4. Although the same approach can be used to prove such formulas for the orthogonal and symplectic groups, this leads to longer calculations which we omit and only state the dimension formulas in Section 5.2.

Primitive idempotents for the symmetric group can be regarded as elements of the Brauer algebra due to the embedding $\mathbb{C}[\mathfrak{S}_m] \subset \mathcal{B}_m(\omega)$. Recall that $h(\lambda)$ denotes the product of hooks of λ as defined in (1.3).

PROPOSITION 1.3.5. *Suppose that λ is a diagram with m boxes and $\mathcal{U}$ is a standard tableau of shape λ. The trace of the corresponding primitive idempotent $e_{\mathcal{U}} \in \mathbb{C}[\mathfrak{S}_m] \subset \mathcal{B}_m(\omega)$ is found by*

$$\operatorname{tr}_{1,\dots,m} e_{\mathcal{U}} = \frac{1}{h(\lambda)} \prod_{(i,j)\in\lambda} (\omega + j - i). \tag{1.56}$$

PROOF. Since for partial traces we have

$$\operatorname{tr}_m\Big(s_{m-1}\, \frac{1}{u - x_{m-1}}\, s_{m-1}\Big) = \operatorname{tr}_{m-1}\Big(\frac{1}{u - x_{m-1}}\Big),$$

we derive from Lemma 1.1.6 a recurrence relation for the rational functions

$$A_m(u) = \operatorname{tr}_m\Big(\frac{1}{u - x_m}\Big)$$

in the form

$$A_m(u) = A_{m-1}(u) + \frac{1}{(u - x_{m-1})^2}\,\big(1 + A_m(u)\big).$$

That is,

$$A_m(u) = \frac{(u - x_{m-1})^2}{(u - x_{m-1})^2 - 1}\, A_{m-1}(u) + \frac{1}{(u - x_{m-1})^2 - 1}.$$

Since for $m = 1$ we have $A_1(u) = \omega/u$, solving the recurrence relation we find

$$A_m(u) = \frac{u + \omega}{u} \prod_{a=1}^{m-1} \frac{(u - x_a)^2}{(u - x_a)^2 - 1} - 1.$$

This calculation and the recurrence formula (1.15) allow us to find the partial trace $\mathrm{tr}_m\, e_{\mathcal{U}}$. Indeed, with the notation used in (1.15), the properties (1.11) of the Jucys–Murphy elements imply

$$\mathrm{tr}_m\, e_{\mathcal{U}} = e_{\mathcal{V}} \Big[(u - c)\, A_m(u)\Big]_{u=c} = (\omega + c)\, e_{\mathcal{V}} \Big[\, \frac{u - c}{u} \prod_{a=1}^{m-1} \frac{(u - c_a)^2}{(u - c_a)^2 - 1}\, \Big]_{u=c},$$

where $c_a = j - i$ denotes the content of the box (i, j) of $\mathcal{U}$ occupied by a and $c = c_m$. The evaluation of the rational function in u is well-defined and it depends only on the shape μ of the standard tableau $\mathcal{V}$ but does not depend on $\mathcal{V}$. The result of the evaluation is easily calculated by taking, say, the row tableau; cf. the proof of Proposition 1.2.3. It gives

$$\mathrm{tr}_m\, e_{\mathcal{U}} = (\omega + c)\, \frac{h(\mu)}{h(\lambda)}\, e_{\mathcal{V}}$$

and so (1.56) follows by an obvious induction. □

1.4. Tensor notation

The endomorphisms of the vector space $\mathbb{C}^N$ (linear maps $\mathbb{C}^N \to \mathbb{C}^N$) form an associative algebra which we denote by $\mathrm{End}\, \mathbb{C}^N$. We let $e_1, \dots, e_N$ denote the vectors of the canonical basis of $\mathbb{C}^N$. Writing the endomorphisms as matrices with respect to this basis allows us to identify $\mathrm{End}\, \mathbb{C}^N$ with the algebra of $N \times N$ matrices over $\mathbb{C}$. The matrix units e_{ij} with $i, j \in \{1, \dots, N\}$ form a basis of $\mathrm{End}\, \mathbb{C}^N$. The corresponding endomorphisms act on the basis vectors by

$$e_{ij} : e_k \mapsto \delta_{jk}\, e_i,$$

where δ_{jk} is the Kronecker delta, $\delta_{jk} = 1$ if $j = k$ and $\delta_{jk} = 0$ otherwise. For the product of the matrix units we have $e_{ij}\, e_{kl} = \delta_{jk}\, e_{il}$. The vector space of tensors

$$(\mathbb{C}^N)^{\otimes m} = \underbrace{\mathbb{C}^N \otimes \mathbb{C}^N \otimes \dots \otimes \mathbb{C}^N}_{m} \tag{1.57}$$

is equipped with the basis formed by the tensor products $e_{i_1} \otimes \dots \otimes e_{i_m}$, where the indices $i_1, \dots, i_m$ run over the set $\{1, \dots, N\}$. The endomorphism algebra of the vector space (1.57) will be identified with the tensor product of the endomorphism algebras

$$\underbrace{\mathrm{End}\, \mathbb{C}^N \otimes \dots \otimes \mathrm{End}\, \mathbb{C}^N}_{m} \tag{1.58}$$

via a natural isomorphism so that an element $A_1 \otimes \ldots \otimes A_m$ of the algebra (1.58) is understood as the endomorphism of $(\mathbb{C}^N)^{\otimes m}$ defined by

$$(A_1 \otimes \ldots \otimes A_m)(v_1 \otimes \ldots \otimes v_m) = A_1(v_1) \otimes \ldots \otimes A_m(v_m), \qquad v_i \in \mathbb{C}^N.$$

We will often consider matrices

$$X = \begin{bmatrix} X_{11} & \ldots & X_{1N} \\ \vdots & & \vdots \\ X_{N1} & \ldots & X_{NN} \end{bmatrix}$$

whose entries X_{ij} belong to a certain unital associative algebra $\mathcal{A}$. The algebra of $N \times N$ matrices with entries in $\mathcal{A}$ is isomorphic to the tensor product algebra $\operatorname{End}\mathbb{C}^N \otimes \mathcal{A}$ so that the matrix X will be identified with the element

$$X = \sum_{i,j=1}^{N} e_{ij} \otimes X_{ij} \in \operatorname{End}\mathbb{C}^N \otimes \mathcal{A}. \tag{1.59}$$

Given a scalar or variable u with values in $\mathcal{A}$, the expression $u + X$ will be understood as the sum $u\,1 + X$, where $u\,1$ is the scalar matrix of the same size as X.

We will work with tensor product algebras of the form

$$\operatorname{End}(\mathbb{C}^N)^{\otimes m} \otimes \mathcal{A} = \underbrace{\operatorname{End}\mathbb{C}^N \otimes \ldots \otimes \operatorname{End}\mathbb{C}^N}_{m} \otimes \mathcal{A}. \tag{1.60}$$

For any $a \in \{1, \ldots, m\}$ we will denote by X_a the element (1.59) associated with the a-th copy of $\operatorname{End}\mathbb{C}^N$ so that

$$X_a = \sum_{i,j=1}^{N} 1^{\otimes(a-1)} \otimes e_{ij} \otimes 1^{\otimes(m-a)} \otimes X_{ij} \in \operatorname{End}(\mathbb{C}^N)^{\otimes m} \otimes \mathcal{A}, \tag{1.61}$$

where

$$1 = \sum_{i=1}^{N} e_{ii} \in \operatorname{End}\mathbb{C}^N$$

is the identity endomorphism.

Given any element

$$C = \sum_{i,j,k,l=1}^{N} c_{ijkl}\, e_{ij} \otimes e_{kl} \in \operatorname{End}\mathbb{C}^N \otimes \operatorname{End}\mathbb{C}^N,$$

for any two indices $a, b \in \{1, \ldots, m\}$ such that $a < b$, we set

$$C_{ab} = \sum_{i,j,k,l=1}^{N} c_{ijkl}\, 1^{\otimes(a-1)} \otimes e_{ij} \otimes 1^{\otimes(b-a-1)} \otimes e_{kl} \otimes 1^{\otimes(m-b)} \in \operatorname{End}(\mathbb{C}^N)^{\otimes m}. \tag{1.62}$$

We will keep the same notation C_{ab} for the element $C_{ab} \otimes 1$ of the algebra (1.60).

By the *trace* we will mean the linear map

$$\operatorname{tr} : \operatorname{End}\mathbb{C}^N \to \mathbb{C}, \qquad e_{ij} \mapsto \delta_{ij}.$$

Furthermore, for any $a \in \{1, \ldots, m\}$ the partial trace tr_a will be understood as the linear map

$$\operatorname{tr}_a : \operatorname{End}(\mathbb{C}^N)^{\otimes m} \to \operatorname{End}(\mathbb{C}^N)^{\otimes(m-1)} \tag{1.63}$$

which acts as the trace map on the a-th copy of $\operatorname{End}\mathbb{C}^N$ and is the identity map on all the remaining copies. The full trace $\operatorname{tr}_{1,\dots,m}$ is the composition $\operatorname{tr}_1\circ\cdots\circ\operatorname{tr}_m$.

The following lemma will be often referred to as the *cyclic property of trace*; cf. (1.50).

LEMMA 1.4.1. *Suppose that two elements*

$$X=\sum e_{i_1j_1}\otimes\dots\otimes e_{i_mj_m}\otimes X^{i_1\dots i_m}_{j_1\dots j_m}\qquad\textit{and}$$

$$Y=\sum e_{i_1j_1}\otimes\dots\otimes e_{i_mj_m}\otimes Y^{i_1\dots i_m}_{j_1\dots j_m}$$

of the algebra (1.60) *satisfy the property*

$$X^{i_1\dots i_m}_{j_1\dots j_m}Y^{k_1\dots k_m}_{l_1\dots l_m}=Y^{k_1\dots k_m}_{l_1\dots l_m}X^{i_1\dots i_m}_{j_1\dots j_m}$$

for all values of the indices. Then

$$\operatorname{tr}_{1,\dots,m}XY=\operatorname{tr}_{1,\dots,m}YX.$$

PROOF. We have

$$\operatorname{tr}_{1,\dots,m}XY=\sum X^{i_1\dots i_m}_{j_1\dots j_m}Y^{j_1\dots j_m}_{i_1\dots i_m}=\sum Y^{j_1\dots j_m}_{i_1\dots i_m}X^{i_1\dots i_m}_{j_1\dots j_m}=\operatorname{tr}_{1,\dots,m}YX,$$

as required. □

1.5. Action of the symmetric group and the Brauer algebra

The symmetric group $\mathfrak{S}_m$ acts on the space (1.57) by permuting the tensor factors. More precisely, introduce the element

$$P=\sum_{i,j=1}^{N}e_{ij}\otimes e_{ji}\in\operatorname{End}\mathbb{C}^N\otimes\operatorname{End}\mathbb{C}^N \tag{1.64}$$

which acts on the vector space $\mathbb{C}^N\otimes\mathbb{C}^N$ by swapping the tensor factors,

$$P(v\otimes w)=w\otimes v,\qquad v,w\in\mathbb{C}^N.$$

Then using the notation (1.62), for the images of the transpositions under the action of $\mathfrak{S}_m$ on the space (1.57) we can write

$$s_{a\,b}\mapsto P_{a\,b},\qquad 1\leqslant a<b\leqslant m, \tag{1.65}$$

with

$$P_{a\,b}=\sum_{i,j=1}^{N}1^{\otimes(a-1)}\otimes e_{ij}\otimes 1^{\otimes(b-a-1)}\otimes e_{ji}\otimes 1^{\otimes(m-b)}. \tag{1.66}$$

The image of an arbitrary element $\sigma\in\mathfrak{S}_m$ under the map (1.65) will often be denoted by P_σ. Note the relations

$$P_{a\,b}X_a=X_bP_{a\,b},\qquad 1\leqslant a<b\leqslant m, \tag{1.67}$$

which hold in the algebra (1.60) for an arbitrary element (1.61). Furthermore,

$$\operatorname{tr}_b P_{a\,b}=\operatorname{tr}_a P_{a\,b}=1. \tag{1.68}$$

We will need the action of the Brauer algebra $\mathcal{B}_m(\omega)$ on the tensor space (1.57) with two particular specializations of the parameter ω. The values $\omega=N$ and $\omega=-N$ (the latter with even $N=2n$) occur in the context of the *Schur–Weyl duality*, where the action of the Brauer algebra centralizers the respective actions of the orthogonal and symplectic groups on the space (1.57). In the orthogonal case, the generators of $\mathcal{B}_m(N)$ act by the rule

$$s_{a\,b} \mapsto P_{a\,b}, \qquad \epsilon_{a\,b} \mapsto Q_{a\,b}, \qquad 1 \leqslant a < b \leqslant m, \tag{1.69}$$

where $P_{a\,b}$ is defined by (1.66), while

$$Q_{a\,b} = \sum_{i,j=1}^{N} 1^{\otimes(a-1)} \otimes e_{ij} \otimes 1^{\otimes(b-a-1)} \otimes e_{i'j'} \otimes 1^{\otimes(m-b)} \tag{1.70}$$

and we use the involution on the set $\{1, \dots, N\}$ defined by $i \mapsto i' = N - i + 1$. We point out the following relations:

$$P_{a\,b}\, Q_{a\,b} = Q_{a\,b}\, P_{a\,b} = Q_{a\,b} \qquad \text{and} \qquad Q_{a\,b}^2 = N\, Q_{a\,b}. \tag{1.71}$$

Our notation for the partial traces on the Brauer algebra defined in (1.49) is consistent with the notation (1.63) in the sense that the following diagram commutes:

$$\begin{array}{ccc} \mathcal{B}_m(N) & \xrightarrow{\ \mathrm{tr}_a\ } & \mathcal{B}_{\{1,\dots,\widehat{a},\dots,m\}}(N) \\ \downarrow & & \downarrow \\ \mathrm{End}\,(\mathbb{C}^N)^{\otimes m} & \xrightarrow{\ \mathrm{tr}_a\ } & \mathrm{End}\,(\mathbb{C}^N)^{\otimes m-1}, \end{array} \tag{1.72}$$

where the vertical arrows denote the homomorphisms defined in (1.69).

In the symplectic case, the action of $\mathcal{B}_m(-N)$ with $N = 2n$ in the space (1.57) is defined by

$$s_{a\,b} \mapsto -P_{a\,b}, \qquad \epsilon_{a\,b} \mapsto -Q_{a\,b}, \qquad 1 \leqslant a < b \leqslant m, \tag{1.73}$$

where $P_{a\,b}$ is defined in (1.66), and

$$Q_{a\,b} = \sum_{i,j=1}^{2n} \varepsilon_i \varepsilon_j\, 1^{\otimes(a-1)} \otimes e_{ij} \otimes 1^{\otimes(b-a-1)} \otimes e_{i'j'} \otimes 1^{\otimes(m-b)} \tag{1.74}$$

with $\varepsilon_i = 1$ for $i = 1, \dots, n$ and $\varepsilon_i = -1$ for $i = n+1, \dots, 2n$. The symplectic analogue of (1.71) reads

$$P_{a\,b}\, Q_{a\,b} = Q_{a\,b}\, P_{a\,b} = -Q_{a\,b} \qquad \text{and} \qquad Q_{a\,b}^2 = 2n\, Q_{a\,b}. \tag{1.75}$$

Both in the orthogonal and symplectic case we have the following analogue of relation (1.68) for partial traces:

$$\mathrm{tr}_b\, Q_{a\,b} = \mathrm{tr}_a\, Q_{a\,b} = 1. \tag{1.76}$$

However, the symplectic analogue of the commutative diagram (1.72) takes a different form,

$$\begin{array}{ccc} \mathcal{B}_m(-N) & \xrightarrow{\ -\mathrm{tr}_a\ } & \mathcal{B}_{\{1,\dots,\widehat{a},\dots,m\}}(-N) \\ \downarrow & & \downarrow \\ \mathrm{End}\,(\mathbb{C}^N)^{\otimes m} & \xrightarrow{\ \mathrm{tr}_a\ } & \mathrm{End}\,(\mathbb{C}^N)^{\otimes m-1}, \end{array} \tag{1.77}$$

where the vertical arrows denote the homomorphisms defined in (1.73) and the top horizontal arrow is the composition of the map (1.49) with multiplication by -1.

1.6. Bibliographical notes

Proposition 1.1.2 is due to Jucys [**85**] and Murphy [**120**]. The version of the fusion procedure given in Proposition 1.1.7 is contained in [**109**, Section 6.4], where more detailed bibliographical notes can also be found. Originated from Jucys [**84**], the procedure was re-discovered by Cherednik [**19**] with detailed proofs given by Nazarov [**124**]. The definition of the algebra $\mathcal{B}_m(\omega)$ is due to Brauer [**14**]. For the presentation in terms of the generators (1.25) see [**10**, Section 5]. The Jucys–Murphy elements (1.26) for the Brauer algebra $\mathcal{B}_m(\omega)$ were introduced in [**123**]. They are also recovered from the corresponding elements of the Birman–Murakami–Wenzl algebras introduced in [**101**] at $q = 1$. The properties (1.27) were established in those papers. Two versions of the fusion procedure for the Brauer algebra were given in [**74**] and [**75**], respectively. The expression for the symmetrizer in terms of the Brauer diagrams given in (1.43) is provided by [**69**, Theorem 4.3]. The product formula (1.46) is contained in [**74**, Remark 3.8]. Its derivation from a fusion procedure was given in [**75**, Example 2.3].

CHAPTER 2

Invariants of symmetric algebras

Our goal in this chapter is to review some constructions of $\mathfrak{g}$-invariants of the symmetric algebra $\mathrm{S}(\mathfrak{g})$, where $\mathfrak{g}$ is a simple Lie algebra of classical type. These invariants will be produced with the use of the primitive idempotents for the symmetric group and Brauer algebra, discussed in Chapter 1. First recall some well-known facts on the invariants; see, for instance, Dixmier [**29**, Section 7.3]. The adjoint action of a simple Lie algebra $\mathfrak{g}$ on itself extends to the symmetric algebra $\mathrm{S}(\mathfrak{g})$ by

$$Y \cdot X_1 \dots X_k = \sum_{i=1}^{k} X_1 \dots [Y, X_i] \dots X_k, \qquad Y, X_i \in \mathfrak{g}.$$

Since any element $Y \in \mathfrak{g}$ acts as a derivation, the subspace of $\mathfrak{g}$-invariants

$$\mathrm{S}(\mathfrak{g})^{\mathfrak{g}} = \{P \in \mathrm{S}(\mathfrak{g}) \mid Y \cdot P = 0 \quad \text{for all} \quad Y \in \mathfrak{g}\} \tag{2.1}$$

is a subalgebra of $\mathrm{S}(\mathfrak{g})$. Let n denote the rank of $\mathfrak{g}$. Then

$$\mathrm{S}(\mathfrak{g})^{\mathfrak{g}} = \mathbb{C}[P_1, \dots, P_n] \tag{2.2}$$

for certain algebraically independent invariants $P_1, \dots, P_n$ whose degrees $d_1, \dots, d_n$ are the exponents of $\mathfrak{g}$ increased by 1.

Fix a Cartan subalgebra $\mathfrak{h}$ of $\mathfrak{g}$ and a triangular decomposition $\mathfrak{g} = \mathfrak{n}_- \oplus \mathfrak{h} \oplus \mathfrak{n}_+$. Consider the projection $\mathrm{S}(\mathfrak{g}) \to \mathrm{S}(\mathfrak{h})$ defined by the decomposition $\mathrm{S}(\mathfrak{g}) = \mathrm{S}(\mathfrak{h}) \oplus \mathrm{J}$, where J is the ideal of $\mathrm{S}(\mathfrak{g})$ generated by $\mathfrak{n}_- \cup \mathfrak{n}_+$. The restriction of the projection to the subalgebra of invariants yields the *Chevalley isomorphism*

$$\varsigma : \mathrm{S}(\mathfrak{g})^{\mathfrak{g}} \to \mathrm{S}(\mathfrak{h})^W, \tag{2.3}$$

where W is the *Weyl group* of the root system of $\mathfrak{g}$ and $\mathrm{S}(\mathfrak{h})^W$ denotes the subalgebra of W-invariant elements in $\mathrm{S}(\mathfrak{h})$.

2.1. Invariants in type A

The vector space $\mathrm{Mat}_N(\mathbb{C})$ of $N \times N$ matrices over $\mathbb{C}$ is equipped with a Lie algebra structure defined by the matrix commutator

$$[X, Y] = XY - YX.$$

This defines the *general linear Lie algebra* $\mathfrak{gl}_N$ which has standard basis elements E_{ij} with the commutation relations

$$[E_{ij}, E_{kl}] = \delta_{kj} E_{il} - \delta_{il} E_{kl}, \qquad i, j, k, l \in \{1, \dots, N\}. \tag{2.4}$$

We thus make a notational distinction between the E_{ij} and the introduced above basis elements e_{ij} of the endomorphism algebra $\mathrm{End}\,\mathbb{C}^N$. The *special linear Lie algebra* $\mathfrak{sl}_N$ (the simple Lie algebra of type A_{N-1}) is defined as the subalgebra of $\mathfrak{gl}_N$ spanned by the matrices whose trace is zero.

We will regard $\mathrm{S}(\mathfrak{gl}_N)$ as the commutative associative algebra with the free generators E_{ij} so that its elements are polynomials in the E_{ij}. Introduce the matrix

$$E = \begin{bmatrix} E_{11} & \dots & E_{1N} \\ \vdots & & \vdots \\ E_{N1} & \dots & E_{NN} \end{bmatrix} \tag{2.5}$$

with entries in $\mathrm{S}(\mathfrak{gl}_N)$ and write

$$\det(u+E) = u^N + C_1 u^{N-1} + \dots + C_N \tag{2.6}$$

for a variable u. Here and below expressions like $u+E$ should be understood as $u1+E$, where 1 is the identity matrix of the same size as E. By the well-known property of the characteristic polynomial, all coefficients are invariant under the adjoint action of the general linear group GL_N on $\mathrm{S}(\mathfrak{gl}_N)$. More precisely, GL_N acts on $\mathfrak{gl}_N$ by the rule

$$\mathrm{Ad}\,\mathbf{g} : X \mapsto \mathbf{g}X\mathbf{g}^{-1}, \qquad X \in \mathfrak{gl}_N, \quad \mathbf{g} \in \mathrm{GL}_N\,. \tag{2.7}$$

This extends to an action of GL_N on the symmetric algebra $\mathrm{S}(\mathfrak{gl}_N)$ so that elements of GL_N act on $\mathrm{S}(\mathfrak{gl}_N)$ as automorphisms. The subalgebra $\mathrm{S}(\mathfrak{gl}_N)^{\mathfrak{gl}_N}$ coincides with the subalgebra of GL_N-invariants in $\mathrm{S}(\mathfrak{gl}_N)$. Calculating the image of the basis element E_{ij} under $\mathrm{Ad}\,\mathbf{g}$ we find that the image of the matrix E is given by

$$\mathrm{Ad}\,\mathbf{g} : E \mapsto \mathbf{g}^t\, E\, (\mathbf{g}^t)^{-1}, \tag{2.8}$$

where $\mathbf{g}^t$ is the transpose of $\mathbf{g}$. Hence the polynomial (2.6) is GL_N-invariant.

Moreover, we also have

$$\mathrm{S}(\mathfrak{gl}_N)^{\mathfrak{gl}_N} = \mathbb{C}[C_1, \dots, C_N] \tag{2.9}$$

so that the coefficients of the characteristic polynomial are algebraically independent generators of the algebra of invariants. This can be seen by calculating the Chevalley images of the C_m. Fix a standard triangular decomposition for $\mathfrak{gl}_N$ so that the Cartan subalgebra $\mathfrak{h}$ is spanned by the elements $E_{11}, \dots, E_{NN}$, while the subalgebras $\mathfrak{n}_-$ and $\mathfrak{n}_+$ are spanned by the elements E_{ij} with $i > j$ and $i < j$, respectively. The Weyl group is isomorphic to the symmetric group $\mathfrak{S}_N$ which acts on $\mathrm{S}(\mathfrak{h})$ by permuting the generators E_{ii}. Therefore, the image of the Chevalley isomorphism is the algebra of symmetric polynomials in the variables $\lambda_1, \dots, \lambda_N$, where $\lambda_i = E_{ii}$. We have

$$\varsigma : \det(u+E) \mapsto (u+\lambda_1)\dots(u+\lambda_N)$$

and so the Chevalley image $\varsigma(C_m)$ coincides with the elementary symmetric polynomial

$$e_m(\lambda_1, \dots, \lambda_N) = \sum_{1 \leqslant p_1 < \dots < p_m \leqslant N} \lambda_{p_1} \dots \lambda_{p_m}. \tag{2.10}$$

The polynomials $e_m(\lambda_1, \dots, \lambda_N)$ with $m = 1, \dots, N$ are algebraically independent generators of the algebra of symmetric polynomials which implies (2.9).

By a similar argument, the traces of all powers of the matrix E belong to the algebra of invariants,

$$T_m = \mathrm{tr}\, E^m \in \mathrm{S}(\mathfrak{gl}_N)^{\mathfrak{gl}_N} \tag{2.11}$$

for all $m \geqslant 0$. Moreover, we get an alternative family of generators,

$$\mathrm{S}(\mathfrak{gl}_N)^{\mathfrak{gl}_N} = \mathbb{C}[T_1, \dots, T_N] \tag{2.12}$$

such that

$$\varsigma : T_m \mapsto \lambda_1^m + \dots + \lambda_N^m.$$

To get a more general family of invariants, recall the action of the symmetric group $\mathfrak{S}_m$ on the tensor product space $(\mathbb{C}^N)^{\otimes m}$ as defined in Section 1.5. The general linear group GL_N acts on this space diagonally,

$$\mathbf{h} : v_1 \otimes \dots \otimes v_m \mapsto \mathbf{h} v_1 \otimes \dots \otimes \mathbf{h} v_m, \qquad v_i \in \mathbb{C}^N, \qquad \mathbf{h} \in \mathrm{GL}_N, \tag{2.13}$$

which can be written as $\mathbf{h} \mapsto \mathbf{h}_1 \dots \mathbf{h}_m$, where

$$\mathbf{h}_a = 1^{\otimes(a-1)} \otimes \mathbf{h} \otimes 1^{\otimes(m-a)}.$$

The action of any element of $\mathfrak{S}_m$ on the vector space $(\mathbb{C}^N)^{\otimes m}$ commutes with the action of any element $\mathbf{h} \in \mathrm{GL}_N$. It suffices to check this property for generators of $\mathfrak{S}_m$ which follows from (1.67):

$$P_{a\,a+1} \mathbf{h}_1 \dots \mathbf{h}_m = \mathbf{h}_1 \dots \mathbf{h}_{a+1} \mathbf{h}_a \dots \mathbf{h}_m P_{a\,a+1} = \mathbf{h}_1 \dots \mathbf{h}_m P_{a\,a+1} \tag{2.14}$$

since $\mathbf{h}_{a+1} \mathbf{h}_a = \mathbf{h}_a \mathbf{h}_{a+1}$. We will discuss the commuting actions of $\mathfrak{S}_m$ and GL_N in more detail in Section 4.4 within the context of the classical *Schur–Weyl duality*.

Let S denote the image of $s \in \mathbb{C}[\mathfrak{S}_m]$ under the map (1.65). We regard S as an element of the algebra

$$\underbrace{\mathrm{End}\,\mathbb{C}^N \otimes \dots \otimes \mathrm{End}\,\mathbb{C}^N}_{m} \otimes \mathrm{S}(\mathfrak{gl}_N) \tag{2.15}$$

by identifying it with $S \otimes 1$. Also, as in (1.59), we regard the matrix E as the element

$$E = \sum_{i,j=1}^{N} e_{ij} \otimes E_{ij} \in \mathrm{End}\,\mathbb{C}^N \otimes \mathrm{S}(\mathfrak{gl}_N).$$

PROPOSITION 2.1.1. *For any $s \in \mathbb{C}[\mathfrak{S}_m]$ the element*

$$\mathrm{tr}_{1,\dots,m}\, S\, E_1 \dots E_m \tag{2.16}$$

belongs to the algebra of invariants $\mathrm{S}(\mathfrak{gl}_N)^{\mathfrak{gl}_N}$.

PROOF. As with the characteristic polynomial, we will be proving that the element (2.16) is invariant with respect to the adjoint action of the general linear group GL_N. Due to (2.8), it suffices to verify that this element remains unchanged if E is replaced by $\mathbf{h} E \mathbf{h}^{-1}$ for any $\mathbf{h} \in \mathrm{GL}_N$. We have

$$\begin{aligned}\mathrm{tr}_{1,\dots,m}\, S\, \mathbf{h}_1 E_1 \mathbf{h}_1^{-1} \dots \mathbf{h}_m E_m \mathbf{h}_m^{-1} &= \mathrm{tr}_{1,\dots,m}\, S\, \mathbf{h}_1 \dots \mathbf{h}_m E_1 \dots E_m \mathbf{h}_1^{-1} \dots \mathbf{h}_m^{-1} \\ &= \mathrm{tr}_{1,\dots,m}\, \mathbf{h}_1^{-1} \dots \mathbf{h}_m^{-1} S\, \mathbf{h}_1 \dots \mathbf{h}_m E_1 \dots E_m = \mathrm{tr}_{1,\dots,m}\, S\, E_1 \dots E_m,\end{aligned}$$

where we used the cyclic property of trace (Lemma 1.4.1) and the fact that the actions of S and $\mathbf{h}$ commute. □

Proposition 2.1.1 is also a consequence of Theorem 4.5.1 below which will be proved by a slightly different argument.

REMARK 2.1.2. A seemingly more general family of invariants is obtained by taking the trace

$$\mathrm{tr}_{1,\dots,m}\, S\, E_1^{k_1} \dots E_m^{k_m} \tag{2.17}$$

with some nonnegative powers k_a. This is verified by the same calculation as in the proof of Proposition 2.1.1. However, these elements can be written in the form (2.16) by applying a 'linearization' procedure so that (2.17) is written as

$$\mathrm{tr}_{1,\dots,m,m+1}\, E_1^{k_1} \dots E_m^{k_m-1}\, E_{m+1}\, P_{m\, m+1}\, S$$

in the first step, due to (1.67) and (1.68). In particular,

$$\mathrm{tr}\, E^m = \mathrm{tr}_{1,\dots,m}\, E_1 \dots E_m\, P_{m-1\, m} \dots P_{12} = \mathrm{tr}_{1,\dots,m}\, E_1 \dots E_m\, P_\sigma,$$

for the cycle $\sigma = (m, m-1, \dots, 1)$; see also Section 4.8 below. □

As in Section 1.1, suppose that $\mathcal{U}$ is a standard tableau of shape $\mu \vdash m$ and let $e_{\mathcal{U}} \in \mathbb{C}[\mathfrak{S}_m]$ be the associated primitive idempotent. We let $\mathcal{E}_{\mathcal{U}}$ denote the image of $e_{\mathcal{U}}$ under the action of the symmetric group $\mathfrak{S}_m$ given by (1.65). Consider the algebra (2.15) and employ the tensor notation of Section 1.4. Set

$$\mathrm{I}_\mu = \mathrm{tr}_{1,\dots,m}\, \mathcal{E}_{\mathcal{U}}\, E_1 \dots E_m. \tag{2.18}$$

This element belongs to $\mathrm{S}(\mathfrak{gl}_N)^{\mathfrak{gl}_N}$ by Proposition 2.1.1 and is independent of the choice of the standard tableau $\mathcal{U}$ of shape μ. The latter property follows from the identity (1.9) in the group algebra of $\mathfrak{S}_m$, since by the cyclic property of trace,

$$\begin{aligned}\mathrm{tr}_{1,\dots,m}\, \mathcal{E}_{\mathcal{U}}\, E_1 \dots E_m &= \frac{1}{m!} \sum_{s\in\mathfrak{S}_m} \mathrm{tr}_{1,\dots,m}\, P_s\, \mathcal{E}_{\mathcal{U}}\, E_1 \dots E_m\, P_s^{-1} \\ &= \frac{1}{m!} \sum_{s\in\mathfrak{S}_m} \mathrm{tr}_{1,\dots,m}\, P_s\, \mathcal{E}_{\mathcal{U}}\, P_s^{-1}\, E_1 \dots E_m = \frac{1}{m!}\, \mathrm{tr}_{1,\dots,m}\, \mathcal{X}_\mu\, E_1 \dots E_m,\end{aligned}$$

where $\mathcal{X}_\mu$ denotes the image of the irreducible character χ_μ under the homomorphism (1.65). For a matrix X with entries in $\mathbb{C}$, its *μ-immanant* is defined by

$$\mathrm{imm}_\mu(X) = \sum_{s\in\mathfrak{S}_m} \chi_\mu(s)\, X_{s(1)\,1} \dots X_{s(m)\,m}.$$

The invariant I_μ defined in (2.18) can be written as a linear combination of the immanants of the principal m-submatrices of E with repeated rows and columns. Moreover, the image of I_μ under the Chevalley isomorphism coincides with the Schur polynomial s_μ,

$$\varsigma : \mathrm{I}_\mu \mapsto s_\mu(\lambda_1, \dots, \lambda_N), \tag{2.19}$$

where

$$s_\mu(\lambda_1, \dots, \lambda_N) = \sum_{\mathrm{sh}(\mathcal{T})=\mu} \prod_{\alpha\in\mu} \lambda_{\mathcal{T}(\alpha)},$$

summed over semistandard tableaux $\mathcal{T}$ of shape μ with entries in $\{1, \dots, N\}$ as defined in Section 1.1, where $\mathcal{T}(\alpha)$ denotes the entry of $\mathcal{T}$ in the box α. The property (2.19) can be verified directly (by a simplified version of the argument used in the proof of Theorem 10.1.2 below), but it is naturally explained within the context of Schur–Weyl duality which we will discuss in Section 4.4. Since the Schur polynomials form a basis of the algebra of symmetric polynomials in $\lambda_1, \dots, \lambda_N$ we come to the following result.

PROPOSITION 2.1.3. *The elements* I_μ *with* μ *running over all diagrams with at most* N *rows form a basis of* $\mathrm{S}(\mathfrak{gl}_N)^{\mathfrak{gl}_N}$. □

REMARK 2.1.4. From another perspective, the assignment $\chi_\mu \mapsto \varsigma(\mathrm{I}_\mu)$ gives rise to the classical *characteristic map* providing an isomorphism between the algebra generated by the irreducible characters of the symmetric groups and the algebra of symmetric functions; see Macdonald [**104**, Section I.7]. □

We point out two particular cases of (2.19) corresponding to the row and column diagrams $\mu = (m)$ and $\mu = (1^m)$ for which direct proofs are straightforward. Denote by $H^{(m)}$ and $A^{(m)}$ the elements of the algebra (2.15) (with the identity components in $\mathrm{S}(\mathfrak{gl}_N)$) which are the respective images of the symmetrizer $h^{(m)}$ and anti-symmetrizer $a^{(m)}$ defined in (1.17) and (1.19) under the map (1.65).

PROPOSITION 2.1.5. *Under the Chevalley isomorphism we have*

$$\varsigma : \mathrm{tr}_{1,\dots,m}\, A^{(m)} E_1 \dots E_m \mapsto e_m(\lambda_1, \dots, \lambda_N), \tag{2.20}$$

$$\varsigma : \mathrm{tr}_{1,\dots,m}\, H^{(m)} E_1 \dots E_m \mapsto h_m(\lambda_1, \dots, \lambda_N), \tag{2.21}$$

where the elementary symmetric polynomial $e_m(\lambda_1, \dots, \lambda_N)$ *is defined in* (2.10) *and*

$$h_m(\lambda_1, \dots, \lambda_N) = \sum_{1 \leqslant p_1 \leqslant \dots \leqslant p_m \leqslant N} \lambda_{p_1} \dots \lambda_{p_m} \tag{2.22}$$

is the m*-th complete symmetric polynomial.*

PROOF. We will regard E as the diagonal matrix $E = \mathrm{diag}[\lambda_1, \dots, \lambda_N]$ and calculate the traces. Applying the operator occurring in (2.20) to a basis vector we find

$$A^{(m)} E_1 \dots E_m (e_{p_1} \otimes \dots \otimes e_{p_m}) = \frac{1}{m!} \lambda_{p_1} \dots \lambda_{p_m} \sum_{\sigma \in \mathfrak{S}_m} \mathrm{sgn}\, \sigma \cdot e_{p_{\sigma(1)}} \otimes \dots \otimes e_{p_{\sigma(m)}}.$$

This shows that the corresponding diagonal matrix element of the operator is nonzero only if all indices $p_1, \dots, p_m$ are distinct. Moreover, all $m!$ permutations of such m-tuple $(p_1, \dots, p_m)$ give the same contribution to the trace and so it is found by (2.10). Similarly, for the operator in (2.21) we get

$$H^{(m)} E_1 \dots E_m (e_{p_1} \otimes \dots \otimes e_{p_m}) = \frac{1}{m!} \lambda_{p_1} \dots \lambda_{p_m} \sum_{\sigma \in \mathfrak{S}_m} e_{p_{\sigma(1)}} \otimes \dots \otimes e_{p_{\sigma(m)}}.$$

Hence, if $(p_1, \dots, p_m)$ is a permutation of the multiset containing α_i entries equal to i for $i = 1, \dots, N$, then the corresponding diagonal matrix element of the operator equals

$$\frac{\alpha_1! \dots \alpha_N!}{m!} \lambda_{p_1} \dots \lambda_{p_m}.$$

Taking into account the number of such permutations, we arrive at (2.21). □

COROLLARY 2.1.6. *Each of the families*

$$\mathrm{tr}_{1,\dots,m}\, A^{(m)} E_1 \dots E_m \qquad \textit{and} \qquad \mathrm{tr}_{1,\dots,m}\, H^{(m)} E_1 \dots E_m$$

with $m = 1, \dots, N$ *is algebraically independent and generates the algebra* $\mathrm{S}(\mathfrak{gl}_N)^{\mathfrak{gl}_N}$.

PROOF. This follows from Proposition 2.1.5 since the elementary and complete symmetric polynomials are algebraically independent generators of the algebra of symmetric polynomials in N variables. □

As another consequence of Proposition 2.1.5, a formula for the coefficients of the characteristic polynomial follows:

$$C_m = \operatorname{tr}_{1,\dots,m} A^{(m)} E_1 \dots E_m, \qquad m = 1, \dots, N.$$

It is also implied by the identity

$$\det(u + E) = \operatorname{tr}_{1,\dots,N} A^{(N)} (u + E_1) \dots (u + E_N)$$

whose generalization (3.29) will be proved below.

2.2. Invariants in types B, C and D

Define the *orthogonal Lie algebras* $\mathfrak{o}_N$ with $N = 2n+1$ and $N = 2n$ (simple Lie algebras of types B_n with $n \geqslant 1$ and D_n with $n \geqslant 3$, respectively) and *symplectic Lie algebra* $\mathfrak{sp}_N$ with $N = 2n$ (simple Lie algebra of type C_n), as subalgebras of $\mathfrak{gl}_N$ spanned by the elements F_{ij},

$$F_{ij} = E_{ij} - E_{j'i'} \qquad \text{and} \qquad F_{ij} = E_{ij} - \varepsilon_i \, \varepsilon_j \, E_{j'i'}, \tag{2.23}$$

respectively, for $\mathfrak{o}_N$ and $\mathfrak{sp}_N$. As before, we use the notation $i' = N - i + 1$, and in the symplectic case we set $\varepsilon_i = 1$ for $i = 1, \dots, n$ and $\varepsilon_i = -1$ for $i = n+1, \dots, 2n$. Unless otherwise stated, we will consider the three cases B, C and D simultaneously and use the notation $\mathfrak{g}_N$ for any of the Lie algebras $\mathfrak{o}_N$ or $\mathfrak{sp}_N$. So we have

$$\mathfrak{g}_N = \{X \in \operatorname{Mat}_N(\mathbb{C}) \mid X + X' = 0\},$$

where we use the transposition defined for any $N \times N$ matrix $X = [X_{ij}]$ with entries in an associative algebra $\mathcal{A}$ by

$$(X')_{ij} = \begin{cases} X_{j'i'} & \text{in the orthogonal case,} \\ \varepsilon_i \, \varepsilon_j \, X_{j'i'} & \text{in the symplectic case.} \end{cases} \tag{2.24}$$

This is the transposition associated with the symmetric or skew-symmetric bilinear form on $\mathbb{C}^N$ defined on the basis vectors by

$$\langle e_i, e_j \rangle = g_{ij}. \tag{2.25}$$

The respective matrices $G = [g_{ij}]$ are antidiagonal and given by

$$g_{ij} = \begin{cases} \delta_{ij'} & \text{in the symmetric case,} \\ \varepsilon_i \, \delta_{ij'} & \text{in the skew-symmetric case,} \end{cases} \tag{2.26}$$

where $N = 2n$ is even in the skew-symmetric case. So (2.24) can be written as

$$X' = G X^t G^{-1},$$

where $X \to X^t$ is the usual matrix transposition, $(X^t)_{ij} = X_{ji}$.

When X is regarded as the element (1.59), this transposition can be understood as the anti-automorphism of the algebra $\operatorname{End} \mathbb{C}^N$ defined on the matrix unites by

$$e_{ij} \mapsto \begin{cases} e_{j'i'} & \text{in the orthogonal case,} \\ \varepsilon_i \, \varepsilon_j \, e_{j'i'} & \text{in the symplectic case.} \end{cases} \tag{2.27}$$

Accordingly, for each $a \in \{1, \dots, m\}$ the a-th transposition on the algebra (1.60) acts as the map (2.27) on the a-th copy of $\operatorname{End} \mathbb{C}^N$ and as the identity map on all other tensor factors.

LEMMA 2.2.1. *Suppose that two elements*

$$X = \sum_{i,j=1}^{N} e_{ij} \otimes X_{ij} \qquad \text{and} \qquad Y = \sum_{i,j=1}^{N} e_{ij} \otimes Y_{ij}$$

of the algebra $\operatorname{End}\mathbb{C}^N \otimes \mathcal{A}$ *satisfy the property*

$$X_{ij}\, Y_{kl} = Y_{kl}\, X_{ij} \qquad \text{for all} \quad i, j, k, l.$$

Then

$$(XY)' = Y'X'.$$

PROOF. In the orthogonal case we have

$$Y'X' = \sum_{i,j,k,l=1}^{N} e_{l'k'}\, e_{j'i'} \otimes Y_{kl}\, X_{ij} = \sum_{i,j,l=1}^{N} e_{l'i'} \otimes Y_{jl}\, X_{ij}.$$

On the other hand,

$$XY = \sum_{i,j,k,l=1}^{N} e_{ij}\, e_{kl} \otimes X_{ij}\, Y_{kl} = \sum_{i,j,l=1}^{N} e_{il} \otimes Y_{jl}\, X_{ij},$$

so that the application of the transposition yields $Y'X'$. The same argument is used in the symplectic case together with the relation $\varepsilon_i^2 = 1$. □

Applying the transposition (2.27) to the a-th and b-th copies of $\operatorname{End}\mathbb{C}^N$ in (1.60) and using Lemma 2.2.1, we derive from (1.67) that for an arbitrary element (1.59) we have

$$X_a' Q_{ab} = X_b Q_{ab} \qquad \text{and} \qquad Q_{ab} X_a = Q_{ab} X_b' \tag{2.28}$$

for all $1 \leqslant a < b \leqslant m$.

Introduce the $N \times N$ matrix $F = [F_{ij}]$ and note that the relations (2.23) read $F = E - E'$, where E is the matrix (2.5). Hence F possesses the symmetry property $F + F' = 0$. Regarding F as a matrix with entries in the symmetric algebra $\mathrm{S}(\mathfrak{g}_N)$, for its characteristic polynomial we find

$$\det(u + F) = \det(u - F') = (-1)^N \det(-u + GF^tG^{-1}) = (-1)^N \det(-u + F).$$

Therefore,

$$\det(u + F) = \begin{cases} u^{2n} + C_1\, u^{2n-2} + \dots + C_n & \text{if} \quad N = 2n, \\ u^{2n+1} + C_1\, u^{2n-1} + \dots + C_n\, u & \text{if} \quad N = 2n+1, \end{cases}$$

for some coefficients $C_i \in \mathrm{S}(\mathfrak{g}_N)$.

The orthogonal group O_N and symplectic group Sp_N are defined as the groups of complex matrices preserving the respective symmetric and skew-symmetric forms (2.25). We will use the notation G_N for either of these groups, so that

$$\mathrm{G}_N = \{\mathbf{g} \in \mathrm{Mat}_N(\mathbb{C}) \mid \mathbf{g}^t G\, \mathbf{g} = G\}.$$

Note that the matrix G has the property $G^{-1} = \pm G$ so that the definition of G_N can be written equivalently as

$$\mathrm{G}_N = \{\mathbf{g} \in \mathrm{Mat}_N(\mathbb{C}) \mid \mathbf{g}'\mathbf{g} = 1\}. \tag{2.29}$$

The group G_N acts on $\mathfrak{g}_N$ by the rule

$$\operatorname{Ad}\mathbf{g} : X \mapsto \mathbf{g} X \mathbf{g}^{-1}, \qquad X \in \mathfrak{g}_N, \quad \mathbf{g} \in \mathrm{G}_N\,. \tag{2.30}$$

For the image of the matrix F we obtain from (2.8) that

$$\mathrm{Ad}\,\mathbf{g} : F \mapsto \mathbf{g}^t\, F\, (\mathbf{g}^t)^{-1}. \tag{2.31}$$

Observe that $\mathbf{h} = \mathbf{g}^t$ belongs to G_N. Hence the same argument as for type A in Section 2.1 shows that all coefficients of the polynomial $\det\,(u + F)$ belong to the subalgebra of invariants $\mathrm{S}(\mathfrak{g}_N)^{\mathrm{G}_N}$. Furthermore, this subalgebra can be described by

$$\mathrm{S}(\mathfrak{g}_N)^{\mathrm{G}_N} = \mathbb{C}[C_1, \dots, C_n], \tag{2.32}$$

which follows by calculating the images of the coefficients C_m under the Chevalley isomorphism. To do the calculation, we make a standard choice of the triangular decomposition for $\mathfrak{g}_N$ so that the Cartan subalgebra $\mathfrak{h}$ is spanned by the elements $F_{11}, \dots, F_{nn}$, while the subalgebras $\mathfrak{n}_-$ and $\mathfrak{n}_+$ are spanned by the elements F_{ij} with $i > j$ and $i < j$, respectively. By setting $\lambda_i = F_{ii}$ for $i = 1, \dots, n$, we have

$$\varsigma : \det\,(u + F) \mapsto \begin{cases} (u - \lambda_1^2) \dots (u - \lambda_n^2) & \text{if} \quad N = 2n, \\ u\,(u - \lambda_1^2) \dots (u - \lambda_n^2) & \text{if} \quad N = 2n + 1. \end{cases}$$

As with type A, the traces of powers $\mathrm{tr}\, F^m$ of the matrix F are also G_N-invariants. They are zero for odd values of m due to the skew-symmetry property of F. Set

$$T_k = \mathrm{tr}\, F^{2k} \in \mathrm{S}(\mathfrak{g}_N)^{\mathrm{G}_N}.$$

We get an alternative family of generators,

$$\mathrm{S}(\mathfrak{g}_N)^{\mathrm{G}_N} = \mathbb{C}[T_1, \dots, T_n]$$

and

$$\varsigma : T_k \mapsto 2\,(\lambda_1^{2k} + \dots + \lambda_n^{2k}).$$

The algebra of invariants (2.32) coincides with $\mathrm{S}(\mathfrak{g}_N)^{\mathfrak{g}_N}$ in types B and C, but is properly contained in $\mathrm{S}(\mathfrak{g}_N)^{\mathfrak{g}_N}$ in type D. If $\mathfrak{g}_N = \mathfrak{o}_{2n}$, then

$$C_n = \det F = (-1)^n\, \big(\mathrm{Pf}\ F\big)^2,$$

where $\mathrm{Pf}\ F$ is the *Pfaffian* defined by[1]

$$\mathrm{Pf}\ F = \frac{1}{2^n n!} \sum_{\sigma \in \mathfrak{S}_{2n}} \mathrm{sgn}\, \sigma \cdot F_{\sigma(1)\,\sigma(2)'} \dots F_{\sigma(2n-1)\,\sigma(2n)'}. \tag{2.33}$$

Indeed, for the even skew-symmetric matrix $[F_{i\,j'}] = FG$ we have the well-known relation $\det FG = \big(\mathrm{Pf}\ F\big)^2$ and note that $\det G = (-1)^n$. The Pfaffian $\mathrm{Pf}\ F$ is found by the expansion

$$\frac{\Phi^n}{n!} = \big(e_1 \wedge \dots \wedge e_{2n}\big) \otimes \mathrm{Pf}\ F, \tag{2.34}$$

with

$$\Phi = \sum_{i<j} (e_i \wedge e_j) \otimes F_{i\,j'} \in \Lambda(\mathbb{C}^{2n}) \otimes \mathrm{S}(\mathfrak{o}_{2n}),$$

[1] This is actually the Pfaffian of the skew-symmetric matrix $[F_{i\,j'}]$. We abuse notation slightly by introducing the Pfaffian of the matrix which is skew-symmetric with respect to the *antidiagonal*.

where $\Lambda(\mathbb{C}^{2n})$ is the exterior algebra of the vector space $\mathbb{C}^{2n}$. Under the map (2.31) we have $FG \mapsto \mathbf{h}FG\mathbf{h}^t$ with $\mathbf{h} = \mathbf{g}^t$. Hence, by using (2.34) we find that the image of Pf F under this map is given by

$$\text{Pf } F \mapsto \det \mathbf{h} \cdot \text{Pf } F.$$

This means that Pf F is invariant under the adjoint action of the special orthogonal group SO_{2n}, and hence under the action of $\mathfrak{o}_{2n}$. However, Pf F does not belong to the algebra (2.32) for $\mathrm{G}_N = \mathrm{O}_{2n}$. Under the Chevalley isomorphism we have

$$\varsigma : \text{Pf } F \mapsto \lambda_1 \dots \lambda_n.$$

Thus, the subalgebra of $\mathfrak{g}_N$-invariants can be described by

$$\mathrm{S}(\mathfrak{g}_N)^{\mathfrak{g}_N} = \begin{cases} \mathbb{C}[C_1, \dots, C_n] & \text{for } \mathfrak{g}_N = \mathfrak{o}_{2n+1}, \ \mathfrak{sp}_{2n}, \\ \mathbb{C}[C_1, \dots, C_{n-1}, \text{Pf } F] & \text{for } \mathfrak{g}_N = \mathfrak{o}_{2n}, \end{cases} \tag{2.35}$$

since the Chevalley images of the corresponding elements are algebraically independent generators of the algebra of W-invariants of $\mathrm{S}(\mathfrak{h})$.

To prove the orthogonal and symplectic analogues of Proposition 2.1.1, consider the action of the Brauer algebra on the tensor product space $(\mathbb{C}^N)^{\otimes m}$ as defined in Section 1.5. The action of the group G_N on this space is obtained from (2.13) by restriction so that

$$\mathbf{h} : v_1 \otimes \dots \otimes v_m \mapsto \mathbf{h}v_1 \otimes \dots \otimes \mathbf{h}v_m, \qquad v_i \in \mathbb{C}^N, \qquad \mathbf{h} \in \mathrm{G}_N,$$

which we write as $\mathbf{h} \mapsto \mathbf{h}_1 \dots \mathbf{h}_m$. The action of any element $s \in \mathcal{B}_m(\omega)$ with the respective values $\omega = N$ or $\omega = -N$ on the vector space $(\mathbb{C}^N)^{\otimes m}$ commutes with that of any element $\mathbf{h} \in \mathrm{G}_N$. Indeed, this was already verified for the generators $s_{a\,a+1}$ in (2.14), whereas for the action of the generators $\epsilon_{a\,a+1}$ we have

$$Q_{a\,a+1}\mathbf{h}_a\,\mathbf{h}_{a+1} = Q_{a\,a+1}\mathbf{h}'_{a+1}\mathbf{h}_{a+1} = Q_{a\,a+1} = \mathbf{h}_a\,\mathbf{h}'_a Q_{a\,a+1} = \mathbf{h}_a\,\mathbf{h}_{a+1}Q_{a\,a+1}$$

where we used (2.29) and the properties

$$Q_{a\,a+1}\mathbf{h}_a = Q_{a\,a+1}\mathbf{h}'_{a+1} \qquad \text{and} \qquad \mathbf{h}'_a Q_{a\,a+1} = \mathbf{h}_{a+1}Q_{a\,a+1}$$

implied by (2.28). Therefore,

$$Q_{a\,a+1}\,\mathbf{h}_1 \dots \mathbf{h}_m = \mathbf{h}_1 \dots \mathbf{h}_m\,Q_{a\,a+1}.$$

We will return to the commuting actions of the Brauer algebra and the group G_N in the context of the *Brauer–Schur–Weyl duality* in Section 5.2.

Let S denote the image of $s \in \mathcal{B}_m(\omega)$ under the respective map (1.69) or (1.73). We regard S as an element of the algebra

$$\underbrace{\mathrm{End}\,\mathbb{C}^N \otimes \dots \otimes \mathrm{End}\,\mathbb{C}^N}_{m} \otimes \mathrm{S}(\mathfrak{g}_N) \tag{2.36}$$

by identifying it with $S \otimes 1$. Also, we regard the matrix F as the element

$$F = \sum_{i,j=1}^{N} e_{ij} \otimes F_{ij} \in \mathrm{End}\,\mathbb{C}^N \otimes \mathrm{S}(\mathfrak{g}_N).$$

PROPOSITION 2.2.2. *For any $s \in \mathcal{B}_m(\omega)$ with $\omega = N$ or $\omega = -N$, respectively, the element*

$$\mathrm{tr}_{1,\dots,m}\, S\, F_1 \dots F_m \tag{2.37}$$

belongs to the algebra of invariants $\mathrm{S}(\mathfrak{g}_N)^{\mathrm{G}_N}$. If m is odd, then the element is zero.

PROOF. Use the same argument as for the proof of Proposition 2.1.1. It relies on the verified above property that the operator S commutes with the product $\mathbf{h}_1 \dots \mathbf{h}_m$ for any $\mathbf{h} \in \mathrm{G}_N$. All the generators C_k in (2.32) are homogeneous polynomials of even degree in the variables F_{ij}. Since the invariant (2.37) is a homogeneous polynomial of an odd degree, it must be zero. □

We will now use Proposition 2.2.2 for some particular choices of the element s. Let $S^{(m)}$ and $A^{(m)}$ denote the respective images of the symmetrizer $s^{(m)}$ and anti-symmetrizer $a^{(m)}$ in the Brauer algebra $\mathcal{B}_m(\omega)$ (defined by equivalent formulas in Section 1.2) under its actions (1.69) (with $\omega = N$) or (1.73) (with $\omega = -N$). Consider the algebra (2.36) and employ the tensor notation of Section 1.4. We have the following analogue of Proposition 2.1.5.

PROPOSITION 2.2.3. *Under the Chevalley isomorphism for* $\mathfrak{g}_N = \mathfrak{o}_N$ *we have*

$$\varsigma : \mathrm{tr}_{1,\dots,2k} A^{(2k)} F_1 \dots F_{2k} \mapsto (-1)^k \, e_k(\lambda_1^2, \dots, \lambda_n^2),$$

and for $\mathfrak{g}_N = \mathfrak{sp}_{2n}$ *we have*

$$\varsigma : \mathrm{tr}_{1,\dots,2k} A^{(2k)} F_1 \dots F_{2k} \mapsto h_k(\lambda_1^2, \dots, \lambda_n^2). \tag{2.38}$$

PROOF. To calculate the traces, we will regard F as the diagonal matrix

$$F = \begin{cases} \mathrm{diag}\,[\lambda_1, \dots, \lambda_n, -\lambda_n, \dots, -\lambda_1] & \text{if} \quad N = 2n, \\ \mathrm{diag}\,[\lambda_1, \dots, \lambda_n, 0, -\lambda_n, \dots, -\lambda_1] & \text{if} \quad N = 2n+1. \end{cases} \tag{2.39}$$

Hence, by (2.20), in the orthogonal case it will be sufficient to verify the relation

$$e_{2k}(\lambda_1, \lambda_2, \dots, -\lambda_2, -\lambda_1) = (-1)^k \, e_k(\lambda_1^2, \dots, \lambda_n^2).$$

It is immediate from the identity

$$\sum_{m=0}^{N} u^{N-m} \, e_m(\lambda_1, \lambda_2, \dots, -\lambda_2, -\lambda_1) = (u + \lambda_1)(u + \lambda_2) \dots (u - \lambda_2)(u - \lambda_1).$$

In the symplectic case the operator $A^{(m)}$ on the tensor product space coincides with the action of the symmetrizer $h^{(m)}$ defined by (1.65), because of the signs in (1.73). Therefore, the claim follow from (2.21) due to the generating function identity

$$\begin{aligned} \sum_{m=0}^{\infty} t^m \, h_m(\lambda_1, \lambda_2, \dots, -\lambda_2, -\lambda_1) &= \frac{1}{(1 - t\lambda_1)(1 - t\lambda_2) \dots (1 + t\lambda_2)(1 + t\lambda_1)} \\ &= \frac{1}{(1 - t^2\lambda_1^2) \dots (1 - t^2\lambda_n^2)}, \end{aligned}$$

thus completing the proof. □

The calculation of the Chevalley images of the invariants associated with the symmetrizer $S^{(m)}$ will require some extra work. Set

$$\gamma_m(\omega) = \frac{\omega + m - 2}{\omega + 2m - 2} \tag{2.40}$$

and consider the orthogonal case first.

PROPOSITION 2.2.4. *If* $\mathfrak{g}_N = \mathfrak{o}_N$ *then under the Chevalley isomorphism we have*

$$\varsigma : \gamma_{2k}(N) \, \mathrm{tr}_{1,\dots,2k} S^{(2k)} F_1 \dots F_{2k} \mapsto h_k(\lambda_1^2, \dots, \lambda_n^2). \tag{2.41}$$

PROOF. Assuming that $F = \mathrm{diag}[\lambda_1, \dots, \lambda_N]$ is a diagonal matrix of the form (2.39) and m is a positive integer we will calculate the trace

$$\mathrm{tr}_{1,\dots,m}\, S^{(m)} F_1 \dots F_m \tag{2.42}$$

by applying the operator to a basis vector $e_{p_1} \otimes \dots \otimes e_{p_m}$. Using the expression (1.41) for the symmetrizer, we can write

$$S^{(m)} = H^{(m)} \sum_{r=0}^{\lfloor m/2 \rfloor} \frac{(-1)^r}{2^r\, r!} \binom{N/2 + m - 2}{r}^{-1} \sum_{a_i < b_i} Q_{a_1 b_1} Q_{a_2 b_2} \dots Q_{a_r b_r}, \tag{2.43}$$

with the second sum taken over the sets of disjoint pairs $\{(a_1, b_1), \dots, (a_r, b_r)\}$ of indices from $\{1, \dots, m\}$. Since $H^{(m)}$ commutes with the sum

$$\sum_{a_i < b_i} Q_{a_1 b_1} Q_{a_2 b_2} \dots Q_{a_r b_r} \tag{2.44}$$

for each fixed value of r, all permutations of the m-tuple $(p_1, \dots, p_m)$ give the same contribution to the trace. Moreover, since $\lambda_{n+1} = 0$ in the case $N = 2n + 1$, it suffices to consider basis vectors of the form

$$\underbrace{e_1 \otimes \dots \otimes e_1}_{q_1} \otimes \underbrace{e_{1'} \otimes \dots \otimes e_{1'}}_{q_{1'}} \otimes \dots \otimes \underbrace{e_n \otimes \dots \otimes e_n}_{q_n} \otimes \underbrace{e_{n'} \otimes \dots \otimes e_{n'}}_{q_{n'}} \tag{2.45}$$

with $q_1 + \dots + q_{n'} = m$. The product $Q_{a_1 b_1} Q_{a_2 b_2} \dots Q_{a_r b_r}$ occurring in (2.44) annihilates this vector, unless for each pair (a_i, b_i) the a_i-th and b_i-th tensor factors are, respectively, e_l and $e_{l'}$ for some l. For such a pair we have

$$Q_{a_i b_i} \big(\dots \otimes e_l \otimes \dots \otimes e_{l'} \otimes \dots \big) = \sum_{l_i = 1}^{N} \big(\dots \otimes e_{l_i} \otimes \dots \otimes e_{l_i'} \otimes \dots \big).$$

Suppose that $s_1, \dots, s_n$ are nonnegative integers such that $s_1 + \dots + s_n = r$ and suppose that for each $l = 1, \dots, n$ the number of the pairs (a_i, b_i) corresponding to l is s_l. Then the number of the products $Q_{a_1 b_1} Q_{a_2 b_2} \dots Q_{a_r b_r}$ with this property can be evaluated as

$$\prod_{l=1}^{n} \binom{q_l}{s_l} \binom{q_{l'}}{s_l} s_l!.$$

Furthermore, due to the subsequent application of the operator $H^{(m)}$ in (2.43), all permutations of the factors in the tensor products of the basis vectors of $\mathbb{C}^N$ give the same contribution to the diagonal matrix element. The same observation applies to the 2^r swaps of the tensor factors $e_{l_i} \otimes e_{l_i'} \mapsto e_{l_i'} \otimes e_{l_i}$ for the values $l_i' = 1, \dots, n$. Clearly, we may ignore the values $l_i = n + 1$ in the case $N = 2n + 1$. Thus, taking into account the number of permutations

$$\frac{r!}{s_1! \dots s_n!}$$

of an r-multiset with the multiplicities $s_1, \dots, s_n$, we find that the diagonal matrix element of the operator

$$H^{(m)} \sum_{a_i < b_i} Q_{a_1 b_1} Q_{a_2 b_2} \dots Q_{a_r b_r} F_1 \dots F_m$$

corresponding to the vector (2.45) equals

$$\frac{q_1!\,q_{1'}!\dots q_n!\,q_{n'}!}{m!} \times 2^r r! \sum \binom{q_1}{s_1}\binom{q_{1'}}{s_1}\dots\binom{q_n}{s_n}\binom{q_{n'}}{s_n}\lambda_1^{q_1}(-\lambda_1)^{q_{1'}}\dots\lambda_n^{q_n}(-\lambda_n)^{q_{n'}},$$

summed over all n-tuples of nonnegative integers s_l such that $s_1+\dots+s_n=r$. Note that the number of different basis vectors obtained by permuting the tensor factors in (2.45) equals

$$\frac{m!}{q_1!\,q_{1'}!\dots q_n!\,q_{n'}!}.$$

The trace (2.42) will be found by taking the sum of the diagonal matrix elements corresponding to all basis vectors and by taking the sum over the parameter r with the coefficients occurring in (2.43). Let us look at the coefficient of a monomial of the form $\lambda_1^{m_1}\dots\lambda_n^{m_n}$ in the resulting expression for the trace. The sum

$$\sum_{q_i+q_{i'}=m_i}(-1)^{q_{i'}}\binom{q_i}{s_i}\binom{q_{i'}}{s_i} \tag{2.46}$$

with a fixed value of s_i can be calculated as follows. Recalling the generating function

$$\sum_{q=0}^{\infty}\binom{q}{s}z^q=\frac{z^s}{(1-z)^{s+1}}$$

we find that the generating function of the coefficients (2.46) is given by

$$\frac{z^{s_i}}{(1-z)^{s_i+1}}\,\frac{(-z)^{s_i}}{(1+z)^{s_i+1}}=\frac{(-1)^{s_i}z^{2s_i}}{(1-z^2)^{s_i+1}}.$$

This formal power series is even so that the sum (2.46) is zero for odd m_i. Therefore, if m is odd then the trace (2.42) is zero, in agreement with Proposition 2.2.2, because at least one power m_i must be odd. If $m=2k$ is even, then we may assume that all powers m_i of the monomial $\lambda_1^{m_1}\dots\lambda_n^{m_n}$ are even, say $m_i=2k_i$. The sum (2.46) then simplifies to $(-1)^{s_i}\binom{k_i}{s_i}$. Hence the coefficient of $\lambda_1^{2k_1}\dots\lambda_n^{2k_n}$ in the trace (2.42) with $m=2k$ equals

$$\sum_{r=0}^{k}(-1)^r\binom{N/2+2k-2}{r}^{-1}\sum_{s_1+\dots+s_n=r}(-1)^{s_1}\binom{k_1}{s_1}\dots(-1)^{s_n}\binom{k_n}{s_n}$$

which clearly coincides with

$$\sum_{r=0}^{k}\binom{N/2+2k-2}{r}^{-1}\binom{k}{r}. \tag{2.47}$$

By the recurrence relation for the binomial coefficients we have the identity

$$\sum_{r=0}^{k}\binom{x-r}{k-r}=\binom{x+1}{k}$$

which holds for an arbitrary variable x. By multiplying both sides by the inverse of $\binom{x}{k}$ we get

$$\sum_{r=0}^{k} \binom{x}{r}^{-1} \binom{k}{r} = \frac{x+1}{x-k+1}. \tag{2.48}$$

Hence the sum (2.47) equals

$$\frac{N+4k-2}{N+2k-2} = \gamma_{2k}(N)^{-1}$$

as required. □

COROLLARY 2.2.5.
(i) *The family*

$$\gamma_{2k}(N)\, \mathrm{tr}_{1,\dots,2k}\, S^{(2k)} F_1 \dots F_{2k} \tag{2.49}$$

with $k = 1, \dots, n$ *is algebraically independent and generates the algebra* $\mathrm{S}(\mathfrak{o}_N)^{\mathfrak{o}_N}$ *for* $N = 2n+1$.

(ii) *The family* (2.49) *with* $k = 1, \dots, n-1$ *together with* Pf F *is algebraically independent and generates the algebra* $\mathrm{S}(\mathfrak{o}_N)^{\mathfrak{o}_N}$ *for* $N = 2n$.

PROOF. Both claims follow from Proposition 2.2.4, since the Chevalley images of the families are algebraically independent generators of the respective algebras of W-invariants $\mathrm{S}(\mathfrak{h})^W$. □

In the symplectic case the symmetrizer $S^{(m)}$ is well-defined for the values of m satisfying $m \leqslant n+1$, while $S^{(m)}$ is *not* defined for $m \geqslant n+2$ as some denominators in the formulas defining $s^{(m)}$ vanish for $\omega = -2n$. In fact, $S^{(n+1)} = 0$, as can be seen from the Howe duality interpretation of the symmetrizer (see Proposition 2.3.2 below) or from the Brauer–Schur–Weyl duality (see decomposition (5.13) in Section 5.2). Nonetheless, the definition of the invariants (2.37) with $S = S^{(m)}$ can be extended to all values $1 \leqslant m \leqslant 2n+1$ by a certain kind of 'analytic continuation' as we demonstrate below in Proposition 2.2.8. We keep using the notation (2.40).

PROPOSITION 2.2.6. *If* $\mathfrak{g}_N = \mathfrak{sp}_{2n}$ *and* $2k \leqslant n$ *then under the Chevalley isomorphism we have*

$$\varsigma : \gamma_{2k}(-2n)\, \mathrm{tr}_{1,\dots,2k}\, S^{(2k)} F_1 \dots F_{2k} \mapsto (-1)^k\, e_k(\lambda_1^2, \dots, \lambda_n^2). \tag{2.50}$$

PROOF. Assume that $F = \mathrm{diag}[\lambda_1, \dots, \lambda_{2n}]$ is a diagonal matrix of the form (2.39) and for a positive integer $m \leqslant n$ calculate the trace

$$\mathrm{tr}_{1,\dots,m}\, S^{(m)} F_1 \dots F_m \tag{2.51}$$

by applying the operator to basis vectors of the tensor product space. We let $H^{(m)}$ denote the image of the element $h^{(m)}$ of the Brauer algebra $\mathcal{B}_m(-2n)$ under the action (1.73). Note that $H^{(m)}$ acts as the anti-symmetrization operator on the tensor product space. Using the expression (1.41) for the symmetrizer, we can write

$$S^{(m)} = H^{(m)} \sum_{r=0}^{\lfloor m/2 \rfloor} \frac{1}{2^r r!} \binom{-n+m-2}{r}^{-1} \sum_{a_i < b_i} Q_{a_1 b_1} Q_{a_2 b_2} \dots Q_{a_r b_r}, \tag{2.52}$$

with the second sum taken over the sets of disjoint pairs $\{(a_1, b_1), \ldots, (a_r, b_r)\}$ of indices from $\{1, \ldots, m\}$. Since the operator $H^{(m)}$ commutes with the sum

$$\sum_{a_i<b_i} Q_{a_1 b_1} Q_{a_2 b_2} \ldots Q_{a_r b_r} \tag{2.53}$$

for each fixed value of r, it will be sufficient to calculate the diagonal matrix elements for the tensor products of distinct basis vectors of the form

$$e_{c_1} \otimes e_{c_1'} \otimes \ldots \otimes e_{c_p} \otimes e_{c_p'} \otimes e_{d_1} \otimes \ldots \otimes e_{d_{m-2p}}, \qquad p = 0, 1, \ldots, \lfloor m/2 \rfloor, \tag{2.54}$$

where $1 \leqslant c_1 < \cdots < c_p \leqslant n$ and $1 \leqslant d_1 < \cdots < d_{m-2p} \leqslant 2n$ with the condition that $d_i + d_j \neq 2n+1$ for all i and j. For $i = 1, \ldots, p$ we have

$$Q_{2i-1\, 2i}(e_{c_i} \otimes e_{c_i'}) = \sum_{l=1}^{n} (e_l \otimes e_{l'} - e_{l'} \otimes e_l).$$

By applying the product of $H^{(m)}$ and the operator (2.53) to a basis vector of the form (2.54) we find that the coefficient of this vector in the expansion equals $2^r \dfrac{r!}{m!} \dbinom{p}{r}$. Since $\lambda_{i'} = -\lambda_i$, the contributions of the basis vectors with $m - 2p > 0$ to the trace in (2.51) add up to zero. In particular, this shows that if m is odd, then the trace (2.51) is zero, in agreement with Proposition 2.2.2. Now suppose that $m = 2k$ is even. We may assume that $p = k$ in (2.54), so that the coefficient of the monomial $\lambda_{c_1}^2 \ldots \lambda_{c_k}^2$ in the trace equals

$$(-1)^k \sum_{r=0}^{k} \binom{-n+2k-2}{r}^{-1} \binom{k}{r} = (-1)^k \frac{-n+2k-1}{-n+k-1} = (-1)^k \gamma_{2k}(-2n)^{-1},$$

due to the identity (2.48), as required. □

Now we would like to extend the construction of the invariants in $\mathrm{S}(\mathfrak{sp}_{2n})$ associated with the symmetrizer $S^{(m)}$ to all values $1 \leqslant m \leqslant 2n+1$.

LEMMA 2.2.7. *If $m \leqslant n$ then for the m-th partial trace we have*

$$\mathrm{tr}_m\, S^{(m)} F_m = -\frac{n-m+1}{m(n-m+2)} \sum_{a=1}^{m-1} F_a\, S^{(m-1)}.$$

PROOF. The left hand side also equals $\mathrm{tr}_m\, F_m S^{(m)}$ by the cyclic property of trace. Take the image of the recurrence relation (1.53) under the homomorphism (1.73) to get

$$S^{(m)} = \frac{1}{m(-2n+2m-4)} \Big(-1 + \sum_{a=1}^{m-1} (P_{a\,m} - Q_{a\,m})\Big) \times \Big(2n - m + 3 + \sum_{a=1}^{m-1} (P_{a\,m} - Q_{a\,m})\Big)\, S^{(m-1)}.$$

Now take into account (1.75) to calculate the partial trace

$$\mathrm{tr}_m F_m\Big(-1+\sum_{a=1}^{m-1}(P_{a\,m}-Q_{a\,m})\Big) \times \Big(2n-m+3+\sum_{a=1}^{m-1}(P_{a\,m}-Q_{a\,m})\Big)\, S^{(m-1)}. \tag{2.55}$$

We have $\mathrm{tr}_m F_m = 0$, and using (1.67), (1.68), (1.76) and (2.28) we find

$$\mathrm{tr}_m F_m P_{a\,m} = \mathrm{tr}_m P_{a\,m} F_a = F_a \qquad \text{and} \qquad \mathrm{tr}_m F_m Q_{a\,m} = \mathrm{tr}_m F_a' Q_{a\,m} = -F_a.$$

Furthermore, if $a, b \in \{1, \dots, m-1\}$ are distinct indices, then

$$\mathrm{tr}_m F_m\,(P_{a\,m}-Q_{a\,m})\,P_{b\,m} = \mathrm{tr}_m P_{b\,m} F_b\,(P_{a\,b}-Q_{a\,b}) = F_b\,(P_{a\,b}-Q_{a\,b}),$$

where we set $P_{a\,b} = P_{b\,a}$ and $Q_{a\,b} = Q_{b\,a}$ for $a > b$. Similarly,

$$\begin{aligned}\mathrm{tr}_m F_m\,(P_{a\,m}-Q_{a\,m})\,Q_{b\,m} &= \mathrm{tr}_m F_m\,(Q_{a\,b}-P_{a\,b})\,Q_{b\,m}\\ &= \mathrm{tr}_m\,(Q_{a\,b}-P_{a\,b})\,F_m Q_{b\,m} = (P_{a\,b}-Q_{a\,b})\,F_b.\end{aligned}$$

By the properties (1.31) of the symmetrizer, we have

$$F_b\,(P_{a\,b}-Q_{a\,b})\,S^{(m-1)} = -F_b\,S^{(m-1)},$$

whereas

$$(P_{a\,b}-Q_{a\,b})\,F_b\,S^{(m-1)} = (F_a P_{a\,b} - Q_{a\,b} F_b)\,S^{(m-1)} = -F_a\,S^{(m-1)}.$$

The last relation holds since

$$Q_{a\,b} F_b\,S^{(m-1)} = -Q_{a\,b} P_{a\,b} F_b\,S^{(m-1)} = Q_{a\,b} F_a\,S^{(m-1)} = -Q_{a\,b} F_b\,S^{(m-1)},$$

so that $Q_{a\,b} F_b\,S^{(m-1)} = 0$. Thus, collecting all the terms, we conclude that (2.55) equals

$$2\,(n-m+1)\sum_{a=1}^{m-1} F_a\,S^{(m-1)}$$

completing the proof. □

PROPOSITION 2.2.8. *The expression*

$$\gamma_m(-2n)\,\mathrm{tr}_{1,\dots,m}\,S^{(m)} F_1 \dots F_m \tag{2.56}$$

admits an equivalent form representing a well-defined element of $\mathrm{S}(\mathfrak{sp}_{2n})^{\mathfrak{sp}_{2n}}$ *for all* $1 \leqslant m \leqslant 2n+1$.

PROOF. Suppose that $m \leqslant n$. By calculating the m-th partial trace with the use of Lemma 2.2.7, and applying the cyclic property of trace, we will write (2.56) as a linear combination of expressions of the form

$$\gamma_{m-1}(-2n)\,\mathrm{tr}_{1,\dots,m-1}\,S^{(m-1)} F_1 \dots F_a^2 \dots F_{m-1}, \qquad a = 1, \dots, m-1.$$

All coefficients of this linear combination are well-defined for all $1 \leqslant m \leqslant 2n+1$. By (1.31) and (1.73), we have

$$S^{(m-1)} = (-1)^{a-1}\,S^{(m-1)}\,P_\sigma$$

for the cycle $\sigma = (1, 2, \dots, a)$. Hence,

$$\begin{aligned}\operatorname{tr}_{1,\dots,m-1} S^{(m-1)} F_1 \dots F_a^2 \dots F_{m-1} &= (-1)^{a-1} \operatorname{tr}_{1,\dots,m-1} S^{(m-1)} P_\sigma F_a^2 F_1 \dots F_{a-1} F_{a+1} \dots F_{m-1}\\ &= (-1)^{a-1} \operatorname{tr}_{1,\dots,m-1} S^{(m-1)} F_1^2 F_2 \dots F_a F_{a+1} \dots F_{m-1} P_\sigma\end{aligned}$$

which equals

$$\operatorname{tr}_{1,\dots,m-1} S^{(m-1)} F_1^2 F_2 \dots F_{m-1}$$

by the cyclic property of trace. Therefore, (2.56) is proportional to the expression

$$\gamma_{m-1}(-2n) \operatorname{tr}_{1,\dots,m-1} S^{(m-1)} F_1^2 F_2 \dots F_{m-1}. \tag{2.57}$$

Now apply Lemma 2.2.7 again for the $(m-1)$-th partial trace. Repeating the previous calculation we will represent (2.57) as a linear combination of two expressions

$$\gamma_{m-2}(-2n) \operatorname{tr}_{1,\dots,m-2} S^{(m-2)} F_1^3 F_2 \dots F_{m-2}$$

and

$$\gamma_{m-2}(-2n) \operatorname{tr}_{1,\dots,m-2} S^{(m-2)} F_1^2 F_2^2 F_3 \dots F_{m-2}$$

so that the coefficients are well-defined for all $1 \leqslant m \leqslant 2n+1$. Continuing with the $(m-2)$-th partial trace, etc., these calculations can be performed l times in total, with any l satisfying $2l \leqslant m+1$ to write (2.56) as a linear combination of expressions of the form

$$\gamma_{m-l}(-2n) \operatorname{tr}_{1,\dots,m-l} S^{(m-l)} F_1^{k_1} \dots F_{m-l}^{k_{m-l}}. \tag{2.58}$$

Observe that (2.58) is well-defined for $m - l \leqslant n$. Hence, if $n+1 \leqslant m \leqslant 2n+1$, then taking $l = m - n$ we get a required expression defined for these values of m. Moreover, every element (2.58) belongs to the subalgebra of invariants $\mathrm{S}(\mathfrak{sp}_{2n})^{\mathfrak{sp}_{2n}}$; cf. Remark 2.1.2.

To complete the proof, we need to show that all possible linear combinations representing the expression (2.56) for $n+1 \leqslant m \leqslant 2n+1$, obtained by the described above procedure, coincide, as elements of the symmetric algebra $\mathrm{S}(\mathfrak{sp}_{2n})$. If the condition $m \leqslant n$ holds, then by (2.32) the invariant (2.56) is a unique polynomial in the generators $C_1, \dots, C_n$,

$$\gamma_m(-2n) \operatorname{tr}_{1,\dots,m} S^{(m)} F_1 \dots F_m = \sum a^{(m)}_{k_1,\dots,k_n} C_1^{k_1} \dots C_n^{k_n},$$

with the summation over nonnegative integers k_a subject to the degree condition $2k_1 + 4k_2 + \dots + 2nk_n = m$. In particular, the invariant is zero if m is odd. Moreover, $k_a = 0$ unless $2a \leqslant m$ so we may write the coefficients as $a^{(m)}_{k_1,\dots,k_n} = a^{(m)}_{k_1,\dots,k_p}$ with $p = \lfloor m/2 \rfloor$. Now suppose that m is fixed and let n vary. The coefficients $a^{(m)}_{k_1,\dots,k_p}$ are rational functions in n and so they are determined by infinitely many values of n with $n \geqslant m$. The first part of the proof demonstrates that the rational functions $a^{(m)}_{k_1,\dots,k_p}$ are defined for all values $n \geqslant (m-1)/2$, that is, for $n \geqslant p$. Thus, we may conclude that since all equivalent expressions for (2.56) obtained in the first part of the proof coincide, as elements of $\mathrm{S}(\mathfrak{sp}_{2n})^{\mathfrak{sp}_{2n}}$ for infinitely many values of n, they must coincide for all values of $n \geqslant (m-1)/2$ for which they are defined. □

We will use expression (2.56) for all values $1 \leqslant m \leqslant 2n+1$, with the understanding that it represents a well-defined element of $\mathrm{S}(\mathfrak{sp}_{2n})^{\mathfrak{sp}_{2n}}$ obtained in the proof of Proposition 2.2.8.

EXAMPLE 2.2.9. For $m = 2$ and $n \geqslant 2$ expression (2.56) takes the form

$$\gamma_2(-2n)\, \mathrm{tr}_{1,2}\, S^{(2)} F_1\, F_2 = \frac{n}{n-1}\, \mathrm{tr}_{1,2} \Big(\frac{1}{2} - \frac{P_{12}}{2} - \frac{Q_{12}}{2n}\Big) F_1\, F_2.$$

Since $\mathrm{tr}\, F = 0$, the properties (1.68) and (1.76) together with (1.67) and (2.28) imply that this expression coincides with $-\frac{1}{2}\,\mathrm{tr}\, F^2$. This is a well-defined element of $\mathrm{S}(\mathfrak{sp}_{2n})^{\mathfrak{sp}_{2n}}$ for all $n \geqslant 1$. □

COROLLARY 2.2.10. *Proposition 2.2.6 is valid for all values $1 \leqslant m \leqslant 2n+1$. Moreover, the family*

$$\gamma_{2k}(-2n)\, \mathrm{tr}_{1,\dots,2k}\, S^{(2k)} F_1 \dots F_{2k} \tag{2.59}$$

with $k = 1, \dots, n$ is algebraically independent and generates the algebra $\mathrm{S}(\mathfrak{sp}_{2n})^{\mathfrak{sp}_{2n}}$.

PROOF. As the proof of Proposition 2.2.8 shows, for a fixed m the polynomial in $C_1, \dots, C_p$ representing the element (2.56) is determined by the values $n \geqslant m$. Hence the claim follows from Proposition 2.2.6. The second part of the corollary holds since the Chevalley images of the elements (2.59) are algebraically independent generators of the algebra $\mathrm{S}(\mathfrak{h})^W$. □

2.3. Symmetrizer and extremal projector

Consider the orthogonal case $\mathfrak{g}_N = \mathfrak{o}_N$ first. We regard the operator $H^{(m)}$ on tensor product space

$$\underbrace{\mathbb{C}^N \otimes \mathbb{C}^N \otimes \dots \otimes \mathbb{C}^N}_{m} \tag{2.60}$$

as the image of the element $h^{(m)} \in \mathcal{B}_m(N)$ under the action (1.65). Denote by $\mathcal{P}_N$ the space of polynomials in variables $z_1, \dots, z_N$ and identify the image of the vector space (2.60) under the operator $H^{(m)}$ with the subspace $\mathcal{P}_N^m$ of homogeneous polynomials of degree m,

$$H^{(m)}(\mathbb{C}^N)^{\otimes m} \cong \mathcal{P}_N^m \tag{2.61}$$

via the isomorphism

$$H^{(m)}(e_{i_1} \otimes \dots \otimes e_{i_m}) \mapsto z_{i_1} \dots z_{i_m}. \tag{2.62}$$

Due to (2.43) we may consider $S^{(m)}$ as an operator on the space $\mathcal{P}_N^m$. This operator commutes with the action of the orthogonal group O_N on this space and so it can be expressed in terms of the action of the generators of the Lie algebra $\mathfrak{sl}_2$ provided by a particular case of the *Howe duality* [**67**, Section 3.4]. To get such an expression, let $\{e, f, h\}$ be the standard basis of $\mathfrak{sl}_2$ with the commutation relations

$$[e, f] = h, \qquad [h, e] = 2e, \qquad [h, f] = -2f.$$

The action of the basis elements on the space $\mathcal{P}_N$ is given by

$$e \mapsto -\frac{1}{2} \sum_{i=1}^{N} \partial_i\, \partial_{i'}, \qquad f \mapsto \frac{1}{2} \sum_{i=1}^{N} z_i\, z_{i'}, \qquad h \mapsto -\frac{N}{2} - \sum_{i=1}^{N} z_i\, \partial_i, \tag{2.63}$$

where ∂_i denotes the partial derivative over z_i. Recall that the *extremal projector* p for the Lie algebra $\mathfrak{sl}_2$ is given by the formula

$$p = 1 + \sum_{r=1}^{\infty} \frac{(-1)^r}{r!\,(h+2)\dots(h+r+1)}\, f^r e^r. \tag{2.64}$$

The projector p is an element of an extension of the universal enveloping algebra $\mathrm{U}(\mathfrak{sl}_2)$ and it possesses the properties

$$ep = pf = 0.$$

The center of the universal enveloping algebra $\mathrm{U}(\mathfrak{sl}_2)$ is generated by the Casimir element

$$C = fe + \frac{h(h+2)}{4}.$$

One easily verifies by induction the identity in $\mathrm{U}(\mathfrak{sl}_2)$

$$f^r e^r = \Big(C - \frac{h(h+2)}{4}\Big)\Big(C - \frac{(h+2)(h+4)}{4}\Big)\dots\Big(C - \frac{(h+2r-2)(h+2r)}{4}\Big)$$

for any $r \geqslant 1$. Therefore the extremal projector (2.64) can be written as

$$p = 1 + \sum_{r=1}^{\infty} \frac{(-1)^r}{r!\,(h+2)\dots(h+r+1)} \times fe\big(fe-(h+2)\big)\dots\big(fe-(r-1)(h+r)\big), \tag{2.65}$$

which is also equivalent to the infinite product expansion

$$p = \prod_{r=1}^{\infty}\Big(1 - \frac{fe}{r(h+r+1)}\Big). \tag{2.66}$$

The basis element h of $\mathfrak{sl}_2$ acts on the space $\mathcal{P}_N^m$ as multiplication by the scalar $-N/2-m$. Furthermore, this space is annihilated by the action of e^r with $r > m/2$. Hence, using (2.64) we may regard p as an operator on $\mathcal{P}_N$. The image of this operator coincides with the subspace of $\mathfrak{sl}_2$-singular vectors.

PROPOSITION 2.3.1. *The operator $S^{(m)}$ on the space $\mathcal{P}_N^m$ coincides with the restriction of the action of the extremal projector p to this space.*

PROOF. We will regard $S^{(m)}$ as the image of the symmetrizer $s^{(m)}$ given by the formula (1.44) under the action (1.69) of the Brauer algebra with $\omega = N$. We have the identity for operators on the space $\mathcal{P}_N^m$,

$$\sum_{a<b} Q_{ab} = -2\,fe.$$

The proof is completed by comparing the formula for $S^{(m)}$ with (2.65). □

Now let $\mathfrak{g}_N = \mathfrak{sp}_{2n}$. The element $H^{(m)} \in \mathcal{B}_m(-2n)$ acts as the anti-symmetrizer on the tensor product space (2.60). The exterior algebra $\Lambda_{2n} = \Lambda(\mathbb{C}^{2n})$ will be regarded as the space of polynomials in the anti-commuting variables $\zeta_1, \dots, \zeta_{2n}$. Denote by Λ_{2n}^m the subspace of homogeneous polynomials of degree m. Identify the vector spaces

$$H^{(m)}(\mathbb{C}^{2n})^{\otimes m} \cong \Lambda_{2n}^m \tag{2.67}$$

via the isomorphism

$$H^{(m)}(e_{i_1} \otimes \dots \otimes e_{i_m}) \mapsto \zeta_{i_1} \wedge \dots \wedge \zeta_{i_m}. \tag{2.68}$$

If the condition $m \leqslant n+1$ holds, then $S^{(m)}$ is a well-defined operator on Λ_{2n}^m. We will regard it as the image of the symmetrizer $s^{(m)}$ given by the formula (1.44) under the action (1.73) of the Brauer algebra with $\omega = -2n$. The operator $S^{(m)}$ commutes with the diagonal action of the symplectic group Sp_{2n} on the space (2.60) and so can be expressed in terms of the action of the generators of the Lie algebra $\mathfrak{sl}_2$ provided by the *skew Howe duality* [**67**, Section 3.8]. The action of the basis elements of $\mathfrak{sl}_2$ on the space Λ_{2n} is given by

$$e \mapsto \sum_{i=1}^{n} \partial_i \wedge \partial_{i'}, \quad f \mapsto -\sum_{i=1}^{n} \zeta_i \wedge \zeta_{i'}, \quad h \mapsto n - \sum_{i=1}^{n} (\zeta_i \wedge \partial_i + \zeta_{i'} \wedge \partial_{i'}), \tag{2.69}$$

where ∂_i denotes the (left) partial derivative over ζ_i and $i' = 2n - i + 1$. Note that the basis element h acts as multiplication by the scalar $n - m$ on Λ_{2n}^m.

PROPOSITION 2.3.2. *If $m \leqslant n+1$ then the operator $S^{(m)}$ on the space Λ_{2n}^m coincides with the restriction of the action of the extremal projector p to this subspace. In particular, $S^{(n+1)} = 0$.*

PROOF. We have the identity for operators on Λ_{2n}^m,

$$\sum_{a<b} Q_{ab} = 2fe.$$

Therefore, to prove the first claim, it remains to compare the formula for $S^{(m)}$ implied by (1.44) with (2.65). Due to the property $ep = 0$ of the extremal projector, the image $p(\Lambda_{2n}^m)$ is a subspace of $\mathfrak{sl}_2$-singular vectors of weight $n-m$. Since Λ_{2n} is a finite-dimensional representation of $\mathfrak{sl}_2$, then the weight of any nonzero singular vector must be a nonnegative integer. This implies that $p(\Lambda_{2n}^m) = 0$ for $m = n+1$ and so $S^{(n+1)} = 0$. □

2.4. Bibliographical notes

In Sections 2.2 and 2.3 we followed [**110**]. Some generalizations of relations (2.41) and (2.50) where the symmetrizer $S^{(2k)}$ is replaced by the idempotents in the Brauer algebra associated with standard tableaux, were proved in [**117**] in a different way. The formula (2.64) for the extremal projector was generalized by Asherova, Smirnov and Tolstoy [**8**] to arbitrary simple Lie algebras. Its properties were further studied by Zhelobenko [**157**], [**159**]. For an interpretation of the infinite product formula (2.66) and its generalizations to the Kac–Moody algebras see his paper [**158**].

CHAPTER 3

Manin matrices

This chapter is devoted to a special class of matrices whose origins go back to Manin [**106**], [**107**]. Some generator matrices associated with the Lie algebra $\mathfrak{gl}_N$, the affine Kac–Moody algebra $\widehat{\mathfrak{gl}}_N$ and the Yangian $\mathrm{Y}(\mathfrak{gl}_N)$ turn out to belong to this class. Consequently, general properties of Manin matrices will be useful in those particular cases. Orthogonal and symplectic analogues of Manin matrices will be discussed below in Section 5.6.

3.1. Definition and basic properties

DEFINITION 3.1.1. A square matrix

$$M = \begin{bmatrix} M_{11} & \dots & M_{1N} \\ \vdots & & \vdots \\ M_{N1} & \dots & M_{NN} \end{bmatrix}$$

with entries in an associative algebra $\mathcal{A}$ over $\mathbb{C}$ is called a *Manin matrix* if the following relations hold

$$M_{ij}\, M_{kl} - M_{kl}\, M_{ij} = M_{kj}\, M_{il} - M_{il}\, M_{kj} \tag{3.1}$$

for all $i, j, k, l \in \{1, \dots, N\}$. □

If the algebra $\mathcal{A}$ is commutative, then any matrix M satisfies (3.1). Therefore, the properties of Manin matrices which we discuss below generalize those of the usual numerical matrices. Observe also that the relations (3.1) imply that the entries of each column of a Manin matrix pairwise commute.

Consider the tensor product algebra

$$\operatorname{End}\mathbb{C}^N \otimes \operatorname{End}\mathbb{C}^N \otimes \mathcal{A} \tag{3.2}$$

and use the notation of Section 1.4.

LEMMA 3.1.2. *Each of following relations provides an equivalent definition of Manin matrices:*

$$(1 - P)\, M_1\, M_2\, (1 + P) = 0, \tag{3.3}$$

$$(1 - P)\, (M_1\, M_2 - M_2\, M_1) = 0, \tag{3.4}$$

$$(M_1\, M_2 - M_2\, M_1)\, (1 + P) = 0. \tag{3.5}$$

PROOF. Note first that the three relations are equivalent to each other by (1.67). We have

$$M_1\, M_2 = \sum_{i,j,k,l=1}^{N} e_{ij} \otimes e_{kl} \otimes M_{ij}\, M_{kl}.$$

Hence, using formula (1.64) for the element P we get

$$P\,M_1\,M_2 = \sum_{i,j,k,l=1}^{N} e_{kj}\otimes e_{il}\otimes M_{ij}\,M_{kl},$$

$$M_1\,M_2\,P = \sum_{i,j,k,l=1}^{N} e_{il}\otimes e_{kj}\otimes M_{ij}\,M_{kl},$$

and

$$P\,M_1\,M_2\,P = \sum_{i,j,k,l=1}^{N} e_{kl}\otimes e_{ij}\otimes M_{ij}\,M_{kl}.$$

Therefore, taking the coefficient of the basis vector $e_{ij}\otimes e_{kl}$ on the left hand side of (3.3) we recover (3.1). □

For any $m \geqslant 1$ consider the tensor product algebra

$$\underbrace{\operatorname{End}\mathbb{C}^N\otimes\ldots\otimes\operatorname{End}\mathbb{C}^N}_{m}\otimes\mathcal{A}. \tag{3.6}$$

As before, we denote by $H^{(m)}$ and $A^{(m)}$ the elements of the algebra (3.6) which are the respective images of the symmetrizer $h^{(m)}$ and anti-symmetrizer $a^{(m)}$ defined in (1.17) and (1.19) under the map (1.65).

LEMMA 3.1.3. *If M is a Manin matrix, then we have the identities in the algebra* (3.6),

$$A^{(m)}M_1\ldots M_m\,A^{(m)} = A^{(m)}M_1\ldots M_m \tag{3.7}$$

and

$$H^{(m)}M_1\ldots M_m\,H^{(m)} = M_1\ldots M_m H^{(m)}. \tag{3.8}$$

PROOF. To prove (3.7) it suffices to show that for any element $\sigma\in\mathfrak{S}_m$ we have

$$A^{(m)}M_1\ldots M_m\,P_\sigma = \operatorname{sgn}\sigma\cdot A^{(m)}M_1\ldots M_m, \tag{3.9}$$

where P_σ denotes the image of σ under the map (1.65). Since the group $\mathfrak{S}_m$ is generated by the adjacent transpositions, it is enough to verify (3.9) for the elements $\sigma = s_a$ with $a = 1,\ldots,m-1$. Hence we only need to consider the case $m = 2$. However, the relation (3.9) with $\sigma = s_1$ reads

$$\frac{1-P}{2}\,M_1\,M_2\,P = -\frac{1-P}{2}\,M_1\,M_2$$

which an equivalent form of (3.3). Similarly, the proof of (3.8) reduces to checking that for any $\sigma\in\mathfrak{S}_m$

$$P_\sigma\,M_1\ldots M_m\,H^{(m)} = M_1\ldots M_m H^{(m)}. \tag{3.10}$$

This follows again from (3.3) written in the form

$$P\,M_1\,M_2\,\frac{1+P}{2} = M_1\,M_2\,\frac{1+P}{2}.$$

□

3.2. Identities and invertibility

We point out some useful recurrence formulas for the symmetrizer and anti-symmetrizer. Note that by (1.20)

$$A^{(m)} = \frac{1}{m} A^{(m-1)} (1 - P_{1\,m} - \dots - P_{m-1\,m}).$$

Multiply both sides by $A^{(m-1)}$ from the right and use the relations

$$A^{(m)} A^{(m-1)} = A^{(m)} \qquad \text{and} \qquad A^{(m-1)} P_{a\,m} A^{(m-1)} = A^{(m-1)} P_{m-1\,m} A^{(m-1)}$$

for $1 \leqslant a < m$ to get the formula

$$A^{(m)} = \frac{1}{m} A^{(m-1)} - \frac{m-1}{m} A^{(m-1)} P_{m-1\,m} A^{(m-1)}. \tag{3.11}$$

Similarly, we derive from (1.18) that

$$H^{(m)} = \frac{1}{m} H^{(m-1)} + \frac{m-1}{m} H^{(m-1)} P_{m-1\,m} H^{(m-1)}. \tag{3.12}$$

From now on we will assume that the associative algebra $\mathcal{A}$ is unital. Given a Manin matrix M with entries in $\mathcal{A}$, introduce the sums of elements of $\mathcal{A}$ by

$$\mathrm{Bos} = 1 + \sum_{m=1}^{\infty} \mathrm{tr}_{1,\dots,m}\, H^{(m)} M_1 \dots M_m, \tag{3.13}$$

$$\mathrm{Ferm} = 1 + \sum_{m=1}^{N} (-1)^m\, \mathrm{tr}_{1,\dots,m}\, A^{(m)} M_1 \dots M_m. \tag{3.14}$$

The following is a Manin matrix version of the *MacMahon Master Theorem.*

THEOREM 3.2.1. *We have the identity*

$$\mathrm{Bos} \times \mathrm{Ferm} = 1.$$

PROOF. It is sufficient to show that for any integer $1 \leqslant m \leqslant N$ we have the identity in the algebra (3.6)

$$\sum_{r=0}^{m} (-1)^{m-r} \mathrm{tr}_{1,\dots,r} H^{(r)} M_1 \dots M_r \times \mathrm{tr}_{r+1,\dots,m} A^{\{r+1,\dots,m\}} M_{r+1} \dots M_m = 0,$$

where $A^{\{r+1,\dots,m\}}$ denotes the anti-symmetrizer in (3.6) over the copies of $\mathrm{End}\,\mathbb{C}^N$ labeled by $r+1,\dots,m$ (with the identity components in the first r copies). The identity can be written as

$$\sum_{r=0}^{m} (-1)^r \mathrm{tr}_{1,\dots,m} H^{(r)} A^{\{r+1,\dots,m\}} M_1 \dots M_m = 0. \tag{3.15}$$

Our next step is to show that the left hand side of (3.15) remains unchanged after the replacement of the product of the symmetrizer and anti-symmetrizer by

$$\begin{aligned} &H^{(r)} A^{\{r+1,\dots,m\}} \\ &\qquad \mapsto \frac{r(m-r+1)}{m} H^{(r)} A^{\{r,\dots,m\}} + \frac{(r+1)(m-r)}{m} H^{(r+1)} A^{\{r+1,\dots,m\}}. \end{aligned} \tag{3.16}$$

Indeed, by (3.11) and (3.12) the right hand side of (3.16) equals

$$\frac{r(m-r+1)}{m} H^{(r)}\Big(\frac{1}{m-r+1} A^{\{r+1,\dots,m\}} - \frac{m-r}{m-r+1} A^{\{r+1,\dots,m\}} P_{r\,r+1} A^{\{r+1,\dots,m\}}\Big)$$
$$+ \frac{(r+1)(m-r)}{m} \Big(\frac{1}{r+1} H^{(r)} + \frac{r}{r+1} H^{(r)} P_{r\,r+1} H^{(r)}\Big) A^{\{r+1,\dots,m\}}.$$

To simplify this expression note that permuting $H^{(r)}$ with $A^{\{r+1,\dots,m\}}$ and using the cyclic property of the trace we obtain the relation

$$\operatorname{tr}_{1,\dots,m} H^{(r)} A^{\{r+1,\dots,m\}} P_{r\,r+1} A^{\{r+1,\dots,m\}} M_1 \dots M_m$$
$$= \operatorname{tr}_{1,\dots,m} H^{(r)} P_{r\,r+1} A^{\{r+1,\dots,m\}} M_1 \dots M_m A^{\{r+1,\dots,m\}}.$$

Now apply (3.7) to write this element as

$$\operatorname{tr}_{1,\dots,m} H^{(r)} P_{r\,r+1} A^{\{r+1,\dots,m\}} M_1 \dots M_m.$$

Similarly, by the cyclic property of the trace, we get

$$\operatorname{tr}_{1,\dots,m} H^{(r)} P_{r\,r+1} H^{(r)} A^{\{r+1,\dots,m\}} M_1 \dots M_m$$
$$= \operatorname{tr}_{1,\dots,m} P_{r\,r+1} A^{\{r+1,\dots,m\}} H^{(r)} M_1 \dots M_m H^{(r)}.$$

Hence, by (3.8) this equals

$$\operatorname{tr}_{1,\dots,m} P_{r\,r+1} A^{\{r+1,\dots,m\}} M_1 \dots M_m H^{(r)}$$
$$= \operatorname{tr}_{1,\dots,m} H^{(r)} P_{r\,r+1} A^{\{r+1,\dots,m\}} M_1 \dots M_m.$$

We have thus verified that the replacement (3.16) does not affect the left hand side of (3.15). On the other hand, (3.15) vanishes after this replacement since we get a telescoping sum equal to zero. □

The summands in (3.13) and (3.14) can be written explicitly in terms of the matrix elements of the matrix M as follows.

PROPOSITION 3.2.2. *For any $1 \leqslant m \leqslant N$ we have*

$$\operatorname{tr}_{1,\dots,m} A^{(m)} M_1 \dots M_m = \sum_{1\leqslant i_1<\dots<i_m\leqslant N} \sum_{\sigma\in\mathfrak{S}_m} \operatorname{sgn}\sigma \cdot M_{i_{\sigma(1)}\, i_1} \dots M_{i_{\sigma(m)}\, i_m}. \tag{3.17}$$

Moreover, for any $m \geqslant 1$ we have

$$\operatorname{tr}_{1,\dots,m} H^{(m)} M_1 \dots M_m = \sum_{1\leqslant i_1\leqslant\dots\leqslant i_m\leqslant N} \frac{1}{\alpha_1! \dots \alpha_N!} \sum_{\sigma\in\mathfrak{S}_m} M_{i_m\, i_{\sigma(m)}} \dots M_{i_1\, i_{\sigma(1)}}, \tag{3.18}$$

where α_i is the multiplicity of index $i \in \{1,\dots,N\}$ in the multiset $\{i_1,\dots,i_m\}$.

PROOF. Write

$$A^{(m)} M_1 \dots M_m = \sum_{I,J} e_{i_1 j_1} \otimes \dots \otimes e_{i_m j_m} \otimes M^{i_1 \dots i_m}_{j_1 \dots j_m}, \tag{3.19}$$

summed over all m-tuples of indices $I = (i_1, \dots, i_m)$ and $J = (j_1, \dots, j_m)$ from $\{1, \dots, N\}$, where $M^{\,i_1\dots i_m}_{\,j_1\dots j_m} \in \mathcal{A}$. For each $a = 1, \dots, m-1$ we have

$$P_{a\,a+1}A^{(m)}M_1\dots M_m = -A^{(m)}M_1\dots M_m = A^{(m)}M_1\dots M_m P_{a\,a+1},$$

where the second equality holds due to (3.9). This implies that the matrix elements $M^{\,i_1\dots i_m}_{\,j_1\dots j_m}$ are skew-symmetric with respect to permutations of the upper indices and of the lower indices. Hence

$$\mathrm{tr}_{1,\dots,m}\, A^{(m)}M_1\dots M_m = \sum_I M^{\,i_1\dots i_m}_{\,i_1\dots i_m} = m! \sum_{1\leqslant i_1<\dots<i_m\leqslant N} M^{\,i_1\dots i_m}_{\,i_1\dots i_m}.$$

Calculating now $M^{\,i_1\dots i_m}_{\,i_1\dots i_m}$ from (3.19) we get (3.17).

Similarly, for the proof of (3.18) write the matrix element decomposition as in (3.19),

$$M_1\dots M_m\, H^{(m)} = \sum_{I,J} e_{i_1j_1}\otimes\dots\otimes e_{i_mj_m}\otimes \tilde{M}^{\,i_1\dots i_m}_{\,j_1\dots j_m}, \tag{3.20}$$

with some elements $\tilde{M}^{\,i_1\dots i_m}_{\,j_1\dots j_m} \in \mathcal{A}$ which are symmetric with respect to permutations of the upper indices and of the lower indices. Using the cyclic property of trace, we get

$$\mathrm{tr}_{1,\dots,m}\, H^{(m)}M_1\dots M_m = \mathrm{tr}_{1,\dots,m}\, M_1\dots M_m\, H^{(m)} = \sum_I \tilde{M}^{\,i_1\dots i_m}_{\,i_1\dots i_m}$$

which equals

$$\sum_{1\leqslant i_1\leqslant\dots\leqslant i_m\leqslant N} \frac{m!}{\alpha_1!\dots\alpha_N!}\,\tilde{M}^{\,i_m\dots i_1}_{\,i_m\dots i_1},$$

where α_i is the multiplicity of index $i \in \{1, \dots, N\}$ in the multiset $\{i_1, \dots, i_m\}$. Calculating $\tilde{M}^{\,i_m\dots i_1}_{\,i_m\dots i_1}$ from (3.20) we come to (3.18). □

Remark 3.2.3. As the proof of Proposition 3.2.2 shows, we may replace the summation conditions in (3.17) by $N \geqslant i_1 > \dots > i_m \geqslant 1$ and those in (3.18) by $N \geqslant i_1 \geqslant \dots \geqslant i_m \geqslant 1$ for the respective formulas to remain valid. □

Consider now the particular case $m = N$ of (3.7). The anti-symmetrizer $A^{(N)}$ projects the space $(\mathbb{C}^N)^{\otimes N}$ to its one-dimensional subspace spanned by the vector

$$\sum_{\sigma\in\mathfrak{S}_N} \mathrm{sgn}\,\sigma\cdot e_{\sigma(1)}\otimes\dots\otimes e_{\sigma(N)}. \tag{3.21}$$

This implies that

$$A^{(N)}M_1\dots M_N\, A^{(N)} = A^{(N)}\mathrm{cdet}\, M$$

for a uniquely determined element $\mathrm{cdet}\, M \in \mathcal{A}$ called the *column-determinant* of the matrix M. By (3.7) we also have

$$A^{(N)}M_1\dots M_N = A^{(N)}\mathrm{cdet}\, M. \tag{3.22}$$

Applying both sides to the vector $e_1\otimes\dots\otimes e_N$ and comparing the coefficients of the vector (3.21) on both sides we get the formula

$$\mathrm{cdet}\, M = \sum_{\sigma\in\mathfrak{S}_N} \mathrm{sgn}\,\sigma\cdot M_{\sigma(1)\,1}\dots M_{\sigma(N)\,N}. \tag{3.23}$$

Due to (1.67), for any element $s \in \mathfrak{S}_N$ we have

$$A^{(N)}M_{s(1)}\dots M_{s(N)} = A^{(N)}\mathrm{cdet}\, M \tag{3.24}$$

which follows by multiplying both sides of (3.22) by P_s from the right. This leads to a generalization of (3.23),

$$\operatorname{cdet} M = \operatorname{sgn} s \sum_{\sigma\in\mathfrak{S}_N} \operatorname{sgn}\sigma \cdot M_{\sigma(1)\, s(1)} \dots M_{\sigma(N)\, s(N)}.$$

Hence, the column-determinant of a Manin matrix shares the usual properties of determinant: if two rows or two columns are swapped then it changes the sign.

REMARK 3.2.4. The column-determinant of any square matrix A over a ring coincides with the *row-determinant*

$$\operatorname{rdet} B = \sum_{\sigma\in\mathfrak{S}_N} \operatorname{sgn}\sigma \cdot B_{1\,\sigma(1)} \dots B_{N\,\sigma(N)} \tag{3.25}$$

of the transposed matrix $B = A^t$. □

Taking $\omega = N$ in Lemma 1.3.3 and using (1.72), we find that

$$\operatorname{tr}_m A^{(m)} = \frac{N-m+1}{m} A^{(m-1)}. \tag{3.26}$$

This implies

$$\operatorname{tr}_{1,\dots,N} A^{(N)} = 1 \tag{3.27}$$

so that we recover the particular $m = N$ case of (3.17) by taking trace in (3.22):

$$\operatorname{tr}_{1,\dots,N} A^{(N)} M_1 \dots M_N = \operatorname{cdet} M. \tag{3.28}$$

Now let u be a formal variable.

COROLLARY 3.2.5. *We have the identities*

$$\operatorname{cdet}(1 + u M) = \sum_{m=0}^{N} u^m \operatorname{tr}_{1,\dots,m} A^{(m)} M_1 \dots M_m, \tag{3.29}$$

$$\big[\operatorname{cdet}(1 - u M)\big]^{-1} = \sum_{m=0}^{\infty} u^m \operatorname{tr}_{1,\dots,m} H^{(m)} M_1 \dots M_m. \tag{3.30}$$

PROOF. By the MacMahon Master Theorem (Theorem 3.2.1), identities (3.29) and (3.30) are equivalent. Identity (3.29) holds since the coefficient of u^m on the left hand side equals the sum of the column-determinants of all principle $m \times m$ submatrices of M, which coincides with the corresponding coefficient on the right hand side due to (3.17). □

Define the *comatrix* for a Manin matrix M as the matrix $\widehat{M}$ with the entries in the algebra $\mathcal{A}$ defined by

$$\widehat{M}_{ij} = (-1)^{i+j} \operatorname{cdet} M^{ji}, \tag{3.31}$$

where M^{ji} is the matrix obtained from M by deleting row j and column i.

LEMMA 3.2.6. *We have the relation*

$$\widehat{M} M = \big(\operatorname{cdet} M\big)\, 1, \tag{3.32}$$

where 1 *denotes the identity matrix.*

PROOF. First observe that the definition (3.31) of the comatrix can be written equivalently in the matrix form as

$$A^{(N)} M_1 \dots M_{N-1} = A^{(N)} \widehat{M}_N. \tag{3.33}$$

Indeed, (3.7) implies

$$A^{(N)} M_1 \dots M_{N-1} = A^{(N)} M_1 \dots M_{N-1} A^{(N-1)}$$

so that the matrix relation (3.33) is equivalent to the equality of the matrix coefficients corresponding to the basis vectors of the form

$$e_1 \otimes \dots \otimes \widehat{e}_i \otimes \dots \otimes e_N \otimes e_j, \qquad i, j \in \{1, \dots, N\}, \tag{3.34}$$

where the hat indicates that the corresponding tensor factor should be skipped. Apply both sides of (3.33) to such a vector and compare the coefficients of the vector (3.21). Using notation (3.19), we get the relation

$$(-1)^{N-j} M^{\,1\dots\widehat{j}\dots N}_{\,1\dots\widehat{i}\dots N} = (-1)^{N-i} \widehat{M}_{ij}$$

which is equivalent to (3.31). Now, using (3.22) and (3.33) we obtain

$$A^{(N)} \mathrm{cdet}\, M = A^{(N)} M_1 \dots M_N = A^{(N)} \widehat{M}_N M_N.$$

On applying both sides to the vectors (3.34) we get (3.32). □

If the algebra $\mathcal{A}$ is not commutative, then we need to distinguish between the left and right inverses of elements of $\mathcal{A}$, as well as between the left and right inverses of matrices over $\mathcal{A}$. Lemma 3.2.6 implies that if $\mathrm{cdet}\, M$ has a left inverse $l(M) \in \mathcal{A}$, then M is a left-invertible matrix with its left inverse found by $l(M)\, \widehat{M}$. On the other hand, if the matrix M has a right inverse $M^{\vee}$, then the element $\mathrm{cdet}\, M$ is right-invertible which can be seen as follows. Consider relation (3.24) with the permutation s such that $s(i) = N - i + 1$ for all $i = 1, \dots, N$,

$$A^{(N)} M_N \dots M_1 = A^{(N)} \mathrm{cdet}\, M. \tag{3.35}$$

It implies

$$A^{(N)} = \mathrm{cdet}\, M\, A^{(N)} M_1^{\vee} \dots M_N^{\vee}.$$

Multiply both sides by $A^{(N)}$ from the right and note that

$$A^{(N)} M_1^{\vee} \dots M_N^{\vee} A^{(N)} = A^{(N)}\, r(M) \tag{3.36}$$

for a certain element $r(M) \in \mathcal{A}$. This implies $\mathrm{cdet}\, M \cdot r(M) = 1$ thus verifying that $\mathrm{cdet}\, M$ is right-invertible.

Note also that due to the associativity of matrix multiplication, if a matrix M has both left and right inverses, then they coincide. Hence, they are equal to a uniquely determined (two-sided) inverse M^{-1} of the matrix M.

PROPOSITION 3.2.7. *If a Manin matrix M is right-invertible and* $\mathrm{cdet}\, M$ *is left-invertible, then M is invertible and M^{-1} is a Manin matrix.*

PROOF. By the above arguments, the assumptions imply that M is an invertible matrix and $\mathrm{cdet}\, M$ is an invertible element of the algebra $\mathcal{A}$. By (3.35) we have (assuming $N \geqslant 2$)

$$(\mathrm{cdet}\, M)^{-1} A^{(N)} M_N \dots M_3 = A^{(N)} M_1^{-1} M_2^{-1}$$

so that the right hand side is unchanged after the multiplication by $-P_{12}$ from the right. Hence, using (1.67) we come to

$$A^{(N)}(M_1^{-1}M_2^{-1} - M_2^{-1}M_1^{-1}) = 0. \tag{3.37}$$

Using (3.26) and taking the partial trace $\mathrm{tr}_{3,\dots,N}$ in (3.37) we get

$$A^{(2)}(M_1^{-1}M_2^{-1} - M_2^{-1}M_1^{-1}) = 0$$

so that M^{-1} is a Manin matrix by (3.4). □

COROLLARY 3.2.8. *Under the assumptions of Proposition* 3.2.7 *we have*

$$\mathrm{cdet}(M^{-1}) = \big(\mathrm{cdet}\, M\big)^{-1}.$$

PROOF. Indeed, the right inverse element $r(M)$ defined in (3.36) coincides with $\mathrm{cdet}(M^{-1})$, since $M^{\vee} = M^{-1}$. □

COROLLARY 3.2.9. *If M is a Manin matrix, then for any nonnegative integer r the following identity holds:*

$$(1-P)\sum_{k+l=r}[M_1^k, M_2^l] = 0, \tag{3.38}$$

where k and l run over nonnegative integers.

PROOF. Clearly, the matrix $1 - tM$ with entries in the algebra $\mathcal{A} \otimes \mathbb{C}[[t]]$ of formal power series in a variable t is also a Manin matrix. By Proposition 3.2.7, so is its inverse, and by (3.4) we have the relation

$$(1-P)\big[(1-tM_1)^{-1}, (1-tM_2)^{-1}\big] = 0.$$

Taking the coefficient of t^r we get (3.38). □

The next theorem provides the *Newton identity* for Manin matrices. As before, M denotes an arbitrary Manin matrix with entries in a unital associative algebra $\mathcal{A}$.

THEOREM 3.2.10. *We have the identity*

$$\partial_t\,\mathrm{cdet}(1+tM) = \mathrm{cdet}(1+tM)\sum_{m=0}^{\infty}(-t)^m\,\mathrm{tr}\, M^{m+1}. \tag{3.39}$$

PROOF. By (3.22) we have

$$A^{(N)}(1+tM_1)\dots(1+tM_N) = A^{(N)}\mathrm{cdet}\,(1+tM).$$

Calculate the derivative of both sides over t:

$$\sum_{a=1}^{N} A^{(N)}(1+tM_1)\dots M_a \dots(1+tM_N) = A^{(N)}\partial_t\,\mathrm{cdet}(1+tM).$$

Replace the factor M_a by $t^{-1}(1+tM_a) - t^{-1}$, then take the trace of both sides over all N copies of $\mathrm{End}\,\mathbb{C}^N$ and use (3.27) to get

$$N t^{-1}\mathrm{cdet}(1+tM) - t^{-1}\sum_{a=1}^{N}\mathrm{tr}_{1,\dots,N}\, A^{(N)}(1+tM_1)\dots\widehat{(1+tM_a)}\dots(1+tM_N)$$
$$= \partial_t\,\mathrm{cdet}(1+tM),$$

where the hat indicates the factor to be skipped. Observe that for each value of a the corresponding term in the sum coincides with the term for $a = N$ which equals

$$\operatorname{tr}_{1,\dots,N} A^{(N)}(1 + tM_1)\dots(1 + tM_{N-1}). \tag{3.40}$$

Indeed, taking the cycle permutation $s = (a, a+1, \dots, N)$ we find that the trace (3.40) equals

$$\begin{aligned}
&\operatorname{sgn} s \cdot \operatorname{tr}_{1,\dots,N} A^{(N)} P_s (1 + tM_1)\dots(1 + tM_{N-1}) \\
&\quad = \operatorname{sgn} s \cdot \operatorname{tr}_{1,\dots,N} A^{(N)}(1 + tM_1)\dots(\widehat{1 + tM_a})\dots(1 + tM_N) P_s \\
&\quad = \operatorname{sgn} s \cdot \operatorname{tr}_{1,\dots,N} P_s A^{(N)}(1 + tM_1)\dots(\widehat{1 + tM_a})\dots(1 + tM_N) \\
&\quad = \operatorname{tr}_{1,\dots,N} A^{(N)}(1 + tM_1)\dots(\widehat{1 + tM_a})\dots(1 + tM_N),
\end{aligned}$$

where we used (1.67) and the cyclic property of trace. Furthermore, using (3.26) and (3.33) we find that (3.40) equals the trace of the comatrix associated with the matrix $1 + tM$. Lemma 3.2.6 implies that this equals $\operatorname{cdet}(1 + tM) \operatorname{tr}(1 + tM)^{-1}$ and so we come to the identity

$$\operatorname{cdet}(1 + tM)\big(N t^{-1} - t^{-1} \operatorname{tr}(1 + tM)^{-1}\big) = \partial_t \operatorname{cdet}(1 + tM).$$

It can be written in the form

$$\operatorname{cdet}(1 + tM) \sum_{m=0}^{\infty} (-t)^m \operatorname{tr} M^{m+1} = \partial_t \operatorname{cdet}(1 + tM),$$

as required. □

The following equivalent form of the Newton identity holds.

COROLLARY 3.2.11. *We have*

$$-\partial_t \big[\operatorname{cdet}(1 + tM)\big]^{-1} = \sum_{m=0}^{\infty} (-t)^m \operatorname{tr} M^{m+1} \cdot \big[\operatorname{cdet}(1 + tM)\big]^{-1}.$$

PROOF. By the Leibniz rule,

$$\begin{aligned}
0 &= \partial_t \big[\operatorname{cdet}(1 + tM) \operatorname{cdet}(1 + tM)^{-1}\big] \\
&= \big[\partial_t \operatorname{cdet}(1 + tM)\big] \operatorname{cdet}(1 + tM)^{-1} + \operatorname{cdet}(1 + tM)\, \partial_t \big[\operatorname{cdet}(1 + tM)^{-1}\big]
\end{aligned}$$

so that the desired identity is implied by Theorem 3.2.10. □

3.3. Bibliographical notes

Definition 3.1.1 goes back to the work of Manin [**106**], [**107**]. Such matrices are also known in the literature as the *right-quantum matrices* (with $q = 1$); see [**54**]. A detailed account of their algebraic properties and applications with an extensive list of references can be found in [**20**] and [**21**]. The MacMahon Master Theorem for Manin matrices (Theorem 3.2.1) originates in [**54**]. More proofs and versions can be found in [**43**], [**44**], [**64**], [**92**] and [**114**]. Our proof follows [**114**], where its super-version is given. All other results, including Lemma 3.1.3, Proposition 3.2.7 and Theorem 3.2.10, are contained in [**20**] and [**21**]; see also [**22**] for their q-analogues. Some of the arguments were simplified by taking advantage of the matrix form of the definition of Manin matrices provided by Lemma 3.1.2. The assumption on $\operatorname{cdet} M$ in Proposition 3.2.7 was erroneously omitted in [**21**, Theorem 1, Section 4.3]; cf. [**22**, Theorem 4.7].

CHAPTER 4

Casimir elements for $\mathfrak{gl}_N$

In the following two chapters we construct 'quantum analogues' of the invariants in the symmetric algebras considered in Chapter 2. Those invariants will be 'lifted' to the universal enveloping algebra $\mathrm{U}(\mathfrak{g})$ thus providing its central elements. As in Chapter 2, we will focus on such elements produced with the use of idempotents for the symmetric group and Brauer algebra. However, keeping in mind affine counterparts of these constructions, we will depart from our reliance on the adjoint action of the associated group. Instead, we will develop matrix techniques by writing the commutation relations of Lie algebras in a matrix form. This will allow us to work within the Lie algebra settings. The same approach will then be applied to affine Kac–Moody algebras in Chapters 7 and 8.

We will rely on some standard facts about simple Lie algebras and their representations which can be found in the books by Dixmier [**29**], Goodman and Wallach [**59**] and Humphreys [**70**]. We start by recalling matrix presentations of Lie algebras.

4.1. Matrix presentations of simple Lie algebras

Suppose that $\mathfrak{g}$ is a finite-dimensional Lie algebra over $\mathbb{C}$ equipped with a nondegenerate symmetric invariant bilinear form $\langle\ ,\ \rangle$. Choose a basis $J^1, \dots, J^d$ of $\mathfrak{g}$ and let $J_1, \dots, J_d$ be its dual with respect to the form so that $\langle J_i, J^k \rangle = \delta_{ik}$. Let π be a faithful representation of $\mathfrak{g}$ afforded by a finite-dimensional vector space V,

$$\pi : \mathfrak{g} \to \operatorname{End} V.$$

Introduce the elements

$$G = \sum_{i=1}^{d} \pi(J^i) \otimes J_i \in \operatorname{End} V \otimes \mathrm{U}(\mathfrak{g})$$

and

$$\Omega = \sum_{i=1}^{d} \pi(J^i) \otimes \pi(J_i) \in \operatorname{End} V \otimes \operatorname{End} V.$$

It is easy to verify that G and Ω are independent of the choice of the basis J^i. In particular,

$$\Omega = \sum_{i=1}^{d} \pi(J_i) \otimes \pi(J^i). \tag{4.1}$$

Consider the tensor product algebra $\operatorname{End} V \otimes \operatorname{End} V \otimes \mathrm{U}(\mathfrak{g})$ and identify Ω with the element $\Omega \otimes 1$. Introduce elements of this algebra by

$$G_1 = \sum_{i=1}^{d} \pi(J^i) \otimes 1 \otimes J_i \qquad \text{and} \qquad G_2 = \sum_{i=1}^{d} 1 \otimes \pi(J^i) \otimes J_i.$$

Write the commutation relations for $\mathfrak{g}$,

$$[J_i, J_j] = \sum_{k=1}^{d} c_{ij}^{k}\, J_k \tag{4.2}$$

with structure coefficients c_{ij}^k. We will regard the universal enveloping algebra $\mathrm{U}(\mathfrak{g})$ as the associative algebra with generators J_i subject to the defining relations (4.2), where the left hand side is understood as the commutator $J_iJ_j - J_jJ_i$.

PROPOSITION 4.1.1. *The defining relations of* $\mathrm{U}(\mathfrak{g})$ *are equivalent to the matrix relation*

$$G_1 G_2 - G_2 G_1 = -\Omega\, G_2 + G_2\, \Omega. \tag{4.3}$$

PROOF. The left hand side of (4.3) reads

$$\sum_{i,j=1}^{d} \pi(J^i) \otimes \pi(J^j) \otimes (J_iJ_j - J_jJ_i).$$

For the right hand side we have

$$- \sum_{i,k=1}^{d} \pi(J^i) \otimes \pi\big([J_i, J^k]\big) \otimes J_k. \tag{4.4}$$

By the invariance of the form, we find

$$\langle [J_i, J^k], J_j \rangle = -\langle J^k, [J_i, J_j] \rangle = -c_{ij}^k.$$

Hence (4.4) equals

$$\sum_{i,j,k=1}^{d} c_{ij}^k\, \pi(J^i) \otimes \pi(J^j) \otimes J_k.$$

Since the representation π is faithful, we conclude that (4.3) is equivalent to the defining relations (4.2) of $\mathrm{U}(\mathfrak{g})$. □

The defining relations (4.3) can be written in an equivalent form

$$G_1 G_2 - G_2 G_1 = \Omega\, G_1 - G_1\, \Omega, \tag{4.5}$$

which is easily verified with the use of (4.1). The element G can be regarded as an $n \times n$ matrix ($n = \dim V$) with entries in $\mathrm{U}(\mathfrak{g})$ so that Proposition 4.1.1 gives a *matrix presentation* of $\mathrm{U}(\mathfrak{g})$. As we will see below in Proposition 4.2.1, single matrix relations (4.3) or (4.5), encoding the commutation relations of $\mathfrak{g}$, are convenient for constructing central elements of $\mathrm{U}(\mathfrak{g})$.

4.2. Harish-Chandra isomorphism

The *center* $\mathrm{Z}(\mathfrak{g})$ of the universal enveloping algebra $\mathrm{U}(\mathfrak{g})$ is defined by

$$\mathrm{Z}(\mathfrak{g}) = \{z \in \mathrm{U}(\mathfrak{g}) \mid zu = uz \quad \text{for all} \quad u \in \mathrm{U}(\mathfrak{g})\}.$$

Any element of the center is called a *Casimir element* for $\mathfrak{g}$. Since $\mathrm{U}(\mathfrak{g})$ is generated by basis elements of $\mathfrak{g}$, for any $z \in \mathrm{U}(\mathfrak{g})$ we have

$$z \in \mathrm{Z}(\mathfrak{g}) \qquad \text{if and only if} \qquad zx = xz \quad \text{for all} \quad x \in \mathfrak{g}.$$

This condition can be restricted further to a subset of elements $x \in \mathfrak{g}$ which generate $\mathfrak{g}$ as a Lie algebra. A family of Casimir elements (often called *Gelfand invariants* following [**55**]) can be produced with the use of the matrix presentations introduced in Section 4.1.

PROPOSITION 4.2.1. *All elements* $\operatorname{tr} G^k$ *with* $k \geqslant 1$ *belong to the center of* $\mathrm{U}(\mathfrak{g})$.

PROOF. Relation (4.3) implies

$$G_1 G_2^k - G_2^k G_1 = \sum_{r=1}^{k} G_2^{r-1}(-\Omega\, G_2 + G_2 \Omega) G_2^{k-r} = -\Omega\, G_2^k + G_2^k \Omega.$$

By taking trace over the second copy of $\operatorname{End} V$ and using its cyclic property (Lemma 1.4.1), we get $[G_1, \operatorname{tr}_2 G_2^k] = 0$ as required. $\square$

The Casimir elements $\operatorname{tr} G^k$ are widely used in representation theory, especially for the Lie algebras $\mathfrak{g}$ of classical types. In those cases one usually takes V to be the first fundamental (or vector) representation; see Section 4.8 and Example 5.3.3 below.

Now suppose that $\mathfrak{g}$ is a simple Lie algebra over $\mathbb{C}$. Choose a Cartan subalgebra $\mathfrak{h}$ of $\mathfrak{g}$ and a triangular decomposition $\mathfrak{g} = \mathfrak{n}_- \oplus \mathfrak{h} \oplus \mathfrak{n}_+$ as in Chapter 2. Recall that the *Harish-Chandra homomorphism*

$$\mathrm{U}(\mathfrak{g})^{\mathfrak{h}} \to \mathrm{U}(\mathfrak{h}) \tag{4.6}$$

is the projection of the $\mathfrak{h}$-centralizer $\mathrm{U}(\mathfrak{g})^{\mathfrak{h}}$ in the universal enveloping algebra to $\mathrm{U}(\mathfrak{h})$ whose kernel is the two-sided ideal $\mathrm{U}(\mathfrak{g})^{\mathfrak{h}} \cap \mathrm{U}(\mathfrak{g})\mathfrak{n}_+$. This ideal coincides with $\mathrm{U}(\mathfrak{g})^{\mathfrak{h}} \cap \mathfrak{n}_-\, \mathrm{U}(\mathfrak{g})$. The restriction of the homomorphism (4.6) to the center $\mathrm{Z}(\mathfrak{g})$ of $\mathrm{U}(\mathfrak{g})$ yields an isomorphism

$$\chi : \mathrm{Z}(\mathfrak{g}) \to \mathrm{U}(\mathfrak{h})^W \tag{4.7}$$

called the *Harish-Chandra isomorphism*, where $\mathrm{U}(\mathfrak{h})^W$ denotes the subalgebra of invariants in $\mathrm{U}(\mathfrak{h})$ with respect to a certain (shifted) action of the Weyl group W of $\mathfrak{g}$. This leads to a description of the center $\mathrm{Z}(\mathfrak{g})$ as an algebra of polynomials,

$$\mathrm{Z}(\mathfrak{g}) = \mathbb{C}[P_1^\circ, \dots, P_n^\circ],$$

for certain algebraically independent central elements $P_1^\circ, \dots, P_n^\circ$, where n is the rank of $\mathfrak{g}$. The universal enveloping algebra $\mathrm{U}(\mathfrak{g})$ is equipped with a canonical filtration so that the associated graded algebra $\operatorname{gr} \mathrm{U}(\mathfrak{g})$ is isomorphic to the symmetric algebra $\mathrm{S}(\mathfrak{g})$. Moreover, the associated graded algebra $\operatorname{gr} \mathrm{Z}(\mathfrak{g})$ is isomorphic to the subalgebra of $\mathfrak{g}$-invariants in $\mathrm{S}(\mathfrak{g})$ as defined in (2.1). For each i denote by P_i the *symbol* of the element P_i°, that is, the image of P_i in the corresponding graded component of $\mathrm{S}(\mathfrak{g})$. Then relation (2.2) holds, and the respective degrees $d_1, \dots, d_n$ of the elements $P_1^\circ, \dots, P_n^\circ$ coincide with the exponents of $\mathfrak{g}$ increased by 1.

Now recall the general linear Lie algebra $\mathfrak{gl}_N$ defined by the commutation relations (2.4). We will regard the basis elements E_{ij} of $\mathfrak{gl}_N$ as generators of the universal enveloping algebra $\mathrm{U}(\mathfrak{gl}_N)$. So we will think of $\mathrm{U}(\mathfrak{gl}_N)$ as the associative algebra with these generators subject to the defining relations

$$E_{ij}\, E_{kl} - E_{kl}\, E_{ij} = \delta_{kj}\, E_{il} - \delta_{il}\, E_{kj}, \qquad i, j, k, l \in \{1, \dots, N\}. \tag{4.8}$$

Taking $\mathcal{A} = \mathrm{U}(\mathfrak{gl}_N)$ in (1.59), combine the elements E_{ij} into the matrix E so that

$$E = \sum_{i,j=1}^{N} e_{ij} \otimes E_{ij} \in \operatorname{End} \mathbb{C}^N \otimes \mathrm{U}(\mathfrak{gl}_N).$$

Consider the tensor product algebra

$$\operatorname{End} \mathbb{C}^N \otimes \operatorname{End} \mathbb{C}^N \otimes \mathrm{U}(\mathfrak{gl}_N) \tag{4.9}$$

and use the notation of Section 1.4. We identify the permutation operator (1.64) with the element $P \otimes 1$ of the algebra (4.9).

PROPOSITION 4.2.2. *The defining relations of the algebra* $\mathrm{U}(\mathfrak{gl}_N)$ *can be written in the form*

$$E_1\, E_2 - E_2\, E_1 = (E_1 - E_2)\, P. \tag{4.10}$$

PROOF. This can be derived from Proposition 4.1.1 and it is also easy to check directly. Namely, compare the coefficients of the basis vectors $e_{ij} \otimes e_{kl} \otimes 1$ on both sides of (4.10). This is equivalent to applying the operators on both sides to the basis vector $e_j \otimes e_l$ of $\mathbb{C}^N \otimes \mathbb{C}^N$ and then comparing the coefficients of the vector $e_i \otimes e_k$. For the left hand side we have

$$\sum_{i,k=1}^{N} e_i \otimes e_k \otimes \big(E_{ij}\, E_{kl} - E_{kl}\, E_{ij}\big),$$

while for the right hand side we get

$$(E_1 - E_2)\, P(e_j \otimes e_l) = (E_1 - E_2)\,(e_l \otimes e_j) = \sum_{i=1}^{N} e_i \otimes e_j \otimes E_{il} - \sum_{k=1}^{N} e_l \otimes e_k \otimes E_{kj}.$$

Equating the coefficients of $e_i \otimes e_k$ we recover the defining relations (4.8). □

The advantage of having the defining relations for $\mathrm{U}(\mathfrak{gl}_N)$ written as a single relation (4.10) for the matrix E will be apparent in Theorem 4.5.1 below, where we produce a family of Casimir elements for the Lie algebra $\mathfrak{gl}_N$.

REMARK 4.2.3. The matrix form of the defining relations given in Proposition 4.2.2 leads to the definition of the *degenerate affine Hecke algebra* $\mathcal{H}_m$ as follows; see [**31**]. First note the relations

$$P_{a\,b}\, E_c = E_c\, P_{a\,b} \qquad \text{and} \qquad P_{a\,b}\, E_a = E_b\, P_{a\,b}, \tag{4.11}$$

which hold in the algebra (1.60) with $\mathcal{A} = \mathrm{U}(\mathfrak{gl}_N)$ for $1 \leqslant a < b \leqslant m$ and $c \neq a, b$. Moreover, by (4.10) for $a < b$ we have

$$E_a\, E_b - E_b\, E_a = E_a\, P_{a\,b} - P_{a\,b}\, E_a. \tag{4.12}$$

It was pointed out in [**3**] that $\mathcal{H}_m$ is isomorphic to the algebra generated by (abstract) elements $E_1, \dots, E_m$ and the group algebra $\mathbb{C}[\mathfrak{S}_m]$ subject to the relations (4.11) and (4.12), where $P_{a\,b}$ is understood as the transposition $s_{a\,b} \in \mathfrak{S}_m$. To see the connection with the definition of [**31**], set $u_a = E_a - x_a$ for $a = 1, \dots, m$, where x_a is the Jucys–Murphy element defined in (1.10). The elements u_a pairwise commute, $u_a u_b = u_b u_a$, while

$$s_a u_a = u_{a+1} s_a + 1 \qquad \text{and} \qquad s_b u_a = u_a s_b \quad \text{if} \quad b \neq a-1, a.$$

The map taking E_a to $u_a + x_a$ and identical on $\mathbb{C}[\mathfrak{S}_m]$ provides an isomorphism between the two presentations of $\mathcal{H}_m$. □

REMARK 4.2.4. The *adjoint action* of the group GL_N of all invertible $N \times N$ matrices over $\mathbb{C}$ on the Lie algebra $\mathfrak{gl}_N$ is defined by (2.7). This extends to a unique action of the group GL_N on $\mathrm{U}(\mathfrak{gl}_N)$ so that each element of the group acts as an automorphism. The center coincides with the subalgebra of invariants under this action, $\mathrm{Z}(\mathfrak{gl}_N) = \mathrm{U}(\mathfrak{gl}_N)^{\mathrm{GL}_N}$. □

EXAMPLE 4.2.5. It is an easy calculation to verify directly or with the use of Remark 4.2.4 that

$$\sum_{i=1}^{N} E_{ii} \qquad \text{and} \qquad \sum_{i,j=1}^{N} E_{ij} E_{ji}$$

are Casimir elements for $\mathfrak{gl}_N$. □

Given an N-tuple of complex numbers $\lambda = (\lambda_1, \dots, \lambda_N)$, the corresponding *irreducible highest weight representation* $L(\lambda)$ of the Lie algebra $\mathfrak{gl}_N$ is generated by a nonzero vector $\xi \in L(\lambda)$ (the *highest vector*) such that

$$E_{ij}\, \xi = 0 \qquad \text{for} \quad 1 \leqslant i < j \leqslant N, \qquad \text{and} \tag{4.13}$$

$$E_{ii}\, \xi = \lambda_i\, \xi \qquad \text{for} \quad 1 \leqslant i \leqslant N. \tag{4.14}$$

Any element $z \in \mathrm{Z}(\mathfrak{gl}_N)$ acts in $L(\lambda)$ by multiplying each vector by a scalar $\chi(z)$. When regarded as a function of the highest weight, $\chi(z)$ is a symmetric polynomial in the variables $l_1, \dots, l_N$, where $l_i = \lambda_i - i + 1$. This provides an equivalent interpretation of the Harish-Chandra isomorphism (4.7) as the mapping $z \mapsto \chi(z)$ defines an algebra isomorphism

$$\chi : \mathrm{Z}(\mathfrak{gl}_N) \to \mathbb{C}[l_1, \dots, l_N]^{\mathfrak{S}_N}, \tag{4.15}$$

where $\mathbb{C}[l_1, \dots, l_N]^{\mathfrak{S}_N}$ denotes the algebra of $\mathfrak{S}_N$-invariant (symmetric) polynomials in $l_1, \dots, l_N$ and l_i is identified with the element $E_{ii} - i + 1 \in \mathrm{U}(\mathfrak{h})$. Thus, the commutative algebra $\mathrm{Z}(\mathfrak{gl}_N)$ can be regarded as an algebra of polynomials in N variables. The preimages of any family of algebraically independent generators of the algebra of symmetric polynomials are algebraically independent generators of $\mathrm{Z}(\mathfrak{gl}_N)$.

REMARK 4.2.6. The shifts in the definition of the variables l_i are determined by the half-sum of the positive roots for the simple Lie algebra $\mathfrak{sl}_N$. Clearly, given any constant $a \in \mathbb{C}$, one could set $l_i = \lambda_i - i + a$ for $i = 1, \dots, N$ to get an alternative definition of the isomorphism (4.15). This extra freedom is explained by the fact that the reductive Lie algebra $\mathfrak{gl}_N$ is isomorphic to the direct sum of $\mathfrak{sl}_N$ and the one-dimensional center spanned by the scalar matrices. Apart from the value $a = 1$ which we use in (4.15), some other common choices of a in the literature include $(N+1)/2$ and N. □

EXAMPLE 4.2.7. For the Harish-Chandra images of the Casimir elements of Example 4.2.5 we have

$$\chi: \sum_{i=1}^{N} E_{ii} \mapsto \sum_{i=1}^{N} l_i + \binom{N}{2},$$

$$\chi: \sum_{i,j=1}^{N} E_{ij} E_{ji} \mapsto \sum_{i=1}^{N} l_i^2 + (N-1) \sum_{i=1}^{N} l_i + \binom{N}{3}.$$

They are found by the application of the Casimir elements to the highest vector ξ in the representation $L(\lambda)$ of $\mathfrak{gl}_N$. □

4.3. Factorial Schur polynomials

Certain particular families of symmetric polynomials will occur as Harish-Chandra images of central elements for the simple Lie algebras of all classical types. To describe them, consider the algebra of symmetric polynomials in the independent variables $x_1, \dots, x_n$ over $\mathbb{C}$ and fix a sequence $a = (a_1, a_2, \dots)$ of complex numbers. The *factorial elementary* and *complete symmetric polynomials* are defined by the respective formulas

$$(4.16) \quad e_k(x_1, \dots, x_n \,|\, a) = \sum_{1 \leqslant p_1 < \dots < p_k \leqslant n} (x_{p_1} - a_{p_1})(x_{p_2} - a_{p_2-1}) \dots (x_{p_k} - a_{p_k-k+1}),$$

and

$$(4.17) \quad h_k(x_1, \dots, x_n \,|\, a) = \sum_{1 \leqslant p_1 \leqslant \dots \leqslant p_k \leqslant n} (x_{p_1} - a_{p_1})(x_{p_2} - a_{p_2+1}) \dots (x_{p_k} - a_{p_k+k-1}),$$

so that $e_k(x_1, \dots, x_n \,|\, a) = 0$ for $k > n$. These polynomials are particular cases of the *factorial* (or *double*) *Schur polynomials* $s_\mu(x_1, \dots, x_n \,|\, a)$ defined as follows. Suppose that μ is a diagram with at most n rows; see Section 1.1. Then

$$(4.18) \quad s_\mu(x_1, \dots, x_n \,|\, a) = \sum_{\mathrm{sh}(\mathcal{T})=\mu} \prod_{\alpha \in \mu} \big(x_{\mathcal{T}(\alpha)} - a_{\mathcal{T}(\alpha)+c(\alpha)}\big),$$

summed over semistandard tableaux $\mathcal{T}$ of shape μ with entries in $\{1, \dots, n\}$, where $\mathcal{T}(\alpha)$ denotes the entry of the tableau $\mathcal{T}$ at the box $\alpha = (i, j)$ of μ and $c(\alpha) = j - i$ is the content of this box. In the particular cases $\mu = (1^k)$ and $\mu = (k)$ the polynomial (4.18) coincides with (4.16) and (4.17), respectively.

When a is specialized to the sequence of zeros, then (4.16) and (4.17) become the respective elementary and complete symmetric polynomials $e_k(x_1, \dots, x_n)$ and $h_k(x_1, \dots, x_n)$, and $s_\mu(x_1, \dots, x_n \,|\, a)$ becomes the Schur polynomial $s_\mu(x_1, \dots, x_n)$. These homogeneous polynomials also coincide with the top degree components of the respective factorial counterparts. This implies, in particular, that the factorial Schur polynomials with μ running over all diagrams with at most n rows form a basis of the algebra of symmetric polynomials $\mathbb{C}[x_1, \dots, x_n]^{\mathfrak{S}_n}$. Some other equivalent definitions of the factorial Schur polynomials (4.18) and their basic properties are discussed in the book by Macdonald [**104**, Section I.3]. Here we derive their vanishing and characterization properties to be used below.

We will suppose that the sequence a is multiplicity-free, that is $a_k \neq a_l$ for all $k \neq l$. Set $x = (x_1, \dots, x_n)$ and for any partition $\lambda = (\lambda_1, \dots, \lambda_n)$ such that $\ell(\lambda) \leqslant n$ introduce the n-tuple of elements of a by

$$a_\lambda = (a_{\lambda_1+n}, \dots, a_{\lambda_n+1}). \tag{4.19}$$

As before we let λ'_j denote the number of boxes in the column j of the diagram λ.

PROPOSITION 4.3.1. *If $\mu \not\subset \lambda$ then $s_\mu(a_\lambda\,|\,a) = 0$. Moreover,*

$$s_\mu(a_\mu\,|\,a) = \prod_{(i,j)\in\mu} \big(a_{\mu_i+n-i+1} - a_{n-\mu'_j+j}\big). \tag{4.20}$$

PROOF. The polynomials $s_\mu(x\,|\,a)$ are symmetric in x, so replacing x with $(x_n, \dots, x_1)$, we may rewrite (4.18) in the form

$$s_\mu(x\,|\,a) = \sum_{\mathrm{sh}(\mathcal{T})=\mu} \prod_{\alpha\in\mu} \big(x_{n-\mathcal{T}(\alpha)+1} - a_{\mathcal{T}(\alpha)+c(\alpha)}\big), \tag{4.21}$$

with the summation over semistandard μ-tableau $\mathcal{T}$ with entries from $\{1, \dots, n\}$. Suppose that $s_\mu(a_\lambda\,|\,a) \neq 0$. Then at least one summand in (4.21) does not vanish for $x = a_\lambda$,

$$\prod_{\alpha\in\mu} \big(a_{\lambda_{n-\mathcal{T}(\alpha)+1}+\mathcal{T}(\alpha)} - a_{\mathcal{T}(\alpha)+c(\alpha)}\big) \neq 0. \tag{4.22}$$

Since the sequence a is multiplicity free, this implies that

$$\lambda_{n-\mathcal{T}(\alpha)+1} \neq c(\alpha) \tag{4.23}$$

for all $\alpha \in \mu$. For the entries of the first row of the tableau $\mathcal{T}$ we have

$$\mathcal{T}(1,1) \leqslant \dots \leqslant \mathcal{T}(1,\mu_1).$$

Note that $c(1,1) = 0$ so that using (4.23) with $\alpha = (1,1)$ we obtain $\lambda_{n-\mathcal{T}(1,1)+1} \geqslant 1$. Furthermore, $c(1,2) = 1$ and since

$$\lambda_{n-\mathcal{T}(1,2)+1} \geqslant \lambda_{n-\mathcal{T}(1,1)+1} \geqslant 1,$$

by using (4.23) with $\alpha = (1,2)$ we get $\lambda_{n-\mathcal{T}(1,2)+1} \geqslant 2$. Continuing in the same manner, we conclude that $\lambda_{n-\mathcal{T}(1,i)+1} \geqslant i$ for all $i = 1, \dots, \mu_1$. On the other hand, for the entries of the i-th column of $\mathcal{T}$ we have

$$\mathcal{T}(1,i) < \dots < \mathcal{T}(\mu'_i, i).$$

Therefore,

$$\lambda_{n-\mathcal{T}(\mu'_i,i)+1} \geqslant \dots \geqslant \lambda_{n-\mathcal{T}(1,i)+1} \geqslant i. \tag{4.24}$$

This means that the diagram λ has at least μ'_i rows of length at least i so that $\lambda'_i \geqslant \mu'_i$. Hence, $\mu \subset \lambda$ thus proving the first part of the proposition.

Now take $\lambda = \mu$ and suppose that (4.22) holds for a certain tableau $\mathcal{T}$. By the above argument, the inequalities (4.24) hold for $\lambda = \mu$ which implies that such tableau $\mathcal{T}$ is determined uniquely by $\mathcal{T}(k,i) = n - \mu'_i + k$ for $k = 1, \dots, \mu'_i$. Thus, there is a unique nonzero summand in (4.21) with $x = a_\mu$. By using the values of the tableau, we come to (4.20). □

The vanishing properties of the factorial Schur polynomials given by Proposition 4.3.1 are characteristic in the sense of the next proposition. We suppose that μ is a diagram with at most n rows.

PROPOSITION 4.3.2. *Let $f(x)$ be a symmetric polynomial of degree not exceeding $|\mu|$. If the top degree component of $f(x)$ coincides with $s_\mu(x)$ and $f(a_\lambda) = 0$ for all λ such that $|\lambda| < |\mu|$, then $f(x) = s_\mu(x\,|\,a)$.*

PROOF. Represent the difference $f(x) - s_\mu(x\,|\,a)$ as a linear combination of the basis polynomials,

$$f(x) - s_\mu(x\,|\,a) = \sum_{\nu \in S} c_\nu\, s_\nu(x\,|\,a), \tag{4.25}$$

where $c_\nu \in \mathbb{C}$ and the sum is taken over the set S of the diagrams ν with $|\nu| < |\mu|$. Let λ run over the same set S. Then $\mu \not\subset \lambda$ and $s_\mu(a_\lambda\,|\,a) = 0$ by Proposition 4.3.1. Hence putting $x = a_\lambda$ in (4.25) for all such λ, we obtain a system of linear equations on the coefficients c_ν of the form

$$\sum_{\nu \in S} c_\nu\, s_\nu(a_\lambda\,|\,a) = 0.$$

Equip the set S with any linear ordering $\prec$ such that $|\lambda| < |\nu|$ implies $\lambda \prec \nu$. By arranging the system of equations according to the ordering $\prec$ we find that the corresponding matrix is triangular due to the vanishing properties of Proposition 4.3.1. The diagonal entries of the matrix are the values $s_\nu(a_\nu\,|\,a)$. All the numbers $\nu_i + n - i + 1$ and $n - \nu'_j + j$ are distinct and so the product in (4.20) is nonzero since the sequence a is multiplicity-free. Thus $s_\nu(a_\nu\,|\,a) \neq 0$ and the system has only the trivial solution $c_\nu = 0$ which means that $f(x) = s_\mu(x\,|\,a)$. $\square$

4.4. Schur–Weyl duality

Recall the action of the symmetric group $\mathfrak{S}_m$ on the tensor product space $(\mathbb{C}^N)^{\otimes m}$ as defined in Section 1.5. The general linear group GL_N acts on this space diagonally; see (2.13). The corresponding action of the Lie algebra $\mathfrak{gl}_N$ is given by

$$X : v_1 \otimes \dots \otimes v_m \mapsto \sum_{a=1}^{m} v_1 \otimes \dots \otimes X\, v_a \otimes \dots \otimes v_m, \qquad v_i \in \mathbb{C}^N, \qquad X \in \mathfrak{gl}_N.$$

In line with our tensor notation, these actions are written as

$$\mathbf{h} \mapsto \mathbf{h}_1 \dots \mathbf{h}_m \qquad \text{and} \qquad X \mapsto X_1 + \dots + X_m,$$

where

$$\mathbf{h}_a = 1^{\otimes (a-1)} \otimes \mathbf{h} \otimes 1^{\otimes (m-a)} \qquad \text{and} \qquad X_a = 1^{\otimes (a-1)} \otimes X \otimes 1^{\otimes (m-a)}.$$

As we verified in Section 2.1, the action of any element $s \in \mathfrak{S}_m$ on the vector space $(\mathbb{C}^N)^{\otimes m}$ commutes with that of any element $\mathbf{h} \in \mathrm{GL}_N$ (and hence with the action of any element $X \in \mathfrak{gl}_N$). By the classical *Schur–Weyl duality*, these actions of $\mathfrak{S}_m$ and GL_N centralize each other. This leads to the multiplicity-free decomposition as a representation of the group $\mathfrak{S}_m \times \mathrm{GL}_N$,

$$(\mathbb{C}^N)^{\otimes m} \cong \bigoplus_{\lambda \vdash m,\ \ell(\lambda) \leqslant N} V_\lambda \otimes L(\lambda), \tag{4.26}$$

where V_λ and $L(\lambda)$ are the respective irreducible representations of $\mathfrak{S}_m$ and GL_N associated with a Young diagram λ which contains $|\lambda| = m$ boxes, and the number of nonzero rows $\ell(\lambda)$ does not exceed N. As a $\mathfrak{gl}_N$-module, $L(\lambda)$ is the highest weight representation with the highest weight $\lambda = (\lambda_1, \dots, \lambda_N)$, where one sets $\lambda_i = 0$ for all $i = \ell(\lambda)+1, \dots, N$.

Let $\mathcal{U}$ be a standard tableau of shape $\lambda \vdash m$ and let $e_{\mathcal{U}} \in \mathbb{C}[\mathfrak{S}_m]$ be the associated primitive idempotent; see Section 1.1. Denote by $\mathcal{E}_{\mathcal{U}}$ the image of $e_{\mathcal{U}}$ under the action of the symmetric group $\mathfrak{S}_m$ given by (1.65). The space $\mathcal{E}_{\mathcal{U}}(\mathbb{C}^N)^{\otimes m}$ is an irreducible representation of GL_N isomorphic to $L(\lambda)$. Therefore, the trace $\mathrm{tr}_{1,\dots,m}\, \mathcal{E}_{\mathcal{U}}\, \mathbf{h}_1 \dots \mathbf{h}_m$ coincides with the character of the representation $L(\lambda)$ evaluated at the element $\mathbf{h}$. This value is given by the Weyl character formula so that the trace equals the Schur polynomial s_μ evaluated at the eigenvalues $h_1, \dots, h_N$ of the matrix $\mathbf{h}$,

$$\mathrm{tr}_{1,\dots,m}\, \mathcal{E}_{\mathcal{U}}\, \mathbf{h}_1 \dots \mathbf{h}_m = s_\mu(h_1, \dots, h_N); \tag{4.27}$$

see also (2.19). In particular, the dimension $\dim L(\lambda)$ coincides with the trace

$$\mathrm{tr}_{1,\dots,m}\, \mathcal{E}_{\mathcal{U}} = s_\mu(1, \dots, 1).$$

By the correspondence (1.72), the trace formula for the idempotents given in Proposition 1.3.5 implies the *Robinson hook dimension formula*

$$\dim L(\lambda) = \frac{1}{h(\lambda)} \prod_{(i,j)\in\lambda} (N + j - i). \tag{4.28}$$

4.5. A general construction of central elements

Now suppose that $s \in \mathbb{C}[\mathfrak{S}_m]$ is an arbitrary element and let S denote its image under the map (1.65). We regard S as an element of the algebra

$$\underbrace{\mathrm{End}\, \mathbb{C}^N \otimes \dots \otimes \mathrm{End}\, \mathbb{C}^N}_{m} \otimes \mathrm{U}(\mathfrak{gl}_N) \tag{4.29}$$

by identifying it with $S \otimes 1$.

THEOREM 4.5.1. *For any $s \in \mathbb{C}[\mathfrak{S}_m]$ and $u_1, \dots, u_m \in \mathbb{C}$ the element*

$$\mathrm{tr}_{1,\dots,m}\, S\, (u_1 + E_1) \dots (u_m + E_m) \tag{4.30}$$

belongs to the center $\mathrm{Z}(\mathfrak{gl}_N)$.

PROOF. Consider the tensor product

$$\mathrm{End}\, \mathbb{C}^N \otimes \mathrm{End}\, (\mathbb{C}^N)^{\otimes m} \otimes \mathrm{U}(\mathfrak{gl}_N) \tag{4.31}$$

with an additional copy of the endomorphism algebra $\mathrm{End}\, \mathbb{C}^N$. We will label the copies of this algebra respectively by $0, 1, \dots, m$. It will be sufficient to show that the following commutator in the algebra (4.31) is zero,

$$\big[E_0, \mathrm{tr}_{1,\dots,m}\, S\, (u_1 + E_1) \dots (u_m + E_m)\big] = 0. \tag{4.32}$$

To this end, note that by (1.67) and Proposition 4.2.2 we can write

$$[E_0, u_a + E_a] = P_{0\,a}(u_a + E_a) - (u_a + E_a)P_{0\,a}$$

so that

$$\begin{aligned}\big[E_0, S\,(u_1+E_1)\dots(u_m+E_m)\big]\\ &= \sum_{a=1}^{m} S\,(u_1+E_1)\dots\big(P_{0\,a}(u_a+E_a)-(u_a+E_a)P_{0\,a}\big)\dots(u_m+E_m)\\ &= S\sum_{a=1}^{m} P_{0\,a}\,(u_1+E_1)\dots(u_m+E_m) - S\,(u_1+E_1)\dots(u_m+E_m)\sum_{a=1}^{m} P_{0\,a},\end{aligned}$$

where we used the observation that $E_0\, S = S\, E_0$ and that $P_{0\,a}$ commutes with E_b for $b \neq a$. Furthermore, the sum of the permutation operators $P_{0\,a}$ commutes with the image of any element of $\mathfrak{S}_m$ under the map (1.65) so that

$$\begin{aligned}\mathrm{tr}_{1,\dots,m}\, S\sum_{a=1}^{m} P_{0\,a}\,(u_1+E_1)\dots(u_m+E_m)\\ &= \mathrm{tr}_{1,\dots,m}\sum_{a=1}^{m} P_{0\,a}\, S\,(u_1+E_1)\dots(u_m+E_m)\\ &= \mathrm{tr}_{1,\dots,m}\, S\,(u_1+E_1)\dots(u_m+E_m)\sum_{a=1}^{m} P_{0\,a},\end{aligned}$$

where the last equality holds by the cyclic property of trace; see Lemma 1.4.1. Its application relies on the fact that two elements of the algebra (4.31) of the form

$$X\otimes 1^{\otimes m}\otimes 1 \qquad \text{and} \qquad 1\otimes 1^{\otimes m}\otimes y$$

commute for any $X \in \mathrm{End}\,\mathbb{C}^N$ and $y \in \mathrm{U}(\mathfrak{gl}_N)$. This proves (4.32). □

REMARK 4.5.2. A slight modification of the above argument shows that all elements of the form

$$\mathrm{tr}_{1,\dots,m}\, S\, E_{a_1}\dots E_{a_k}$$

with arbitrary parameters $a_i \in \{1,\dots,m\}$ also belong to $\mathrm{Z}(\mathfrak{gl}_N)$. □

In what follows we consider particular choices of the parameters $u_1,\dots,u_m$ and the element $s \in \mathbb{C}[\mathfrak{S}_m]$ in Theorem 4.5.1. The next lemma will allow us to rely on the properties of Manin matrices discussed in Chapter 3.

LEMMA 4.5.3. *Suppose that α and β are elements of a certain unital associative algebra $\mathcal{D}$ which satisfy the relation*

$$\alpha\beta-\beta\alpha=\beta^2. \tag{4.33}$$

Then the matrix $\alpha+E\,\beta$ with entries in the algebra $\mathrm{U}(\mathfrak{gl}_N)\otimes\mathcal{D}$ is a Manin matrix.

PROOF. Set $M=\alpha+E\,\beta$ (we abbreviate notation by omitting the tensor product signs). Using Proposition 4.2.2 we find

$$\begin{aligned}M_1\,M_2-M_2\,M_1 &= \big(\alpha+E_1\,\beta\big)\big(\alpha+E_2\,\beta\big)-\big(\alpha+E_2\,\beta\big)\big(\alpha+E_1\,\beta\big)\\ &= (E_1\,E_2-E_2\,E_1)\,\beta^2-(E_1-E_2)(\alpha\beta-\beta\alpha)=(E_1-E_2)(P-1)\,\beta^2.\end{aligned}$$

Hence, on multiplying this element from the right by $1+P$ we get 0, and the claim follows by using the equivalent definition (3.5) of Manin matrices. □

Examples of elements α and β satisfying (4.33) are provided by

$$\alpha = -\partial_t, \quad \beta = t^{-1} \qquad \text{and} \qquad \alpha = u\, e^{-\partial_u}, \quad \beta = e^{-\partial_u}.$$

In the first example we can take $\mathcal{D}$ to be the algebra of polynomial differential operators of the form

$$a_0 + a_1\, \partial_t + \cdots + a_k\, \partial_t^k, \qquad k \geqslant 0,$$

where each coefficient a_i is a Laurent polynomial in t. In the second example we take $\mathcal{D}$ to be generated by all polynomials in u and an additional element $e^{-\partial_u}$, subject to the relations

$$e^{-\partial_u} f(u) = f(u-1)\, e^{-\partial_u} \tag{4.34}$$

for any polynomial $f(u)$.

LEMMA 4.5.4. *Let $u_1, \dots, u_m$ be complex parameters and let S be the image of an element s of the center of the group algebra $\mathbb{C}[\mathfrak{S}_m]$ under the homomorphism* (1.65)*. For any permutations $\sigma, \tau \in \mathfrak{S}_m$ we have the identity*

$$\mathrm{tr}_{1,\dots,m}\, S\, (E_{\sigma(1)} + u_{\tau(1)}) \dots (E_{\sigma(m)} + u_{\tau(m)}) = \mathrm{tr}_{1,\dots,m}\, S\, (E_1 + u_1) \dots (E_m + u_m).$$

PROOF. By the cyclic property of trace, the right hand side equals

$$\begin{aligned}\mathrm{tr}_{1,\dots,m}\, & P_\sigma S\, (E_1 + u_1) \dots (E_m + u_m) P_\sigma^{-1} \\ &= \mathrm{tr}_{1,\dots,m}\, S\, P_\sigma\, (E_1 + u_1) \dots (E_m + u_m) P_\sigma^{-1} \\ &= \mathrm{tr}_{1,\dots,m}\, S\, (E_{\sigma(1)} + u_1) \dots (E_{\sigma(m)} + u_m),\end{aligned}$$

where we also used (1.67). Thus, it suffices to verify the identity in the case where σ is the identity permutation. We may assume that τ is an adjacent transposition $s_a = (a\ a+1)$. By Proposition 4.2.2,

$$(E_a + u_{a+1})(E_{a+1} + u_a) - (E_{a+1} + u_a)(E_a + u_{a+1}) = P_{a\,a+1}\, E_{a+1} - E_{a+1}\, P_{a\,a+1}.$$

Since $S\, P_{a\,a+1} = P_{a\,a+1}\, S$, the claim follows from the cyclic property of trace and the first part of the proof. □

In particular, Lemma 4.5.4 holds for the symmetrizer and anti-symmetrizer $S = H^{(m)}$ and $S = A^{(m)}$.

4.6. Capelli determinant

Take $S = A^{(m)}$ with $1 \leqslant m \leqslant N$ in Theorem 4.5.1 and specialize the parameters by

$$u_a = u - a + 1, \qquad a = 1, \dots, m,$$

for a variable u. Then the element (4.30) becomes a polynomial in u of degree m,

$$\mathrm{tr}_{1,\dots,m}\, A^{(m)}\, (u + E_1) \dots (u + E_m - m + 1), \tag{4.35}$$

whose coefficients are Casimir elements for $\mathfrak{gl}_N$.

PROPOSITION 4.6.1. *The Harish-Chandra images of the coefficients of the polynomial* (4.35) *are found by*

$$\begin{aligned}\chi : \mathrm{tr}_{1,\dots,m}\, & A^{(m)}\, (u + E_1) \dots (u + E_m - m + 1) \\ & \mapsto \sum_{1 \leqslant i_1 < \dots < i_m \leqslant N} (u + \lambda_{i_1}) \cdots (u + \lambda_{i_m} - m + 1).\end{aligned}$$

PROOF. Use Lemma 4.5.3 to apply Proposition 3.2.2 to the Manin matrix $M = (u+E)e^{-\partial_u}$. By (4.34) the left hand side of (3.17) can be written as

$$\operatorname{tr}_{1,\dots,m} A^{(m)}(u+E_1)e^{-\partial_u}\dots(u+E_m)e^{-\partial_u}$$
$$= \operatorname{tr}_{1,\dots,m} A^{(m)}(u+E_1)\dots(u+E_m-m+1)e^{-m\partial_u}.$$

Now consider the product $M_{i_{\sigma(1)}i_1}\dots M_{i_{\sigma(m)}i_m}$ which occurs as a summand in the right hand side of (3.17). It is clear from (4.13) that its application to the highest vector ξ of a highest weight representation $L(\lambda)$ yields zero unless σ is the identity permutation in $\mathfrak{S}_m$. In that case, by (4.14)

$$M_{i_1 i_1}\dots M_{i_m i_m}\,\xi = (u+E_{i_1i_1})e^{-\partial_u}\dots(u+E_{i_mi_m})e^{-\partial_u}\,\xi$$
$$= (u+\lambda_{i_1})\dots(u+\lambda_{i_m}-m+1)e^{-m\partial_u}\,\xi.$$

Taking the sum over the indices $1 \leqslant i_1 < \dots < i_m \leqslant N$ we get the desired formula for the Harish-Chandra image of the polynomial (4.35). □

Due to Proposition 4.6.1 all coefficients of the polynomial in u given by

$$\sum_{1\leqslant i_1<\dots<i_m\leqslant N}(u+\lambda_{i_1})\cdots(u+\lambda_{i_m}-m+1) \tag{4.36}$$

are symmetric polynomials in the variables $l_i = \lambda_i - i + 1$ with $i = 1,\dots,N$. In particular, taking $u = 0$ in (4.36) we get the factorial elementary symmetric polynomial

$$e_m(l_1,\dots,l_N\,|\,a) = \sum_{1\leqslant i_1<\dots<i_m\leqslant N}(l_{i_1}+i_1-1)\cdots(l_{i_m}+i_m-m),$$

associated with the parameter sequence $a = (a_i)$, $a_i = -i+1$; see (4.16).

We use the column-determinant (3.23) to define the *Capelli determinant* by

$$C(u) = \operatorname{cdet}\begin{bmatrix} u+E_{11} & E_{12} & \dots & E_{1N}\\ E_{21} & u+E_{22}-1 & \dots & E_{2N}\\ \vdots & \vdots & & \vdots\\ E_{N1} & E_{N2} & \dots & u+E_{NN}-N+1\end{bmatrix}. \tag{4.37}$$

This is a polynomial in u,

$$C(u) = u^N + C_1^\circ u^{N-1} + \dots + C_N^\circ, \qquad C_k^\circ \in \mathrm{U}(\mathfrak{gl}_N). \tag{4.38}$$

Note that the symbols $C_k \in \mathrm{S}(\mathfrak{gl}_N)$ of the elements C_k° coincide with the coefficients of the characteristic polynomial $\det(u+E)$; see (2.6). Relation (3.28) and the proof of Proposition 4.6.1 imply the properties of $C(u)$ given by the next corollary.

COROLLARY 4.6.2. *The Capelli determinant coincides with the polynomial in u defined in (4.35) with $m = N$,*

$$C(u) = \operatorname{tr}_{1,\dots,N} A^{(N)}(u+E_1)\dots(u+E_N-N+1)$$

so that all coefficients C_k° are Casimir elements for $\mathfrak{gl}_N$. Moreover, under the Harish-Chandra isomorphism (4.15) we have

$$\chi: C(u) \to (u+l_1)\dots(u+l_N).$$

In particular, the elements $C_1^\circ,\dots,C_N^\circ$ are algebraically independent generators of the center $\mathrm{Z}(\mathfrak{gl}_N)$ of the universal enveloping algebra $\mathrm{U}(\mathfrak{gl}_N)$. □

The last property holds since the elementary symmetric polynomials are algebraically independent generators of the algebra $\mathbb{C}[l_1,\dots,l_N]^{\mathfrak{S}_N}$.

4.7. Permanent-type elements

Now take $S = H^{(m)}$ with $m \geqslant 1$ in Theorem 4.5.1 and specialize the parameters by

$$u_a = u + a - 1, \qquad a = 1,\dots,m,$$

for a variable u. The element (4.30) becomes a polynomial in u of degree m:

$$\operatorname{tr}_{1,\dots,m} H^{(m)}\,(u + E_1)\dots(u + E_m + m - 1). \tag{4.39}$$

Its coefficients are Casimir elements for $\mathfrak{gl}_N$.

PROPOSITION 4.7.1. *The Harish-Chandra images of the coefficients of the polynomial* (4.39) *are found by*

$$\chi : \operatorname{tr}_{1,\dots,m} H^{(m)}\,(u + E_1)\dots(u + E_m + m - 1) \mapsto \sum_{1\leqslant i_1\leqslant\dots\leqslant i_m\leqslant N} (u + \lambda_{i_1})\cdots(u + \lambda_{i_m} + m - 1).$$

PROOF. We use again Proposition 3.2.2 and Lemma 4.5.3. The left hand side of (3.18) can be written for the Manin matrix $M = (u + E + m - 1)\,e^{-\partial_u}$ as

$$\begin{aligned}\operatorname{tr}_{1,\dots,m} H^{(m)}&(u + E_1 + m - 1)\,e^{-\partial_u}\dots(u + E_m + m - 1)\,e^{-\partial_u}\\ &= \operatorname{tr}_{1,\dots,m} H^{(m)}\,(u + E_1 + m - 1)\dots(u + E_m)\,e^{-m\,\partial_u}\\ &= \operatorname{tr}_{1,\dots,m} H^{(m)}\,(u + E_1)\dots(u + E_m + m - 1)\,e^{-m\,\partial_u},\end{aligned}$$

where the second equality holds due to Lemma 4.5.4. For $\sigma \in \mathfrak{S}_m$ consider the product $M_{i_m\, i_{\sigma(m)}}\dots M_{i_1\, i_{\sigma(1)}}$ which occurs as a summand on the right hand side of (3.18). By (4.13) its application to the highest vector ξ of a highest weight representation $L(\lambda)$ yields zero unless σ belongs to the stabilizer of the multiset $\{i_1,\dots,i_m\}$. In this case, by (4.14) we find

$$\begin{aligned}M_{i_m\, i_m}\dots M_{i_1\, i_1}\,\xi &= (u + E_{i_m i_m} + m - 1)\,e^{-\partial_u}\dots(u + E_{i_1 i_1} + m - 1)\,e^{-\partial_u}\,\xi\\ &= (u + \lambda_{i_m} + m - 1)\dots(u + \lambda_{i_1})\,e^{-m\,\partial_u}\,\xi.\end{aligned}$$

Taking into account the number $\alpha_1!\dots\alpha_N!$ of permutations which stabilize the multiset $\{i_1,\dots,i_m\}$, we obtain the desired formula. $\square$

Due to Proposition 4.7.1 all coefficients of the polynomial in u given by

$$\sum_{1\leqslant i_1\leqslant\dots\leqslant i_m\leqslant N} (u + \lambda_{i_1})\cdots(u + \lambda_{i_m} + m - 1)$$

are symmetric polynomials in the variables $l_i = \lambda_i - i + 1$ with $i = 1,\dots,N$. In particular, taking $u = 0$ we get the factorial complete symmetric polynomial

$$h_m(l_1,\dots,l_N\,|\,a) = \sum_{1\leqslant i_1\leqslant\dots\leqslant i_m\leqslant N} (l_{i_1} + i_1 - 1)\cdots(l_{i_m} + i_m + m - 2),$$

associated with the parameter sequence $a = (a_i)$, $a_i = -i + 1$; see (4.17).

4.8. Gelfand invariants

Take $S = P_\sigma$ in Theorem 4.5.1, where $\sigma = (m, m-1, \dots, 1)$ is a long cycle in the symmetric group $\mathfrak{S}_m$. We have

$$P_\sigma = P_{m-1\,m} \dots P_{2\,3} P_{1\,2}. \tag{4.40}$$

We also set $u_a = 0$ for all $a = 1, \dots, m$. Then by (1.68) and the cyclic property of trace, (4.30) becomes the Casimir element $\operatorname{tr} E^m$ known as the *Gelfand invariant*. Indeed, note that by (1.67)

$$\begin{aligned} &\operatorname{tr}_{1,\dots,m} E_1 \dots E_m P_{m-1\,m} \dots P_{2\,3} P_{1\,2} \\ &\qquad\qquad = \operatorname{tr}_{1,\dots,m} E_1 \dots E_{m-1} P_{m-1\,m} E_{m-1} P_{m-2\,m-1} \dots P_{2\,3} P_{1\,2}. \end{aligned}$$

Since $\operatorname{tr}_m P_{m-1\,m} = 1$, applying the partial trace tr_m we bring the expression to the form

$$\operatorname{tr}_{1,\dots,m-1} E_1 \dots E_{m-1}^2 P_{m-2\,m-1} \dots P_{2\,3} P_{1\,2}$$

and then continue by induction to get

$$\operatorname{tr}_{1,\dots,m} P_\sigma E_1 \dots E_m = \operatorname{tr} E^m \tag{4.41}$$

which is therefore a Casimir element for $\mathfrak{gl}_N$ for any $m \geqslant 1$.

A more direct way to come to this conclusion is to use the following straightforward consequence of Proposition 4.2.2:

$$E_1 E_2^m - E_2^m E_1 = P_{12} E_2^m - E_2^m P_{12},$$

then take the partial trace tr_2 and apply Lemma 1.4.1; cf. Proposition 4.2.1.

The next theorem is the *Newton identity* for the Gelfand invariants. The argument will be similar to the proof of Theorem 3.2.10 but we need to use a difference operator instead of the derivative. We use the Capelli determinant $C(u)$; see (4.38).

THEOREM 4.8.1. *We have the identity*

$$1 + \sum_{m=0}^{\infty} \frac{(-1)^m \operatorname{tr} E^m}{(u-N+1)^{m+1}} = \frac{C(u+1)}{C(u)}.$$

PROOF. By Corollary 4.6.2,

$$\operatorname{tr}_{1,\dots,N} A^{(N)} (u + E_1) \dots (u + E_N - N + 1) = C(u).$$

Hence, by Lemma 4.5.4 we also have

$$\operatorname{tr}_{1,\dots,N} A^{(N)} (u + E_1) \dots (u + E_{N-1} - N + 2)(u + E_N + 1) = C(u+1).$$

Therefore,

$$\begin{aligned} C(u+1) - C(u) &= N \operatorname{tr}_{1,\dots,N} A^{(N)} (u + E_1) \dots (u + E_{N-1} - N + 2) \\ &= N \operatorname{tr}_{1,\dots,N} A^{(N)} C(u) (u + E_N - N + 1)^{-1}. \end{aligned}$$

Applying the conjugation by P_{1N} we can write the right hand side as

$$N \operatorname{tr}_{1,\dots,N} A^{(N)} C(u) (u + E_1 - N + 1)^{-1}.$$

Finally, calculate the partial trace $\operatorname{tr}_{2,\dots,N}$ by using (3.26) to get

$$C(u+1) - C(u) = C(u) \sum_{m=0}^{\infty} \frac{(-1)^m \operatorname{tr} E^m}{(u-N+1)^{m+1}},$$

as required. ∎

COROLLARY 4.8.2. *The Harish-Chandra images of the Gelfand invariants are found by*

$$1+\sum_{m=0}^{\infty}\frac{(-1)^m\,\chi(\operatorname{tr}E^m)}{(u-N+1)^{m+1}}=\prod_{i=1}^{N}\frac{u+l_i+1}{u+l_i}. \tag{4.42}$$

Equivalently, setting $\bar{l}_i=\lambda_i-i+N$ *we have*

$$\chi(\operatorname{tr}E^m)=\sum_{k=1}^{N}\bar{l}_k^{\,m}\,\frac{(\bar{l}_1-\bar{l}_k+1)\dots(\bar{l}_N-\bar{l}_k+1)}{(\bar{l}_1-\bar{l}_k)\dots\wedge\dots(\bar{l}_N-\bar{l}_k)}, \tag{4.43}$$

where the symbol $\wedge$ *indicates that the zero factor is skipped.*

PROOF. Relation (4.42) follows from Corollary 4.6.2 and Theorem 4.8.1. Setting $v=u-N+1$ we can write this relation in the form

$$1+\sum_{m=0}^{\infty}(-1)^m\,\chi(\operatorname{tr}E^m)\,v^{-m-1}=\prod_{i=1}^{N}\frac{v+\bar{l}_i+1}{v+\bar{l}_i}.$$

Now use a partial fraction decomposition for the right hand side,

$$\prod_{i=1}^{N}\frac{v+\bar{l}_i+1}{v+\bar{l}_i}=1+\frac{a_1}{v+\bar{l}_1}+\dots+\frac{a_N}{v+\bar{l}_N}.$$

The constants a_k are found by multiplying both sides by $v+\bar{l}_k$ and setting $v=-\bar{l}_k$. This gives (4.43) by expanding the rational functions into power series in v^{-1}. □

The Harish-Chandra images of certain modified Gelfand invariants will be calculated in Corollary 13.4.3 below.

4.9. Quantum immanants

As in Section 1.1, suppose that $\mathcal{U}$ is a standard tableau of shape $\mu\vdash m$ and let $e_{\mathcal{U}}\in\mathbb{C}[\mathfrak{S}_m]$ be the associated primitive idempotent. By $c_a=c_a(\mathcal{U})$ we denote the content $c(\alpha)=j-i$ of the box $\alpha=(i,j)$ in $\mathcal{U}$ occupied by $a\in\{1,\dots,m\}$. As before, we let $\mathcal{E}_{\mathcal{U}}$ denote the image of $e_{\mathcal{U}}$ under the action of the symmetric group $\mathfrak{S}_m$ given by (1.65). The operator $\mathcal{E}_{\mathcal{U}}$ is zero if the length $\ell(\mu)$ of μ exceeds N. So we will assume that $\ell(\mu)\leqslant N$. Now specialize the parameters in Theorem 4.5.1 by

$$u_a=u+c_a,\qquad a=1,\dots,m$$

for a variable u and take $S=\mathcal{E}_{\mathcal{U}}$. Denote the corresponding Casimir element by

$$\mathbb{S}_\mu(u)=\operatorname{tr}_{1,\dots,m}\mathcal{E}_{\mathcal{U}}\,(u+E_1+c_1)\dots(u+E_m+c_m). \tag{4.44}$$

PROPOSITION 4.9.1. *The polynomial* $\mathbb{S}_\mu(u)$ *is independent of the choice of the standard tableau* $\mathcal{U}$ *of shape* μ. *Moreover, the Harish-Chandra images of its coefficients are found by*

$$\chi:\mathbb{S}_\mu(u)\mapsto\sum_{\operatorname{sh}(\mathcal{T})=\mu}\prod_{\alpha\in\mu}\big(u+\lambda_{\mathcal{T}(\alpha)}+c(\alpha)\big),$$

summed over semistandard tableau $\mathcal{T}$ *of shape* μ *with entries in* $\{1,\dots,N\}$.

We postpone the proof till Chapter 10 as Proposition 4.9.1 will follow from a more general result on Yangian characters; see Remark 10.1.5(i). In the particular cases where μ is the column or row-diagram, the statement reduces to Propositions 4.6.1 and 4.7.1, as $\mathbb{S}_\mu(u)$ coincides with the polynomials (4.35) and (4.39), respectively.

The Harish-Chandra image of the polynomial $\mathbb{S}_\mu(u)$ is a symmetric polynomial in the variables $l_i = \lambda_i - i + 1$ with $i = 1, \dots, N$. The Casimir elements $\mathbb{S}_\mu = \mathbb{S}_\mu(0)$ given by

$$\mathbb{S}_\mu = \mathrm{tr}_{1,\dots,m}\, \mathcal{E}_\mathcal{U}\, (E_1 + c_1) \dots (E_m + c_m) \tag{4.45}$$

are called the *quantum immanants*. Their images are the factorial Schur polynomials

$$s_\mu(l_1, \dots, l_N \,|\, a) = \sum_{\mathrm{sh}(\mathcal{T})=\mu} \prod_{\alpha \in \mu} \big(l_{\mathcal{T}(\alpha)} + \mathcal{T}(\alpha) + c(\alpha) - 1\big),$$

associated with the parameter sequence $a = (a_i)$, $a_i = -i + 1$; see (4.18). The following is a quantum version of Proposition 2.1.3.

COROLLARY 4.9.2. *The quantum immanants $\mathbb{S}_\mu$ with μ running over all diagrams with at most N rows form a basis of* $\mathrm{Z}(\mathfrak{gl}_N)$.

PROOF. The factorial Schur polynomial $s_\mu(l_1, \dots, l_N \,|\, a)$ is a non-homogeneous symmetric polynomial in $l_1, \dots, l_N$ whose top degree component coincides with the Schur polynomial $s_\mu(l_1, \dots, l_N)$. Since the Schur polynomials form a basis of the algebra of symmetric polynomials so do their factorial counterparts. The claim follows due to the Harish-Chandra isomorphism (4.15). □

REMARK 4.9.3. Our definition of the quantum immanants (4.45) is slightly different from the original one due to Okounkov [**128**], where they are given by

$$\mathrm{tr}_{1,\dots,m}\, \mathcal{E}_\mathcal{U}\, (E_1 - c_1) \dots (E_m - c_m). \tag{4.46}$$

These elements possess certain stability properties which motivated their definition. To make a connection between the two families, consider the automorphism of the algebra $\mathrm{U}(\mathfrak{gl}_N)$ defined by

$$\phi : E_{ij} \to -E_{ji}, \tag{4.47}$$

so that ϕ takes E to the negative transpose matrix $-E^t$. Then

$$\phi : \mathbb{S}_\mu \mapsto \mathrm{tr}_{1,\dots,m}\, \mathcal{E}_\mathcal{U}\, (-E_1^t + c_1) \dots (-E_m^t + c_m).$$

It follows by induction from (1.14) that the element $\mathcal{E}_\mathcal{U}$ is stable under the transposition applied simultaneously to all m copies of $\mathrm{End}\, \mathbb{C}^N$ in (4.29). Since this operation does not affect the trace, we obtain

$$\phi : \mathbb{S}_\mu \mapsto (-1)^m\, \mathrm{tr}_{1,\dots,m}\, \mathcal{E}_\mathcal{U}\, (E_1 - c_1) \dots (E_m - c_m).$$

On the other hand, by twisting the action of $\mathfrak{gl}_N$ on the finite-dimensional irreducible representation $L(\lambda_1, \dots, \lambda_N)$ by the automorphism (4.47) we obtain a representation isomorphic to $L(-\lambda_N, \dots, -\lambda_1)$. Hence, the shifted variables $l_i = \lambda_i - i + 1$ get mapped by

$$l_i \mapsto -l_{N-i+1} - N + 1, \qquad i = 1, \dots, N.$$

The conclusion is that the Harish-Chandra image of the Casimir element (4.46) coincides with the polynomial $s_\mu(l_1 + N - 1, \dots, l_N + N - 1 \,|\, a)$; see also [**109**, Theorem 7.4.6]. □

4.10. Bibliographical notes

Matrix presentations of simple Lie algebras were used by Gould [**60**] in relation with characteristic identities. In a different way, they also emerge in the work of Drinfeld [**32**] in the context of quantum groups and can be obtained by restricting the R-matrix presentations of the Yangians $\mathrm{Y}(\mathfrak{g})$ to the subalgebras $\mathrm{U}(\mathfrak{g})$. A detailed exposition of properties of the factorial Schur functions can be found e.g. in the paper by Macdonald [**103**]. Propositions 4.3.1 and 4.3.2 are due to Okounkov [**128**]. More details on the use of the idempotents for the derivation of the Robinson formula (4.28) can be found in [**117**]; see also [**34**] and [**104**, Section I.3, Example 4] for other proofs. The first direct proof that the coefficients of the Capelli determinant $C(u)$ in (4.38) are Casimir elements was given by Howe and Umeda [**68**]. The images of the Gelfand invariants under the Harish-Chandra isomorphism (Corollary 4.8.2) were first found by Perelomov and Popov [**133**]. This corollary implies the Newton identity of Theorem 4.8.1. Direct proofs of this theorem were given by Umeda [**148**] and Itoh [**76**]; see also [**109**, Chapter 7] for its derivation with the use of the Yangians and more references. The matrices of the form $\alpha + E\beta$ as in Lemma 4.5.3 are among principle examples of Manin matrices; see [**21**]. The quantum immanants and related higher Capelli identities are due to Okounkov [**128**]; see also Nazarov [**124**] and Okounkov and Olshanski [**129**].

CHAPTER 5

Casimir elements for $\mathfrak{o}_N$ and $\mathfrak{sp}_N$

Extending the arguments of Chapter 4, we will now use the Brauer–Schur–Weyl duality between the Brauer algebra and the classical groups in types B, C and D to construct Casimir elements for the orthogonal and symplectic Lie algebras. As with type A, our approach will be based on matrix presentations of these Lie algebras provided by Proposition 4.1.1.

5.1. Harish-Chandra isomorphism

Recall from Section 2.2 that the Lie subalgebra of $\mathfrak{gl}_N$ spanned by the elements

$$F_{ij} = E_{ij} - E_{j'i'}, \qquad i,j = 1,\dots,N, \tag{5.1}$$

is the orthogonal Lie algebra $\mathfrak{o}_N$. As before, we use the notation $i' = N - i + 1$. If $N = 2n+1$ is odd with $n \geqslant 1$, then this is a simple Lie algebra of type B_n, and if $N = 2n$ is even with $n \geqslant 3$, then this is a simple Lie algebra of type D_n. The Lie algebra $\mathfrak{o}_4$ is semi-simple, it is isomorphic to the direct sum $\mathfrak{sl}_2 \oplus \mathfrak{sl}_2$. In both odd and even cases, (4.8) implies the commutation relations

$$F_{ij}\, F_{kl} - F_{kl}\, F_{ij} = \delta_{kj}\, F_{il} - \delta_{il}\, F_{kl} - \delta_{ki'}\, F_{j'l} + \delta_{j'l}\, F_{ki'}, \tag{5.2}$$

for all $i,j,k,l \in \{1,\dots,N\}$. Note also the symmetry relation

$$F_{ij} + F_{j'i'} = 0, \qquad i,j \in \{1,\dots,N\}.$$

Similarly, if $N = 2n$ is even, then the Lie subalgebra of $\mathfrak{gl}_{2n}$ spanned by the elements

$$F_{ij} = E_{ij} - \varepsilon_i \varepsilon_j\, E_{j'i'}, \qquad i,j = 1,\dots,2n, \tag{5.3}$$

is the symplectic Lie algebra $\mathfrak{sp}_{2n}$ which is a simple Lie algebra of type C_n. We set $\varepsilon_i = 1$ for $i = 1,\dots,n$ and $\varepsilon_i = -1$ for $i = n+1,\dots,2n$. We have the commutation relations

$$F_{ij}\, F_{kl} - F_{kl}\, F_{ij} = \delta_{kj}\, F_{il} - \delta_{il}\, F_{kl} - \varepsilon_i \varepsilon_j \big(\delta_{ki'}\, F_{j'l} - \delta_{j'l}\, F_{ki'}\big) \tag{5.4}$$

for all $i,j,k,l \in \{1,\dots,2n\}$. Note also the symmetry relation

$$F_{ij} + \varepsilon_i \varepsilon_j\, F_{j'i'} = 0, \qquad i,j \in \{1,\dots,2n\}.$$

We will use the uniform notation $\mathfrak{g}_N$ for the Lie algebra $\mathfrak{o}_N$ (with $N = 2n$ or $N = 2n+1$) or $\mathfrak{sp}_N$ (with $N = 2n$). Combine the elements F_{ij} into the matrix $F = [F_{ij}]$ and identify it with the element

$$F = \sum_{i,j=1}^{N} e_{ij} \otimes F_{ij} \in \operatorname{End} \mathbb{C}^N \otimes \mathrm{U}(\mathfrak{g}_N). \tag{5.5}$$

Consider the tensor product algebra

$$\operatorname{End} \mathbb{C}^N \otimes \operatorname{End} \mathbb{C}^N \otimes \mathrm{U}(\mathfrak{g}_N)$$

and use the operators $P = P_{12}$ and $Q = Q_{12}$ defined in (1.64), (1.70) and (1.74) for the respective cases. Observe that the application of the transposition (2.27) to the operator P with respect to either copy of the endomorphism algebra yields the operator Q.

PROPOSITION 5.1.1. *The defining relations of the algebra* $\mathrm{U}(\mathfrak{g}_N)$ *can be written in the form*

$$F_1\, F_2 - F_2\, F_1 = (P - Q)\, F_2 - F_2\, (P - Q) \tag{5.6}$$

together with the relation $F + F' = 0$.

PROOF. Clearly, $F + F' = 0$ is equivalent to the symmetry relations. One way to proceed with (5.6) is to derive it from the general Proposition 4.1.1. We will use Lemma 2.2.1 instead. Write (4.10) in the form

$$E_1\, E_2 - E_2\, E_1 = P\, E_2 - E_2\, P.$$

The application of the transposition (2.27) to the first copy of $\operatorname{End}\mathbb{C}^N$ in (4.9) gives

$$E_1'\, E_2 - E_2\, E_1' = Q\, E_2 - E_2\, Q,$$

while its application to the second copy with the use of Lemma 2.2.1 gives

$$E_1\, E_2' - E_2'\, E_1 = E_2'\, Q - Q\, E_2'.$$

Now applying the transposition to the first copy of $\operatorname{End}\mathbb{C}^N$ in the latter relation we come to

$$E_1'\, E_2' - E_2'\, E_1' = E_2'\, P - P\, E_2'.$$

These four relations imply

$$\begin{aligned} F_1\, F_2 - F_2\, F_1 &= (E_1 - E_1')(E_2 - E_2') - (E_2 - E_2')(E_1 - E_1') \\ &= P\, E_2 - E_2\, P - Q\, E_2 + E_2\, Q - E_2'\, Q + Q\, E_2' + E_2'\, P - P\, E_2' \\ &= (P - Q)\, F_2 - F_2\, (P - Q), \end{aligned}$$

as claimed. □

By (2.28), for the matrix F we have

$$Q\, F_1 + Q\, F_2 = 0 \qquad \text{and} \qquad F_1\, Q + F_2\, Q = 0. \tag{5.7}$$

Together with (1.67) this implies an equivalent form of relations (5.6),

$$F_1\, F_2 - F_2\, F_1 = F_1\, (P - Q) - (P - Q)\, F_1.$$

REMARK 5.1.2. The matrix form of the defining relations for $\mathrm{U}(\mathfrak{g}_N)$ as provided by Proposition 5.1.1 leads to the definition of a Brauer-type analogue of the degenerate affine Hecke algebra; cf. Remark 4.2.3. Consider the algebra $\mathcal{F}_m(\omega)$ generated by the Brauer algebra $\mathcal{B}_m(\omega)$ and the (abstract) elements $F_1, \dots, F_m$ subject to the relations

$$P_{a\,b}\, F_c = F_c\, P_{a\,b}, \qquad Q_{a\,b}\, F_c = F_c\, Q_{a\,b},$$

where $1 \leqslant a < b \leqslant m$ and $c \neq a, b$,

$$P_{a\,b}\, F_a = F_b\, P_{a\,b}, \qquad Q_{a\,b}\, F_a + Q_{a\,b}\, F_b = 0 \qquad \text{and} \qquad F_a\, Q_{a\,b} + F_b\, Q_{a\,b} = 0,$$

together with

$$F_a\, F_b - F_b\, F_a = F_a\, (P_{a\,b} - Q_{a\,b}) - (P_{a\,b} - Q_{a\,b})\, F_a,$$

where $P_{a\,b}$ and $Q_{a\,b}$ are understood as the respective elements $s_{a\,b}$ and $\epsilon_{a\,b}$ of $\mathcal{B}_m(\omega)$. The quotient of $\mathcal{F}_m(\omega)$ by some additional relations is known as the *Nazarov–Wenzl algebra* and was originally introduced in [**123**]. □

Given any n-tuple of complex numbers $\lambda = (\lambda_1, \dots, \lambda_n)$, the corresponding *irreducible highest weight representation* $L(\lambda)$ of the Lie algebra $\mathfrak{g}_N$ is generated by a nonzero vector $\xi \in L(\lambda)$ (the *highest vector*) such that

$$\begin{aligned} F_{ij}\,\xi &= 0 \qquad \text{for} \quad 1 \leqslant i < j \leqslant N, \qquad \text{and} \\ F_{ii}\,\xi &= \lambda_i\,\xi \qquad \text{for} \quad 1 \leqslant i \leqslant n. \end{aligned}$$

As before, we denote by $\mathrm{Z}(\mathfrak{g}_N)$ the center of the universal enveloping algebra $\mathrm{U}(\mathfrak{g}_N)$. Any element $z \in \mathrm{Z}(\mathfrak{g}_N)$ acts in $L(\lambda)$ by multiplying each vector by a scalar $\chi(z)$. When regarded as a function of the highest weight, $\chi(z)$ is a polynomial in $\lambda_1, \dots, \lambda_n$ with certain symmetry properties arising from the action of the corresponding Weyl group W. Introduce shifted variables $l_1, \dots, l_n$ by $l_i = \lambda_i + \rho_i$, where the ρ_i are the coordinates of the half-sum of the positive roots of $\mathfrak{g}_N$ so that $\rho_i = n - i + \varepsilon$ with

$$\varepsilon = \begin{cases} 0 & \text{for} \quad \mathfrak{g}_N = \mathfrak{o}_{2n}, \\ \frac{1}{2} & \text{for} \quad \mathfrak{g}_N = \mathfrak{o}_{2n+1}, \\ 1 & \text{for} \quad \mathfrak{g}_N = \mathfrak{sp}_{2n}. \end{cases} \tag{5.8}$$

In types B_n and C_n the polynomial $\chi(z)$ in $l_1, \dots, l_n$ is invariant under any permutations of the variables and under the changes of signs $l_i \mapsto -l_i$ of any subset of the variables. In type D_n the polynomial $\chi(z)$ in $l_1, \dots, l_n$ is invariant under any permutations of the variables and under the changes of signs $l_i \mapsto -l_i$ of any subset containing an even number of the variables. In all three cases, the mapping $z \mapsto \chi(z)$ defines an algebra isomorphism

$$\chi : \mathrm{Z}(\mathfrak{g}_N) \to \mathbb{C}[l_1, \dots, l_n]^W \tag{5.9}$$

to the algebra of invariant polynomials. This provides an equivalent interpretation of the Harish-Chandra isomorphism defined in (4.7), where l_i is identified with the element $F_{ii} + \rho_i \in \mathrm{U}(\mathfrak{h})$. Note also that in types B_n and C_n the image of the isomorphism coincides with the algebra of symmetric polynomials in the variables $l_1^2, \dots, l_n^2$,

$$\mathbb{C}[l_1, \dots, l_n]^W = \mathbb{C}[l_1^2, \dots, l_n^2]^{\mathfrak{S}_n}.$$

In type D_n, the image of the isomorphism is generated by the algebra of symmetric polynomial in the variables $l_1^2, \dots, l_n^2$ and the polynomial $l_1 \dots l_n$; cf. (2.35).

Recall from Section 2.2 that the classical group $\mathrm{G}_N = \mathrm{O}_N$ or $\mathrm{G}_N = \mathrm{Sp}_N$ is defined as the group of complex matrices preserving the respective form (2.25). The *adjoint action* of the group G_N on $\mathfrak{g}_N$ is defined by (2.30). This extends to a unique action of G_N on $\mathrm{U}(\mathfrak{g}_N)$ so that each element of the group acts as an automorphism. The subalgebra of invariants $\mathrm{U}(\mathfrak{g}_N)^{\mathrm{G}_N}$ under this action coincides with the center $\mathrm{Z}(\mathfrak{g}_N)$ in types B_n and C_n, and is properly contained in the center in type D_n. In all three cases, the restriction of (5.9) induces the isomorphism

$$\chi : \mathrm{U}(\mathfrak{g}_N)^{\mathrm{G}_N} \to \mathbb{C}[l_1^2, \dots, l_n^2]^{\mathfrak{S}_n}. \tag{5.10}$$

We will need the factorial elementary and complete symmetric polynomials as given in (4.16) and (4.17) for the particular sequences a defined by

$$a = (\varepsilon^2, (\varepsilon + 1)^2, (\varepsilon + 2)^2, \dots), \tag{5.11}$$

where ε is introduced in (5.8), so that $a_i = (\varepsilon + i - 1)^2$.

Note that any element $z \in \mathrm{U}(\mathfrak{g}_N)^{\mathrm{G}_N}$ is uniquely determined by the eigenvalues $\chi(z)$ in the irreducible modules $L(\lambda)$, where $\lambda = (\lambda_1, \dots, \lambda_n)$ runs over the set of partitions with $\ell(\lambda) \leqslant n$. Hence, we come to the following characterization properties of Casimir elements.

PROPOSITION 5.1.3. *Suppose that k is a nonnegative integer and $z \in \mathrm{U}(\mathfrak{g}_N)^{\mathrm{G}_N}$ is an element of degree $\leqslant 2k$ with respect to the canonical filtration, which vanishes in each representation $L(\lambda)$ with $|\lambda| < k$. If the homogeneous component of $\chi(z)$ of degree $2k$ coincides with $e_k(\lambda_1^2, \dots, \lambda_n^2)$ or $h_k(\lambda_1^2, \dots, \lambda_n^2)$ then $\chi(z)$ equals $e_k(l_1^2, \dots, l_n^2 \,|\, a)$ or $h_k(l_1^2, \dots, l_n^2 \,|\, a)$, respectively.*

PROOF. We will use Proposition 4.3.2 for two special choices of the partition μ by taking $\mu = (1^k)$ and $\mu = (k)$. By (5.10), the image $\chi(z)$ is a symmetric polynomial in the variables $x = (l_1^2, \dots, l_n^2)$. Due to our choice of the sequence a in (5.11), for any given partition λ for the n-tuple (4.19) we have $a_\lambda = (l_1^2, \dots, l_n^2)$. Therefore, the symmetric polynomial $\chi(z)$ satisfies the respective assumptions of Proposition 4.3.2 and the claim follows. □

5.2. Brauer–Schur–Weyl duality

Finite-dimensional irreducible representations of the orthogonal group O_N are parameterized by all diagrams λ with the property $\lambda'_1 + \lambda'_2 \leqslant N$, where λ'_j denotes the number of boxes in the column j of λ. The corresponding representation will be denoted by $\overline{L}(\lambda)$. Let λ^* be the diagram obtained from λ by replacing the first column with the column containing $N - \lambda'_1$ boxes. If $N = 2n+1$ and $\lambda'_1 \leqslant n$ then the associated representation of the Lie algebra $\mathfrak{o}_N$ in the space $\overline{L}(\lambda)$ is irreducible and isomorphic to the representation $L(\lambda)$ whose highest weight $(\lambda_1, \dots, \lambda_n)$ coincides with λ; if $\lambda'_1 > n$ then the associated representation of $\mathfrak{o}_N$ is isomorphic to $L(\lambda^*)$.

If $N = 2n$ and $\lambda'_1 < n$ then the associated representation of the Lie algebra $\mathfrak{o}_N$ in the space $\overline{L}(\lambda)$ is irreducible and isomorphic to $L(\lambda)$, while for $\lambda'_1 > n$ the associated representation of $\mathfrak{o}_N$ is isomorphic to $L(\lambda^*)$. If $N = 2n$ and $\lambda'_1 = n$ then the associated representation of $\mathfrak{o}_N$ in $\overline{L}(\lambda)$ is isomorphic to the direct sum of two irreducible representations $L(\lambda)$ and $L(\lambda^\circ)$ with $\lambda^\circ = (\lambda_1, \dots, \lambda_{n-1}, -\lambda_n)$.

Finite-dimensional irreducible representations of the symplectic group Sp_N with $N = 2n$ are parameterized by partitions λ whose lengths do not exceed n. The associated representation of the Lie algebra $\mathfrak{sp}_N$ is irreducible and isomorphic to $L(\lambda)$.

As G_N is a subgroup of GL_N, the action of G_N on the tensor product space $(\mathbb{C}^N)^{\otimes m}$ can be defined by the restriction of the diagonal action of GL_N defined in (2.13). The centralizer of this action in the endomorphism algebra $\mathrm{End}\,(\mathbb{C}^N)^{\otimes m}$ coincides with the homomorphic image of the Brauer algebra $\mathcal{B}_m(\omega)$ with the parameter ω specialized to N and $-N$, respectively, in the orthogonal and symplectic case, where the homomorphism is defined in Section 1.5. This implies the tensor product decomposition analogous to (4.26): for $\mathrm{G}_N = \mathrm{O}_N$ we have

$$(\mathbb{C}^N)^{\otimes m} \cong \bigoplus_{f=0}^{\lfloor m/2 \rfloor} \bigoplus_{\substack{\lambda \vdash m-2f \\ \lambda'_1 + \lambda'_2 \leqslant N}} V_\lambda \otimes \overline{L}(\lambda), \tag{5.12}$$

where V_λ and $\overline{L}(\lambda)$ are the respective irreducible representations of $\mathcal{B}_m(N)$ and O_N associated with the diagram λ.

Similarly, in the symplectic case with $N = 2n$,

$$(\mathbb{C}^N)^{\otimes m} \cong \bigoplus_{f=0}^{\lfloor m/2 \rfloor} \bigoplus_{\substack{\lambda \vdash m-2f \\ \lambda_1 \leqslant n}} V_\lambda \otimes L(\lambda'), \tag{5.13}$$

where V_λ and $L(\lambda')$ are the respective irreducible representations of $\mathcal{B}_m(-N)$ and Sp_N associated with λ and λ'.

Now suppose that λ is a partition of m with $\lambda'_1 \leqslant n$ in the orthogonal case and $\lambda_1 \leqslant n$ in the symplectic case. Let $\mathcal{U}$ be a standard tableau of shape λ which we also regard as an updown tableau. We let $e_{\mathcal{U}}$ be the associated primitive idempotent of the respective Brauer algebra $\mathcal{B}_m(N)$ or $\mathcal{B}_m(-N)$; see Section 1.2. Denote by $\mathcal{E}_{\mathcal{U}}$ the image of $e_{\mathcal{U}}$ under the respective action of the Brauer algebra as defined in Section 1.5. Denote by $\mathcal{V}$ the standard tableau obtained from $\mathcal{U}$ by deleting the box occupied by m and let μ be the shape of $\mathcal{V}$. By the decompositions (5.12) and (5.13), the subspace $\mathcal{E}_{\mathcal{U}}(\mathbb{C}^N)^{\otimes m}$ is an irreducible representation of G_N isomorphic to $\overline{L}(\lambda)$ in the orthogonal case and to $L(\lambda')$ in the symplectic case. Therefore, the trace $\mathrm{tr}_{1,\dots,m}\,\mathcal{E}_{\mathcal{U}}$ equals the dimension of the respective representation $\overline{L}(\lambda)$ or $L(\lambda')$. Thus, by calculating the trace we can recover the well-known hook dimension formulas for these representations; cf. the proof of (4.28). We omit the calculations and only state the formulas. For $\mathrm{G}_N = \mathrm{O}_N$ we set

$$D(\lambda) = \frac{1}{h(\lambda)} \prod_{(i,j)\in\lambda} \big(N - 1 + d(i,j)\big),$$

where

$$d(i,j) = \begin{cases} \lambda_i + \lambda_j - i - j + 1 & \text{if} \quad i \leqslant j, \\ -\lambda'_i - \lambda'_j + i + j - 1 & \text{if} \quad i > j. \end{cases}$$

We have the relation

$$\mathrm{tr}_m\,\mathcal{E}_{\mathcal{U}} = \mathcal{E}_{\mathcal{V}}\,\frac{D(\lambda)}{D(\mu)}$$

which implies

$$\dim \overline{L}(\lambda) = D(\lambda). \tag{5.14}$$

Similarly, if $\mathrm{G}_N = \mathrm{Sp}_N$, for any diagram ρ with at most n rows set

$$D(\rho) = \frac{1}{h(\rho)} \prod_{(i,j)\in\rho} \big(N + 1 + d(i,j)\big),$$

where the parameters $d(i,j)$ are now defined by

$$d(i,j) = \begin{cases} \rho_i + \rho_j - i - j + 1 & \text{if} \quad i > j, \\ -\rho'_i - \rho'_j + i + j - 1 & \text{if} \quad i \leqslant j. \end{cases}$$

Now we have

$$\mathrm{tr}_m\,\mathcal{E}_{\mathcal{U}} = \mathcal{E}_{\mathcal{V}}\,\frac{D(\lambda')}{D(\mu')}$$

so that

$$\dim L(\rho) = D(\rho). \tag{5.15}$$

5.3. A general construction of central elements

Suppose that $s \in \mathcal{B}_m(\omega)$ is an arbitrary element, where the parameter ω of the Brauer algebra is specialized by

$$\omega = \begin{cases} N & \text{for} \quad \mathfrak{g}_N = \mathfrak{o}_N, \\ -N & \text{for} \quad \mathfrak{g}_N = \mathfrak{sp}_N. \end{cases} \tag{5.16}$$

We let S denote the respective image of s under the map (1.69) or (1.73), which is also regarded as the element $S \otimes 1$ of the algebra

$$\underbrace{\operatorname{End}\mathbb{C}^N \otimes \ldots \otimes \operatorname{End}\mathbb{C}^N}_{m} \otimes \mathrm{U}(\mathfrak{g}_N). \tag{5.17}$$

The following is an analogue of Theorem 4.5.1 for the orthogonal and symplectic Lie algebras and it is proved by a similar argument.

THEOREM 5.3.1. *For any $s \in \mathcal{B}_m(\omega)$ and $u_1, \ldots, u_m \in \mathbb{C}$ the element*

$$\operatorname{tr}_{1,\ldots,m} S\,(u_1 + F_1) \ldots (u_m + F_m) \tag{5.18}$$

belongs to the center $\mathrm{Z}(\mathfrak{g}_N)$.

PROOF. Take the tensor product

$$\operatorname{End}\mathbb{C}^N \otimes \operatorname{End}(\mathbb{C}^N)^{\otimes m} \otimes \mathrm{U}(\mathfrak{g}_N) \tag{5.19}$$

with the copies of the endomorphism algebra $\operatorname{End}\mathbb{C}^N$ labeled by $0, 1, \ldots, m$. It is sufficient to show that

$$\big[F_0, \operatorname{tr}_{1,\ldots,m} S\,(u_1 + F_1) \ldots (u_m + F_m)\big] = 0 \tag{5.20}$$

in the algebra (5.19). By Proposition 5.1.1 we can write

$$[F_0, u_a + F_a] = (P_{0\,a} - Q_{0\,a})(u_a + F_a) - (u_a + F_a)(P_{0\,a} - Q_{0\,a})$$

so that

$$\begin{aligned}\big[F_0,\, S\,(u_1 + F_1) \ldots (u_m + F_m)\big] = S \sum_{a=1}^{m} (P_{0\,a} - Q_{0\,a})\,(u_1 + F_1) \ldots (u_m + F_m) \\ - S\,(u_1 + F_1) \ldots (u_m + F_m) \sum_{a=1}^{m} (P_{0\,a} - Q_{0\,a}).\end{aligned}$$

By the properties of the Jucys–Murphy elements (1.26), the sum $\sum_{a=1}^{m}(P_{0\,a} - Q_{0\,a})$ commutes with S, so that the trace $\operatorname{tr}_{1,\ldots,m}$ of the above commutator vanishes due to the cyclic property of trace. □

REMARK 5.3.2. Although we will not use it below, we point out a more general family of Casimir elements for $\mathfrak{g}_N$. Using the assumptions of Theorem 5.3.1, consider the element

$$\operatorname{tr}_{1,\ldots,m} S\,(u_1 + F_1) R_1 (u_2 + F_2) \ldots R_{m-1}(u_m + F_m), \tag{5.21}$$

where

$$R_a = r_a + r_{a\,a+1} Q_{a\,a+1} + \cdots + r_{a\,m} Q_{a\,m}, \qquad a = 1, \ldots, m-1,$$

for some complex numbers r_a and $r_{a\,b}$. Applying the same approach as in the proof of Theorem 5.3.1, we can see that the element (5.21) belongs to the center $\mathrm{Z}(\mathfrak{g}_N)$. One extra step in the argument is the use of the relations

$$Q_{c\,a}(P_{0\,a}-Q_{0\,a}) = -Q_{c\,a}(P_{0\,c}-Q_{0\,c}) \qquad \text{and}$$
$$(P_{0\,a}-Q_{0\,a})Q_{a\,b} = -(P_{0\,b}-Q_{0\,b})Q_{a\,b}$$

which hold for $c < a < b$.

Slightly modifying the proof of the theorem in a different direction, we can show that all elements of the form

$$\mathrm{tr}_{1,\dots,m}\, S\, F_{a_1}\dots F_{a_k} \tag{5.22}$$

with arbitrary parameters $a_i \in \{1,\dots,m\}$ also belong to $\mathrm{Z}(\mathfrak{g}_N)$. □

EXAMPLE 5.3.3. As in Section 4.8, take $S = P_\sigma$ in Theorem 5.3.1, where $\sigma = (m, m-1,\dots,1)$ is a long cycle (see (4.40)) and set $u_a = 0$ for all $a = 1,\dots,m$. Repeating the argument of Section 4.8 we get

$$\mathrm{tr}_{1,\dots,m}\, P_\sigma\, F_1\dots F_m = \mathrm{tr}\, F^m$$

which is therefore a Casimir element for $\mathfrak{g}_N$ for any $m \geqslant 1$. It is known as the *Gelfand invariant* for the Lie algebra $\mathfrak{g}_N$. In fact, it belongs to the algebra of invariants $\mathrm{U}(\mathfrak{g}_N)^{\mathrm{G}_N}$. The Harish-Chandra image $\chi(\mathrm{tr}\, F^m)$ is a symmetric polynomial in $l_1^2,\dots,l_n^2$; see (5.10). If $m = 2k$ is even, then the top degree component of the polynomial $\chi(\mathrm{tr}\, F^{2k})$ coincides with $2\,(l_1^{2k}+\dots+l_n^{2k})$. Therefore, the Gelfand invariants

$$\mathrm{tr}\, F^2,\ \mathrm{tr}\, F^4,\dots,\mathrm{tr}\, F^{2n}$$

are algebraically independent generators of the algebra $\mathrm{U}(\mathfrak{g}_N)^{\mathrm{G}_N}$.

There are analogues of the Newton identities (cf. Theorem 4.8.1) relating the Gelfand invariants with Capelli-type determinants. However, the known proofs of these identities rely on some other methods as compared to Theorem 4.8.1; see Section 5.7 for references. □

The following is an analogue of Lemma 4.5.4 for the orthogonal and symplectic Lie algebras.

LEMMA 5.3.4. *Let $u_1,\dots,u_m$ be complex parameters and let S be the image of an element s of the center of the Brauer algebra $\mathcal{B}_m(\omega)$ (with ω specialized as in* (5.16)*), under the respective homomorphism* (1.69) *or* (1.73)*. For any permutations $\sigma,\tau \in \mathfrak{S}_m$ we have the identity*

$$\mathrm{tr}_{1,\dots,m}\, S\,(F_{\sigma(1)}+u_{\tau(1)})\dots(F_{\sigma(m)}+u_{\tau(m)}) = \mathrm{tr}_{1,\dots,m}\, S\,(F_1+u_1)\dots(F_m+u_m).$$

PROOF. We argue as in the proof of Lemma 4.5.4, except that the commutation relations are now implied by Proposition 5.1.1 to give

$$\begin{aligned}(F_a+u_{a+1})(F_{a+1}+u_a) &- (F_{a+1}+u_a)(F_a+u_{a+1})\\ &= (P_{a\,a+1}-Q_{a\,a+1})F_{a+1} - F_{a+1}(P_{a\,a+1}-Q_{a\,a+1}).\end{aligned}$$

This leads to the desired relation. □

5.4. Symmetrizer and anti-symmetrizer for $\mathfrak{o}_N$

Here we formulate analogues of Proposition 4.6.1 and 4.7.1 for the Lie algebras $\mathfrak{g}_N = \mathfrak{o}_N$ with $N = 2n+1$ or $N = 2n$ (the symplectic case will be considered in the next section). Unlike $\mathfrak{gl}_N$, the proofs of these analogues are more involved due to the lack of Manin matrix techniques. Although there are certain analogues of Manin matrices in types B, C and D (see Section 5.6 below), they do not seem to suit well the calculations of Harish-Chandra images. Our proofs will instead rely on the properties of factorial symmetric polynomials; see Proposition 5.1.3.

Recall that the symmetrizer $s^{(m)}$ and anti-symmetrizer $a^{(m)}$ in the Brauer algebra $\mathcal{B}_m(\omega)$ are defined Section 1.2, and $S^{(m)}$ and $A^{(m)}$ are the respective images of these elements under the action of the corresponding Brauer algebra in the tensor spaces, defined by (1.69) with $\omega = N$. We will use the function $\gamma_m(\omega)$ of the parameter ω defined by (2.40).

The factorial elementary and complete symmetric polynomials defined in (4.16) and (4.17) will be associated with the sequence of parameters a given in (5.11). Take even values $m = 2k$ and set $S = S^{(2k)}$ with $k \geqslant 1$ in Theorem 5.3.1. By the cyclic property of trace we can place $S^{(2k)}$ as the last factor in (5.18). We choose a particular specialization of the parameters $u_1, \dots, u_{2k}$ as follows.

PROPOSITION 5.4.1. *For any $k \geqslant 1$ the image of the Casimir element*

$$\gamma_{2k}(N)\,\mathrm{tr}_{1,\dots,2k}\,(F_1 + k - 1)\dots(F_{2k} - k)\,S^{(2k)} \in \mathrm{Z}(\mathfrak{o}_N) \tag{5.23}$$

under the Harish-Chandra isomorphism coincides with $h_k(l_1^2, \dots, l_n^2\,|\,a)$.

PROOF. It will be convenient for the arguments below to interpret the tensor notation in a certain dual way. Namely, we will write the tensor factors in (5.17) in a different order, by placing $\mathrm{U}(\mathfrak{g}_N)$ first:

$$\mathrm{U}(\mathfrak{g}_N) \otimes \underbrace{\mathrm{End}\,\mathbb{C}^N \otimes \dots \otimes \mathrm{End}\,\mathbb{C}^N}_{m}. \tag{5.24}$$

Accordingly, we swap the factors in (5.5) so that now

$$F_a = \sum_{i,j=1}^{N} F_{ij} \otimes 1^{\otimes(a-1)} \otimes e_{ij} \otimes 1^{\otimes(m-a)}. \tag{5.25}$$

This does not affect the notation for the action of the Brauer algebra. In particular, Theorem 5.3.1 holds in the same form.

Denote the Casimir element (5.23) by D_k. We will use Proposition 5.1.3 and start by showing that D_k vanishes in all representations $L(\lambda)$ of $\mathfrak{o}_N$, where the partitions λ satisfy $|\lambda| < k$. We employ a realization of $L(\lambda)$ in tensor spaces as follows. Let $r = |\lambda|$. Consider the action of the Lie algebra $\mathfrak{o}_N$ on $\mathbb{C}^N$ defined by

$$F_{ij} \mapsto -e_{ji} + e_{i'j'}, \qquad 1 \leqslant i, j \leqslant N,$$

so that this representation is isomorphic to $L(1, 0, \dots, 0)$. The space of tensors $(\mathbb{C}^N)^{\otimes r}$ then also becomes a representation of $\mathfrak{o}_N$. With the notation (5.25), under the corresponding homomorphism

$$\varphi : \mathrm{U}(\mathfrak{o}_N) \otimes \mathrm{End}\,(\mathbb{C}^N)^{\otimes 2k} \to \mathrm{End}\,(\mathbb{C}^N)^{\otimes r} \otimes \mathrm{End}\,(\mathbb{C}^N)^{\otimes 2k}$$

we have

$$\varphi(F_a) = \sum_{b=1}^{r} \left(-P_{b\,r+a} + Q_{b\,r+a}\right), \qquad a = 1, \dots, 2k. \tag{5.26}$$

By (5.12), the decomposition of $(\mathbb{C}^N)^{\otimes r}$ into a direct sum of irreducible representations of $\mathfrak{o}_N$ contains $L(\lambda)$ with a nonzero multiplicity. Hence, the desired vanishing condition of Proposition 5.1.3 will follow if we show that

$$\big(\varphi(F_1) + k - 1\big) \dots \big(\varphi(F_{2k}) - k\big)\, S'^{(2k)} = 0, \tag{5.27}$$

where $S'^{(2k)}$ denotes the image of the symmetrizer in the Brauer algebra $\mathcal{B}_{2k}(N)$ acting on the last $2k$ copies of the tensor product space $(\mathbb{C}^N)^{\otimes(r+2k)}$. Due to (5.26) and relations (1.31), the desired identity (5.27) can be written in the form

$$\big(-Y_{r+1} + k - 1\big) \dots \big(-Y_{r+2k} + k - 1\big)\, S'^{(2k)} = 0, \tag{5.28}$$

where Y_{r+b} with $b = 1, \dots, 2k$ are the images of the respective Jucys–Murphy elements $y_{r+b} - (\omega - 1)/2$ given in (1.26) under the action of the Brauer algebra $\mathcal{B}_{r+2k}(N)$ in $(\mathbb{C}^N)^{\otimes(r+2k)}$. To prove the identity (5.28) note that the two operators $(-Y_{r+1} + k - 1) \dots (-Y_{r+2k} + k - 1)$ and $S'^{(2k)}$ on the vector space $(\mathbb{C}^N)^{\otimes(r+2k)}$ commute with the action of O_N. We will show that their images have zero intersection. To describe the image of the first operator, represent the vector space as the direct sum of irreducible representations of O_N,

$$(\mathbb{C}^N)^{\otimes(r+2k)} = \sum_{l=0}^{\lfloor \frac{r}{2} \rfloor + k} \sum_{\nu \vdash r+2k-2l} \sum_{\mathcal{U}} \mathcal{E}_{\mathcal{U}} (\mathbb{C}^N)^{\otimes(r+2k)},$$

where the last sum is taken over all updown tableaux $\mathcal{U} = (\Lambda_1, \dots, \Lambda_{r+2k})$ of shape ν associated with the Brauer algebra $\mathcal{B}_{r+2k}(N)$. We claim that if $\mathcal{U}$ is an updown tableau of shape $\nu = \Lambda_{r+2k}$ with $\nu_1 \geqslant k$ then

$$\big(-Y_{r+1} + k - 1\big) \dots \big(-Y_{r+2k} + k - 1\big)\, \mathcal{E}_{\mathcal{U}} = 0. \tag{5.29}$$

Indeed, since $r < k$, there exists a pair of diagrams $(\Lambda_{r+a}, \Lambda_{r+a+1})$ such that $a \in \{0, 1, \dots, 2k-1\}$ with the property that the second diagram is obtained from the first by adding the box $\alpha = (1, k)$. The content of this box is $c(\alpha) = (\omega - 1)/2 + k - 1$ so that (5.29) follows from relations (1.27). Thus, as a representation of O_N, the image

$$\big(-Y_{r+1} + k - 1\big) \dots \big(-Y_{r+2k} + k - 1\big)\, (\mathbb{C}^N)^{\otimes(r+2k)}$$

is contained in a direct sum of representations $\overline{L}(\nu)$ with $\nu_1 < k$.

On the other hand, the operator $S'^{(2k)}$ projects the vector space $(\mathbb{C}^N)^{\otimes(r+2k)}$ onto the tensor product $(\mathbb{C}^N)^{\otimes r} \otimes \overline{L}(2k, 0, \dots, 0)$ of representations of O_N. For the irreducible decomposition of the tensor product of representations of O_N we have

$$\mathbb{C}^N \otimes \overline{L}(\mu) \cong \bigoplus_{\mu^*} \overline{L}(\mu^*), \tag{5.30}$$

where μ^* is obtained from μ by adding or removing one box. This property is equivalent to the branching rule for the Brauer algebras. Hence, applying (5.30) consecutively to the tensor factors of $(\mathbb{C}^N)^{\otimes r} \otimes \overline{L}(2k, 0, \dots, 0)$, we find that, as a representation of O_N, the image $S'^{(2k)}(\mathbb{C}^N)^{\otimes(r+2k)}$ is contained in a direct sum of representations $\overline{L}(\nu)$ with $\nu_1 \geqslant k+1$. Indeed, the maximum number of times a box can be removed from the row diagram $(2k)$ is $r < k$. This completes the proof of (5.28) and hence of (5.27).

The leading term of the symmetric polynomial $\chi(D_k)$ was calculated in Proposition 2.2.4. It coincides with $h_k(\lambda_1^2, \dots, \lambda_n^2)$ by (2.41), so that the proof is completed by the application of Proposition 5.1.3. $\square$

For a different proof of Proposition 5.4.1 see Section 13.4.

Now we use Theorem 5.3.1 again and consider more general Casimir elements. Let u be a variable and set

$$D_m(u) = \gamma_m(N)\, \mathrm{tr}_{1,\dots,m} \Big(F_1 + u + \frac{m-1}{2}\Big)\Big(F_2 + u + \frac{m-3}{2}\Big) \\ \times \dots \times \Big(F_m + u - \frac{m-1}{2}\Big)\, S^{(m)}$$

so that $D_m(u)$ is a polynomial in u whose coefficients are Casimir elements for $\mathfrak{o}_N$. To give an alternative formula for such coefficients, we will use the extended algebra $\mathrm{U}(\mathfrak{o}_N) \otimes \mathcal{A}$, where $\mathcal{A}$ is the algebra of differential operators in a variable z whose coefficients are Laurent polynomials in z.

COROLLARY 5.4.2. *For the images under the Harish-Chandra isomorphism we have*

$$\chi : D_m(u) \mapsto \sum_{r=0}^{\lfloor \frac{m}{2} \rfloor} \binom{N+m-2}{m-2r} h_r(l_1^2, \dots, l_n^2\,|\,a) \prod_{i=0}^{m-2r-1} \Big(u - \frac{m-1}{2} + r + i\Big)$$

and

$$\chi : \gamma_m(N)\, \mathrm{tr}_{1,\dots,m} \big(-\partial_z + F_1\, z^{-1}\big) \dots \big(-\partial_z + F_m\, z^{-1}\big)\, S^{(m)} \\ \mapsto \sum_{r=0}^{\lfloor \frac{m}{2} \rfloor} \binom{N+m-2}{m-2r} h_r(l_1^2, \dots, l_n^2\,|\,a)\, z^{-2r} \big(-\partial_z + r\, z^{-1}\big)^{m-2r}.$$

PROOF. Observe that by the definition of trace, $D_m(u)$ is stable under the transposition (2.24) applied simultaneously to each of the m copies of the algebra $\mathrm{End}\,\mathbb{C}^N$. This simultaneous transposition is an anti-automorphism of the algebra $\mathrm{End}\,(\mathbb{C}^N)^{\otimes m}$ which preserves all elements P_{ab} and Q_{ab}. Hence the symmetrizer $S^{(m)}$ is also stable under this transposition, as can be easily seen from the formulas of Section 1.2, for instance, (1.46). On the other hand, since $F' = -F$, using Lemma 5.3.4, for the image of $D_m(u)$ under this anti-automorphism we find

$$\begin{aligned} D_m(u) &= \gamma_m(N)\, \mathrm{tr}_{1,\dots,m} \Big(-F_1 + u + \frac{m-1}{2}\Big) \dots \Big(-F_m + u - \frac{m-1}{2}\Big)\, S^{(m)} \\ &= (-1)^m\, \gamma_m(N)\, \mathrm{tr}_{1,\dots,m} \Big(F_1 - u - \frac{m-1}{2}\Big) \dots \Big(F_m - u + \frac{m-1}{2}\Big)\, S^{(m)} \\ &= (-1)^m\, \gamma_m(N)\, \mathrm{tr}_{1,\dots,m} \Big(F_1 - u + \frac{m-1}{2}\Big) \dots \Big(F_m - u - \frac{m-1}{2}\Big)\, S^{(m)} \end{aligned}$$

which coincides with $(-1)^m\, D_m(-u)$. This shows that the polynomials $D_{2k}(u)$ are even, while the polynomials $D_{2k-1}(u)$ are odd; in particular, $D_{2k-1}(0) = 0$. The value $\chi\big(D_{2k}(1/2)\big) = \chi\big(D_{2k}(-1/2)\big)$ is found in Proposition 5.4.1. These values agree with the Harish-Chandra images provided in the statement of the corollary. Hence, the proof will be completed if we show that the polynomials $D_m(u)$ and the polynomials which are claimed to be their Harish-Chandra images satisfy the same recurrence relations. We show first that

$$D_m(u+1/2) - D_m(u-1/2) = (N+m-2)\, D_{m-1}(u) \tag{5.31}$$

for all $m \geqslant 1$, where we set $D_0(u) = 1$. By Lemma 5.3.4 we can write

$$D_m(u+1/2) = \gamma_m(N)\, \mathrm{tr}_{1,\dots,m} \Big(F_1 + u + \frac{m}{2} - 1\Big) \\ \times \dots \times \Big(F_{m-1} + u - \frac{m}{2} - 1\Big)\Big(F_m + u + \frac{m}{2}\Big)\, S^{(m)},$$

so that

$$D_m(u+1/2) - D_m(u-1/2) \\ = m\, \gamma_m(N)\, \mathrm{tr}_{1,\dots,m} \Big(F_1 + u + \frac{m}{2} - 1\Big) \dots \Big(F_{m-1} + u - \frac{m}{2} - 1\Big)\, S^{(m)}.$$

Due to Lemma 1.3.2, the partial trace of the symmetrizer is found by

$$\mathrm{tr}_m\, S^{(m)} = \frac{(N+m-3)(N+2m-2)}{m\,(N+2m-4)}\, S^{(m-1)} \tag{5.32}$$

thus verifying (5.31). It is straightforward to check that the same relation is satisfied by the alleged images $\chi(D_m(u))$ as stated in the corollary.

To prove the second part, note that by the relation

$$z^m \big(-\partial_z + F_1\, z^{-1}\big) \dots \big(-\partial_z + F_m\, z^{-1}\big) = \big(-z\,\partial_z + F_1 + m - 1\big) \dots \big(-z\,\partial_z + F_m\big)$$

the polynomial $D_m(u)$ can be written in the form

$$z^m\, \gamma_m(N)\, \mathrm{tr}_{1,\dots,m} \big(-\partial_z + F_1\, z^{-1}\big) \dots \big(-\partial_z + F_m\, z^{-1}\big)\, S^{(m)}$$

with the replacement of $-z\,\partial_z$ by $u - (m-1)/2$. Similarly, for any $k \geqslant 0$,

$$z^k \big(-\partial_z + r\, z^{-1}\big)^k = \big(-z\,\partial_z + r + k - 1\big) \dots \big(-z\,\partial_z + r\big),$$

so that the second relation follows from the first. □

Now we will prove analogues of Proposition 5.4.1 and Corollary 5.4.2, where the role of the symmetrizer $S^{(m)}$ is taken by the anti-symmetrizer $A^{(m)}$. First, set $S = A^{(2k)}$ in Theorem 5.3.1.

PROPOSITION 5.4.3. *For any $1 \leqslant k \leqslant n$ the image of the Casimir element*

$$\mathrm{tr}_{1,\dots,2k}\, (F_1 - k + 1) \dots (F_{2k} + k)\, A^{(2k)} \in \mathrm{Z}(\mathfrak{o}_N) \tag{5.33}$$

under the Harish-Chandra isomorphism coincides with $(-1)^k\, e_k(l_1^2, \dots, l_n^2\,|\,a)$.

PROOF. Denote the Casimir element (5.33) by C_k. We use Proposition 5.1.3 and show that C_k vanishes in all representations $L(\lambda)$ of $\mathfrak{o}_N$, where the partitions λ satisfy $|\lambda| < k$. Using the same realization of $L(\lambda)$ as in the proof of Proposition 5.4.1, we come to showing the following analogue of (5.27):

$$\big(\varphi(F_1) - k + 1\big) \dots \big(\varphi(F_{2k}) + k\big)\, A'^{(2k)} = 0. \tag{5.34}$$

Here the homomorphism φ is defined in (5.26) and $A'^{(2k)}$ denotes the image of the anti-symmetrizer in the Brauer algebra $\mathcal{B}_{2k}(N)$ acting of the last $2k$ copies of the tensor product space $(\mathbb{C}^N)^{\otimes(r+2k)}$ with $r = |\lambda|$. By (1.48), the identity (5.34) can be written in the form

$$\big(-Y_{r+1} - k + 1\big) \dots \big(-Y_{r+2k} - k + 1\big)\, A'^{(2k)} = 0.$$

To verify this relation, we show exactly as in the proof of Proposition 5.4.1 that the image of the operator $(-Y_{r+1} - k + 1) \dots (-Y_{r+2k} - k + 1)$ on the vector space $(\mathbb{C}^N)^{\otimes(r+2k)}$ is contained in a direct sum of representations $\overline{L}(\nu)$ of O_N with $\ell(\nu) < k$, while the image of the operator $A'^{(2k)}$ is contained in a direct sum of representations $\overline{L}(\nu)$ with $\ell(\nu) \geqslant k+1$.

Finally, by Proposition 2.2.3, the leading term of the symmetric polynomial $\chi(C_k)$ coincides with $(-1)^k e_k(\lambda_1^2, \dots, \lambda_n^2)$. □

For $m = 1, \dots, N$ consider the polynomials in u whose coefficients are Casimir elements for $\mathfrak{o}_N$ given by

$$C_m(u) = \mathrm{tr}_{1,\dots,m} \Big(F_1 + u + \frac{m-1}{2}\Big)\Big(F_2 + u + \frac{m-3}{2}\Big) \dots \Big(F_m + u - \frac{m-1}{2}\Big) A^{(m)}.$$

COROLLARY 5.4.4. *For the images under the Harish-Chandra isomorphism we have*

$$\chi : C_m(u) \mapsto \sum_{r=0}^{\lfloor \frac{m}{2} \rfloor} (-1)^r \binom{N-2r}{m-2r} e_r(l_1^2, \dots, l_n^2 \,|\, a) \prod_{i=0}^{m-2r-1} \Big(u - \frac{m-1}{2} + r + i\Big)$$

and

$$\chi : \mathrm{tr}_{1,\dots,m} \big(-\partial_z + F_1 z^{-1}\big) \dots \big(-\partial_z + F_m z^{-1}\big) A^{(m)}$$

$$\mapsto \sum_{r=0}^{\lfloor \frac{m}{2} \rfloor} (-1)^r \binom{N-2r}{m-2r} e_r(l_1^2, \dots, l_n^2 \,|\, a)\, z^{-2r} \big(-\partial_z + r z^{-1}\big)^{m-2r}.$$

PROOF. We argue as in the proof of Corollary 5.4.2. Using Lemma 5.3.4 we show first that $C_m(u) = (-1)^m C_m(-u)$ so that $C_{2k-1}(0) = 0$. Moreover, $\chi\big(C_{2k}(1/2)\big) = \chi\big(C_{2k}(-1/2)\big)$ is found from Proposition 5.4.3. As a next step, we verify that

$$C_m(u + 1/2) - C_m(u - 1/2) = (N - m + 1)\, C_{m-1}(u), \qquad m \geqslant 1, \tag{5.35}$$

where $C_0(u) = 1$. This follows easily from Lemma 5.3.4, and the formula (3.26) for the partial trace of the anti-symmetrizer, thus verifying (5.35). The same relation is satisfied by the polynomials which are claimed to be the images $\chi\big(C_m(u)\big)$ as stated in the corollary. The second part follows from the first; cf. the proof of Corollary 5.4.2. □

In the case $\mathfrak{g}_N = \mathfrak{o}_{2n}$ introduce the (noncommutative) *Pfaffian* Pf F as the element of the universal enveloping algebra $\mathrm{U}(\mathfrak{o}_{2n})$ defined by the formula

$$\mathrm{Pf}\, F = \frac{1}{2^n n!} \sum_{\sigma \in \mathfrak{S}_{2n}} \mathrm{sgn}\, \sigma \cdot F_{\sigma(1)\,\sigma(2)'} \dots F_{\sigma(2n-1)\,\sigma(2n)'}. \tag{5.36}$$

The Pfaffian belongs to the center $\mathrm{Z}(\mathfrak{o}_{2n})$ and its image under the Harish-Chandra isomorphism is found by

$$\chi : \mathrm{Pf}\, F \mapsto l_1 \dots l_n.$$

EXAMPLE 5.4.5. Since $F_{ij} + F_{j'i'} = 0$, for $n = 1$ we have[1]

$$\mathrm{Pf}\, F = F_{11}, \qquad \chi : \mathrm{Pf}\, F \mapsto l_1.$$

---

[1]We thus use an obvious version of the Harish-Chandra isomorphism for the abelian Lie algebra $\mathfrak{o}_2$.

For $n = 2$ we obtain

$$\begin{aligned}\operatorname{Pf} F &= \frac{1}{2}\,(F_{13}F_{31} - F_{12}F_{21} + F_{11}F_{22} + F_{31}F_{13} - F_{21}F_{12} + F_{22}F_{11})\\ &= F_{11}F_{22} - F_{21}F_{12} + F_{31}F_{13} + F_{22}\end{aligned}$$

and $\chi : \operatorname{Pf} F \mapsto \lambda_1\lambda_2 + \lambda_2$ so that the image coincides with $l_1 l_2$. □

COROLLARY 5.4.6. *If $\mathfrak{g}_N = \mathfrak{o}_{2n+1}$ then the elements of each family* (5.23) *and* (5.33) *with $k = 1, \dots, n$ are algebraically independent generators of the center* $\mathrm{Z}(\mathfrak{o}_{2n+1})$.

If $\mathfrak{g}_N = \mathfrak{o}_{2n}$ then $\operatorname{Pf} F$ *together with the elements of either family* (5.23) *or* (5.33) *with $k = 1, \dots, n-1$ are algebraically independent generators of the center* $\mathrm{Z}(\mathfrak{o}_{2n})$.

PROOF. By Propositions 5.4.1 and 5.4.3, the images of the elements of each family under the isomorphism (5.9) are algebraically independent generators of the algebra of invariant polynomials $\mathbb{C}[l_1, \dots, l_n]^W$. □

5.5. Symmetrizer and anti-symmetrizer for $\mathfrak{sp}_N$

Now we let $\mathfrak{g}_N = \mathfrak{sp}_N$ with $N = 2n$. Note that the image of the symmetrizer $h^{(m)} \in \mathbb{C}[\mathfrak{S}_m]$ under the action (1.73) coincides with the *anti-symmetrization operator* on the tensor space $(\mathbb{C}^{2n})^{\otimes m}$. Therefore, the image of $h^{(m)}$ is the zero operator for all $m > 2n$. On the other hand, as we pointed out in Section 2.2, the image $S^{(m)}$ of the symmetrizer $s^{(m)} \in \mathcal{B}_m(\omega)$ under the action (1.73) with $\omega = -2n$ is well-defined provided that $m \leqslant n+1$ and $S^{(n+1)} = 0$; see Proposition 2.3.2. The operator $S^{(m)}$ is not defined for $m \geqslant n+2$ since some denominators in the formulas defining $s^{(m)}$ vanish for $\omega = -2n$. Consider the expression

$$\gamma_m(-2n)\,\mathrm{tr}_{1,\dots,m}\, S^{(m)}\,(u_1 + F_1)\dots(u_m + F_m) \tag{5.37}$$

obtained by taking $S = S^{(m)}$ in (5.18) and multiplying it by the coefficient (2.40) with $\omega = -2n$. For any fixed m regard (5.37) as a function of n which is defined for all integer values $n \geqslant m$. Our next goal is to show that (5.37) is well-defined for all integer values $n \geqslant (m-1)/2$ (that is, it has 'removable singularities' at the extra values on n). In other words, for a given n, the expression (5.37) can be evaluated for all m such that $1 \leqslant m \leqslant 2n+1$; cf. Proposition 2.2.8.

LEMMA 5.5.1. *Suppose that $n \geqslant m$. For any $1 \leqslant a < b \leqslant m$ we have the following relations in the algebra* (5.17):

$$S^{(m)}\big(F_a F_b - F_b F_a\big) = S^{(m)}\big(F_a - F_b\big) \tag{5.38}$$

and

$$\big(F_a F_b - F_b F_a\big)S^{(m)} = -\big(F_a - F_b\big)S^{(m)}. \tag{5.39}$$

PROOF. By Proposition 5.1.1 we have

$$F_a F_b - F_b F_a = F_a\,(P_{ab} - Q_{ab}) - (P_{ab} - Q_{ab})\,F_a.$$

Therefore, both relations follow from (1.31), (1.67) and (1.73) with the use of the identities

$$S^{(m)} F_a Q_{ab} = 0 \qquad \text{and} \qquad Q_{ab} F_a S^{(m)} = 0. \tag{5.40}$$

The latter hold due to (1.75) and (5.7); cf. the proof of Lemma 2.2.7:

$$S^{(m)}F_a Q_{ab} = -S^{(m)}F_b Q_{ab} = S^{(m)}F_b P_{ab} Q_{ab} = S^{(m)}P_{ab}F_a Q_{ab} = -S^{(m)}F_a Q_{ab},$$

and the same calculation applies to verify the second identity in (5.40). □

LEMMA 5.5.2. *Suppose that $n \geqslant m$. For the m-th partial trace in the algebra (5.17) we have*

$$\operatorname{tr}_m S^{(m)}F_m = -\frac{n-m+1}{m(n-m+2)}\sum_{a=1}^{m-1} F_a\, S^{(m-1)}.$$

PROOF. Apply the same argument as for the symmetric algebra counterpart of this relation; see Lemma 2.2.7. □

PROPOSITION 5.5.3. *The expression (5.37) admits an equivalent form which represents a well-defined element of $\mathrm{Z}(\mathfrak{sp}_{2n})$ for all $n \geqslant (m-1)/2$.*

PROOF. Expanding the product $(u_1+F_1)\dots(u_m+F_m)$, we come to verifying the claim for expressions of the form

$$\frac{1}{n-m+1}\operatorname{tr}_{1,\dots,m} S^{(m)}F_{a_1}\dots F_{a_k}, \tag{5.41}$$

where $1 \leqslant a_1 < \dots < a_k \leqslant m$. For all $l = 0,1,\dots,\lceil m/2\rceil$ consider more general expressions

$$\frac{1}{n-m+l+1}\operatorname{tr}_{1,\dots,m-l} S^{(m-l)}F_{b_1}\dots F_{b_p}F_{a_1}\dots F_{a_k}, \tag{5.42}$$

where all indices $b_1,\dots,b_p$ belong to the set $\{1,2,\dots,\min(l,m-l)\}$ and

$$l+1 \leqslant a_1 < \dots < a_k \leqslant m-l.$$

Note that (5.41) is a particular case of (5.42) with $l=0$. We will prove by (reverse) induction on l that each expression (5.42) admits an equivalent form which is well-defined for all $n \geqslant (m-1)/2$. The induction base is the case $l = \lceil m/2\rceil$, where the claim holds trivially. Indeed, we then have $m = 2l$ or $m = 2l-1$, and the condition $n \geqslant (m-1)/2$ implies $n \geqslant l$ or $n \geqslant l-1$, respectively. In both cases the condition $n \geqslant m-l$ holds so that the expression (5.42) is well-defined (the empty product corresponding to the value $k=0$ is understood as being equal to 1).

Now consider an expression of the form (5.42) with $l < \lceil m/2\rceil$. Suppose first that $a_k < m-l$. Note the following consequence of Lemma 1.3.2 and the diagram (1.77):

$$\operatorname{tr}_m S^{(m)} = \frac{(n-m+1)(2n-m+3)}{m(n-m+2)}\, S^{(m-1)}. \tag{5.43}$$

Calculating the partial trace tr_{m-l} by using this relation, we can write (5.42) as

$$\frac{2n-m+l+3}{(m-l)(n-m+l+2)}\operatorname{tr}_{1,\dots,m-l-1} S^{(m-l-1)}F_{b_1}\dots F_{b_p}F_{a_1}\dots F_{a_k}.$$

Up to the constant factor, this expression is of the form (5.42) with the parameter l replaced by $l+1$. Hence, the claim follows by the induction hypothesis. Now suppose that $a_k = m - l$. Applying Lemma 5.5.2 together with the cyclic property of trace, we can write (5.42) as

$$-\frac{1}{(m-l)(n-m+l+2)}\operatorname{tr}_{1,\dots,m-l-1} S^{(m-l-1)} F_{b_1}\dots F_{b_p} \times F_{a_1}\dots F_{a_{k-1}} \sum_{a=1}^{m-l-1} F_a.$$

We will use Lemma 5.5.1 to show that this expression equals a linear combination of expressions of the form (5.42) with the parameter l replaced by $l+1$ so that the induction hypothesis could be applied. For the summands with $a = 1,\dots,l$ this can be done as follows. First apply the cyclic property of trace to move the factor $S^{(m-l-1)}$ to the right-most position, then transform the product

$$F_{a_1}\dots F_{a_{k-1}} F_a\, S^{(m-l-1)}$$

by making a repeated use of (5.39) to move F_a to the leftmost position and do the same with all the terms of the form F_a occurring as a result of the use of relation (5.39). Note that this relation is applicable to permute the factors F_{a_i} and F_a because the symmetrizer can be written as the product $S^{(m-l-1)} = S^{(h)} S^{(m-l-1)}$ for an appropriate value $h < m - l - 1$.

Furthermore, for any value $a \in \{l+1,\dots,m-l-1\}$ we can have at most two factors of the form F_a in the product. Suppose that $a_i = a$ for some subscript $i \in \{1,\dots,k-1\}$ (otherwise the argument is simplified in an obvious way). First, apply a conjugation by an appropriate permutation operator P_σ (together with (1.31)) and the cyclic property of trace to bring the corresponding product under the trace to the form

$$S^{(m-l-1)} F_{b_1}\dots F_{b_p} F_{a_2}\dots F_{a_i} F_{l+1} F_{a_{i+1}}\dots F_{a_{k-1}} F_{l+1}.$$

Namely, we use either the cycle $\sigma = (a_1, a_2, \dots, a_i, l+1)$ if $a_1 > l+1$, or the cycle $\sigma = (a_2,\dots,a_i,l+1)$ if $a_1 = l+1$. Then, arguing as above, use (5.38) repeatedly to move F_{l+1} to the position next after F_{b_p} and do the same with all the terms of the form F_{l+1} occurring as a result of the use of relation (5.38). Finally, use the cyclic property of trace again and transform the resulting products of the form

$$F_{b_1}\dots F_{b_p}\dots F_{a_{k-1}} F_{l+1}\, S^{(m-l-1)}$$

by using (5.39) repeatedly to move F_{l+1} and all terms of the form F_{l+1} occurring in the process to the position just after the term of the form F_{l+1} or to the position next after F_{b_p} if the term of the form F_{l+1} is not present.

To show that the possible expressions representing (5.37) by the above procedure are well-defined for $n+1 \leqslant m \leqslant 2n+1$, we need to verify that they coincide, as elements of $\mathrm{U}(\mathfrak{sp}_{2n})$. First observe that all these expressions belong to the center $\mathrm{Z}(\mathfrak{sp}_{2n})$; see (5.22) in Remark 5.3.2. If the condition $m \leqslant n$ holds, then the central element (5.37) is a unique polynomial in the Gelfand invariants $T_1,\dots,T_n$ with $T_k = \operatorname{tr} F^{2k}$ (see Example 5.3.3) and so equals

$$\sum a^{(m)}_{k_1,\dots,k_n} T_1^{k_1}\dots T_n^{k_n},$$

summed over nonnegative integers k_a such that $2k_1 + 4k_2 + \dots + 2nk_n \leqslant m$, where the coefficients $a^{(m)}_{k_1,\dots,k_n}$ are polynomials in $u_1,\dots,u_m$. In particular, $k_a = 0$ unless $2a \leqslant m$ so we may write the coefficients as $a^{(m)}_{k_1,\dots,k_n} = a^{(m)}_{k_1,\dots,k_p}$ with $p = \lfloor m/2 \rfloor$.

Now suppose that m is fixed and let n vary. By the definition of the symmetrizer $S^{(m)}$, each coefficient $a^{(m)}_{k_1,\dots,k_p}$ is a rational function in n. Hence, it is determined by infinitely many values of n with $n \geqslant m$. Due to the first part of the proof, the rational functions $a^{(m)}_{k_1,\dots,k_p}$ are defined for all values $n \geqslant p$. Since all equivalent expressions for (5.37) obtained in the first part of the proof coincide for infinitely many values of n, they must coincide for all values of $n \geqslant (m-1)/2$ for which they are defined. □

In what follows, we will use notation (5.37) for all values $1 \leqslant m \leqslant 2n+1$, assuming that it represents a well-defined element of $\mathrm{Z}(\mathfrak{sp}_{2n})$ obtained in the proof of Proposition 5.5.3.

Recall the factorial symmetric functions (4.16) and (4.17) which are associated with the sequence $a = (a_i)$ defined in (5.11), so that in the symplectic case $a_i = i^2$ for $i \geqslant 1$. The factorial elementary symmetric polynomials $e_k(l_1^2,\dots,l_n^2\,|\,a)$ with $k = 1,\dots,n$ are algebraically independent generators of the algebra $\mathbb{C}[l_1^2,\dots,l_n^2]^{\mathfrak{S}_n}$. Therefore, the Harish-Chandra image of the Casimir element (5.37) can be written as a unique linear combination

$$\sum c^{(m)}_{k_1,\dots,k_p} \prod_{r=1}^{p} e_r(l_1^2,\dots,l_n^2\,|\,a)^{k_p},$$

with the summation taken over nonnegative integers k_a satisfying the condition $2k_1+4k_2+\dots+2pk_p \leqslant m$, where $p = \lfloor m/2 \rfloor$ and the coefficients $c^{(m)}_{k_1,\dots,k_p}$ are polynomials in $u_1,\dots,u_m$. The coefficients of these polynomials are rational functions in n. Since a rational function is uniquely determined by its values at infinitely many points, the Harish-Chandra image of the Casimir element (5.37) is uniquely determined by its values for $n \geqslant m$.

Take even values $m = 2k$ and set $S = S^{(2k)}$ with $k \geqslant 1$ in Theorem 5.3.1 and consider a particular specialization of the parameters $u_1,\dots,u_{2k}$ as follows.

PROPOSITION 5.5.4. *For any $1 \leqslant k \leqslant n$ the image of the Casimir element*

$$\gamma_{2k}(-2n)\,\mathrm{tr}_{1,\dots,2k}\,(F_1-k+1)\dots(F_{2k}+k)\,S^{(2k)} \in \mathrm{Z}(\mathfrak{sp}_{2n}) \tag{5.44}$$

under the Harish-Chandra isomorphism coincides with $(-1)^k\,e_k(l_1^2,\dots,l_n^2\,|\,a)$.

PROOF. As pointed out above, it will be sufficient to prove the statement for any fixed k under the assumption that $n \geqslant 2k$. We will use the same argument as in the proof of Proposition 5.4.1. Let D_k denote the Casimir element (5.44). Relying on Proposition 5.1.3 we will show that D_k vanishes in all representations $L(\lambda)$ of Sp_{2n}, where the partitions λ satisfy $|\lambda| < k$. Set $r = |\lambda|$ and consider the action of the Lie algebra $\mathfrak{sp}_{2n}$ on $\mathbb{C}^{2n}$ defined by

$$F_{ij} \mapsto -e_{ji} + \varepsilon_i\varepsilon_j\, e_{i'j'}, \qquad 1 \leqslant i,j \leqslant 2n.$$

Under the corresponding representation in the tensor space

$$\varphi : \mathrm{U}(\mathfrak{sp}_{2n}) \otimes \mathrm{End}\,(\mathbb{C}^{2n})^{\otimes 2k} \to \mathrm{End}\,(\mathbb{C}^{2n})^{\otimes r} \otimes \mathrm{End}\,(\mathbb{C}^{2n})^{\otimes 2k}$$

we have

$$\varphi(F_a) = \sum_{b=1}^{r} \big(-P_{b\,r+a} + Q_{b\,r+a}\big), \qquad a = 1,\dots,2k. \tag{5.45}$$

The vanishing condition of Proposition 5.1.3 will be verified if we show that if $|\lambda| < k$ then

$$(\varphi(F_1) - k + 1) \dots (\varphi(F_{2k}) + k) \, S'^{(2k)} = 0, \tag{5.46}$$

where $S'^{(2k)}$ denotes the image of the symmetrizer in the Brauer algebra $\mathcal{B}_{2k}(-2n)$ acting of the last $2k$ copies of the tensor product space $(\mathbb{C}^{2n})^{\otimes(r+2k)}$. Due to (5.45) and relations (1.31), the desired identity (5.46) can be written in the form

$$(Y_{r+1} - k + 1) \dots (Y_{r+2k} - k + 1) \, S'^{(2k)} = 0, \tag{5.47}$$

where Y_{r+b} with $b = 1, \dots, 2k$ are the images of the respective Jucys–Murphy elements $y_{r+b} - (\omega - 1)/2$ given in (1.26) under the action (1.73) of the Brauer algebra $\mathcal{B}_{r+2k}(-2n)$ in $(\mathbb{C}^{2n})^{\otimes(r+2k)}$. The identity (5.47) is verified by the same argument as in the proof of Proposition 5.4.1. The leading term of $\chi(D_k)$ coincides with $(-1)^k e_k(\lambda_1^2, \dots, \lambda_n^2)$ as shown in Proposition 2.2.6; see (2.50). □

A different proof of Proposition 5.5.4 will follow from the results of Section 13.4.

For $1 \leqslant m \leqslant 2n+1$ consider the polynomials in a variable u whose coefficients are Casimir elements for $\mathfrak{sp}_{2n}$ given by

$$D_m(u) = \gamma_m(-2n) \operatorname{tr}_{1,\dots,m} \Big(F_1 + u + \frac{m-1}{2}\Big)\Big(F_2 + u + \frac{m-3}{2}\Big) \\ \times \dots \times \Big(F_m + u - \frac{m-1}{2}\Big) \, S^{(m)}.$$

COROLLARY 5.5.5. *For the images under the Harish-Chandra isomorphism we have*

$$\chi : D_m(u) \mapsto \sum_{r=0}^{\lfloor \frac{m}{2} \rfloor} (-1)^r \binom{2n-2r+1}{m-2r} e_r(l_1^2, \dots, l_n^2 \,|\, a) \prod_{i=0}^{m-2r-1} \Big(u - \frac{m-1}{2} + r + i\Big)$$

and

$$\chi : \gamma_m(-2n) \operatorname{tr}_{1,\dots,m} \big(-\partial_z + F_1 z^{-1}\big) \dots \big(-\partial_z + F_m z^{-1}\big) \, S^{(m)} \\ \mapsto \sum_{r=0}^{\lfloor \frac{m}{2} \rfloor} (-1)^r \binom{2n-2r+1}{m-2r} e_r(l_1^2, \dots, l_n^2 \,|\, a) \, z^{-2r} \big(-\partial_z + r z^{-1}\big)^{m-2r}.$$

PROOF. Exactly as in the proof of Corollary 5.4.2, we use Lemma 5.3.4 to verify the relations $D_m(u) = (-1)^m D_m(-u)$ and

$$D_m(u + 1/2) - D_m(u - 1/2) = (2n - m + 2) \, D_{m-1}(u), \qquad m \geqslant 1,$$

with $D_0(u) = 1$. Since the same relation is satisfied by the polynomials which are the desired images $\chi\big(D_m(u)\big)$, the statement follows from Proposition 5.5.4. □

Now let $S = A^{(2k)}$ in Theorem 5.3.1 with $k \geqslant 1$.

PROPOSITION 5.5.6. *For any $k \geqslant 1$ the image of the Casimir element*

$$\operatorname{tr}_{1,\dots,2k} (F_1 + k - 1) \dots (F_{2k} - k) \, A^{(2k)} \in \mathrm{Z}(\mathfrak{sp}_{2n}) \tag{5.48}$$

under the Harish-Chandra isomorphism coincides with $h_k(l_1^2, \dots, l_n^2 \,|\, a)$.

PROOF. Denote the Casimir element (5.48) by C_k. We use Proposition 5.1.3 and show that C_k vanishes in all representations $L(\lambda)$ of Sp_{2n}, where the partitions λ satisfy $|\lambda| < k$. Using the same realization of $L(\lambda)$ as in the proof of Proposition 5.5.4, we come to verifying the following analogue of (5.46):

$$\big(\varphi(F_1) + k - 1\big) \dots \big(\varphi(F_{2k}) - k\big)\, A'^{(2k)} = 0.$$

We show that the image of the operator $\big(-Y_{r+1} + k - 1\big) \dots \big(-Y_{r+2k} + k - 1\big)$ on the vector space $(\mathbb{C}^{2n})^{\otimes (r+2k)}$ is contained in a direct sum of representations $L(\nu)$ of Sp_{2n} with $\nu_1 < k$, while the image of the operator $A'^{(2k)}$ is contained in a direct sum of representations $L(\nu)$ with $\nu_1 \geqslant k+1$. The proof is completed by taking into account Proposition 2.2.3; the leading term of $\chi(C_k)$ coincides with $h_k(\lambda_1^2, \dots, \lambda_n^2)$ by (2.38). □

Now consider the polynomials in u given by

$$C_m(u) = \mathrm{tr}_{1,\dots,m} \Big(F_1 + u + \frac{m-1}{2}\Big)\Big(F_2 + u + \frac{m-3}{2}\Big) \dots \Big(F_m + u - \frac{m-1}{2}\Big)\, A^{(m)}.$$

Their coefficients are Casimir elements for $\mathfrak{sp}_{2n}$.

COROLLARY 5.5.7. *For the images under the Harish-Chandra isomorphism we have*

$$\chi : C_m(u) \mapsto \sum_{r=0}^{\lfloor \frac{m}{2} \rfloor} \binom{2n+m-1}{m-2r} h_r(l_1^2, \dots, l_n^2 \,|\, a) \prod_{i=0}^{m-2r-1} \Big(u - \frac{m-1}{2} + r + i\Big)$$

and

$$\chi : \mathrm{tr}_{1,\dots,m} \big(-\partial_z + F_1 z^{-1}\big) \dots \big(-\partial_z + F_m z^{-1}\big)\, A^{(m)}$$
$$\mapsto \sum_{r=0}^{\lfloor \frac{m}{2} \rfloor} \binom{2n+m-1}{m-2r} h_r(l_1^2, \dots, l_n^2 \,|\, a)\, z^{-2r} \big(-\partial_z + r\, z^{-1}\big)^{m-2r}.$$

PROOF. Lemma 5.3.4 implies $C_m(u) = (-1)^m C_m(-u)$ so that for the odd values $m = 2k-1$ we have $C_{2k-1}(0) = 0$. The values $\chi\big(C_{2k}(1/2)\big) = \chi\big(C_{2k}(-1/2)\big)$ are provided by Proposition 5.5.6. Furthermore, we verify that

$$C_m(u + 1/2) - C_m(u - 1/2) = (2n + m - 1)\, C_{m-1}(u), \qquad m \geqslant 1, \tag{5.49}$$

where $C_0(u) = 1$. This follows from Lemma 5.3.4 together with the relation for the partial trace of the operator $A^{(m)}$ implied by (1.54) and (1.77),

$$\mathrm{tr}_m\, A^{(m)} = \frac{2n + m - 1}{m}\, A^{(m-1)}.$$

This verifies (5.49). The same relation is satisfied by the polynomials which are claimed to be the images $\chi\big(C_m(u)\big)$ as stated in the corollary. The second part follows from the first as in the proof of Corollary 5.4.2. □

COROLLARY 5.5.8. *The elements of each family* (5.44) *and* (5.48) *with* $k = 1, \dots, n$ *are algebraically independent generators of the center* $\mathrm{Z}(\mathfrak{sp}_{2n})$.

PROOF. The images of the elements of each family under the Harish-Chandra isomorphism (5.9) are algebraically independent generators of the algebra of symmetric polynomials $\mathbb{C}[l_1^2, \dots, l_n^2]^{\mathfrak{S}_n}$. □

5.6. Manin matrices in types B, C and D

Our calculations of the Harish-Chandra images in Propositions 4.6.1 and 4.7.1 were based on properties of Manin matrices described in Proposition 3.2.2. On the other hand, the orthogonal and symplectic counterparts of these results stated in Sections 5.4 and 5.5 were derived in a different way, by using the Brauer–Schur–Weyl duality. Nevertheless, it is possible to define natural analogues of Manin matrices in types B, C and D and make a connection with the generator matrices F similar to Lemma 4.5.3.

In the following definition which we make by analogy with Lemma 3.1.2, we use tensor product algebra (3.2) and assume that M is a $N \times N$ matrix with entries in an associative algebra $\mathcal{A}$. As before, we suppose that even values $N = 2n$ are associated with types C_n or D_n, whereas odd values $N = 2n + 1$ are associated with type B_n.

DEFINITION 5.6.1. A matrix M is called a *Manin matrix of type B or D* if it satisfies

$$(1 - P)\, M_1\, M_2 \Big(\frac{1+P}{2} - \frac{Q}{N}\Big) = 0. \tag{5.50}$$

A matrix M is called a *Manin matrix of type C* if it satisfies

$$\Big(\frac{1-P}{2} - \frac{Q}{2n}\Big) M_1\, M_2\, (1 + P) = 0. \tag{5.51}$$

Note that if $n = 1$ in the symplectic case, then $P + Q = 1$ so that condition (5.51) holds trivially for any matrix M.

We point out the following as an analogue of Lemma 3.1.3 which is verified in the same way. As before, we denote by $H^{(m)}$ and $S^{(m)}$ the respective images of the element $h^{(m)}$ defined in (1.17) and the element $s^{(m)}$ given by equivalent formulas in Section 1.2, with respect to the homomorphism (1.69) in the orthogonal case and the homomorphism (1.73) in the symplectic case. The identities hold in the algebra (3.6).

LEMMA 5.6.2. *If M is a Manin matrix of type B or D then*

$$H^{(m)} M_1 \dots M_m\, S^{(m)} = M_1 \dots M_m\, S^{(m)}. \tag{5.52}$$

Moreover, for any permutation $\sigma \in \mathfrak{S}_m$ we have

$$M_{\sigma(1)} \dots M_{\sigma(m)}\, S^{(m)} = M_1 \dots M_m\, S^{(m)}. \tag{5.53}$$

If M is a Manin matrix of type C then

$$S^{(m)} M_1 \dots M_m\, H^{(m)} = S^{(m)} M_1 \dots M_m. \tag{5.54}$$

Moreover, for any permutation $\sigma \in \mathfrak{S}_m$ we have

$$S^{(m)} M_{\sigma(1)} \dots M_{\sigma(m)} = S^{(m)} M_1 \dots M_m. \tag{5.55}$$

PROOF. The proof of (5.52) and (5.53) reduces to checking that for any $\sigma \in \mathfrak{S}_m$

$$P_\sigma\, M_1 \dots M_m\, S^{(m)} = M_1 \dots M_m S^{(m)}. \tag{5.56}$$

It is sufficient to verify this property for the adjacent transpositions $\sigma = s_a$ with $a = 1, \dots, m-1$ and so it will follow from the particular case $m = 2$. In this case (5.56) follows from the definition (5.50) which reads $P\, M_1\, M_2\, S^{(2)} = M_1\, M_2\, S^{(2)}$.

Similarly, for the proof of (5.54) and (5.55) we need to verify that

$$S^{(m)} M_1 \dots M_m P_\sigma = \operatorname{sgn} \sigma \cdot S^{(m)} M_1 \dots M_m,$$

for any $\sigma \in \mathfrak{S}_m$. This reduces to the particular case $m = 2$ in the same way as for types B and D. In that case the property is equivalent to the definition (5.51) written in the form $S^{(2)} M_1 M_2 P = -S^{(2)} M_1 M_2$. □

Suppose that α and β are elements of a unital associative algebra $\mathcal{D}$ which satisfy the relation (4.33) and recall the matrix F defined in (5.5) which is associated with the Lie algebra $\mathfrak{g}_N$ of type B, C or D.

LEMMA 5.6.3. *The matrix $\alpha + F\beta$ with entries in the algebra $\mathrm{U}(\mathfrak{g}_N) \otimes \mathcal{D}$ is a Manin matrix of type B, C or D, respectively.*

PROOF. Set $M = \alpha + F\beta$. Using Proposition 5.1.1 and (1.67) we find

$$\begin{aligned} M_1 M_2 - M_2 M_1 &= \big(\alpha + F_1 \beta\big)\big(\alpha + F_2 \beta\big) - \big(\alpha + F_2 \beta\big)\big(\alpha + F_1 \beta\big) \\ &= (F_1 F_2 - F_2 F_1)\,\beta^2 - (F_1 - F_2)(\alpha\beta - \beta\alpha) \\ &= \big(F_1 (P-1) + F_2 (Q - P + 1) - Q F_2\big)\,\beta^2. \end{aligned}$$

In the cases B and D, we need to verify that $(M_1 M_2 - M_2 M_1) S^{(2)} = 0$. This follows from (1.31) and the relation $Q F_2 S^{(2)} = 0$. The latter holds since

$$Q F_2 S^{(2)} = Q P F_2 S^{(2)} = Q F_1 S^{(2)} = Q F_2' S^{(2)} = -Q F_2 S^{(2)},$$

where we used (2.28). Similarly, in the case C we write

$$M_1 M_2 - M_2 M_1 = \big(F_2 Q + (P - Q + 1) F_2 - (P+1) F_1\big)\,\beta^2$$

and verify that $S^{(2)} (M_1 M_2 - M_2 M_1) = 0$. □

As we will see in Chapter 8, certain generator matrices for the affine Kac–Moody algebras of types B, C and D are Manin matrices of the respective types. It would be interesting to extend the general algebraic properties described in Chapter 3 to the Manin matrices of those types.

5.7. Bibliographical notes

For general results outlined in Sections 5.1 and 5.2 see [**29**], [**59**], [**70**] and [**153**]. The hook dimension formulas for representations of O_N and Sp_N were given in [**34**]; another proof can be found in [**117**] together with a version of the characteristic map for the Brauer algebra. The exposition in Sections 5.4 and 5.5 follows [**73**]. Propositions 5.4.1 and 5.5.4 were also proved in [**111**] in a different way. A Capelli-type determinant for $\mathfrak{o}_N$ was given by Wachi [**150**]; see also Itoh [**77**] for its relationship with the Casimir elements (5.33) and Itoh [**78**] for the corresponding results concerning the permanent-type Casimir elements (5.48) in the symplectic case. A determinant of different type was found as an application of the Olshanski twisted Yangians; see [**109**, Chapter 7] for a detailed exposition and more constructions of Casimir elements, including analogues of the Newton identities of Theorem 4.8.1 and their equivalence to the Perelomov–Popov formulas for the Harish-Chandra images of the Gelfand invariants. These formulas were also proved in [**25**] with the use of the Schur–Weyl duality.

CHAPTER 6

Feigin–Frenkel center

Now we introduce the principal object of the book, the *center of the affine vertex algebra at the critical level.* Its structure was described by a theorem of Feigin and Frenkel [**39**] and hence is also known as the *Feigin–Frenkel center.* We begin by reviewing basic definitions of the vertex algebra theory and establish key properties of the center following the books by Frenkel [**46**], Frenkel and Ben-Zvi [**47**] and Kac [**87**]. Our goal will be to extend the R-matrix techniques developed in Chapters 4 and 5 to the affine Kac–Moody algebras and apply these techniques to construct explicit generators of the center for all classical types. This will allow us to give a direct proof of the Feigin–Frenkel theorem in those cases. The vertex algebra structure on the center will then be used to construct *Sugawara operators* for the affine Kac–Moody algebras $\widehat{\mathfrak{g}}$. These operators are affine analogues of the Casimir elements for the classical Lie algebras. They generate the center of a completed universal enveloping algebra of $\widehat{\mathfrak{g}}$ at the critical level.

6.1. Center of a vertex algebra

Let V be a vector space over $\mathbb{C}$. Consider formal Laurent series in a variable z with coefficients in the endomorphism algebra $\operatorname{End} V$. A series of the form

$$\sum_{n\in\mathbb{Z}} c_n z^{-n-1} \in \operatorname{End} V[[z, z^{-1}]]$$

is called a *field*, if for any $v \in V$ there exists an integer $N \geqslant 0$ such that $c_n v = 0$ for all $n \geqslant N$.

DEFINITION 6.1.1. A *vertex algebra* is a vector space V (the *space of states*) with the additional data $(Y, T, \mathbf{1})$, where $\mathbf{1}$ is the *vacuum vector* $\mathbf{1} \in V$, the *translation* T is an operator $T : V \to V$ and the *state-field correspondence* Y is a linear map

$$Y : V \to \operatorname{End} V[[z, z^{-1}]]$$

such that the image of any element $a \in V$ is a field, $Y : a \mapsto a(z)$,

$$a(z) = \sum_{n\in\mathbb{Z}} a_{(n)} z^{-n-1}, \qquad a_{(n)} \in \operatorname{End} V.$$

These data must satisfy the following axioms:

(1) $\mathbf{1}(z) = \mathrm{id}_V$,

(2) $a(z)\mathbf{1}$ is a power series in z and $a(z)\mathbf{1}\big|_{z=0} = a$ for each $a \in V$,

(3) $T\mathbf{1} = 0$,

(4) $\big[T, a(z)\big] = \partial_z a(z)$ for each $a \in V$,

(5) for any states $a, b \in V$ there exists a nonnegative integer N such that $(z - w)^N [a(z), b(w)] = 0$. □

The field $a(z)$ is often written in the form $a(z) = Y(a, z)$, and the endomorphisms $a_{(n)}$ are called the *Fourier coefficients* of $a(z)$. The span in $\operatorname{End} V$ of all Fourier coefficients $a_{(n)}$ of all fields $a(z)$ is a Lie subalgebra of $\operatorname{End} V$. The commutator is given by

$$[a_{(m)}, b_{(k)}] = \sum_{n\geqslant 0} \binom{m}{n} (a_{(n)}\, b)_{(m+k-n)}. \tag{6.1}$$

The axioms imply that the operator T acts by the rule $T(a) = a_{(-2)}\mathbf{1}$ so it is determined by the map Y. This implies a more general formula

$$T^n(a) = n!\, a_{(-n-1)}\mathbf{1}, \qquad n \geqslant 0. \tag{6.2}$$

For every integer n the corresponding *n-th product* $(a, b) \mapsto a_{(n)}b$ equips V with an algebra structure which need not be associative or commutative.

A *subalgebra* of a vertex algebra V is a subspace U of V containing the vacuum vector $\mathbf{1}$, such that $a_{(n)}U \subset U$ for all $a \in U$ and all $n \in \mathbb{Z}$. In particular, U is T-invariant; that is, $T(U) \subset U$. A subalgebra U is itself a vertex algebra; the state-field correspondence is obtained by the restriction of the endomorphisms so that

$$a(z) = \sum_{n\in\mathbb{Z}} a_{(n)}\big|_U\, z^{-n-1} \qquad \text{for any} \quad a \in U.$$

A vertex algebra V is called *commutative* (or *holomorphic*) if $a_{(n)} = 0$ for all $n \geqslant 0$ and all $a \in V$; that is, each field $a(z)$ is a power series in z. Equivalently, V is commutative if and only if $[a(z), b(w)] = 0$ for all $a, b \in V$.

PROPOSITION 6.1.2. *Suppose that V is a commutative vertex algebra. Then the (-1)-product makes V into a unital commutative associative algebra.*

PROOF. Write $ab = a_{(-1)}b$ for $a, b \in V$. The axioms give $\mathbf{1}a = a\mathbf{1} = a$ for any $a \in V$. Furthermore, since $[a(z), b(w)] = 0$ we have

$$a(z)b(w)c = b(w)a(z)c \qquad \text{for any} \quad a, b, c \in V.$$

Setting $z = w = 0$ we get $a(bc) = b(ac)$. By taking $c = \mathbf{1}$ we conclude that the product is commutative. Therefore, we also have $a(cb) = (ac)b$ so that the product is associative. □

DEFINITION 6.1.3. The *center* of a vertex algebra V is the subspace

$$\mathfrak{z}(V) = \{b \in V \mid a_{(n)}\, b = 0 \qquad \text{for all} \quad a \in V \quad \text{and all} \quad n \geqslant 0\}.$$

Equivalently, $b \in \mathfrak{z}(V)$ if and only if $[Y(a, z), Y(b, w)] = 0$ for all $a \in V$. □

The equivalence of the definitions is implied by the commutator formula (6.1).

PROPOSITION 6.1.4. *The center $\mathfrak{z}(V)$ of a vertex algebra V is a commutative vertex algebra.*

PROOF. We show first that $\mathfrak{z}(V)$ is a subalgebra of V. Clearly, $\mathbf{1} \in \mathfrak{z}(V)$. Furthermore, suppose that $b, c \in \mathfrak{z}(V)$ and $a \in V$. Then for $n \geqslant 0$ and arbitrary $m \in \mathbb{Z}$ we have

$$a_{(n)}b_{(m)}c = b_{(m)}a_{(n)}c + [a_{(n)}, b_{(m)}]c = 0.$$

Therefore, $b_{(m)}c \in \mathfrak{z}(V)$ so that $\mathfrak{z}(V)$ is a subalgebra of V. Finally, if $a \in \mathfrak{z}(V)$, then $a_{(n)} = 0$ for all $n \geqslant 0$, as an element of $\operatorname{End}\mathfrak{z}(V)$. Hence the vertex algebra $\mathfrak{z}(V)$ is commutative. □

In particular, $\mathfrak{z}(V)$ is T-invariant which can also be easily checked directly. Due to Proposition 6.1.2, the center of any vertex algebra is a unital commutative associative algebra.

DEFINITION 6.1.5. A *Lie conformal algebra* is a $\mathbb{C}[\partial]$-module R endowed with a $\mathbb{C}$-linear map

$$R \otimes R \to \mathbb{C}[\lambda] \otimes R, \qquad a \otimes b \mapsto [a_\lambda b],$$

called the λ-*bracket*, which satisfies the following axioms:

(1) $[\partial a_\lambda b] = -\lambda [a_\lambda b], \qquad [a_\lambda \partial b] = (\lambda + \partial)[a_\lambda b] \qquad$ (*sesquilinearity*),
(2) $[b_\lambda a] = -[a_{-\lambda-\partial} b] \qquad$ (*skewsymmetry*),
(3) $\big[a_\lambda [b_\mu c]\big] = \big[[a_\lambda b]_{\lambda+\mu} c\big] + \big[b_\mu [a_\lambda c]\big] \qquad$ (*Jacobi identity*). □

It can be derived from the axioms listed in Definition 6.1.1 that a vertex algebra V is a Lie conformal algebra with $\partial = T$ and the λ-bracket

$$[a_\lambda b] = \sum_{n=0}^{\infty} \frac{\lambda^n}{n!} a_{(n)} b, \qquad a, b \in V. \tag{6.3}$$

In terms of the λ-bracket on V, Definition 6.1.3 can be written as

$$\mathfrak{z}(V) = \{ b \in V \mid [a_\lambda b] = 0 \quad \text{for all} \quad a \in V \}.$$

6.2. Affine vertex algebras

Let $\mathfrak{g}$ be a simple Lie algebra over $\mathbb{C}$ equipped with a standard symmetric invariant bilinear form $\langle\ ,\ \rangle$ defined as the *normalized Killing form*

$$\langle X, Y \rangle = \frac{1}{2h^\vee} \operatorname{tr}\big(\operatorname{ad} X \operatorname{ad} Y\big), \tag{6.4}$$

where $h^\vee$ is the *dual Coxeter number* for $\mathfrak{g}$. The normalization is determined by the condition that the square length of the longest root equals 2. The corresponding *affine Kac–Moody algebra* $\widehat{\mathfrak{g}}$ is defined as the central extension

$$\widehat{\mathfrak{g}} = \mathfrak{g}[t, t^{-1}] \oplus \mathbb{C}K, \tag{6.5}$$

where $\mathfrak{g}[t, t^{-1}]$ is the Lie algebra of Laurent polynomials in t with coefficients in $\mathfrak{g}$. For any $r \in \mathbb{Z}$ and $X \in \mathfrak{g}$ we will write $X[r] = X\, t^r$. The commutation relations of the Lie algebra $\widehat{\mathfrak{g}}$ have the form

$$\big[X[r], Y[s]\big] = [X, Y][r+s] + r\, \delta_{r,-s} \langle X, Y \rangle\, K, \qquad X, Y \in \mathfrak{g},$$

and the element K is central in $\widehat{\mathfrak{g}}$.

For any $\kappa \in \mathbb{C}$ introduce the *universal enveloping algebra* $\mathrm{U}_\kappa(\widehat{\mathfrak{g}})$ *at the level* κ as the quotient of $\mathrm{U}(\widehat{\mathfrak{g}})$ by the ideal generated by $K - \kappa$. Similarly, we will say that a $\widehat{\mathfrak{g}}$-module V is *at the level* κ if K acts on V as multiplication by the scalar κ. The *vacuum module at the level* κ over $\widehat{\mathfrak{g}}$ is the quotient

$$V_\kappa(\mathfrak{g}) = \mathrm{U}_\kappa(\widehat{\mathfrak{g}}) / \mathrm{I}, \tag{6.6}$$

where I is the left ideal of $\mathrm{U}_\kappa(\widehat{\mathfrak{g}})$ generated by $\mathfrak{g}[t]$. By the Poincaré–Birkhoff–Witt theorem, the vacuum module is isomorphic to the universal enveloping algebra $\mathrm{U}\big(t^{-1}\mathfrak{g}[t^{-1}]\big)$, as a vector space. We will equip this vector space with a vertex algebra structure by using its $\widehat{\mathfrak{g}}$-module structure; see Definition 6.1.1. Introduce the data $(Y, T, \mathbf{1})$ as follows. The vacuum vector is $1 \in V_\kappa(\mathfrak{g})$ and the translation operator

$$T : V_\kappa(\mathfrak{g}) \to V_\kappa(\mathfrak{g}), \tag{6.7}$$

is determined by the properties

$$T : 1 \mapsto 0 \qquad \text{and} \qquad \big[T, X[r]\big] = -r\, X[r-1], \quad X \in \mathfrak{g}, \quad r < 0,$$

where $X[r]$ is understood as the operator of multiplication by $X[r]$. The state-field correspondence Y is defined by setting $Y(1, z) = \mathrm{id}$,

$$Y(J^a[-1], z) = \sum_{r \in \mathbb{Z}} J^a[r]\, z^{-r-1}, \tag{6.8}$$

where $J^1, \dots, J^d$ is a basis of $\mathfrak{g}$, and $J^a[r]$ is regarded as an element of $\operatorname{End} V_\kappa(\mathfrak{g})$ via the action of $\widehat{\mathfrak{g}}$. The map Y is extended to the whole of $V_\kappa(\mathfrak{g})$ with the use of *normal ordering*. The normally ordered product of fields

$$a(z) = \sum_{r \in \mathbb{Z}} a_{(r)} z^{-r-1} \qquad \text{and} \qquad b(w) = \sum_{r \in \mathbb{Z}} b_{(r)} w^{-r-1}$$

is the formal Laurent series

$$:a(z)\, b(w): = a(z)_+\, b(w) + b(w)\, a(z)_-, \tag{6.9}$$

where

$$a(z)_+ = \sum_{r<0} a_{(r)} z^{-r-1} \qquad \text{and} \qquad a(z)_- = \sum_{r \geqslant 0} a_{(r)} z^{-r-1}.$$

This definition extends to an arbitrary number of fields with the convention that the normal ordering is read from right to left. For instance,

$$:a(z)\, b(w)\, c(v): = \ :a(z)\, \big(:b(w)\, c(v):\big):.$$

Then, denoting the field in (6.8) by $J^a(z)$ we have

$$\begin{aligned} &Y(J^{a_1}[-r_1-1] \dots J^{a_m}[-r_m-1], z) \\ &\qquad = \frac{1}{r_1! \dots r_m!} :\partial_z^{r_1} J^{a_1}(z) \dots \partial_z^{r_m} J^{a_m}(z): \end{aligned} \tag{6.10}$$

for any $r_1, \dots, r_m \geqslant 0$. One can verify that the map Y is well defined and that all axioms in Definition 6.1.1 are satisfied by the data. So $V_\kappa(\mathfrak{g})$ is a vertex algebra which is called the (*universal*) *affine vertex algebra*. Its Lie conformal algebra structure is determined by the λ-bracket (6.3). In particular,

$$\big[X[-1]_\lambda Y[-1]\big] = [X, Y][-1] + \lambda \langle X, Y \rangle\, \kappa, \qquad X, Y \in \mathfrak{g}.$$

By the results of Section 6.1, the center $\mathfrak{z}\big(V_\kappa(\mathfrak{g})\big)$ is a unital commutative associative algebra. It can be shown that this algebra is trivial (coincides with $\mathbb{C}\,1$) unless the level is *critical*, $\kappa = -h^\vee$. In what follows we will be concerned with the structure of the center at the critical level.

DEFINITION 6.2.1. The *Feigin–Frenkel center* $\mathfrak{z}(\widehat{\mathfrak{g}})$ is the center of the vertex algebra $V_{-h^\vee}(\mathfrak{g})$. Any element of $\mathfrak{z}(\widehat{\mathfrak{g}})$ is called a *Segal–Sugawara vector*. □

By the definition of the state-field correspondence map, $\mathfrak{z}(\widehat{\mathfrak{g}})$ coincides with the subspace of $\mathfrak{g}[t]$-invariants on the vacuum module,

$$\mathfrak{z}(\widehat{\mathfrak{g}}) = \{v \in V_{-h^\vee}(\mathfrak{g}) \mid \mathfrak{g}[t]\, v = 0\}.$$

Identifying the vector space $V_{-h^\vee}(\mathfrak{g})$ with $\mathrm{U}\left(t^{-1}\mathfrak{g}[t^{-1}]\right)$, we get a vector space embedding

$$\mathfrak{z}(\widehat{\mathfrak{g}}) \hookrightarrow \mathrm{U}\left(t^{-1}\mathfrak{g}[t^{-1}]\right)$$

so that Segal–Sugawara vectors can be viewed as elements of $\mathrm{U}\left(t^{-1}\mathfrak{g}[t^{-1}]\right)$. It is immediate from the axioms of vertex algebra that the (-1)-product on $\mathfrak{z}(\widehat{\mathfrak{g}})$ coincides with the product in the universal enveloping algebra. In other words, $\mathfrak{z}(\widehat{\mathfrak{g}})$ can be regarded as a subalgebra of $\mathrm{U}\left(t^{-1}\mathfrak{g}[t^{-1}]\right)$. Moreover, by Propositions 6.1.2 and 6.1.4 this subalgebra is commutative.

The same algebra structure on $\mathfrak{z}(\widehat{\mathfrak{g}})$ can be defined equivalently as follows. As before, let I denote the left ideal of $\mathrm{U}_{-h^\vee}(\widehat{\mathfrak{g}})$ generated by $\mathfrak{g}[t]$ and let Norm I be its normalizer,

$$\mathrm{Norm}\ \mathrm{I} = \{v \in \mathrm{U}_{-h^\vee}(\widehat{\mathfrak{g}}) \mid \mathrm{I}\, v \subset \mathrm{I}\}.$$

The normalizer is a subalgebra of $\mathrm{U}_{-h^\vee}(\widehat{\mathfrak{g}})$, and I is a two-sided ideal of Norm I. The Feigin–Frenkel center $\mathfrak{z}(\widehat{\mathfrak{g}})$ coincides with the associative algebra defined as the quotient

$$\mathfrak{z}(\widehat{\mathfrak{g}}) = \mathrm{Norm}\ \mathrm{I}\,/\,\mathrm{I}.$$

Another way to interpret the algebra structure on $\mathfrak{z}(\widehat{\mathfrak{g}})$ is provided by the isomorphism

$$\mathfrak{z}(\widehat{\mathfrak{g}}) \cong \mathrm{End}_{\widehat{\mathfrak{g}}}\, V_{-h^\vee}(\mathfrak{g}).$$

Namely, any element $a \in \mathfrak{z}(\widehat{\mathfrak{g}})$ gives rise to the $\widehat{\mathfrak{g}}$-endomorphism of the vacuum module $V_{-h^\vee}(\mathfrak{g})$ which is determined by the condition $1 \mapsto a\,1$.

EXAMPLE 6.2.2. The *canonical Segal–Sugawara vector* $S \in \mathfrak{z}(\widehat{\mathfrak{g}})$ is given by

$$S = \sum_{a=1}^{d} J_a[-1]J^a[-1], \tag{6.11}$$

where $J_1, \dots, J_d$ is the basis of $\mathfrak{g}$ dual to $J^1, \dots, J^d$ with respect to the form (6.4). The element S is independent of the choice of the dual bases. Since $\mathfrak{g} = [\mathfrak{g}, \mathfrak{g}]$, to verify that S is a Segal–Sugawara vector it suffices to show that $J_i[0]\,S = 0$ and $J_i[1]\,S = 0$ for all i. The first condition is essentially equivalent to the property that

$$C = \sum_{a=1}^{d} J_a\, J^a$$

is a Casimir element for $\mathfrak{g}$. The second condition is implied by the fact that the eigenvalue of C in the adjoint representation equals $2h^\vee$; that is,

$$\sum_{a=1}^{d} \left[J_a, [J^a, X]\right] = 2h^\vee X$$

for any $X \in \mathfrak{g}$. □

EXAMPLE 6.2.3. Take $\mathfrak{g} = \mathfrak{sl}_2$ with its standard basis e, f, h and the commutation relations

$$[e, f] = h, \qquad [h, e] = 2e, \qquad [h, f] = -2f. \tag{6.12}$$

The only nonzero values of the form (6.4) on the basis elements are given by

$$\langle e, f\rangle = 1 \qquad \text{and} \qquad \langle h, h\rangle = 2.$$

Hence the dual basis is $f, e, h/2$ and so the Segal–Sugawara vector (6.11) is written as

$$S = e[-1]\,f[-1] + f[-1]\,e[-1] + \frac{1}{2}\,h[-1]^2.$$

All translations $T^r S$ with $r \geqslant 0$ turn out to be algebraically independent, and every element of the Feigin–Frenkel center $\mathfrak{z}(\widehat{\mathfrak{sl}}_2)$ is a polynomial in the Segal–Sugawara vectors $T^r S$. This is a particular case of the general result (6.15); cf. Example 7.1.6 in the next chapter. □

REMARK 6.2.4. By a theorem of Rybnikov [**140**], the commutative subalgebra $\mathfrak{z}(\widehat{\mathfrak{g}})$ of $\mathrm{U}\left(t^{-1}\mathfrak{g}[t^{-1}]\right)$ coincides with the centralizer of the canonical Segal–Sugawara vector (6.11). In particular, this subalgebra is maximal commutative. □

6.3. Feigin–Frenkel theorem

We will regard the translation T on the affine vertex algebra as the derivation of the algebra $\mathrm{U}\left(t^{-1}\mathfrak{g}[t^{-1}]\right)$ whose action on the generators is given by

$$T : X[r] \mapsto -r\,X[r-1], \qquad X \in \mathfrak{g}, \quad r < 0. \tag{6.13}$$

The algebra $\mathrm{U}\left(t^{-1}\mathfrak{g}[t^{-1}]\right)$ is equipped with the grading where the homogeneous components are eigenspaces of another derivation D defined on the generators by

$$D : X[r] \mapsto -r\,X[r], \qquad X \in \mathfrak{g}, \quad r < 0. \tag{6.14}$$

The subalgebra $\mathfrak{z}(\widehat{\mathfrak{g}})$ of $\mathrm{U}\left(t^{-1}\mathfrak{g}[t^{-1}]\right)$ is invariant with respect to each of the derivations T and D. In particular, all homogeneous components of any Segal–Sugawara vector belong to $\mathfrak{z}(\widehat{\mathfrak{g}})$.

Let n denote the rank of the simple Lie algebra $\mathfrak{g}$. A set of homogeneous elements $S_1, \dots, S_n \in \mathfrak{z}(\widehat{\mathfrak{g}})$ is called a *complete set of Segal–Sugawara vectors* if all elements $T^r S_l$ are algebraically independent, and every Segal–Sugawara vector is a polynomial in these elements. In other words, $\mathfrak{z}(\widehat{\mathfrak{g}})$ is then the algebra of polynomials

$$\mathfrak{z}(\widehat{\mathfrak{g}}) = \mathbb{C}[T^r S_l \mid l = 1, \dots, n, \quad r \geqslant 0]. \tag{6.15}$$

We can now state the Feigin–Frenkel theorem in the following form.

THEOREM 6.3.1. *A complete set of Segal–Sugawara vectors exists for any* $\mathfrak{g}$.

We aim to produce complete sets of homogeneous Segal–Sugawara vectors for the simple Lie algebras $\mathfrak{g}$ of the classical types A, B, C and D and hence give a direct proof of Theorem 6.3.1 in those cases. As a first step, consider a 'classical limit' of the vacuum module $V_{-h^\vee}(\mathfrak{g})$ and its $\mathfrak{g}[t]$-invariants. Regard $V_{-h^\vee}(\mathfrak{g})$ as a $\mathfrak{g}[t]$-module obtained by restriction of the action of $\widehat{\mathfrak{g}}$ to the subalgebra $\mathfrak{g}[t]$. Use the canonical filtration on the universal enveloping algebra $\mathrm{U}\left(t^{-1}\mathfrak{g}[t^{-1}]\right)$ to equip the vector space $V_{-h^\vee}(\mathfrak{g})$ with an ascending filtration $\mathrm{U}_0 \subset \mathrm{U}_1 \subset \dots$. The subspace U_p is spanned by all monomials of the form

$$X_{i_1}[-r_1] \dots X_{i_s}[-r_s], \qquad r_1, \dots, r_s \geqslant 1, \qquad s \leqslant p, \tag{6.16}$$

with $X_{i_1}, \dots, X_{i_s} \in \mathfrak{g}$. Clearly, $\mathfrak{g}[t]\,\mathrm{U}_p \subset \mathrm{U}_p$ so that the filtration is preserved by the action of $\mathfrak{g}[t]$. Therefore, the associated graded space

$$\mathrm{gr}\,V_{-h^\vee}(\mathfrak{g}) = \bigoplus_{p \geqslant 0} \mathrm{U}_p \,/\, \mathrm{U}_{p-1}, \qquad \mathrm{U}_{-1} = \{0\}, \tag{6.17}$$

becomes a $\mathfrak{g}[t]$-module. As a vector space, $\mathrm{gr}\,V_{-h^\vee}(\mathfrak{g})$ will be identified with the symmetric algebra $\mathrm{S}\left(t^{-1}\mathfrak{g}[t^{-1}]\right)$. The action of $\mathfrak{g}[t]$ on $\mathrm{S}\left(t^{-1}\mathfrak{g}[t^{-1}]\right)$ is obtained by extending the adjoint representation of $\mathfrak{g}[t]$ on $\mathfrak{g}[t,t^{-1}]/\mathfrak{g}[t] \cong t^{-1}\mathfrak{g}[t^{-1}]$ to the symmetric algebra. More explicitly, take a monomial of the form (6.16), regarding it as an element of the symmetric algebra and let $Y[r] \in \mathfrak{g}[t]$ so that $r \geqslant 0$. Then

$$(6.18)\quad Y[r]\cdot X_{i_1}[-r_1]\dots X_{i_s}[-r_s] = \sum_{a=1,\ r_a>r}^{s} X_{i_1}[-r_1]\dots \big[Y,X_{i_a}\big][r-r_a]\dots X_{i_s}[-r_s].$$

This shows that $Y[r]$ acts as a derivation on the algebra $\mathrm{S}\left(t^{-1}\mathfrak{g}[t^{-1}]\right)$. In particular, this implies that the subspace

$$\mathrm{S}\left(t^{-1}\mathfrak{g}[t^{-1}]\right)^{\mathfrak{g}[t]} = \left\{v \in \mathrm{S}\left(t^{-1}\mathfrak{g}[t^{-1}]\right) \mid \mathfrak{g}[t]\,v = 0\right\}$$

of $\mathfrak{g}[t]$-invariants in $\mathrm{S}\left(t^{-1}\mathfrak{g}[t^{-1}]\right)$ is closed under multiplication. Such invariants can be produced by the following procedure. Choose a basis $J_1,\dots,J_d$ of $\mathfrak{g}$ and suppose that

$$(6.19)\qquad P = P(J_1,\dots,J_d)$$

is an element of the symmetric algebra $\mathrm{S}(\mathfrak{g})$. Introduce formal power series in a variable z with coefficients in $t^{-1}\mathfrak{g}[t^{-1}]$ by

$$J_i(z) = \sum_{r=0}^{\infty} J_i[-r-1]\,z^r, \qquad i = 1,\dots,d.$$

Now make the substitution into (6.19) and expand into a power series of z,

$$P\big(J_1(z),\dots,J_d(z)\big) = \sum_{r=0}^{\infty} P_{(-r-1)}\,z^r.$$

The coefficients $P_{(-r-1)}$ are elements of the symmetric algebra $\mathrm{S}\left(t^{-1}\mathfrak{g}[t^{-1}]\right)$ and they can also be found as follows. Consider the derivation

$$(6.20)\qquad T : \mathrm{S}\left(t^{-1}\mathfrak{g}[t^{-1}]\right) \to \mathrm{S}\left(t^{-1}\mathfrak{g}[t^{-1}]\right)$$

induced by (6.13) so that it acts on the generators by $T : Y[-r] \mapsto rY[-r-1]$ for $r \geqslant 1$ and $Y \in \mathfrak{g}$. Then

$$(6.21)\qquad P_{(-r-1)} = \frac{T^r}{r!}\,P\big(J_1[-1],\dots,J_d[-1]\big).$$

When written in the equivalent form

$$P\big(J_1(z),\dots,J_d(z)\big) = e^{zT} P\big(J_1[-1],\dots,J_d[-1]\big),$$

this is essentially the statement that the commutative algebra $\mathrm{S}\left(t^{-1}\mathfrak{g}[t^{-1}]\right)$ equipped with the derivation T is a commutative vertex algebra; cf. (6.2).

LEMMA 6.3.2. *If the element $P \in \mathrm{S}(\mathfrak{g})$ is $\mathfrak{g}$-invariant, then all elements $P_{(-r-1)}$ with $r \geqslant 0$ are $\mathfrak{g}[t]$-invariants in $\mathrm{S}\left(t^{-1}\mathfrak{g}[t^{-1}]\right)$.*

PROOF. As a Lie algebra, $\mathfrak{g}[t]$ is generated by elements of the form $Y[0]$ and $Y[1]$ with $Y \in \mathfrak{g}$. Hence it suffices to show that $P_{(-r-1)}$ vanishes under the action of $Y[0]$ and $Y[1]$. We have the relations for the derivations in $\mathrm{S}\left(t^{-1}\mathfrak{g}[t^{-1}]\right)$, which are easily verified on generators,

$$Y[0]T = TY[0] \qquad\text{and}\qquad Y[1]T = TY[1] + Y[0].$$

Therefore, the claim follows from (6.21) by induction on r. □

The subalgebra of invariants is described by the following theorem.

THEOREM 6.3.3. *If $P_1, \dots, P_n$ are algebraically independent generators of the algebra $\mathrm{S}(\mathfrak{g})^{\mathfrak{g}}$ of $\mathfrak{g}$-invariants in $\mathrm{S}(\mathfrak{g})$, then the elements $P_{1\,(-r-1)}, \dots, P_{n\,(-r-1)}$ with $r = 0, 1, \dots$ are algebraically independent generators of the algebra of $\mathfrak{g}[t]$-invariants in $\mathrm{S}\,(t^{-1}\mathfrak{g}[t^{-1}])$.*

PROOF. The adjoint $\mathfrak{g}$-module $\mathfrak{g}$ is isomorphic to the coadjoint module $\mathfrak{g}^*$. An isomorphism is given by the linear map taking any element $X \in \mathfrak{g}$ to the functional $f_X \in \mathfrak{g}^*$ such that $f_X(Y) = \langle X, Y\rangle$ for $Y \in \mathfrak{g}$, with the invariant bilinear form defined in (6.4). Therefore, the $\mathfrak{g}[t]$-module $\mathrm{S}\,(t^{-1}\mathfrak{g}[t^{-1}])$ can be identified with $\mathrm{S}\,(t^{-1}\mathfrak{g}^*[t^{-1}])$ and we will work with the latter for notational convenience. Furthermore, the $\mathfrak{g}[t]$-module $t^{-1}\mathfrak{g}^*[t^{-1}] \cong \mathfrak{g}^*[t, t^{-1}]/\mathfrak{g}^*[t]$ is isomorphic to the (restricted) dual $\mathfrak{g}[t]^*$ of the adjoint module $\mathfrak{g}[t]$. To verify this claim, keep the basis $J_1, \dots, J_d$ of $\mathfrak{g}$ and let $J_1^*, \dots, J_d^*$ be the dual basis of $\mathfrak{g}^*$. We can represent any element of $\mathfrak{g}[t]^*$ as a finite linear combination of the dual basis vectors $J_i[r]^*$ with $i = 1, \dots, d$ and $r \geqslant 0$ such that $J_i[r]^*\big(J_j[s]\big) = \delta_{ij}\,\delta_{rs}$. For any $s \geqslant 0$ the isomorphism takes the basis vector $J_j[s]^* \in \mathfrak{g}[t]^*$ to $J_j^*[-s-1] \in t^{-1}\mathfrak{g}^*[t^{-1}]$. Indeed, for the action of the Lie algebra $\mathfrak{g}[t]$ we have

$$J_i[r]\, J_j[s]^* = -\sum_{k=1}^{d} c_{ik}^{j}\, J_k[-r+s]^* \qquad \text{if} \quad r \leqslant s$$

and $J_i[r]\, J_j[s]^* = 0$ otherwise, where c_{ij}^k are the structure constants of $\mathfrak{g}$ with respect to the basis $J_1, \dots, J_d$. This agrees with the action of $J_i[r]$ on the basis vector $J_j^*[-s-1]$ of $t^{-1}\mathfrak{g}^*[t^{-1}]$ and the claim follows.

Hence, the $\mathfrak{g}[t]$-module $\mathrm{S}\,(t^{-1}\mathfrak{g}^*[t^{-1}])$ is isomorphic to $\mathrm{S}\,(\mathfrak{g}[t]^*)$. The latter can be naturally identified, as a $\mathfrak{g}[t]$-module, with the algebra of polynomial functions $\mathrm{Fun}\,\mathfrak{g}[t]$ on $\mathfrak{g}[t]$. Using this identification, we will be proving the theorem in the following equivalent form. Suppose that $Q_1, \dots, Q_n$ are algebraically independent generators of the algebra $(\mathrm{Fun}\,\mathfrak{g})^{\mathfrak{g}}$ of $\mathfrak{g}$-invariants in $\mathrm{Fun}\,\mathfrak{g}$. Write an arbitrary element $X \in \mathfrak{g}$ in coordinates as $X = x_1 J_1 + \dots + x_d J_d$ with $x_i \in \mathbb{C}$ so that each generator Q_i can be regarded as a polynomial in the variables $x_1, \dots, x_d$. Making the substitutions

$$x_i(z) = \sum_{r=0}^{\infty} x_{ir}\, z^r, \qquad i = 1, \dots, d,$$

for certain variables x_{ir}, define polynomials $Q_{i\,(r)}$ by the expansions

$$Q_i\big(x_1(z), \dots, x_d(z)\big) = \sum_{r=0}^{\infty} Q_{i\,(r)}\, z^r. \tag{6.22}$$

Each $Q_{i\,(r)}$ is a polynomial function on the Lie algebra $\mathfrak{g}[t]$ whose elements are understood as finite linear combinations

$$\sum_{i=1}^{d} \sum_{r \geqslant 0} x_{i\,r}\, J_i[r].$$

It follows from Lemma 6.3.2 and can also be checked directly, that all polynomials $Q_{i\,(r)}$ with $i = 1, \dots, n$ and $r \geqslant 0$ belong to the algebra $\big(\mathrm{Fun}\,\mathfrak{g}[t]\big)^{\mathfrak{g}[t]}$ of $\mathfrak{g}[t]$-invariants in $\mathrm{Fun}\,\mathfrak{g}[t]$. We will be proving that they are algebraically independent generators of $\big(\mathrm{Fun}\,\mathfrak{g}[t]\big)^{\mathfrak{g}[t]}$.

Observe that the polynomial $Q_{i\,(r)}$ does not depend on the variables $x_{j\,s}$ with $s > r$. If there is an algebraic dependence between the polynomials $Q_{i\,(r)}$, it can only involve finitely many of them. Furthermore, any given $\mathfrak{g}[t]$-invariant in the algebra $\operatorname{Fun}\mathfrak{g}[t]$ does not depend on the variables $x_{i\,s}$ with large enough values of s. Therefore, it will be sufficient to prove that for any nonnegative integer m the polynomials $Q_{i\,(r)}$ with $i = 1, \dots, n$ and $r = 0, \dots, m$ are algebraically independent generators of the algebra of invariants $(\operatorname{Fun}\mathfrak{g}_m)^{\mathfrak{g}_m}$, where $\mathfrak{g}_m$ denotes the quotient of $\mathfrak{g}[t]$ by the ideal $t^{m+1}\mathfrak{g}[t]$. Moreover, it is clear from the way the polynomials $Q_{i\,(r)}$ are defined, that it is enough to establish these properties for one particular family of generators $Q_1, \dots, Q_n$ of $(\operatorname{Fun}\mathfrak{g})^{\mathfrak{g}}$.

Choose a regular nilpotent element $e \in \mathfrak{g}$. By the Jacobson–Morozov theorem, it can be completed to an $\mathfrak{sl}_2$-subalgebra of $\mathfrak{g}$ with the standard basis elements e, f, h as in (6.12). Identify $\mathfrak{g}$ with the constant term subalgebra of $\mathfrak{g}_m$ and consider the affine subspace

$$\mathfrak{k}_m = e + \mathfrak{g}^f + \mathfrak{g}^f\, t + \dots + \mathfrak{g}^f\, t^m$$

of $\mathfrak{g}_m$, where $\mathfrak{g}^f$ denotes the centralizer of f in $\mathfrak{g}$. Let

$$\rho : (\operatorname{Fun}\mathfrak{g}_m)^{\mathfrak{g}_m} \to \operatorname{Fun}\mathfrak{k}_m \tag{6.23}$$

be the homomorphism defined by restricting polynomial functions from $\mathfrak{g}_m$ to the affine subspace $\mathfrak{k}_m$. By the classical work of Kostant [**94**], there exist algebraically independent generators $Q_1, \dots, Q_n$ of $(\operatorname{Fun}\mathfrak{g})^{\mathfrak{g}}$ and a basis $U_1, \dots, U_n$ of the centralizer $\mathfrak{g}^f$ such that for all $j = 1, \dots, n$ and all n-tuples $(c_1, \dots, c_n)$ of complex numbers we have

$$Q_j(e + c_1 U_1 + \dots + c_n U_n) = c_j. \tag{6.24}$$

The algebra $\operatorname{Fun}\mathfrak{k}_m$ is freely generated by the coordinate functions $p_{ir} : u \mapsto a_{ir}$, with $i = 1, \dots, n$ and $r = 0, \dots, m$, where

$$u = e + \sum_{i=1}^{n}\sum_{r=0}^{m} a_{ir}\, U_i\, t^r \in \mathfrak{k}_m. \tag{6.25}$$

By the property (6.24) of the generators of $(\operatorname{Fun}\mathfrak{g})^{\mathfrak{g}}$, for any value of t in $\mathbb{C}$ we have

$$Q_j(u) = a_{j0} + a_{j1}\, t + \dots + a_{jm}\, t^m.$$

On the other hand, the definition (6.22) implies

$$Q_j(u) = Q_{j\,(0)} + Q_{j\,(1)}\, t + \dots + Q_{j\,(m)}\, t^m,$$

where the polynomials $Q_{j\,(r)}$ are evaluated at the respective coefficients of the powers of t in (6.25). We may thus conclude that all coordinate functions p_{ir} are contained in the image of the homomorphism ρ defined in (6.23) and so this homomorphism is surjective.

As a final step, we will verify that the homomorphism ρ is injective. Consider the adjoint group G_m associated with $\mathfrak{g}_m$. The algebra $(\operatorname{Fun}\mathfrak{g}_m)^{\mathfrak{g}_m}$ coincides with the algebra of G_m-invariants $(\operatorname{Fun}\mathfrak{g}_m)^{\mathrm{G}_m}$. The injectivity of the homomorphism $\rho : (\operatorname{Fun}\mathfrak{g}_m)^{\mathrm{G}_m} \to \operatorname{Fun}\mathfrak{k}_m$ will follow if we show that $(\operatorname{Ad}\mathrm{G}_m)(\mathfrak{k}_m)$ is dense in $\mathfrak{g}_m$; that is, the map

$$\phi : \mathrm{G}_m \times \mathfrak{k}_m \to \mathfrak{g}_m, \qquad (g, x) \mapsto (\operatorname{Ad} g)(x)$$

is dominant. By [**143**, Theorem 4.3.6(i)], it will be sufficient to verify that the differential $d\phi_{(\mathrm{id}, e)}$ at the point (id, e) is surjective. Identifying the tangent spaces to $\mathfrak{g}_m$ and $\mathfrak{k}_m$ at e with $\mathfrak{g}_m$ and $\mathfrak{g}^f + \mathfrak{g}^f t + \dots + \mathfrak{g}^f t^m$, respectively, we can write the differential in the form

$$d\phi_{(\mathrm{id},e)} : \mathfrak{g}_m \oplus \big(\mathfrak{g}^f + \mathfrak{g}^f t + \dots + \mathfrak{g}^f t^m\big) \to \mathfrak{g}_m, \qquad (x, v) \mapsto [x, e] + v.$$

However, the decomposition $\mathfrak{g} = [\mathfrak{g}, e] \oplus \mathfrak{g}^f$ is well-known due to Kostant [**94**]. This implies that

$$\mathfrak{g}_m = [\mathfrak{g}_m, e] + \mathfrak{g}^f t + \dots + \mathfrak{g}^f t^m$$

so that the differential is surjective. The proof is now complete as we showed that the map (6.23) is an isomorphism. □

In Chapters 7 and 8 we will construct certain elements $S_1, \dots, S_n$ of the algebra $\mathfrak{z}(\widehat{\mathfrak{g}})$ in an explicit form for each Lie algebra $\mathfrak{g}$ of classical type. We will prove that the families are complete sets of Segal–Sugawara vectors by relying on Theorem 6.3.3 as follows. Consider the symbols $\overline{S}_1, \dots, \overline{S}_n$ in the associated graded space as defined in (6.17). It will be possible to regard them as elements of $\mathrm{S}(\mathfrak{g})$ via the embedding $\mathfrak{g} \hookrightarrow t^{-1}\mathfrak{g}[t^{-1}]$ such that $X \mapsto X[-1]$. Then verify that they are algebraically independent generators of $\mathrm{S}(\mathfrak{g})^{\mathfrak{g}}$ and apply Theorem 6.3.3. This will imply by a standard argument that all elements $T^r S_l$ with $l = 1, \dots, n$ and $r \geqslant 0$ are algebraically independent and generate $\mathfrak{z}(\widehat{\mathfrak{g}})$. Indeed, if there was an algebraic dependence between the elements $T^r S_l$, then the symbols $T^r \overline{S}_l$ would also be algebraically dependent. Furthermore, suppose that a certain nonzero element $S \in \mathfrak{z}(\widehat{\mathfrak{g}})$ of the minimal possible degree is not a polynomial in the $T^r S_l$. Consider its symbol $\overline{S} \in \mathrm{S}\,\big(t^{-1}\mathfrak{g}[t^{-1}]\big)^{\mathfrak{g}[t]}$ and write it as a polynomial $P(T^r\overline{S}_l)$ in the generators $T^r\overline{S}_l$. The difference $S - P(T^r S_l)$ is an element of $\mathfrak{z}(\widehat{\mathfrak{g}})$ whose degree is less than the degree of S and it is not a polynomial in the $T^r S_l$. This makes a contradiction.

REMARK 6.3.4. In the case where $\mathfrak{g}$ is an arbitrary simple Lie algebra, the elements $S_1, \dots, S_n$ of any complete set of Segal–Sugawara vectors are 'liftings' to the algebra $\mathrm{U}\,\big(t^{-1}\mathfrak{g}[t^{-1}]\big)$ of certain algebraically independent generators of the algebra $\mathrm{S}(\mathfrak{g})^{\mathfrak{g}}$ in the sense described above. In particular, the respective degrees of the homogeneous elements $S_1, \dots, S_n$ must coincide with the degrees $d_1, \dots, d_n$ as defined in (2.2); see [**46**, Proposition 4.3.3]. For the classical types this will follow from our constructions of the Segal–Sugawara vectors. □

EXAMPLE 6.3.5. Theorem 6.3.3 clearly extends to the reductive Lie algebra $\mathfrak{gl}_N$ and holds in the same form. Recall the generators of the algebra of $\mathfrak{gl}_N$-invariants in $\mathrm{S}(\mathfrak{gl}_N)$ introduced in Section 2.1. Consider the matrix $E(z) = [E_{ij}(z)]$ with

$$E_{ij}(z) = \sum_{r=0}^{\infty} E_{ij}[-r-1]\, z^r$$

and expand

$$\det\,\big(u + E(z)\big) = u^N + \sum_{m=1}^{N} u^{N-m} \sum_{r=0}^{\infty} C_{m\,(-r-1)}\, z^r, \tag{6.26}$$

where $u + E(z)$ is understood as $u1 + E(z)$, where 1 is the $N \times N$ identity matrix. Then the elements $C_{m\,(-r-1)}$ with $m = 1, \dots, N$ and $r \geqslant 0$ are algebraically independent generators of $\mathrm{S}\left(t^{-1}\mathfrak{gl}_N[t^{-1}]\right)^{\mathfrak{gl}_N[t]}$. Similarly, writing

$$\mathrm{tr}\, E(z)^m = \sum_{r=0}^{\infty} T_{m\,(-r-1)}\, z^r \tag{6.27}$$

we obtain another algebraically independent family of generators $T_{m\,(-r-1)}$ with $m = 1, \dots, N$ and $r \geqslant 0$. □

EXAMPLE 6.3.6. Recall that for $\mathfrak{g} = \mathfrak{o}_N$ and $\mathfrak{sp}_N$ generators of the subalgebra of $\mathfrak{g}$-invariants in the symmetric algebra $\mathrm{S}(\mathfrak{g})$ were produced in Section 2.2. Consider the matrix $F(z) = [F_{ij}(z)]$ with

$$F_{ij}(z) = \sum_{r=0}^{\infty} F_{ij}[-r-1]\, z^r$$

and expand

$$\det\left(u + F(z)\right) = u^N + \sum_{m=1}^{n} u^{N-2m} \sum_{r=0}^{\infty} C_{m\,(-r-1)}\, z^r.$$

Moreover, in the case of $\mathfrak{o}_{2n}$ consider the Pfaffian $\mathrm{Pf}\, F(z)$ defined by the formula (2.33), where the elements F_{ij} should be respectively replaced by $F_{ij}(z)$, and expand it as a power series in z,

$$\mathrm{Pf}\, F(z) = \sum_{r=0}^{\infty} \mathrm{Pf}_{(-r-1)}\, z^r.$$

By Theorem 6.3.3, the elements $C_{m\,(-r-1)}$ with $m = 1, \dots, n$ and $r \geqslant 0$ are algebraically independent generators of $\mathrm{S}\left(t^{-1}\mathfrak{g}[t^{-1}]\right)^{\mathfrak{g}[t]}$ for $\mathfrak{g} = \mathfrak{sp}_{2n}$ and $\mathfrak{g} = \mathfrak{o}_{2n+1}$, while the elements $C_{m\,(-r-1)}$ with $m = 1, \dots, n-1$ together with $\mathrm{Pf}_{(-r-1)}$ for $r \geqslant 0$ are algebraically independent generators of $\mathrm{S}\left(t^{-1}\mathfrak{o}_{2n}[t^{-1}]\right)^{\mathfrak{o}_{2n}[t]}$.

Similarly, writing

$$\mathrm{tr}\, F(z)^{2m} = \sum_{r=0}^{\infty} T_{m\,(-r-1)}\, z^r$$

we obtain another algebraically independent family of generators $T_{m\,(-r-1)}$ with $m = 1, \dots, n$ and $r \geqslant 0$ in the cases $\mathfrak{g} = \mathfrak{sp}_{2n}$ and $\mathfrak{g} = \mathfrak{o}_{2n+1}$. In the case $\mathfrak{g} = \mathfrak{o}_{2n}$ the family is comprised of the elements $T_{m\,(-r-1)}$ with $m = 1, \dots, n-1$ together with $\mathrm{Pf}_{(-r-1)}$ for $r \geqslant 0$. □

6.4. Affine symmetric functions

A natural class of affine analogues of symmetric polynomials emerges by the application of a Chevalley-type projection to elements of the algebra of $\mathfrak{g}[t]$-invariants in $\mathrm{S}\left(t^{-1}\mathfrak{g}[t^{-1}]\right)$. Similar to the Chevalley projection (2.3) applied to the $\mathfrak{g}$-invariants in $\mathrm{S}(\mathfrak{g})$, the images in the affine case for the classical types turn out to be related to these affine symmetric polynomials as implied by Examples 6.3.5 and 6.3.6.

Given a symmetric polynomial $P(\lambda)$ in the family of variables $\lambda = (\lambda_1, \dots, \lambda_N)$, replace each variable λ_i by the respective formal power series

$$\lambda_i(z) = \sum_{r=0}^{\infty} \lambda_i[-r-1]\, z^r,$$

where the $\lambda_i[-r-1]$ are independent variables, and write

$$P\big(\lambda_1(z),\dots,\lambda_N(z)\big)=\sum_{r=0}^{\infty}P_r\,z^r,$$

where the coefficients P_r are polynomials in the variables $\lambda_i[-r-1]$. Equivalently, P_r is found as the derivative

$$P_r=\frac{T^rP}{r!},$$

where $P=P(\lambda)$ is regarded as a polynomial in the variables $\lambda_i[-1]=\lambda_i$ and the derivation T acts on the variables by the rule

$$T:\lambda_i[-r]\mapsto r\,\lambda_i[-r-1],\qquad i=1,\dots,N,\quad r\geqslant 1.$$

DEFINITION 6.4.1. Denote by $\Lambda^{\mathrm{aff}}(N)$ the subalgebra of the algebra of polynomials in the variables $\lambda_i[-r-1]$ generated by all coefficients P_r associated with all symmetric polynomials P. Any element of $\Lambda^{\mathrm{aff}}(N)$ is called an *affine symmetric polynomial.* □

It is clear that the algebra $\Lambda^{\mathrm{aff}}(N)$ is generated by the coefficients P_r associated to any family $\{P\}$ of generators of the algebra of symmetric polynomials $\mathbb{C}[\lambda]^{\mathfrak{S}_N}$. Returning to Example 6.3.5, consider the image of the determinant (6.26) under the affine extension of the Chevalley projection,

$$\varsigma: E_{ij}[-r-1]\mapsto \delta_{ij}\,\lambda_i[-r-1].$$

We have

$$\varsigma:\det\big(u+E(z)\big)\mapsto\big(u+\lambda_1(z)\big)\dots\big(u+\lambda_N(z)\big)$$

so that

$$\varsigma: C_{m\,(-r-1)}\mapsto\frac{T^r}{r!}\sum_{1\leqslant p_1<\dots<p_m\leqslant N}\lambda_{p_1}[-1]\dots\lambda_{p_m}[-1].$$

It is not difficult to verify that all these images with $r\geqslant 0$ are algebraically independent generators of the algebra $\Lambda^{\mathrm{aff}}(N)$ (this can be deduced from Lemma 12.3.2 in Chapter 12). Another family can be obtained by using the invariants given in (6.27). Namely, such generators are associated with the symmetric power sums $\lambda_1^m+\dots+\lambda_N^m$ and are given by

$$\sum_{i=1}^{N}\sum_{r_1+\dots+r_m=r}\lambda_i[-r_1-1]\dots\lambda_i[-r_m-1],\qquad m\geqslant 1,\quad r\geqslant 0,\tag{6.28}$$

where the second sum is taken over the m-tuples $(r_1,\dots,r_m)$ of nonnegative integers. We get an algebraically independent family of generators of $\Lambda^{\mathrm{aff}}(N)$ by restricting m to the values $1,\dots,N$. Thus, in type A we get the following affine version of the Chevalley isomorphism (2.3):

$$\varsigma:\mathrm{S}\big(t^{-1}\mathfrak{gl}_N[t^{-1}]\big)^{\mathfrak{gl}_N[t]}\to\Lambda^{\mathrm{aff}}(N).$$

In this form, its extension to other types is straightforward, although it would be interesting to get an independent description of the image of ς inside the algebra of all polynomials, possibly, as a 'classical limit' of the affine Harish-Chandra isomorphism; see Chapter 13.

Setting

$$\deg\lambda_i[-r]=r,\qquad r\geqslant 1,$$

defines a grading on the algebra of polynomials in the $\lambda_i[-r]$. The subalgebra $\Lambda^{\text{aff}}(N)$ inherits the grading so that we have the direct sum decomposition

$$\Lambda^{\text{aff}}(N) = \bigoplus_{k\geqslant 0} \Lambda^{\text{aff}}(N)^k,$$

where $\Lambda^{\text{aff}}(N)^k$ denotes the subspace of $\Lambda^{\text{aff}}(N)$ spanned by homogeneous elements of degree k and we set $\Lambda^{\text{aff}}(N)^0 := \mathbb{C}$. We let $H_N(q)$ denote the corresponding Hilbert–Poincaré series

$$H_N(q) = \sum_{k=0}^{\infty} \dim \Lambda^{\text{aff}}(N)^k \, q^k.$$

By a *plane partition over the N-strip* we will understand a finite sequence of Young diagrams (or partitions) $\lambda^{(1)} \supset \cdots \supset \lambda^{(r)}$ such that $\lambda^{(1)}$ contains at most N rows. Such a plane partition can be viewed as an array formed by unit cubes, the i-th level of the array has the shape $\lambda^{(i)}$.

PROPOSITION 6.4.2. *The dimension* $\dim \Lambda^{\text{aff}}(N)^k$ *coincides with the number of plane partitions over the N-strip containing exactly k unit cubes. Equivalently, the Hilbert–Poincaré series $H_N(q)$ is given by*

$$\prod_{m=1}^{N} \prod_{r\geqslant m} (1-q^r)^{-1} = \prod_{r=1}^{\infty} (1-q^r)^{-\min(r,N)}. \tag{6.29}$$

PROOF. Consider the family of algebraically independent generators of the algebra $\Lambda^{\text{aff}}(N)$ given in (6.28). For each $m = 1, \dots, N$ this family contains generators of degrees $m, m+1, \dots$ so the formula for the Hilbert–Poincaré series $H_N(q)$ follows. It is well-known that the generating function for the plane partitions over the N-strip is given by (6.29); see, e.g., [**104**, Section I.5]. □

6.5. From Segal–Sugawara vectors to Casimir elements

Here we point out a connection between the Feigin–Frenkel center $\mathfrak{z}(\widehat{\mathfrak{g}})$ and the center $\mathrm{Z}(\mathfrak{g})$ of the universal enveloping algebra $\mathrm{U}(\mathfrak{g})$. Given a variable z, consider the evaluation homomorphism

$$\varrho : \mathrm{U}\big(t^{-1}\mathfrak{g}[t^{-1}]\big) \to \mathrm{U}(\mathfrak{g}) \otimes \mathbb{C}[z^{-1}], \qquad X[r] \mapsto X\, z^r, \tag{6.30}$$

for any $X \in \mathfrak{g}$ and $r < 0$, where we suppress the tensor product sign. Recall the derivation T of the algebra $\mathrm{U}\big(t^{-1}\mathfrak{g}[t^{-1}]\big)$ defined in (6.13). The following property is easily verified on monomials in the generators $X[r]$:

$$\varrho(TS) = -\partial_z\, \varrho(S), \tag{6.31}$$

which holds for any $S \in \mathrm{U}\big(t^{-1}\mathfrak{g}[t^{-1}]\big)$.

PROPOSITION 6.5.1. *The image of the Feigin–Frenkel center $\mathfrak{z}(\widehat{\mathfrak{g}})$ under the homomorphism ϱ is contained in the tensor product $\mathrm{Z}(\mathfrak{g}) \otimes \mathbb{C}[z^{-1}]$.*

PROOF. In the algebra $\mathrm{U}_{-h^\vee}(\widehat{\mathfrak{g}})$ we have

$$X[0]\, S = \big[X[0], S\big] + S\, X[0]$$

for any $X \in \mathfrak{g}$ and $S \in \mathrm{U}\big(t^{-1}\mathfrak{g}[t^{-1}]\big)$. Taking this relation modulo the left ideal I generated by $\mathfrak{g}[t]$, we find that the condition $S \in \mathfrak{z}(\widehat{\mathfrak{g}})$ implies that $\big[X[0], S\big] = 0$. Therefore $\big[X, \varrho(S)\big] = 0$ in $\mathrm{U}(\mathfrak{g}) \otimes \mathbb{C}[z^{-1}]$. This shows that $\varrho(S)$ is contained in $\mathrm{Z}(\mathfrak{g}) \otimes \mathbb{C}[z^{-1}]$. □

If the variable z takes a particular nonzero value in $\mathbb{C}$, then the same formula (6.30) defines a homomorphism

$$\varrho_z : \mathrm{U}\big(t^{-1}\mathfrak{g}[t^{-1}]\big) \to \mathrm{U}(\mathfrak{g}), \qquad X[r] \mapsto X\, z^r, \tag{6.32}$$

where we indicate the dependence on $z \in \mathbb{C}$.

PROPOSITION 6.5.2. *Suppose that z is a nonzero complex number. Then the image of the Feigin–Frenkel center $\mathfrak{z}(\widehat{\mathfrak{g}})$ under the homomorphism ϱ_z coincides with $\mathrm{Z}(\mathfrak{g})$. Moreover, if $S_1, \dots, S_n$ is a complete set of Segal–Sugawara vectors, then $\varrho_z(S_1), \dots, \varrho_z(S_n)$ are algebraically independent generators of $\mathrm{Z}(\mathfrak{g})$.*

PROOF. As we pointed out in Remark 6.3.4, elements $S_1, \dots, S_n$ of any complete set of Segal–Sugawara vectors are associated with certain algebraically independent generators of the algebra $\mathrm{S}(\mathfrak{g})^{\mathfrak{g}}$ of the respective degrees $d_1, \dots, d_n$. More precisely, by Proposition 6.5.1 for the images $\varrho_z(S_1), \dots, \varrho_z(S_n)$ we must have

$$\varrho_z(S_i) = S_i^{\circ}\, z^{-d_i}, \qquad i = 1, \dots, n,$$

for certain elements $S_1^{\circ}, \dots, S_n^{\circ}$ of $\mathrm{Z}(\mathfrak{g})$. Moreover, the symbols of these elements in the algebra $\mathrm{S}(\mathfrak{g})^{\mathfrak{g}}$ are its algebraically independent generators. This shows that $S_1^{\circ}, \dots, S_n^{\circ}$ are algebraically independent generators of $\mathrm{Z}(\mathfrak{g})$. The first part of the proposition also follows. □

In particular, the image of the homomorphism ϱ_z does not depend on $z \in \mathbb{C}$ which can also be derived from (6.31) by using the T-invariance of $\mathfrak{z}(\widehat{\mathfrak{g}})$, independently of Proposition 6.5.2.

6.6. Center of the completed universal enveloping algebra

We will now discuss Casimir elements for the affine Kac–Moody algebras $\widehat{\mathfrak{g}}$ associated with simple Lie algebras $\mathfrak{g}$ as defined in (6.5). The universal enveloping algebra $\mathrm{U}(\widehat{\mathfrak{g}})$ needs to be completed to contain such elements. As in Section 6.2, fix an eigenvalue $\kappa \in \mathbb{C}$ of the central element K of $\widehat{\mathfrak{g}}$ and consider the universal enveloping algebra $\mathrm{U}_\kappa(\widehat{\mathfrak{g}})$ at the level κ. We would like the action of the completed algebra to be defined on a certain natural category of smooth modules. A $\widehat{\mathfrak{g}}$-module V at the level κ is called *smooth*, if for any $v \in V$ there exists a nonnegative integer p such that $t^p\mathfrak{g}[t]\, v = 0$, where $t^p\mathfrak{g}[t]$ is the subalgebra of $\widehat{\mathfrak{g}}$ spanned by all elements $X[r]$ with $X \in \mathfrak{g}$ and $r \geqslant p$.

Introduce a linear topology on $\mathrm{U}_\kappa(\widehat{\mathfrak{g}})$ by using the neighborhood basis for 0 formed by the left ideals I_p of $\mathrm{U}_\kappa(\widehat{\mathfrak{g}})$ generated by $t^p\mathfrak{g}[t]$ for all $p \geqslant 0$. The *completed universal enveloping algebra* $\widetilde{\mathrm{U}}_\kappa(\widehat{\mathfrak{g}})$ is the completion of $\mathrm{U}_\kappa(\widehat{\mathfrak{g}})$ with respect to this topology. Equivalently, $\widetilde{\mathrm{U}}_\kappa(\widehat{\mathfrak{g}})$ can be defined as the inverse limit

$$\widetilde{\mathrm{U}}_\kappa(\widehat{\mathfrak{g}}) = \varprojlim \mathrm{U}_\kappa(\widehat{\mathfrak{g}})/I_p. \tag{6.33}$$

The description of the center of the completed algebra $\widetilde{\mathrm{U}}_\kappa(\widehat{\mathfrak{g}})$ is a principal theme of the book by Frenkel [**46**] which stems from the original work of Feigin and Frenkel [**39**]. As with the center of the affine vertex algebra $V_\kappa(\mathfrak{g})$, the center of $\widetilde{\mathrm{U}}_\kappa(\widehat{\mathfrak{g}})$ turns out to be trivial unless the value of κ is critical, $\kappa = -h^\vee$; cf. Section 6.2. We refer the reader to the book [**46**] for a detailed discussion of the center $\mathrm{Z}(\widehat{\mathfrak{g}})$ of the algebra $\widetilde{\mathrm{U}}_{-h^\vee}(\widehat{\mathfrak{g}})$ at the critical level and its relationship with the geometry of *opers*. On the algebraic side, central elements of the completed algebra

$\widetilde{U}_{-h^\vee}(\widehat{\mathfrak{g}})$ are obtained from elements of the Feigin–Frenkel center $\mathfrak{z}(\widehat{\mathfrak{g}})$ by employing the vertex algebra structure. To outline the construction [**46**, Sections 3.2 and 4.3], recall from Section 6.2 that the application of the state-field correspondence map Y to an element $S \in \mathfrak{z}(\widehat{\mathfrak{g}})$ yields a field

$$Y(S,z) = \sum_{r \in \mathbb{Z}} S_{(r)} z^{-r-1}, \tag{6.34}$$

where all coefficients $S_{(r)}$ are operators on the vacuum module $V_{-h^\vee}(\mathfrak{g})$. We will reinterpret (6.34) in such a way that instead of the operators $S_{(r)}$ we will get corresponding elements $S_{[r]}$ of the center $\mathrm{Z}(\widehat{\mathfrak{g}})$ known as *Sugawara operators.* Consider the Laurent series $J^a(z) = Y(J^a[-1], z)$ defined in (6.8),

$$J^a(z) = \sum_{r \in \mathbb{Z}} J^a[r]\, z^{-r-1}.$$

We will now regard $J^a(z)$ as a series with coefficients in the universal enveloping algebra $\mathrm{U}_{-h^\vee}(\widehat{\mathfrak{g}})$. Furthermore, finite linear combinations of the coefficients of the series

$$: \partial_z^{r_1} J^{a_1}(z) \dots \partial_z^{r_m} J^{a_m}(z) :$$

can be regarded as elements of the completed universal enveloping algebra $\widetilde{U}_{-h^\vee}(\widehat{\mathfrak{g}})$ defined in (6.33). The definition of the state-field correspondence map is now interpreted in such a way that for any $a \in V_{-h^\vee}(\mathfrak{g})$ the coefficients $a_{[n]}$ of the series

$$Y[a,z] = \sum_{n \in \mathbb{Z}} a_{[n]} z^{-n-1} \tag{6.35}$$

defined by the same rule (6.10), are understood as elements of $\widetilde{U}_{-h^\vee}(\widehat{\mathfrak{g}})$. The commutator formula (6.1) is now replaced by

$$\left[a_{[m]}, b_{[k]}\right] = \sum_{n \geqslant 0} \binom{m}{n} \left(a_{(n)}\, b\right)_{[m+k-n]}.$$

This formula implies that if a belongs to the Feigin–Frenkel center $\mathfrak{z}(\widehat{\mathfrak{g}})$, then all coefficients $a_{[n]}$ are Sugawara operators; that is, they belong to the center $\mathrm{Z}(\widehat{\mathfrak{g}})$ of the completed algebra $\widetilde{U}_{-h^\vee}(\widehat{\mathfrak{g}})$. Furthermore, we have the following.

PROPOSITION 6.6.1. *Suppose that $S_1, \dots, S_n \in \mathfrak{z}(\widehat{\mathfrak{g}})$ is a complete set of Segal–Sugawara vectors. Then the Sugawara operators*

$$\{S_{1\,[r]}, \dots, S_{n\,[r]} \mid r \in \mathbb{Z}\}$$

are topological generators of $\mathrm{Z}(\widehat{\mathfrak{g}})$. □

Therefore, complete sets of Segal–Sugawara vectors for classical Lie algebras $\mathfrak{g}$ which we will produce in Theorems 7.1.4, 8.1.9 and 8.3.8 below, will provide respective topological generators of the completed universal enveloping algebras; see Chapters 7 and 8 for the formulas. These Sugawara operators will reappear in Chapter 15, where we will calculate their eigenvalues in the Wakimoto modules over $\widehat{\mathfrak{g}}$ of the critical level.

6.7. Bibliographical notes

Our proof of the key Theorem 6.3.3 follows Raïs and Tauvel [**134**] with some modifications as in the paper by Brown and Brundan [**15**]. A different proof is due to Beilinson and Drinfeld (see [**46**, Theorem 3.4.2] for detailed arguments) and Eisenbud and Frenkel [**119**, Proposition A.1]. A generalization of this theorem involving centralizers of nilpotent elements was proved by Panyushev, Premet and Yakimova [**131**]. Prior to the description of the algebra of invariants $\mathfrak{z}(\widehat{\mathfrak{g}})$ of the vacuum module $V_{-h^\vee}(\mathfrak{g})$ for an arbitrary simple Lie algebra $\mathfrak{g}$ in [**39**], particular cases of the Feigin–Frenkel theorem were proved independently by Goodman and Wallach [**58**] (type A) and Hayashi [**65**] (types A, B and C). A geometric proof of the theorem for all $\mathfrak{g}$ was given recently by Raskin [**135**]. The Feigin–Frenkel theorem was generalized by Arakawa [**5**] who showed that the center of the W-algebra at the critical level, associated with an arbitrary nilpotent element $f \in \mathfrak{g}$ coincides with $\mathfrak{z}(\widehat{\mathfrak{g}})$. A super version of affine symmetric polynomials was introduced in [**112**]. Proofs and more details on the center of the completed universal enveloping algebra (Section 6.6) can be found in the book by Frenkel [**46**].

CHAPTER 7

Generators in type A

In this chapter we will produce a few complete sets of Segal–Sugawara vectors for $\mathfrak{gl}_N$. We will thus get a direct proof of the Feigin–Frenkel theorem (Theorem 6.3.1) in type A. We will rely on Theorem 6.3.3 in a way indicated in the comments following the proof of that theorem in Chapter 6. We will also connect the Segal–Sugawara vectors to Casimir elements for $\mathfrak{gl}_N$ and reproduce some constructions of Chapter 4 by applying the general approach explained in Section 6.5. Furthermore, we will use Proposition 6.6.1 to give explicit formulas for generators of the center of the completed universal enveloping algebra of $\widehat{\mathfrak{gl}}_N$ at the critical level.

7.1. Segal–Sugawara vectors

In our constructions of Segal–Sugawara vectors in all classical types, we will use the extended Lie algebra $\widehat{\mathfrak{g}} \oplus \mathbb{C}\tau$ where the additional element τ satisfies the commutation relations

$$[\tau, X[r]] = -r\, X[r-1], \qquad [\tau, K] = 0. \tag{7.1}$$

Instead of the simple Lie algebra $\mathfrak{sl}_N$ of type A we will work with the reductive Lie algebra $\mathfrak{gl}_N$. We extend the form (6.4) to the invariant symmetric bilinear form on $\mathfrak{gl}_N$ which can be written as

$$\langle X,\, Y\rangle = \operatorname{tr} XY - \frac{1}{N} \operatorname{tr} X \operatorname{tr} Y, \qquad X, Y \in \mathfrak{gl}_N. \tag{7.2}$$

Note that the kernel of the form is spanned by the element $E_{11} + \dots + E_{NN}$, and its restriction to the subalgebra $\mathfrak{sl}_N$ is given by

$$\langle X,\, Y\rangle = \operatorname{tr} XY, \qquad X, Y \in \mathfrak{sl}_N.$$

The affine Kac–Moody algebra $\widehat{\mathfrak{gl}}_N = \mathfrak{gl}_N[t, t^{-1}] \oplus \mathbb{C}K$ has the commutation relations

$$\big[E_{ij}[r], E_{kl}[s]\big] = \delta_{kj}\, E_{i\,l}[r+s] - \delta_{i\,l}\, E_{kj}[r+s] + r\,\delta_{r,-s}\, K\Big(\delta_{kj}\,\delta_{i\,l} - \frac{\delta_{ij}\,\delta_{kl}}{N}\Big), \tag{7.3}$$

and the element K is central. The critical level $-N$ coincides with the negative of the dual Coxeter number for $\mathfrak{sl}_N$. With the above definition of the bilinear form on $\mathfrak{gl}_N$ the Feigin–Frenkel center $\mathfrak{z}(\widehat{\mathfrak{gl}}_N)$ coincides with both the algebra of $\mathfrak{gl}_N[t]$-invariants and $\mathfrak{sl}_N[t]$-invariants of the vacuum module,

$$\mathfrak{z}(\widehat{\mathfrak{gl}}_N) = \{v \in V_{-N}(\mathfrak{gl}_N) \mid \mathfrak{gl}_N[t]\, v = 0\} = \{v \in V_{-N}(\mathfrak{gl}_N) \mid \mathfrak{sl}_N[t]\, v = 0\}.$$

For any $r \in \mathbb{Z}$ combine the elements $E_{ij}[r]$ into the matrix $E[r]$ so that

$$E[r] = \sum_{i,j=1}^{N} e_{ij} \otimes E_{ij}[r] \in \operatorname{End} \mathbb{C}^N \otimes \mathrm{U}(\widehat{\mathfrak{gl}}_N).$$

For each $a \in \{1, \dots, m\}$ introduce the element $E[r]_a$ of the algebra

$$\underbrace{\operatorname{End}\mathbb{C}^N \otimes \dots \otimes \operatorname{End}\mathbb{C}^N}_{m} \otimes \mathrm{U} \tag{7.4}$$

by

$$E[r]_a = \sum_{i,j=1}^{N} 1^{\otimes(a-1)} \otimes e_{ij} \otimes 1^{\otimes(m-a)} \otimes E_{ij}[r], \tag{7.5}$$

where U stands for the universal enveloping algebra of $\widehat{\mathfrak{gl}}_N \oplus \mathbb{C}\tau$.

PROPOSITION 7.1.1. *The defining relations of the algebra* $\mathrm{U}(\widehat{\mathfrak{gl}}_N)$ *can be written in the form*

$$E[r]_1\, E[s]_2 - E[s]_2\, E[r]_1 = \big(E[r+s]_1 - E[r+s]_2\big)\, P + r\,\delta_{r,-s}\, K\Big(P - \frac{1}{N}\Big), \tag{7.6}$$

where both sides are elements of the algebra (7.4) *with* $m = 2$.

PROOF. This is immediate from (7.3) by taking the tensor products of $e_{ij} \otimes e_{kl}$ with both sides and then applying the summation over i, j, k, l. □

Recall equivalent definitions of Manin matrices as given by Lemma 3.1.2.

LEMMA 7.1.2. *The matrix* $\mathcal{E} = \tau + E[-1] = \big[\delta_{ij}\tau + E_{ij}[-1]\big]$ *with the entries in the universal enveloping algebra of* $\widehat{\mathfrak{gl}}_N \oplus \mathbb{C}\tau$ *is a Manin matrix.*

PROOF. This is quite similar to the proof of Lemma 4.5.3. Using Proposition 7.1.1 we find

$$\begin{aligned}
\mathcal{E}_1\, \mathcal{E}_2 - \mathcal{E}_2\, \mathcal{E}_1 &= \big(\tau + E[-1]_1\big)\big(\tau + E[-1]_2\big) - \big(\tau + E[-1]_2\big)\big(\tau + E[-1]_1\big) \\
&= E[-1]_1\, E[-1]_2 - E[-1]_2\, E[-1]_1 - E[-2]_1 + E[-2]_2 \\
&= \big(E[-2]_1 - E[-2]_2\big)(P - 1).
\end{aligned}$$

On multiplying this element from the right by $1 + P$ we get 0, and the claim follows by using the equivalent definition (3.5) of Manin matrices. □

As before, we let $H^{(m)}$ and $A^{(m)}$ denote the elements of the algebra (7.4) (with the identity components in U) which are the respective images of the symmetrizer $h^{(m)}$ and anti-symmetrizer $a^{(m)}$ defined in (1.17) and (1.19) under the map (1.65). Define the elements $\phi_{m\,a}, \psi_{m\,a}, \theta_{m\,a} \in V_{-N}(\mathfrak{gl}_N) \cong \mathrm{U}\big(t^{-1}\mathfrak{gl}_N[t^{-1}]\big)$ by the expansions

$$\operatorname{tr}_{1,\dots,m} A^{(m)} \mathcal{E}_1 \dots \mathcal{E}_m = \phi_{m\,0}\, \tau^m + \phi_{m\,1}\, \tau^{m-1} + \dots + \phi_{m\,m}, \tag{7.7}$$

$$\operatorname{tr}_{1,\dots,m} H^{(m)} \mathcal{E}_1 \dots \mathcal{E}_m = \psi_{m\,0}\, \tau^m + \psi_{m\,1}\, \tau^{m-1} + \dots + \psi_{m\,m}, \tag{7.8}$$

and

$$\operatorname{tr} \mathcal{E}^m = \theta_{m\,0}\, \tau^m + \theta_{m\,1}\, \tau^{m-1} + \dots + \theta_{m\,m}. \tag{7.9}$$

Note that all elements $\phi_{m\,a}$ are zero for $m > N$ since the anti-symmetrizer $A^{(m)}$ vanishes in this case. Since $\mathcal{E}$ is a Manin matrix by Lemma 7.1.2, taking $m = N$ in (7.7) and applying (3.28) we get

$$\operatorname{tr}_{1,\dots,N} A^{(N)} \mathcal{E}_1 \dots \mathcal{E}_N = \operatorname{cdet} \mathcal{E}, \tag{7.10}$$

where the column-determinant is given by (3.23). Expand it as a polynomial in τ,

$$\operatorname{cdet}\mathcal{E} = \tau^N + \phi_1\,\tau^{N-1} + \cdots + \phi_N \tag{7.11}$$

so that $\phi_{N\,a} = \phi_a$ for $a = 1, \dots, N$. The identity (3.29) implies the expansion of the noncommutative characteristic polynomial,

$$\operatorname{cdet}(u + \mathcal{E}) = \sum_{m=0}^{N} u^{N-m}\,\operatorname{tr}_{1,\dots,m} A^{(m)} \mathcal{E}_1 \dots \mathcal{E}_m,$$

where u is a variable. Hence replacing τ by $u + \tau$ in (7.11), we obtain more general relations

$$\phi_{m\,a} = \binom{N-a}{m-a}\,\phi_a, \qquad 0 \leqslant a \leqslant m \leqslant N. \tag{7.12}$$

In particular, $\phi_{m\,m} = \phi_m$ for $m = 1, \dots, N$.

THEOREM 7.1.3. *All elements $\phi_{m\,a}$, $\psi_{m\,a}$ and $\theta_{m\,a}$ belong to the Feigin–Frenkel center $\mathfrak{z}(\widehat{\mathfrak{gl}}_N)$.*

PROOF. By Lemma 7.1.2, it is sufficient to verify the claim for only one of the three families. The corresponding statements for the two remaining families will then follow from the MacMahon Master Theorem (Theorem 3.2.1) and the Newton identity (Theorem 3.2.10) together with the identities of Corollary 3.2.5. Still we will give direct arguments for two families $\psi_{m\,a}$ and $\theta_{m\,a}$, as they use different matrix calculations. The direct proof for the elements $\phi_{m\,a}$ is quite similar to the one for $\psi_{m\,a}$.

PROOF FOR THE ELEMENTS $\psi_{m\,a}$. It will be enough to verify that these elements are annihilated by the operators $E_{ij}[0]$ and $E_{ij}[1]$ with $1 \leqslant i, j \leqslant N$ in the vacuum module $V_{-N}(\mathfrak{gl}_N)$. Consider the tensor product algebra

$$\underbrace{\operatorname{End}\mathbb{C}^N \otimes \dots \otimes \operatorname{End}\mathbb{C}^N}_{m+1} \otimes \mathrm{U}$$

with the $m + 1$ copies of $\operatorname{End}\mathbb{C}^N$ labeled by $0, 1, \dots, m$. The required annihilation properties can then be written in the equivalent form

$$E[0]_0\,\operatorname{tr}_{1,\dots,m} H^{(m)} \mathcal{E}_1 \dots \mathcal{E}_m = 0 \quad \text{and} \quad E[1]_0\,\operatorname{tr}_{1,\dots,m} H^{(m)} \mathcal{E}_1 \dots \mathcal{E}_m = 0 \tag{7.13}$$

modulo the left ideal of U generated by $\mathfrak{gl}_N[t]$ and $K + N$. To verity the first relation, note that by (1.67) and (7.6) we have

$$[E[0]_0, \mathcal{E}_a] = P_{0\,a}\,\mathcal{E}_a - \mathcal{E}_a\,P_{0\,a}. \tag{7.14}$$

Hence

$$\begin{aligned} E[0]_0\,\operatorname{tr}_{1,\dots,m} H^{(m)} \mathcal{E}_1 \dots \mathcal{E}_m &= \sum_{a=1}^{m} \operatorname{tr}_{1,\dots,m} H^{(m)} \mathcal{E}_1 \dots \big(P_{0\,a}\,\mathcal{E}_a - \mathcal{E}_a\,P_{0\,a}\big) \dots \mathcal{E}_m \\ &= \operatorname{tr}_{1,\dots,m} H^{(m)} \big(P_{0\,1} + \dots + P_{0\,m}\big)\,\mathcal{E}_1 \dots \mathcal{E}_m \\ &\quad - \operatorname{tr}_{1,\dots,m} H^{(m)}\,\mathcal{E}_1 \dots \mathcal{E}_m\,\big(P_{0\,1} + \dots + P_{0\,m}\big). \end{aligned}$$

The sum $P_{0\,1} + \dots + P_{0\,m}$ commutes with the symmetrizer $H^{(m)}$ so that the expression is equal to zero by the cyclic property of trace.

To verify the second property in (7.13) use the relation implied by (7.6):

$$[E[1]_0, \mathcal{E}_a] = E[0]_0 + P_{0\,a}\, E[0]_a - E[0]_a\, P_{0\,a} + K\,(P_{0\,a} - 1/N). \tag{7.15}$$

We have

$$E[1]_0\, \mathrm{tr}_{1,\dots,m}\, H^{(m)} \mathcal{E}_1 \dots \mathcal{E}_m = \sum_{a=1}^{m} \mathrm{tr}_{1,\dots,m}\, H^{(m)} \mathcal{E}_1 \dots \mathcal{E}_{a-1}$$
$$\times \big(E[0]_0 + P_{0\,a}\, E[0]_a - E[0]_a\, P_{0\,a} + K\,(P_{0\,a} - 1/N)\big)\, \mathcal{E}_{a+1} \dots \mathcal{E}_m.$$

Applying (7.14) we can write this expression as

$$\sum_{1\leqslant a<b\leqslant m} \mathrm{tr}_{1,\dots,m}\, H^{(m)} \mathcal{E}_1 \dots \widehat{\mathcal{E}}_a \dots \big(P_{0\,b}\, \mathcal{E}_b - \mathcal{E}_b\, P_{0\,b}\big) \dots \mathcal{E}_m$$
$$+ \sum_{1\leqslant a<b\leqslant m} \mathrm{tr}_{1,\dots,m}\, H^{(m)}\, P_{0\,a}\, \mathcal{E}_1 \dots \widehat{\mathcal{E}}_a \dots \big(P_{ab}\, \mathcal{E}_b - \mathcal{E}_b\, P_{ab}\big) \dots \mathcal{E}_m$$
$$- \sum_{1\leqslant a<b\leqslant m} \mathrm{tr}_{1,\dots,m}\, H^{(m)} \mathcal{E}_1 \dots \widehat{\mathcal{E}}_a \dots \big(P_{ab}\, \mathcal{E}_b - \mathcal{E}_b\, P_{ab}\big) \dots \mathcal{E}_m\, P_{0\,a}$$
$$+ K \sum_{a=1}^{m} \mathrm{tr}_{1,\dots,m}\, H^{(m)} (P_{0\,a} - 1/N)\, \mathcal{E}_1 \dots \widehat{\mathcal{E}}_a \dots \mathcal{E}_m,$$

where the hats indicate that the corresponding symbols should be skipped. We now transform the expression by applying conjugations by elements of $\mathfrak{S}_m$ and using the cyclic property of trace together with the properties of the symmetrizer $H^{(m)}$. For $1 \leqslant a < b \leqslant m$ we have the relations $P_{a\,b}\, H^{(m)} = H^{(m)}\, P_{a\,b} = H^{(m)}$ as well as

$$H^{(m)}\, P_{0\,a}\, P_{a\,b} = H^{(m)}\, P_{0\,b} \qquad \text{and} \qquad P_{a\,b} P_{0\,a}\, H^{(m)} = P_{0\,b}\, H^{(m)}.$$

Hence, the first sum in the expression equals

$$\sum_{1\leqslant a<b\leqslant m} \mathrm{tr}_{1,\dots,m}\, H^{(m)} \mathcal{E}_1 \dots \widehat{\mathcal{E}}_a \dots \big(P_{0\,b}\, \mathcal{E}_b - \mathcal{E}_b\, P_{0\,b}\big) \dots \mathcal{E}_m$$
$$= \sum_{1\leqslant a<b\leqslant m} \mathrm{tr}_{1,\dots,m}\, H^{(m)} P_{m-1\,m} \cdots P_{a\,a+1}\, \mathcal{E}_1 \dots \widehat{\mathcal{E}}_a \dots \big(P_{0\,b}\, \mathcal{E}_b - \mathcal{E}_b\, P_{0\,b}\big) \dots \mathcal{E}_m$$

which we can write as

$$\sum_{1\leqslant a<b\leqslant m} \mathrm{tr}_{1,\dots,m}\, H^{(m)}\, \mathcal{E}_1 \dots \big(P_{0\,b-1}\, \mathcal{E}_{b-1} - \mathcal{E}_{b-1}\, P_{0\,b-1}\big) \dots \mathcal{E}_{m-1}\, P_{m-1\,m} \cdots P_{a\,a+1}$$
$$= \sum_{a=1}^{m-1} a\, \mathrm{tr}_{1,\dots,m}\, \big(H^{(m)} P_{0\,a} - P_{0\,a}\, H^{(m)}\big)\, \mathcal{E}_1 \dots \mathcal{E}_{m-1}.$$

Applying similar transformations to the remaining sums we will get

$$E[1]_0\, \mathrm{tr}_{1,\dots,m}\, H^{(m)} \mathcal{E}_1 \dots \mathcal{E}_m = \sum_{a=1}^{m-1} 2a\, \mathrm{tr}_{1,\dots,m}\, H^{(m)} P_{0\,a}\, \mathcal{E}_1 \dots \mathcal{E}_{m-1}$$
$$- \binom{m}{2} \mathrm{tr}_{1,\dots,m}\, \big(H^{(m)}\, P_{0\,m} + P_{0\,m}\, H^{(m)}\big)\, \mathcal{E}_1 \dots \mathcal{E}_{m-1}$$
$$+ m K\, \mathrm{tr}_{1,\dots,m}\, H^{(m)} \big(P_{0\,m} - 1/N\big)\, \mathcal{E}_1 \dots \mathcal{E}_{m-1}.$$

Taking $\omega = N$ in Lemma 1.3.4 and using (1.72), we find that

$$\operatorname{tr}_m H^{(m)} = \frac{N+m-1}{m} H^{(m-1)}.$$

Furthermore, arguing as in the proof of Lemma 1.3.4, and using (1.68) we find

$$\begin{aligned}\operatorname{tr}_m H^{(m)} P_{0\,m} &= \frac{1}{m} \operatorname{tr}_m H^{(m-1)} \Big(1 + \sum_{a=1}^{m-1} P_{a\,m}\Big) P_{0\,m}\\ &= \frac{1}{m} \operatorname{tr}_m H^{(m-1)} P_{0\,m} \Big(1 + \sum_{a=1}^{m-1} P_{0\,a}\Big) = \frac{1}{m} H^{(m-1)} \Big(1 + \sum_{a=1}^{m-1} P_{0\,a}\Big).\end{aligned}$$

By the cyclic property of trace (or a similar calculation), $\operatorname{tr}_m P_{0\,m} H^{(m)}$ is given by the same formula. Therefore,

$$\begin{aligned}E[1]_0 \operatorname{tr}_{1,\dots,m} H^{(m)} \mathcal{E}_1 \dots \mathcal{E}_m &= \frac{N+m-1}{m} \sum_{a=1}^{m-1} 2a \operatorname{tr}_{1,\dots,m-1} H^{(m-1)} P_{0\,a}\, \mathcal{E}_1 \dots \mathcal{E}_{m-1}\\ &+ (K-m+1) \operatorname{tr}_{1,\dots,m-1} H^{(m-1)} \Big(1 + \sum_{a=1}^{m-1} P_{0\,a}\Big) \mathcal{E}_1 \dots \mathcal{E}_{m-1}\\ &- \frac{K(N+m-1)}{N} \operatorname{tr}_{1,\dots,m-1} H^{(m-1)}\, \mathcal{E}_1 \dots \mathcal{E}_{m-1}.\end{aligned}$$

Finally, using the property (3.10) of Manin matrices and Lemma 7.1.2 we find that for any $1 \leqslant a < b \leqslant m-1$,

$$\begin{aligned}\operatorname{tr}_{1,\dots,m-1} H^{(m-1)} P_{0\,a}\, \mathcal{E}_1 \dots \mathcal{E}_{m-1} &= \operatorname{tr}_{1,\dots,m-1} P_{0\,a}\, \mathcal{E}_1 \dots \mathcal{E}_{m-1} H^{(m-1)}\\ &= \operatorname{tr}_{1,\dots,m-1} P_{0\,a} P_{a\,b}\, \mathcal{E}_1 \dots \mathcal{E}_{m-1} H^{(m-1)}\\ &= \operatorname{tr}_{1,\dots,m-1} P_{a\,b} P_{0\,b}\, \mathcal{E}_1 \dots \mathcal{E}_{m-1} H^{(m-1)}\\ &= \operatorname{tr}_{1,\dots,m-1} H^{(m-1)} P_{0\,b}\, \mathcal{E}_1 \dots \mathcal{E}_{m-1}.\end{aligned}$$

Thus,

$$\begin{aligned}&E[1]_0 \operatorname{tr}_{1,\dots,m} H^{(m)} \mathcal{E}_1 \dots \mathcal{E}_m\\ &\qquad = (K+N) \operatorname{tr}_{1,\dots,m-1} H^{(m-1)} \Big(\sum_{a=1}^{m-1} P_{0\,a} - \frac{m-1}{N}\Big) \mathcal{E}_1 \dots \mathcal{E}_{m-1}\end{aligned}$$

which is zero, since $K + N = 0$ at the critical level.

PROOF FOR THE ELEMENTS $\theta_{m\,a}$. It is sufficient to verify that for all i, j

$$E_{ij}[0] \operatorname{tr} \mathcal{E}^m = E_{ij}[1] \operatorname{tr} \mathcal{E}^m = 0 \tag{7.16}$$

in the $\widehat{\mathfrak{gl}}_N$-module $V_{-N}(\mathfrak{gl}_N) \otimes \mathbb{C}[\tau]$ with the trivial action on $\mathbb{C}[\tau]$. Consider the tensor product algebra

$$\operatorname{End} \mathbb{C}^N \otimes \operatorname{End} \mathbb{C}^N \otimes \operatorname{End} \mathbb{C}^N \otimes \mathrm{U}$$

with the copies of $\operatorname{End} \mathbb{C}^N$ labeled by $0, 1, 2$, where U stands for the universal enveloping algebra of $\widehat{\mathfrak{gl}}_N \oplus \mathbb{C}\tau$. Relations (7.16) can now be written in the matrix form as

$$E[0]_0 \operatorname{tr}_1 \mathcal{E}_1^m = 0 \qquad \text{and} \qquad E[1]_0 \operatorname{tr}_1 \mathcal{E}_1^m = 0 \tag{7.17}$$

modulo the left ideal of U generated by $\mathfrak{gl}_N[t]$ and $K+N$. Note the identity

$$[E[0]_0, \mathcal{E}_1^m] = P_{01}\mathcal{E}_1^m - \mathcal{E}_1^m P_{01}, \qquad m = 0, 1, 2, \dots, \tag{7.18}$$

which follows from (7.14):

$$[E[0]_0, \mathcal{E}_1^m] = \sum_{r=1}^{m} \mathcal{E}_1^{r-1}[E[0]_0, \mathcal{E}_1]\,\mathcal{E}_1^{m-r} = \sum_{r=1}^{m} \mathcal{E}_1^{r-1}[P_{01}, \mathcal{E}_1]\,\mathcal{E}_1^{m-r} = [P_{01}, \mathcal{E}_1^m].$$

Now the first relation in (7.17) follows by taking the trace tr_1 on both sides of (7.18) and using its cyclic property. For the proof of the second relation in (7.17), use (7.15) to write

$$[E[1]_0, \mathcal{E}_1^m] = \sum_{i=1}^{m} \mathcal{E}_1^{i-1}\Big(E[0]_0 + P_{01}E[0]_1 - E[0]_1P_{01} + KP_{01} - \frac{K}{N}\Big)\,\mathcal{E}_1^{m-i}. \tag{7.19}$$

Applying (7.18) and the relation $\mathcal{E}_1^{m-i} = \operatorname{tr}_2 \mathcal{E}_2^{m-i}P_{12}$ we can rewrite (7.19) modulo the left ideal of U generated by $\mathfrak{gl}_N[t]$ as

$$\begin{aligned}[E[1]_0, \mathcal{E}_1^m] = \sum_{i=1}^{m} \Big(\mathcal{E}_1^{i-1}[P_{01}, \mathcal{E}_1^{m-i}] + K\mathcal{E}_1^{i-1}P_{01}\mathcal{E}_1^{m-i}\Big) - \frac{mK}{N}\,\mathcal{E}_1^{m-1}\\ + \operatorname{tr}_2 \sum_{i=1}^{m} \mathcal{E}_1^{i-1}\big(P_{01}E[0]_1 - E[0]_1P_{01}\big)\,\mathcal{E}_2^{m-i}P_{12}.\end{aligned}$$

Now transform the last summand using (7.18) to get

$$\begin{aligned}&\operatorname{tr}_2 \sum_{i=1}^{m} \mathcal{E}_1^{i-1}\big(P_{01}E[0]_1 - E[0]_1P_{01}\big)\,\mathcal{E}_2^{m-i}P_{12}\\ &\qquad = \operatorname{tr}_2 \sum_{i=1}^{m} \mathcal{E}_1^{i-1}P_{01}\,[P_{12}, \mathcal{E}_2^{m-i}]\,P_{12} - \operatorname{tr}_2 \sum_{i=1}^{m} \mathcal{E}_1^{i-1}\,[P_{12}, \mathcal{E}_2^{m-i}]\,P_{01}P_{12}.\end{aligned}$$

Taking into account the relation $\operatorname{tr}_2 P_{02} = 1$, we can simplify this to

$$\sum_{i=1}^{m-1} \Big(N\,\mathcal{E}_1^{i-1}P_{01}\mathcal{E}_1^{m-i} - \mathcal{E}_1^{i-1}P_{01}\operatorname{tr}\mathcal{E}^{m-i} + \mathcal{E}_1^{i-1}\mathcal{E}_0^{m-i}\Big) - (m-1)\,\mathcal{E}_1^{m-1}.$$

Combining all the terms and taking the trace tr_1 we derive

$$\begin{aligned}[E[1]_0, \operatorname{tr}_1 \mathcal{E}_1^m] = (K+N)\Big(\mathcal{E}_0^{m-1} - \frac{m}{N}\operatorname{tr}\mathcal{E}^{m-1}\Big) + (K+N-m+2)\,\mathcal{E}_0^{m-1}\\ + (K+N+1)\operatorname{tr}_1 \sum_{i=2}^{m-1} P_{01}\,\mathcal{E}_0^{i-1}\mathcal{E}_1^{m-i} - \operatorname{tr}_1 \sum_{i=2}^{m-1} \big[\mathcal{E}_0^{i-1}, \mathcal{E}_1^{m-i}\big].\end{aligned}$$

Finally, we use Lemma 7.1.2 and Corollary 3.2.9 to write

$$\sum_{i=2}^{m-1} \big[\mathcal{E}_0^{i-1}, \mathcal{E}_1^{m-i}\big] = \sum_{i=2}^{m-1} P_{01}\,\big[\mathcal{E}_0^{i-1}, \mathcal{E}_1^{m-i}\big].$$

By calculating the trace tr_1 on both sides of this relation we obtain

$$\operatorname{tr}_1 \sum_{i=2}^{m-1} \big[\mathcal{E}_0^{i-1}, \mathcal{E}_1^{m-i}\big] = \operatorname{tr}_1 \sum_{i=2}^{m-1} P_{01}\,\mathcal{E}_0^{i-1}\mathcal{E}_1^{m-i} - (m-2)\,\mathcal{E}_0^{m-1}.$$

Taking this into account, we get

$$\big[E[1]_0, \operatorname{tr}_1 \mathcal{E}_1^m\big] = (K+N)\Big(2\,\mathcal{E}_0^{m-1} - \frac{m}{N}\operatorname{tr}\mathcal{E}^{m-1} + \operatorname{tr}_1 \sum_{i=2}^{m-1} P_{01}\,\mathcal{E}_0^{i-1}\mathcal{E}_1^{m-i}\Big).$$

This expression vanishes in the vacuum module $V_{-N}(\mathfrak{gl}_N)$ as $K+N=0$. □

One more proof of Theorem 7.1.3 will be given in Section 10.5. The Feigin–Frenkel theorem (Theorem 6.3.1) for type A is implied by the following.

THEOREM 7.1.4. *Each of the families*

$$\phi_1,\dots,\phi_N, \qquad \psi_{11},\dots,\psi_{NN} \qquad \textit{and} \qquad \theta_{11},\dots,\theta_{NN}$$

is a complete set of Segal–Sugawara vectors for $\mathfrak{gl}_N$.

PROOF. Recall that the graded space $\operatorname{gr} V_{-N}(\mathfrak{gl}_N)$ defined in (6.17) is identified with the symmetric algebra $\mathrm{S}\left(t^{-1}\mathfrak{gl}_N[t^{-1}]\right)$. The symbols of the elements of each family in the symmetric algebra coincide with the images of certain elements of the symmetric algebra $\mathrm{S}(\mathfrak{gl}_N)$ under the embedding $\mathrm{S}(\mathfrak{gl}_N) \hookrightarrow \mathrm{S}\left(t^{-1}\mathfrak{gl}_N[t^{-1}]\right)$ taking $X \in \mathfrak{gl}_N$ to $X[-1]$. Due to Theorems 6.3.3 and 7.1.3 we only need to verify that the corresponding elements of $\mathrm{S}(\mathfrak{gl}_N)$ are algebraically independent generators of the algebra of invariants $\mathrm{S}(\mathfrak{gl}_N)^{\mathfrak{gl}_N}$; see the arguments following the proof of Theorem 6.3.3. However, this was already done in Section 2.1, where these invariants were produced and their Chevalley images calculated. The respective families of invariants corresponding to the Segal–Sugawara vectors in the formulation of the theorem are given in (2.6), (2.20), (2.21) and (2.11). □

EXAMPLE 7.1.5. We have

$$\phi_1 = \sum_{i=1}^{N} E_{ii}[-1],$$

$$\phi_2 = \sum_{1\leqslant i<j\leqslant N} \big(E_{ii}[-1]\,E_{jj}[-1] - E_{ji}[-1]\,E_{ij}[-1]\big) + \sum_{i=1}^{N}(i-1)\,E_{ii}[-2].$$

EXAMPLE 7.1.6. The definition (7.9) gives

$$\theta_{11} = \operatorname{tr} E[-1],$$

$$\theta_{22} = \operatorname{tr} E[-1]^2 + \operatorname{tr} E[-2],$$

$$\theta_{33} = \operatorname{tr} E[-1]^3 + 2\operatorname{tr} E[-1]\,E[-2] + \operatorname{tr} E[-2]\,E[-1] + 2\operatorname{tr} E[-3],$$

$$\begin{aligned}\theta_{44} = \operatorname{tr} E[-1]^4 &+ 3\operatorname{tr} E[-1]^2 E[-2] + 2\operatorname{tr} E[-1]\,E[-2]\,E[-1] + \operatorname{tr} E[-2]\,E[-1]^2 \\ &+ 6\operatorname{tr} E[-1]\,E[-3] + 2\operatorname{tr} E[-3]\,E[-1] + 3\operatorname{tr} E[-2]^2 + 6\operatorname{tr} E[-4].\end{aligned}$$

These expressions can be written in a simpler form with the use of the relations

$$\operatorname{tr} E[r]\,E[s] = \operatorname{tr} E[s]\,E[r] \qquad \text{and} \qquad \operatorname{tr} E[r]\,E[s]^2 = \operatorname{tr} E[s]^2 E[r] \tag{7.20}$$

where r and s are arbitrary negative integers. Indeed, by (7.6),

$$E[r]_1\,E[s]_2 - E[s]_2\,E[r]_1 = \big(E[r+s]_1 - E[r+s]_2\big)\,P_{12}.$$

Multiplying both sides by P_{12} from the right and taking trace tr_{12} we get the first relation in (7.20). Similarly,

$$E[r]_1\, E[s]_2^2 - E[s]_2^2\, E[r]_1 = \big(E[r+s]_1 - E[r+s]_2\big)\, P_{12}\, E[s]_2 \\ + E[s]_2 \big(E[r+s]_1 - E[r+s]_2\big)\, P_{12}$$

and so, multiplying by P_{12} from the right and using the first relation we find

$$\operatorname{tr} E[r]\, E[s]^2 - \operatorname{tr} E[s]^2 E[r] = N \operatorname{tr} E[r+s]\, E[s] - \operatorname{tr} E[r+s] \operatorname{tr} E[s] \\ + \operatorname{tr} E[s] \operatorname{tr} E[r+s] - N \operatorname{tr} E[s]\, E[r+s] = 0$$

thus implying the second relation in (7.20). Recall that the Feigin–Frenkel center $\mathfrak{z}(\widehat{\mathfrak{gl}}_N)$ is T-invariant; see (6.7). For $k \geqslant 1$ we have

$$T^k\, \theta_{1\,1} = k! \operatorname{tr} E[-k-1].$$

Furthermore, by (7.20)

$$T \operatorname{tr} E[-1]^2 = 2 \operatorname{tr} E[-1] E[-2],$$
$$T^2 \operatorname{tr} E[-1]^2 = 4 \operatorname{tr} E[-1] E[-3] + 2 \operatorname{tr} E[-2]^2,$$

and

$$T \operatorname{tr} E[-1]^3 = 2 \operatorname{tr} E[-1]^2 E[-2] + \operatorname{tr} E[-1] E[-2] E[-1].$$

Taking the above relations into account, we can conclude that all elements

$$\operatorname{tr} E[-1], \qquad \operatorname{tr} E[-1]^2, \qquad \operatorname{tr} E[-1]^3, \qquad \operatorname{tr} E[-1]^4 - \operatorname{tr} E[-2]^2 \tag{7.21}$$

are Segal–Sugawara vectors for $\mathfrak{gl}_N$. In particular, the following are complete sets of Segal–Sugawara vectors:

for $\mathfrak{gl}_2$: $\operatorname{tr} E[-1], \quad \operatorname{tr} E[-1]^2$

for $\mathfrak{gl}_3$: $\operatorname{tr} E[-1], \quad \operatorname{tr} E[-1]^2, \quad \operatorname{tr} E[-1]^3$

for $\mathfrak{gl}_4$: $\operatorname{tr} E[-1], \quad \operatorname{tr} E[-1]^2, \quad \operatorname{tr} E[-1]^3, \quad \operatorname{tr} E[-1]^4 - \operatorname{tr} E[-2]^2.$ □

7.2. Sugawara operators in type A

Introduce the Laurent series in z with coefficients in $\mathrm{U}_{-N}(\widehat{\mathfrak{gl}}_N)$ by

$$E_{ij}(z) = \sum_{r \in \mathbb{Z}} E_{ij}[r]\, z^{-r-1}, \qquad i,j = 1, \dots, N. \tag{7.22}$$

We adopt the matrix notation as in Section 7.1 and combine the series $E_{ij}(z)$ into the matrix $E(z)$ so that

$$E(z) = \sum_{i,j=1}^{N} e_{ij} \otimes E_{ij}(z).$$

For $a \in \{1, \dots, m\}$ we extend the notation (7.5) to the elements $E(z)_a$ of the algebra (7.4), where U now stands for the algebra of polynomial differential operators over z with coefficients in $\widetilde{\mathrm{U}}_{-N}(\widehat{\mathfrak{gl}}_N)[[z, z^{-1}]]$.

As explained in Section 6.6, the application of the state-field correspondence map (6.35) to the coefficients of the polynomials in (7.7)–(7.9), leads to the definition of the Laurent series $\phi_{m\,a}(z), \psi_{m\,a}(z), \theta_{m\,a}(z)$ given by the following expansions with the usual reading of normal ordering from right to left:

$$
\begin{aligned}
(7.23) \qquad :\mathrm{tr}_{1,\dots,m}\, A^{(m)}\big(\partial_z + E(z)_1\big)\dots\big(\partial_z + E(z)_m\big): \\
= \phi_{m\,0}(z)\,\partial_z^m + \dots + \phi_{m\,m}(z),
\end{aligned}
$$

$$
\begin{aligned}
(7.24) \qquad :\mathrm{tr}_{1,\dots,m}\, H^{(m)}\big(\partial_z + E(z)_1\big)\dots\big(\partial_z + E(z)_m\big): \\
= \psi_{m\,0}(z)\,\partial_z^m + \dots + \psi_{m\,m}(z),
\end{aligned}
$$

and

$$
:\mathrm{tr}\,\big(\partial_z + E(z)\big)^m: \; = \theta_{m\,0}(z)\,\partial_z^m + \dots + \theta_{m\,m}(z). \tag{7.25}
$$

Moreover, following (7.11) define the series $\phi_a(z)$ by

$$
:\mathrm{cdet}\,\big(\partial_z + E(z)\big): \; = \partial_z^N + \phi_1(z)\,\partial_z^{N-1} + \dots + \phi_N(z). \tag{7.26}
$$

By the relations between the Segal–Sugawara vectors pointed out in Section 7.1, we have

$$
\phi_{N\,a}(z) = \phi_a(z), \qquad a = 1, \dots, N
$$

and

$$
\phi_{m\,a}(z) = \binom{N-a}{m-a}\,\phi_a(z), \qquad 0 \leqslant a \leqslant m \leqslant N;
$$

see (7.12). In particular, $\phi_{m\,m}(z) = \phi_m(z)$ for $m = 1, \dots, N$. The next theorem is implied by Proposition 6.6.1 together with Theorems 7.1.3 and 7.1.4.

THEOREM 7.2.1. *All coefficients of the Laurent series $\phi_{m\,a}(z)$, $\psi_{m\,a}(z)$ and $\theta_{m\,a}(z)$ belong to the center of the algebra $\widetilde{\mathrm{U}}_{-N}(\widehat{\mathfrak{gl}}_N)$. Moreover, the coefficients of each family of the Laurent series*

$$
\phi_1(z), \dots, \phi_N(z), \qquad \psi_{1\,1}(z), \dots, \psi_{N\,N}(z) \qquad \text{and} \qquad \theta_{1\,1}(z), \dots, \theta_{N\,N}(z)
$$

are topological generators of the center of $\widetilde{\mathrm{U}}_{-N}(\widehat{\mathfrak{gl}}_N)$. □

EXAMPLE 7.2.2. Consider the first few Segal–Sugawara vectors for $\mathfrak{gl}_N$ given in (7.21). The corresponding Laurent series will have the form

$$
\mathrm{tr}\,E(z), \qquad :\mathrm{tr}\,E(z)^2:, \qquad :\mathrm{tr}\,E(z)^3:, \qquad :\mathrm{tr}\,E(z)^4: - :\mathrm{tr}\,\big(\partial_z E(z)\big)^2:.
$$

We have

$$
\mathrm{tr}\,E(z) = \sum_{r\in\mathbb{Z}} \sum_{i=1}^{N} E_{ii}[r]\, z^{-r-1}
$$

so that all elements $E_{11}[r] + \dots + E_{NN}[r]$ are central in $\widetilde{\mathrm{U}}_{-N}(\widehat{\mathfrak{gl}}_N)$ (in fact, these elements are central in $\widetilde{\mathrm{U}}_\kappa(\widehat{\mathfrak{gl}}_N)$ for any level κ). Furthermore,

$$
:\mathrm{tr}\,E(z)^2: \; = \; :\sum_{i,j=1}^{N} E_{ij}(z)E_{ji}(z): = \sum_{i,j=1}^{N} \Big(E_{ij}(z)_+ E_{ji}(z) + E_{ji}(z)E_{ij}(z)_-\Big).
$$

Hence for each $p \in \mathbb{Z}$ the coefficient of z^{-p-2} in this series is a central element of $\widetilde{\mathrm{U}}_{-N}(\widehat{\mathfrak{gl}}_N)$ given by

$$\sum_{i,j=1}^{N} \Big(\sum_{r<0} E_{ij}[r] E_{ji}[p-r] + \sum_{r \geqslant 0} E_{ji}[p-r] E_{ij}[r] \Big).$$

Similar explicit formulas for central elements of $\widetilde{\mathrm{U}}_{-N}(\widehat{\mathfrak{gl}}_N)$ of higher order can be obtained by taking the coefficients of all powers of z in the expanded normally ordered traces; for instance,

$$\begin{aligned} : \operatorname{tr} E(z)^3 : \; = \; : \sum_{i,j,k=1}^{N} E_{ij}(z) E_{jk}(z) E_{ki}(z) : \\ = \sum_{i,j,k=1}^{N} \Big(E_{ij}(z)_+ E_{jk}(z)_+ E_{ki}(z) + E_{ij}(z)_+ E_{ki}(z) E_{jk}(z)_- \\ + E_{jk}(z)_+ E_{ki}(z) E_{ij}(z)_- + E_{ki}(z) E_{jk}(z)_- E_{ij}(z)_- \Big). \end{aligned}$$

We conclude this section by pointing out how the Casimir elements for $\mathfrak{gl}_N$ constructed in Chapter 4 can be reproduced from elements of the Feigin–Frenkel center $\mathfrak{z}(\widehat{\mathfrak{gl}}_N)$ (see Section 6.5) or from the above Sugawara operators.

Consider the Segal–Sugawara vectors for $\mathfrak{gl}_N$ provided by the expansion (7.7); see Theorem 7.1.3. The images of the coefficients of the polynomial (7.7) under the homomorphism ϱ defined in (6.30) can be written as

$$\operatorname{tr}_{1,\dots,m} A^{(m)} \big(-\partial_z + E_1 z^{-1} \big) \dots \big(-\partial_z + E_m z^{-1} \big), \tag{7.27}$$

where the image of τ is understood as the differential operator $-\partial_z$. Using the relation $\partial_z z = z \partial_z + 1$ and setting $u = -\partial_z z$ we can write

$$\begin{aligned} \operatorname{tr}_{1,\dots,m} A^{(m)} \big(-\partial_z + E_1 z^{-1} \big) \dots \big(-\partial_z + E_m z^{-1} \big) z^m \\ = \operatorname{tr}_{1,\dots,m} A^{(m)} \big(u + E_1 \big) \dots \big(u + E_m - m + 1 \big) \end{aligned}$$

thus reproducing the Casimir elements for $\mathfrak{gl}_N$ provided by (4.35). In particular, taking the image of the column-determinant (7.11) under ϱ we recover the Capelli determinant (4.37):

$$\operatorname{cdet} \big(-\partial_z + E z^{-1} \big) z^N = C(u).$$

The permanent-type Casimir elements (4.39) and the Gelfand invariants (4.41) are recovered from the respective Segal–Sugawara vectors given in (7.8) and (7.9).

By an equivalent approach, take the images of the Sugawara operators given above, in the quotient of the completed universal enveloping algebra by the left ideal generated by $\mathfrak{gl}_N[t]$. These images can be regarded as elements of the Feigin–Frenkel center $\mathfrak{z}(\widehat{\mathfrak{gl}}_N)$. For instance, for the left hand side of (7.23) we get

$$\operatorname{tr}_{1,\dots,m} A^{(m)} \big(\partial_z + E(z)_{+1} \big) \dots \big(\partial_z + E(z)_{+m} \big), \tag{7.28}$$

where we set

$$E(z)_+ = \sum_{i,j=1}^{N} e_{ij} \otimes E_{ij}(z)_+, \qquad E_{ij}(z)_+ = \sum_{r<0} E_{ij}[r] \, z^{-r-1}.$$

All coefficients of the power series (7.28) are elements of $\mathfrak{z}(\widehat{\mathfrak{gl}}_N)$. For a nonzero $a \in \mathbb{C}$ apply the evaluation homomorphism ϱ_a defined in (6.32) to these coefficients. We have

$$\varrho_a : E_{ij}(z)_+ \mapsto -E_{ij}(z-a)^{-1}.$$

Now shift the variable by $z \mapsto z+a$ so that the image of (7.28) under ϱ_a will coincide with (7.27), up to a sign. Similarly, by using the Sugawara operators (7.24)–(7.26) we recover the respective Casimir elements considered in Chapter 4.

7.3. Bibliographical notes

Explicit formulas for generators of the Feigin–Frenkel center in type A were first given in the preprint by Chervov and Talalaev [**24**] in a form close to (7.26). That work was inspired by Talalaev's construction of explicit higher Gaudin Hamiltonians [**145**]. A direct proof of Theorem 7.1.4 for the elements ϕ_a was given by Chervov and the author [**23**]. In the proof of Theorem 7.1.3 we followed the matrix approach which was used in [**110**] and in the paper by Ragoucy and the author [**114**]. The latter work is concerned with the Lie superalgebras $\mathfrak{gl}_{m|n}$ and provides a super-analogue of Theorem 7.1.3. A geometric description of the Segal–Sugawara vectors in type A was given in the work by Kamgarpour [**90**] in the context of the local geometric Langlands correspondence.

CHAPTER 8

Generators in types B, C and D

We will now construct some analogues of the Segal–Sugawara vectors of Chapter 7 for the orthogonal and symplectic Lie algebras. As for type A, we thus get a direct proof of the Feigin–Frenkel theorem (Theorem 6.3.1). Then we give explicit formulas for generators of the center of the completed universal enveloping algebra at the critical level as provided by Proposition 6.6.1.

8.1. Segal–Sugawara vectors in types B and D

We will keep the notation for the generators of the orthogonal Lie algebra $\mathfrak{o}_N$ introduced in Chapter 5, where $N = 2n$ or $N = 2n+1$. That is, $\mathfrak{o}_N$ is the subalgebra of $\mathfrak{gl}_N$ spanned by the elements F_{ij} defined in (5.1).

Now consider the affine Kac–Moody algebra $\widehat{\mathfrak{o}}_N = \mathfrak{o}_N[t, t^{-1}] \oplus \mathbb{C}K$ as defined in (6.5) and set $F_{ij}[r] = F_{ij}t^r$ for any $r \in \mathbb{Z}$. The dual Coxeter number for the Lie algebra $\mathfrak{o}_N$ is given by

$$h^\vee = N - 2 \tag{8.1}$$

so that the normalized Killing form (6.4) can be written as

$$\langle X, Y\rangle = \frac{1}{2}\,\mathrm{tr}\, XY, \qquad X, Y \in \mathfrak{o}_N.$$

REMARK 8.1.1. Formula (8.1) provides correct values of the dual Coxeter number for $\mathfrak{o}_N$ only for $N \geqslant 5$; cf. [**86**, Exercise 7.10]. Indeed, the Lie algebras $\mathfrak{o}_3$ and $\mathfrak{sl}_2$ are isomorphic, so that for $N = 3$ we should have $h^\vee = 2$. Moreover, the Lie algebra $\mathfrak{o}_4$ is isomorphic to the direct sum $\mathfrak{sl}_2 \oplus \mathfrak{sl}_2$ and it is not simple. On the other hand, the invariants of the vacuum modules constructed below will still make sense for all values $N \geqslant 3$, where the values of $h^\vee$ should be redefined as in (8.1) for the cases $N = 3$ and $N = 4$. □

By (5.2), the affine Kac–Moody algebra $\widehat{\mathfrak{o}}_N = \mathfrak{o}_N[t, t^{-1}] \oplus \mathbb{C}K$ has the commutation relations

$$\begin{aligned}\big[F_{ij}[r], F_{kl}[s]\big] &= \delta_{kj}\, F_{il}[r+s] - \delta_{il}\, F_{kj}[r+s] \\ &\quad - \delta_{k i'}\, F_{j'l}[r+s] + \delta_{j'l}\, F_{k i'}[r+s] + r\,\delta_{r,-s}\, K\big(\delta_{kj}\,\delta_{il} - \delta_{k i'}\,\delta_{j'l}\big),\end{aligned} \tag{8.2}$$

where the element K is central and the symmetry property $F_{ij}[r] + F_{j'i'}[r] = 0$ holds. For any $r \in \mathbb{Z}$ we regard the $N \times N$ matrix $F[r] = \big[F_{ij}[r]\big]$ as the element

$$F[r] = \sum_{i,j=1}^{N} e_{ij} \otimes F_{ij}[r] \in \mathrm{End}\,\mathbb{C}^N \otimes \mathrm{U}(\widehat{\mathfrak{o}}_N).$$

It has the symmetry property $F[r]+F[r]'=0$; see (2.24). For each $a\in\{1,\dots,m\}$ introduce the element $F[r]_a$ of the algebra

$$\underbrace{\mathrm{End}\,\mathbb{C}^N\otimes\ldots\otimes\mathrm{End}\,\mathbb{C}^N}_{m}\otimes\mathrm{U}(\widehat{\mathfrak{o}}_N) \tag{8.3}$$

by

$$F[r]_a=\sum_{i,j=1}^{N}1^{\otimes(a-1)}\otimes e_{ij}\otimes 1^{\otimes(m-a)}\otimes F_{ij}[r]. \tag{8.4}$$

By analogy with (5.6), it is immediate from (8.2) that the defining relations of the algebra $\mathrm{U}(\widehat{\mathfrak{o}}_N)$ can be written in the form

$$\begin{aligned}&F[r]_1\,F[s]_2-F[s]_2\,F[r]_1\\&\qquad=(P-Q)\,F[r+s]_2-F[r+s]_2\,(P-Q)+r\,\delta_{r,-s}\,(P-Q)\,K,\end{aligned} \tag{8.5}$$

where both sides are elements of the algebra (8.3) with $m=2$; cf. Proposition 7.1.1.

In the case where $N=2n$ is even, by analogy with (5.36) introduce the (noncommutative) *Pfaffian* $\mathrm{Pf}\,F[-1]$ as an element of the universal enveloping algebra $\mathrm{U}\left(t^{-1}\mathfrak{o}_N[t^{-1}]\right)$ by the formula

$$\mathrm{Pf}\,F[-1]=\frac{1}{2^n n!}\sum_{\sigma\in\mathfrak{S}_{2n}}\mathrm{sgn}\,\sigma\cdot F_{\sigma(1)\,\sigma(2)'}[-1]\ldots F_{\sigma(2n-1)\,\sigma(2n)'}[-1]. \tag{8.6}$$

EXAMPLE 8.1.2. For $n=2$ we have

$$\begin{aligned}\mathrm{Pf}\,F[-1]=\frac{1}{2}\Big(&F_{13}[-1]\,F_{31}[-1]-F_{12}[-1]\,F_{21}[-1]+F_{11}[-1]\,F_{22}[-1]\\&+F_{31}[-1]\,F_{13}[-1]-F_{21}[-1]\,F_{12}[-1]+F_{22}[-1]\,F_{11}[-1]\Big)\end{aligned}$$

which simplifies to

$$\begin{aligned}\mathrm{Pf}\,F[-1]&=F_{11}[-1]\,F_{22}[-1]-F_{21}[-1]\,F_{12}[-1]+F_{31}[-1]\,F_{13}[-1]+F_{22}[-2]\\&=F_{11}[-1]\,F_{22}[-1]-F_{12}[-1]\,F_{21}[-1]+F_{13}[-1]\,F_{31}[-1]-F_{22}[-2].\end{aligned}$$

□

The formula (8.6) for the Pfaffian is associated with the presentation of the orthogonal Lie algebra $\mathfrak{o}_{2n}$ corresponding to the symmetric bilinear form (2.25) with the matrix $G=[\delta_{ij'}]$; see (2.26). It will be convenient to have an equivalent expression for the Pfaffian in the *canonical* presentation of $\mathfrak{o}_{2n}$ by skew-symmetric matrices. To this end, fix a square matrix A of size $2n$ such that $AA^t=G$. Introduce new generators F°_{ij} of $\mathfrak{o}_{2n}$ by the formula

$$F^\circ=A^{-1}FA,\qquad F^\circ=[F^\circ_{ij}], \tag{8.7}$$

where $F=[F_{ij}]$ is the generator matrix for $\mathfrak{o}_{2n}$ which is also regarded as the element (5.5). Similar to Proposition 5.1.1, we have the following defining relations for $\mathrm{U}(\mathfrak{o}_{2n})$ in terms of the matrix F°:

$$F^\circ_1\,F^\circ_2-F^\circ_2\,F^\circ_1=(P-P^t)\,F^\circ_2-F^\circ_2\,(P-P^t) \tag{8.8}$$

together with the relation $F^\circ+F^{\circ\, t}=0$, where

$$P^t:=P^{t_1}=P^{t_2}=\sum_{i,j=1}^{N}e_{ij}\otimes e_{ij}\in\mathrm{End}\,\mathbb{C}^N\otimes\mathrm{End}\,\mathbb{C}^N.$$

Indeed, observe first that the mapping $E \mapsto A^{-1}EA$ defines an automorphism of $\mathrm{U}(\mathfrak{gl}_{2n})$ so that (4.10) is satisfied by the matrix $E^\circ = A^{-1}EA$. This is easily verified with the use of (1.67). On the other hand, since

$$F = E - E' = E - GE^tG^{-1},$$

the definition (8.7) is equivalent to $F^\circ = E^\circ - E^{\circ\, t}$. Hence F° is skew-symmetric with respect to the standard transposition t, and relation (8.8) follows by the same argument as in the proof of Proposition 5.1.1 where t is used instead of the transposition (2.24).

Accordingly, the defining relations for the universal enveloping algebra $\mathrm{U}(\widehat{\mathfrak{o}}_{2n})$ take the form

$$\begin{aligned}(8.9)\quad & F^\circ[r]_1\, F^\circ[s]_2 - F^\circ[s]_2\, F^\circ[r]_1\\ &= (P - P^t)\, F^\circ[r+s]_2 - F^\circ[r+s]_2\,(P - P^t) + r\,\delta_{r,-s}\,(P - P^t)\,K,\end{aligned}$$

together with the symmetry property $F^\circ[r] + F^\circ[r]^t = 0$, where for all $r \in \mathbb{Z}$ the generator matrix is given by

$$F^\circ[r] = A^{-1}F[r]A, \qquad F^\circ[r] = \big[F^\circ_{ij}[r]\big].$$

It terms of generators, (8.9) is written as

$$\begin{aligned}(8.10)\quad \big[F^\circ_{ij}[r], F^\circ_{kl}[s]\big] &= \delta_{kj}\, F^\circ_{il}[r+s] - \delta_{il}\, F^\circ_{kj}[r+s]\\ &\quad - \delta_{ki}\, F^\circ_{jl}[r+s] + \delta_{jl}\, F^\circ_{ki}[r+s] + r\,\delta_{r,-s}\, K\big(\delta_{kj}\,\delta_{il} - \delta_{ki}\,\delta_{jl}\big).\end{aligned}$$

In particular, the elements $F^\circ_{ij}[r]$ and $F^\circ_{kl}[s]$ commute if the indices i, j, k, l are distinct. Therefore, the Pfaffian of the matrix $F^\circ[-1]$ can be defined by the usual formula

$$(8.11)\qquad \mathrm{Pf}\ F^\circ[-1] = \sum_\sigma \operatorname{sgn}\sigma \cdot F^\circ_{\sigma(1)\,\sigma(2)}[-1] \dots F^\circ_{\sigma(2n-1)\,\sigma(2n)}[-1],$$

summed over the elements σ of the subset $\mathcal{A}_{2n} \subset \mathfrak{S}_{2n}$ which consists of the permutations with the properties $\sigma(2k-1) < \sigma(2k)$ for all $k = 1, \dots, n$ and $\sigma(1) < \sigma(3) < \dots < \sigma(2n-1)$.

LEMMA 8.1.3. *We have the relation*

$$\mathrm{Pf}\ F[-1] = \det A \cdot \mathrm{Pf}\ F^\circ[-1].$$

PROOF. The Pfaffian $\mathrm{Pf}\ F[-1]$ is found by the expansion

$$(8.12)\qquad \frac{\Psi^n}{n!} = \big(e_1 \wedge \dots \wedge e_{2n}\big) \otimes \mathrm{Pf}\ F[-1],$$

with

$$\Psi = \sum_{i<j} (e_i \wedge e_j) \otimes F_{ij'}[-1] \in \Lambda(\mathbb{C}^{2n}) \otimes \mathrm{U}(\widehat{\mathfrak{o}}_{2n}),$$

where $\Lambda(\mathbb{C}^{2n})$ is the exterior algebra of the vector space $\mathbb{C}^{2n}$. Note the matrix relation $\big[F_{ij'}[-1]\big] = F[-1]G = AF^\circ[-1]A^t$. Therefore, introducing the entries of the matrix $A = [a_{ij}]$, we get

$$\Psi = \frac{1}{2}\sum_{i,j=1}^{2n} (e_i \wedge e_j) \otimes \sum_{k,l=1}^{2n} a_{ik}\, a_{jl}\, F^\circ_{kl}[-1] = \frac{1}{2}\sum_{k,l=1}^{2n} (e^\circ_k \wedge e^\circ_l) \otimes F^\circ_{kl}[-1],$$

where we set

$$e^\circ_k = \sum_{i=1}^{2n} a_{ik}\, e_i.$$

So we may conclude that

$$\frac{\Psi^n}{n!} = \big(e^\circ_1 \wedge \dots \wedge e^\circ_{2n}\big) \otimes \mathrm{Pf}\ F^\circ[-1] = \big(e_1 \wedge \dots \wedge e_{2n}\big) \otimes \det A \cdot \mathrm{Pf}\ F^\circ[-1].$$

It remains to compare this with (8.12). □

PROPOSITION 8.1.4. *The Pfaffian* Pf $F[-1]$ *belongs to the Feigin–Frenkel center* $\mathfrak{z}(\widehat{\mathfrak{o}}_{2n})$.

PROOF. In view of Lemma 8.1.3, it is enough to prove that

$$F^\circ_{ij}[0]\ \mathrm{Pf}\ F^\circ[-1] = 0 \qquad \text{and} \qquad F^\circ_{ij}[1]\ \mathrm{Pf}\ F^\circ[-1] = 0 \tag{8.13}$$

for all $i, j \in \{1, \dots, 2n\}$ in the vacuum module $V_{-h^\vee}(\mathfrak{o}_{2n})$. Note that for any permutation $\pi \in \mathfrak{S}_{2n}$ the mapping

$$F^\circ_{ij}[r] \mapsto F^\circ_{\pi(i)\,\pi(j)}[r], \qquad K \mapsto K$$

defines an automorphism of the Lie algebra $\widehat{\mathfrak{o}}_{2n}$. Moreover, the image of Pf $F^\circ[-1]$ under its extension to $\mathrm{U}(\widehat{\mathfrak{o}}_{2n})$ coincides with $\operatorname{sgn}\pi \cdot \mathrm{Pf}\ F^\circ[-1]$. Hence, it will be sufficient to verify (8.13) for $i = 1$ and $j = 2$. Observe that $F^\circ_{12}[0]$ commutes with the summands in (8.11) corresponding to permutations σ such that $\sigma(1) = 1$ and $\sigma(2) = 2$. Suppose now that $\sigma \in \mathcal{A}_{2n}$ and $\sigma(2) > 2$. Then $\sigma(3) = 2$ and $\sigma(4) > 2$. In $V_{-h^\vee}(\mathfrak{o}_{2n})$ we have

$$\begin{aligned} F^\circ_{12}[0]\, F^\circ_{1\,\sigma(2)}[-1]\, F^\circ_{2\,\sigma(4)}[-1] \dots F^\circ_{\sigma(2n-1)\,\sigma(2n)}[-1] \\ = -\, F^\circ_{2\,\sigma(2)}[-1]\, F^\circ_{2\,\sigma(4)}[-1] \dots F^\circ_{\sigma(2n-1)\,\sigma(2n)}[-1] \\ + F^\circ_{1\,\sigma(2)}[-1]\, F^\circ_{1\,\sigma(4)}[-1] \dots F^\circ_{\sigma(2n-1)\,\sigma(2n)}[-1]. \end{aligned}$$

Set $i = \sigma(2)$ and $j = \sigma(4)$. Note that the permutation $\sigma' = \sigma\,(2\,4)$ also belongs to the subset $\mathcal{A}_{2n}$, and $\operatorname{sgn}\sigma' = -\operatorname{sgn}\sigma$. Applying (8.10) we get

$$-F^\circ_{2\,i}[-1]\, F^\circ_{2\,j}[-1] + F^\circ_{1\,i}[-1]\, F^\circ_{1\,j}[-1] + F^\circ_{2\,j}[-1]\, F^\circ_{2\,i}[-1] - F^\circ_{1\,j}[-1]\, F^\circ_{1\,i}[-1] = 0.$$

This implies that the terms in the expansion of $F^\circ_{12}[0]$ Pf $F^\circ[-1]$ corresponding to the pairs of the form (σ, σ') cancel pairwise, thus proving the first relation in (8.13). Now we verify that

$$F^\circ_{12}[1]\ \mathrm{Pf}\ F^\circ[-1] = 0. \tag{8.14}$$

Consider first the summands in (8.11) with $\sigma(1) = 1$ and $\sigma(2) = 2$. In $V_{-h^\vee}(\mathfrak{o}_{2n})$ we have

$$\begin{aligned} F^\circ_{12}[1]\, F^\circ_{1\,2}[-1]\, F^\circ_{\sigma(3)\,\sigma(4)}[-1] \dots F^\circ_{\sigma(2n-1)\,\sigma(2n)}[-1] \\ = -K\, F^\circ_{\sigma(3)\,\sigma(4)}[-1] \dots F^\circ_{\sigma(2n-1)\,\sigma(2n)}[-1]. \end{aligned}$$

Furthermore, let $\tau \in \mathcal{A}_{2n}$ with $\tau(2) > 2$. Then $\tau(3) = 2$ and $\tau(4) > 2$. We have

$$\begin{aligned} F^\circ_{12}[1]\, F^\circ_{1\,\tau(2)}[-1]\, F^\circ_{2\,\tau(4)}[-1] \dots F^\circ_{\tau(2n-1)\,\tau(2n)}[-1] \\ = -F^\circ_{2\,\tau(2)}[0]\, F^\circ_{2\,\tau(4)}[-1] \dots F^\circ_{\tau(2n-1)\,\tau(2n)}[-1] \\ = F^\circ_{\tau(2)\,\tau(4)}[-1] \dots F^\circ_{\tau(2n-1)\,\tau(2n)}[-1]. \end{aligned}$$

Suppose now that $\sigma \in \mathcal{A}_{2n}$ with $\sigma(1) = 1$ and $\sigma(2) = 2$ is fixed and calculate the coefficient of the monomial

$$F^{\circ}_{\sigma(3)\,\sigma(4)}[-1] \dots F^{\circ}_{\sigma(2n-1)\,\sigma(2n)}[-1]$$

in the expansion of $F^{\circ}_{12}[1]$ Pf $F^{\circ}[-1]$. A contribution to this coefficient can come from the terms in Pf $F^{\circ}[-1]$ corresponding to permutations $\tau \in \mathcal{A}_{2n}$ of the form

$$\tau = \big[1, \sigma(2k-1), 2, \sigma(2k), \sigma(3), \sigma(4), \dots, \widehat{\sigma}(2k-1), \widehat{\sigma}(2k), \dots, \sigma(2n-1), \sigma(2n)\big]$$

and

$$\tau = \big[1, \sigma(2k), 2, \sigma(2k-1), \sigma(3), \sigma(4), \dots, \widehat{\sigma}(2k-1), \widehat{\sigma}(2k), \dots, \sigma(2n-1), \sigma(2n)\big]$$

for $k = 2, 3, \dots, n$, where the hats indicate the numbers to be skipped. These permutations can be represented as the compositions $\tau = \sigma \circ \rho$ for the respective odd and even permutations ρ written in the cycle notation as

$$\rho = (2k, 2k-2, \dots, 4)(2k-1, 2k-3, \dots, 3, 2)$$

and

$$\rho = (2k, 2k-2, \dots, 4, 2k-1, 2k-3, \dots, 3, 2).$$

Therefore, $\operatorname{sgn}\tau = -\operatorname{sgn}\sigma$ and $\operatorname{sgn}\tau = \operatorname{sgn}\sigma$ in the first and second case, respectively. Hence, taking into account the skew-symmetry property of the matrix $F^{\circ}[-1]$ we find that

$$F^{\circ}_{12}[1] \text{ Pf } F^{\circ}[-1] = (-K - 2n + 2) \sum_{\sigma} \operatorname{sgn}\sigma \cdot F^{\circ}_{\sigma(3)\,\sigma(4)}[-1] \dots F^{\circ}_{\sigma(2n-1)\,\sigma(2n)}[-1],$$

summed over $\sigma \in \mathcal{A}_{2n}$ with $\sigma(1) = 1$ and $\sigma(2) = 2$. Since $K + 2n - 2 = 0$ at the critical level, we get (8.14). □

We keep the notation $S^{(m)}$ for the image of the symmetrizer $s^{(m)} \in \mathcal{B}_m(\omega)$ as defined in Section 1.2, under the action of the Brauer algebra $\mathcal{B}_m(N)$ in the tensor space; see (1.69). We also regard $S^{(m)}$ as an element of the algebra (8.3), identifying it with $S^{(m)} \otimes 1$. We will use the constant $\gamma_m(N)$ defined in (2.40) so that

$$\gamma_m(N) = \frac{N + m - 2}{N + 2m - 2}.$$

The critical level of the vacuum module $V_{\kappa}(\mathfrak{o}_N)$ corresponds to $\kappa = -h^{\vee}$; see Remark 8.1.1. We will work with the extended Lie algebra $\widehat{\mathfrak{o}}_N \oplus \mathbb{C}\tau$ where the commutation relations for τ are defined in (7.1). Define the elements $\phi_{m\,a}$ of the vacuum module $V_{-h^{\vee}}(\mathfrak{o}_N) \cong \mathrm{U}\big(t^{-1}\mathfrak{o}_N[t^{-1}]\big)$ by the expansion

$$\gamma_m(N) \operatorname{tr}_{1,\dots,m} S^{(m)} \mathcal{F}_1 \dots \mathcal{F}_m = \phi_{m\,0}\, \tau^m + \phi_{m\,1}\, \tau^{m-1} + \dots + \phi_{m\,m}, \tag{8.15}$$

where $\mathcal{F} = \tau + F[-1] = \big[\delta_{ij}\tau + F_{ij}[-1]\big]$ is the $N \times N$ matrix with entries in the universal enveloping algebra of $\widehat{\mathfrak{o}}_N \oplus \mathbb{C}\tau$.

Recall the Manin matrices of types B and D (of odd and even sizes, respectively) as introduced in Definition 5.6.1.

LEMMA 8.1.5. *The matrix $\mathcal{F}$ is a Manin matrix of type B or D, respectively. Moreover, for any $1 \leqslant a < b \leqslant m$ and any $r, s < 0$ we have the following relations in the algebra* (8.3):

$$S^{(m)}\big(F[r]_a\, F[s]_b - F[s]_b\, F[r]_a\big) = -S^{(m)}\big(F[r+s]_a - F[r+s]_b\big) \tag{8.16}$$

and

$$\big(F[r]_a\, F[s]_b - F[s]_b\, F[r]_a\big) S^{(m)} = \big(F[r+s]_a - F[r+s]_b\big) S^{(m)}. \tag{8.17}$$

PROOF. We start by verifying (8.16). Due to (8.5), we can write

$$F[r]_a\, F[s]_b - F[s]_b\, F[r]_a = (P_{ab} - Q_{ab})\, F[r+s]_b - F[r+s]_b\, (P_{ab} - Q_{ab}).$$

Furthermore, $S^{(m)}(P_{ab} - Q_{ab}) = S^{(m)}$ by (1.31), while

$$S^{(m)} F[r+s]_b\, P_{ab} = S^{(m)} P_{ab}\, F[r+s]_a = S^{(m)} F[r+s]_a.$$

Now (8.16) is implied by the first of the following two general identities which hold for all $r \in \mathbb{Z}$:

$$S^{(m)} F[r]_a\, Q_{a\,b} = 0 \qquad \text{and} \qquad Q_{a\,b} F[r]_a\, S^{(m)} = 0. \tag{8.18}$$

To verify (8.18), for the first relation write

$$\begin{aligned} S^{(m)} F[r]_a\, Q_{a\,b} = S^{(m)} F[r]_a\, P_{a\,b} Q_{a\,b} &= S^{(m)} P_{a\,b}\, F[r]_b\, Q_{a\,b} \\ &= S^{(m)} F[r]_b\, Q_{a\,b} = -S^{(m)} F[r]_a\, Q_{a\,b}, \end{aligned}$$

where we used (1.31) again, together with the first of the identities

$$F[r]_a\, Q_{a\,b} + F[r]_b\, Q_{a\,b} = 0 \quad \text{and} \quad Q_{a\,b} F[r]_a + Q_{a\,b} F[r]_b = 0.$$

These are immediate from (2.28) and the skew-symmetry of the matrix $F[r]$. The proof of (8.17) is quite similar. Finally, we have

$$\mathcal{F}_1 \mathcal{F}_2 - \mathcal{F}_2 \mathcal{F}_1 = F[-1]_1\, F[-1]_2 - F[-1]_2\, F[-1]_1 - F[-2]_1 + F[-2]_2.$$

Hence, applying (8.17) with $m = 2$ we get

$$\big(\mathcal{F}_1 \mathcal{F}_2 - \mathcal{F}_2 \mathcal{F}_1\big)\, S^{(2)} = 0$$

so that $\mathcal{F}$ satisfies (5.50), as required. □

THEOREM 8.1.6. *All elements $\phi_{m\,a}$ belong to the Feigin–Frenkel center $\mathfrak{z}(\widehat{\mathfrak{o}}_N)$.*

PROOF. It will be sufficient to verify that for all i, j

$$F_{ij}[0]\, \mathrm{tr}_{1,\dots,m}\, S^{(m)} \mathcal{F}_1 \dots \mathcal{F}_m = F_{ij}[1]\, \mathrm{tr}_{1,\dots,m}\, S^{(m)} \mathcal{F}_1 \dots \mathcal{F}_m = 0 \tag{8.19}$$

in the $\widehat{\mathfrak{o}}_N$-module $V_{-h^\vee}(\mathfrak{o}_N) \otimes \mathbb{C}[\tau]$ with the trivial action on $\mathbb{C}[\tau]$. Consider the tensor product algebra

$$\underbrace{\mathrm{End}\,\mathbb{C}^N \otimes \dots \otimes \mathrm{End}\,\mathbb{C}^N}_{m+1} \otimes \mathrm{U} \tag{8.20}$$

with $m+1$ copies of $\mathrm{End}\,\mathbb{C}^N$ labeled by $0, 1, \dots, m$, where U stands for the universal enveloping algebra of the Lie algebra $\widehat{\mathfrak{o}}_N \oplus \mathbb{C}\tau$. Relations (8.19) can be written in the equivalent form

$$F[0]_0\, \mathrm{tr}_{1,\dots,m}\, S^{(m)} \mathcal{F}_1 \dots \mathcal{F}_m = 0 \quad \text{and} \quad F[1]_0\, \mathrm{tr}_{1,\dots,m}\, S^{(m)} \mathcal{F}_1 \dots \mathcal{F}_m = 0 \tag{8.21}$$

modulo the left ideal of U generated by $\mathfrak{o}_N[t]$ and $K + h^\vee$. To verity the first relation, note that by (8.5) we have

$$[F[0]_0, \mathcal{F}_a] = \Phi_{0\,a}\, \mathcal{F}_a - \mathcal{F}_a\, \Phi_{0\,a}, \tag{8.22}$$

where we set $\Phi = P - Q$. Hence

$$F[0]_0\, \mathrm{tr}_{1,\dots,m}\, S^{(m)} \mathcal{F}_1 \dots \mathcal{F}_m = \sum_{a=1}^{m} \mathrm{tr}_{1,\dots,m}\, S^{(m)} \mathcal{F}_1 \dots \big(\Phi_{0\,a}\, \mathcal{F}_a - \mathcal{F}_a\, \Phi_{0\,a}\big) \dots \mathcal{F}_m$$

which equals

$$\mathrm{tr}_{1,\dots,m}\, S^{(m)} \sum_{a=1}^{m} \Phi_{0\,a} \mathcal{F}_1 \dots \mathcal{F}_m - \mathrm{tr}_{1,\dots,m}\, S^{(m)}\, \mathcal{F}_1 \dots \mathcal{F}_m \sum_{a=1}^{m} \Phi_{0\,a}.$$

The sum $\sum_{a=1}^{m} \Phi_{0\,a}$ commutes with the action of the symmetrizer $S^{(m)}$ so that the expression is equal to zero by the cyclic property of trace.

To prove the second relation in (8.21) use the following consequence of (8.5):

$$[F[1]_0, \mathcal{F}_a] = F[0]_0 + \Phi_{0\,a}\, F[0]_a - F[0]_a\, \Phi_{0\,a} + \Phi_{0\,a}\, K.$$

We have

$$\begin{aligned} F[1]_0\, \mathrm{tr}_{1,\dots,m}\, S^{(m)} \mathcal{F}_1 \dots \mathcal{F}_m = \sum_{a=1}^{m} & \mathrm{tr}_{1,\dots,m}\, S^{(m)} \mathcal{F}_1 \dots \mathcal{F}_{a-1} \\ & \times \Big(F[0]_0 + \Phi_{0\,a}\, F[0]_a - F[0]_a\, \Phi_{0\,a} + \Phi_{0\,a}\, K \Big) \mathcal{F}_{a+1} \dots \mathcal{F}_m. \end{aligned}$$

Applying (8.22) we can write this expression as

$$\begin{aligned} & \sum_{1 \leqslant a < b \leqslant m} \mathrm{tr}_{1,\dots,m}\, S^{(m)} \mathcal{F}_1 \dots \widehat{\mathcal{F}}_a \dots \Big(\Phi_{0\,b}\, \mathcal{F}_b - \mathcal{F}_b\, \Phi_{0\,b} \Big) \dots \mathcal{F}_m \\ & + \sum_{1 \leqslant a < b \leqslant m} \mathrm{tr}_{1,\dots,m}\, S^{(m)} \Phi_{0\,a}\, \mathcal{F}_1 \dots \widehat{\mathcal{F}}_a \dots \Big(\Phi_{a\,b}\, \mathcal{F}_b - \mathcal{F}_b\, \Phi_{a\,b} \Big) \dots \mathcal{F}_m \\ & - \sum_{1 \leqslant a < b \leqslant m} \mathrm{tr}_{1,\dots,m}\, S^{(m)} \mathcal{F}_1 \dots \widehat{\mathcal{F}}_a \dots \Big(\Phi_{a\,b}\, \mathcal{F}_b - \mathcal{F}_b\, \Phi_{a\,b} \Big) \dots \mathcal{F}_m\, \Phi_{0\,a} \\ & + K \sum_{a=1}^{m} \mathrm{tr}_{1,\dots,m}\, S^{(m)} \Phi_{0\,a}\, \mathcal{F}_1 \dots \widehat{\mathcal{F}}_a \dots \mathcal{F}_m, \end{aligned}$$

where the hats indicate that the corresponding symbols should be skipped. Now we simplify it by using the cyclic property of trace, conjugations by permutation operators and relations (1.31). For the first sum we have

$$\begin{aligned} & \sum_{1 \leqslant a < b \leqslant m} \mathrm{tr}_{1,\dots,m}\, S^{(m)} \mathcal{F}_1 \dots \widehat{\mathcal{F}}_a \dots \Phi_{0\,b}\, \mathcal{F}_b \dots \mathcal{F}_m \\ & \qquad = \sum_{1 \leqslant a < b \leqslant m} \mathrm{tr}_{1,\dots,m}\, S^{(m)} P_{m-1\, m} \dots P_{a\, a+1}\, \mathcal{F}_1 \dots \widehat{\mathcal{F}}_a \dots \Phi_{0\,b}\, \mathcal{F}_b \dots \mathcal{F}_m \\ & \qquad = \sum_{1 \leqslant a < b \leqslant m} \mathrm{tr}_{1,\dots,m}\, S^{(m)} \Phi_{0\, b-1}\, \mathcal{F}_1 \dots \mathcal{F}_{m-1}\, P_{m-1\, m} \dots P_{a\, a+1} \\ & \qquad = \sum_{a=1}^{m-1} a\, \mathrm{tr}_{1,\dots,m}\, S^{(m)} \Phi_{0\,a}\, \mathcal{F}_1 \dots \mathcal{F}_{m-1}. \end{aligned}$$

Transforming all remaining sums in a similar way, we get

$$
\begin{aligned}
F[1]_0 \operatorname{tr} S^{(m)} \mathcal{F}_1 \dots \mathcal{F}_m = & \sum_{a=1}^{m-1} a \operatorname{tr}_{1,\dots,m} S^{(m)} \Phi_{0\,a} \mathcal{F}_1 \dots \mathcal{F}_{m-1} \\
& - \sum_{a=1}^{m-1} a \operatorname{tr}_{1,\dots,m} \Phi_{0\,a} S^{(m)} \mathcal{F}_1 \dots \mathcal{F}_{m-1} \\
& + \sum_{a=1}^{m-1} a \operatorname{tr}_{1,\dots,m} S^{(m)} \Phi_{0\,m} \Phi_{a\,m} \mathcal{F}_1 \dots \mathcal{F}_{m-1} \\
& + \sum_{a=1}^{m-1} a \operatorname{tr}_{1,\dots,m} \Phi_{a\,m} \Phi_{0\,m} S^{(m)} \mathcal{F}_1 \dots \mathcal{F}_{m-1} \\
& + \big(mK - m(m-1)\big) \operatorname{tr}_{1,\dots,m} S^{(m)} \Phi_{0\,m} \mathcal{F}_1 \dots \mathcal{F}_{m-1}.
\end{aligned}
$$

Furthermore, note that

$$S^{(m)} \Phi_{0\,m} P_{a\,m} = S^{(m)} P_{a\,m} \Phi_{0\,a} = S^{(m)} \Phi_{0\,a},$$

and

$$S^{(m)} \Phi_{0\,m} Q_{a\,m} = 0. \tag{8.23}$$

Relation (8.23) holds since

$$S^{(m)} \Phi_{0\,m} Q_{a\,m} = S^{(m)} \Phi_{0\,m} P_{a\,m} Q_{a\,m} = S^{(m)} \Phi_{0\,a} Q_{a\,m} = -S^{(m)} \Phi_{0\,m} Q_{a\,m},$$

where we used (2.28) and observed that Φ is skew-symmetric with respect to the transposition (2.27) applied to the first of second copy of $\operatorname{End}\mathbb{C}^N$. Therefore, simplifying the expressions further we get

$$
\begin{aligned}
F[1]_0 \operatorname{tr}_{1,\dots,m} S^{(m)} \mathcal{F}_1 \dots \mathcal{F}_m = & \sum_{a=1}^{m-1} 2a \operatorname{tr}_{1,\dots,m} S^{(m)} \Phi_{0\,a} \mathcal{F}_1 \dots \mathcal{F}_{m-1} \\
& + \big(mK - m(m-1)\big) \operatorname{tr}_{1,\dots,m} S^{(m)} \Phi_{0\,m} \mathcal{F}_1 \dots \mathcal{F}_{m-1}.
\end{aligned} \tag{8.24}
$$

As a next step, calculate the partial trace tr_m on the right hand side with respect to the m-th copy of $\operatorname{End}\mathbb{C}^N$ in (8.20) with the use of the following lemma.

LEMMA 8.1.7. *We have*

$$\operatorname{tr}_m S^{(m)} \Phi_{0\,m} = \frac{N+2m-2}{m\,(N+2m-4)}\, S^{(m-1)} \sum_{a=1}^{m-1} \Phi_{0\,a}.$$

PROOF. Taking the image of the recurrence relation (1.53) under the homomorphism (1.69) we get

$$S^{(m)} = \frac{1}{m(N+2m-4)} \Big(1 + \sum_{a=1}^{m-1} \Phi_{a\,m}\Big) \Big(N + m - 3 + \sum_{a=1}^{m-1} \Phi_{a\,m}\Big) S^{(m-1)}.$$

Now use relations (1.31) and (1.71) to bring this to the form

$$S^{(m)} = \left(1 + \sum_{a=1}^{m-1} P_{a\,m} - \frac{2}{N+2m-4} \Big(\sum_{a=1}^{m-1} Q_{a\,m} + \sum_{1\leqslant a<b\leqslant m-1} P_{a\,m} Q_{b\,m} \Big) \right) \frac{S^{(m-1)}}{m}.$$

Using the cyclic property of trace we come to calculating the partial trace

$$\operatorname{tr}_m \Phi_{0m}\Bigg(1+\sum_{a=1}^{m-1} P_{am}-\frac{2}{N+2m-4}\Big(\sum_{a=1}^{m-1} Q_{am}+\sum_{1\leqslant a<b\leqslant m-1} P_{am}Q_{bm}\Big)\Bigg)\, S^{(m-1)}.$$

We have $\operatorname{tr}_m \Phi_{0m}=0$, while

$$\operatorname{tr}_m \Phi_{0m} P_{am}=\Phi_{0a} \qquad \text{and} \qquad \operatorname{tr}_m \Phi_{0m} Q_{am}=-\Phi_{0a}.$$

Moreover,

$$\operatorname{tr}_m \Phi_{0m} P_{am} Q_{bm}=\operatorname{tr}_m \Phi_{0m} Q_{ab} P_{am}=\operatorname{tr}_m Q_{ab}\Phi_{0m} P_{am}=Q_{ab}\Phi_{0a}.$$

However, $Q_{ab}\Phi_{0a}S^{(m-1)}=0$, which is a version of (8.23) and is verified in the same way. So the desired expression for $\operatorname{tr}_m S^{(m)}\Phi_{0m}$ follows. □

By using the formula (5.32) for the partial trace $\operatorname{tr}_m S^{(m)}$ and Lemma 8.1.7 we come to the relation

$$\begin{aligned} F[1]_0 \operatorname{tr}_{1,\dots,m} S^{(m)}\mathcal{F}_1\dots\mathcal{F}_m &= \frac{N+2m-2}{m\,(N+2m-4)} \\ &\times\Big((N+m-3)\sum_{a=1}^{m-1} 2\,a \operatorname{tr}_{1,\dots,m-1} S^{(m-1)}\Phi_{0a}\mathcal{F}_1\dots\mathcal{F}_{m-1} \\ &\qquad + m\,(K-m+1)\sum_{a=1}^{m-1}\operatorname{tr}_{1,\dots,m-1} S^{(m-1)}\Phi_{0a}\mathcal{F}_1\dots\mathcal{F}_{m-1}\Big). \end{aligned}$$

Now we need one more lemma.

LEMMA 8.1.8. *For any* $1\leqslant a<b\leqslant m$ *we have*

$$\operatorname{tr}_{1,\dots,m} S^{(m)}\Phi_{0a}\mathcal{F}_1\dots\mathcal{F}_m=\operatorname{tr}_{1,\dots,m} S^{(m)}\Phi_{0b}\mathcal{F}_1\dots\mathcal{F}_m.$$

PROOF. By Lemma 8.1.5, the matrix $\mathcal{F}$ is a Manin matrix of type B or D, respectively so that by the property (5.53) we find that

$$\begin{aligned} \operatorname{tr}_{1,\dots,m} S^{(m)}\Phi_{0a}\mathcal{F}_1\dots\mathcal{F}_m &= \operatorname{tr}_{1,\dots,m}\Phi_{0a}\mathcal{F}_1\dots\mathcal{F}_m S^{(m)} \\ &= \operatorname{tr}_{1,\dots,m}\Phi_{0a}P_{ab}\mathcal{F}_1\dots\mathcal{F}_m S^{(m)} \end{aligned}$$

where we used the cyclic property of trace. Furthermore, this expression equals

$$\operatorname{tr}_{1,\dots,m} P_{ab}\Phi_{0b}\mathcal{F}_1\dots\mathcal{F}_m S^{(m)}=\operatorname{tr}_{1,\dots,m}\Phi_{0b}\mathcal{F}_1\dots\mathcal{F}_m S^{(m)}P_{ab}$$

which coincides with

$$\operatorname{tr}_{1,\dots,m} S^{(m)}\Phi_{0b}\mathcal{F}_1\dots\mathcal{F}_m$$

by the property (1.31) of the symmetrizer. □

Thus, due to Lemma 8.1.8 (applied with m replaced by $m-1$),

$$\begin{aligned} F[1]_0 &\operatorname{tr}_{1,\dots,m} S^{(m)}\mathcal{F}_1\dots\mathcal{F}_m \\ &= (K+N-2)\,\frac{N+2m-2}{N+2m-4}\sum_{a=1}^{m-1}\operatorname{tr}_{1,\dots,m-1} S^{(m-1)}\Phi_{0a}\mathcal{F}_1\dots\mathcal{F}_{m-1} \end{aligned}$$

which is zero since $K+N-2=0$ at the critical level. This completes the proof of the theorem. □

We will now prove the Feigin–Frenkel theorem (Theorem 6.3.1) for the orthogonal Lie algebras.

THEOREM 8.1.9. *The family $\phi_{2\,2}, \phi_{4\,4}, \dots, \phi_{2n\,2n}$ is a complete set of Segal–Sugawara vectors for $\mathfrak{o}_{2n+1}$ and the family $\phi_{2\,2}, \phi_{4\,4}, \dots, \phi_{2n-2\,2n-2}, \operatorname{Pf} F[-1]$ is a complete set of Segal–Sugawara vectors for $\mathfrak{o}_{2n}$.*

PROOF. The symbols of the elements of each family coincide with the images of certain elements of the symmetric algebra $\mathrm{S}(\mathfrak{o}_N)$ under the embedding $\mathrm{S}(\mathfrak{o}_N) \hookrightarrow \mathrm{S}\left(t^{-1}\mathfrak{o}_N[t^{-1}]\right)$ taking $X \in \mathfrak{o}_N$ to $X[-1]$. Due to Theorems 6.3.3 and 8.1.6 and Proposition 8.1.4, we only need to verify that the corresponding elements of $\mathrm{S}(\mathfrak{o}_N)$ are algebraically independent generators of the algebra of invariants $\mathrm{S}(\mathfrak{o}_N)^{\mathfrak{o}_N}$; see the argument after the proof of Theorem 6.3.3. However, this was already pointed out in Section 2.2; see Corollary 2.2.5. □

We conclude this section by providing equivalent expressions for the Segal–Sugawara vectors $\phi_{m\,a}$ in the context of Howe duality. In the notation of Section 2.3, introduce the element $\mathcal{F}^{(m)} \in \operatorname{End} \mathcal{P}_N^m \otimes \mathrm{U}\left(\widehat{\mathfrak{o}}_N \oplus \mathbb{C}\tau\right)$ by setting

$$\mathcal{F}^{(m)} : z_{j_1} \dots z_{j_m} \mapsto \sum_{i_1 \leqslant \dots \leqslant i_m} z_{i_1} \dots z_{i_m} \otimes \mathcal{F}^{i_1 \dots i_m}_{j_1 \dots j_m},$$

where

$$\mathcal{F}^{i_1 \dots i_m}_{j_1 \dots j_m} = \frac{1}{\alpha_1! \dots \alpha_N!\, m!} \sum_{\sigma, \pi \in \mathfrak{S}_m} \mathcal{F}_{i_{\sigma(1)} j_{\pi(1)}} \dots \mathcal{F}_{i_{\sigma(m)} j_{\pi(m)}}$$

and α_i is the multiplicity of i in the multiset $\{i_1, \dots, i_m\}$. Recall that the extremal projector p defined by the equivalent formulas (2.64) and (2.66) is regarded as an element of $\operatorname{End} \mathcal{P}_N^m$ via the action (2.63).

COROLLARY 8.1.10. *The Segal–Sugawara vectors $\phi_{m\,a}$ can be found from the expansion*

$$\gamma_m(N) \operatorname{tr} p\, \mathcal{F}^{(m)} = \phi_{m\,0}\, \tau^m + \phi_{m\,1}\, \tau^{m-1} + \dots + \phi_{m\,m}$$

with the trace taken over the subspace of $\mathfrak{sl}_2$-singular vectors in $\mathcal{P}_N^m$.

PROOF. Recalling the formula (2.43) for the symmetrizer and using the cyclic property of trace, we can write the left hand side of (8.15) in the form

$$\gamma_m(N) \operatorname{tr}_{1,\dots,m} S^{(m)} \mathcal{F}_1 \dots \mathcal{F}_m = \gamma_m(N) \operatorname{tr}_{1,\dots,m} S^{(m)} H^{(m)} \mathcal{F}_1 \dots \mathcal{F}_m H^{(m)}.$$

Taking into account the isomorphism (2.61), we may regard the product

$$H^{(m)} \mathcal{F}_1 \dots \mathcal{F}_m H^{(m)}$$

as an operator on the vector space $\mathcal{P}_N^m$ with coefficients in $\mathrm{U}\left(\widehat{\mathfrak{o}}_N \oplus \mathbb{C}\tau\right)$. This operator coincides with $\mathcal{F}^{(m)}$. It remains to note that the operator $S^{(m)}$ acts on $\mathcal{P}_N^m$ as the extremal projector p by Proposition 2.3.1. □

8.2. Low degree invariants in trace form

As we showed in Theorem 7.1.3, all coefficients of the polynomial $\operatorname{tr} \mathcal{E}^m$ defined in (7.9) are Segal–Sugawara vectors for $\mathfrak{gl}_N$. Their analogues for the orthogonal and symplectic Lie algebras are not known. In particular, the coefficients of the polynomial $\operatorname{tr} \mathcal{F}^m$ are *not*, in general, Segal–Sugawara vectors for $\mathfrak{o}_N$. In this section we aim to construct examples of such vectors in trace form; cf. Example 7.1.6. This can be done by writing the Segal–Sugawara vectors $\phi_{m\,a}$ provided by Theorem 8.1.6 in terms of the trace-type invariants

$$\operatorname{tr} F[r_1]\dots F[r_k], \qquad r_1,\dots,r_k<0.$$

We will produce such expressions for $\phi_{2\,2}$ and $\phi_{4\,4}$. Using (1.46) and (1.69) we get

$$\phi_{2\,2}=\frac{N}{N+2}\operatorname{tr}_{1,2}\Big(\frac{1}{2}+\frac{P_{12}}{2}-\frac{Q_{12}}{N}\Big)\big(F[-1]_1F[-1]_2+F[-2]_2\big).$$

Furthermore, by (1.67) and (1.68),

$$\operatorname{tr}_{1,2}P_{12}F[-1]_1F[-1]_2=\operatorname{tr}_{1,2}F[-1]_2P_{12}F[-1]_2=\operatorname{tr}_2F[-1]_2F[-1]_2=\operatorname{tr}F[-1]^2.$$

Similarly, since $F[-1]'+F[-1]=0$, by (1.76) and (2.28) we have

$$\begin{aligned}\operatorname{tr}_{1,2}Q_{12}F[-1]_1F[-1]_2&=-\operatorname{tr}_{1,2}Q_{12}F[-1]_2F[-1]_2\\&=-\operatorname{tr}_2F[-1]_2F[-1]_2=-\operatorname{tr}F[-1]^2.\end{aligned}$$

Taking into account that $\operatorname{tr}F[r]=0$ for any r, we conclude that

$$\phi_{2\,2}=\frac{1}{2}\operatorname{tr}F[-1]^2=\frac{1}{2}\sum_{i,j=1}^{N}F_{ij}[-1]F_{ji}[-1].$$

Thus, $\phi_{2\,2}$ is proportional to the canonical Segal–Sugawara vector (6.11).

To calculate $\phi_{4\,4}$, note that the constant term of the polynomial

$$\big(\tau+F[-1]_1\big)\big(\tau+F[-1]_2\big)\big(\tau+F[-1]_3\big)\big(\tau+F[-1]_4\big)$$

in τ is found by

$$\begin{aligned}&F[-1]_1F[-1]_2F[-1]_3F[-1]_4+F[-1]_1F[-1]_2F[-2]_4+F[-1]_1F[-2]_3F[-1]_4\\&\quad+F[-1]_1F[-1]_3F[-2]_4+F[-2]_2F[-1]_3F[-1]_4+F[-1]_2F[-2]_3F[-1]_4\\&\quad+F[-1]_2F[-1]_3F[-2]_4+2F[-1]_1F[-3]_4+2F[-1]_2F[-3]_4+2F[-1]_3F[-3]_4\\&\quad+2F[-3]_3F[-1]_4+2F[-2]_3F[-2]_4+F[-2]_2F[-2]_4+6F[-4]_4.\end{aligned}$$

Using the properties (1.31) of the symmetrizer together with the cyclic property of trace and applying conjugations by appropriate permutation operators we can write $\phi_{4\,4}$ in the form

$$\begin{aligned}\phi_{4\,4}=\frac{N+2}{N+6}&\operatorname{tr}_{1,2,3,4}S^{(4)}\Big(F[-1]_1F[-1]_2F[-1]_3F[-1]_4+3F[-1]_1F[-1]_2F[-2]_3\\&+2F[-1]_1F[-2]_2F[-1]_3+F[-2]_1F[-1]_2F[-1]_3+6F[-1]_1F[-3]_2\\&\qquad+3F[-2]_1F[-2]_2+2F[-3]_1F[-1]_2+6F[-4]_1\Big).\end{aligned}$$

Applying Lemma 8.1.5 we get

$$\begin{aligned}\phi_{4\,4}=\frac{N+2}{N+6}\operatorname{tr}_{1,2,3,4}S^{(4)}\Big(&F[-1]_1F[-1]_2F[-1]_3F[-1]_4+6F[-1]_1F[-1]_2F[-2]_3\\&+8F[-1]_1F[-3]_2+3F[-2]_1F[-2]_2+6F[-4]_1\Big).\end{aligned}$$

As a next step, calculate the partial trace tr_4 by using the following general lemma.

LEMMA 8.2.1. *Suppose that X is a certain linear combination of products of elements $F[s]_b$ of the algebra* (8.3) *with $s<0$ and $1\leqslant b\leqslant m-1$. Then for any $r<0$ we have*

$$\mathrm{tr}_{1,\dots,m}\, S^{(m)} X\, F[r]_m = \frac{N+2m-2}{m(N+2m-4)}\, \mathrm{tr}_{1,\dots,m-1}\, S^{(m-1)} X \sum_{a=1}^{m-1} F[r]_a.$$

PROOF. This is essentially a version of Lemma 8.1.7, where Φ_{0m} should be replaced with $F[r]_m$. It is verified by the same argument. □

Lemmas 1.3.2 and 8.2.1 with $m=4$ allow us to bring the expression for the Segal–Sugawara vector ϕ_{44} to the form

$$\begin{aligned}\phi_{44} = \frac{N+2}{4(N+4)}\, \mathrm{tr}_{1,2,3}\, S^{(3)} \Big(& F[-1]_1F[-1]_2F[-1]_3\big(F[-1]_1+F[-1]_2+F[-1]_3\big)\\ &+6(N+1)\,F[-1]_1F[-1]_2F[-2]_3+8(N+1)\,F[-1]_1F[-3]_2\\ &+3(N+1)\,F[-2]_1F[-2]_2+6(N+1)\,F[-4]_1\Big).\end{aligned}$$

For the expression in the first line we can write

$$\begin{aligned}&\mathrm{tr}_{1,2,3}\, S^{(3)}\, F[-1]_1F[-1]_2F[-1]_3\Big(F[-1]_1+F[-1]_2+F[-1]_3\Big)\\ &\quad= \mathrm{tr}_{1,2,3}\, S^{(3)}\Big(F[-1]_1F[-1]_2F[-1]_3+F[-1]_2F[-1]_1F[-1]_3\\ &\qquad\qquad+F[-1]_2F[-1]_3F[-1]_1\Big)F[-1]_1.\end{aligned}$$

Now use (8.16) repeatedly to move the factor $F[-1]_1$ occurring in the second and third terms within the brackets to the leftmost position. After that, use the cyclic property of trace to write $S^{(3)}$ as the last factor to be able to use (8.17) to move the second factor $F[-1]_1$ to the left. This brings the expression to the form

$$\begin{aligned}\mathrm{tr}_{1,2,3}\, S^{(3)}\Big(&3\,F[-1]_1^2F[-1]_2F[-1]_3+3\,F[-1]_1F[-1]_2F[-2]_3\\ &-6\,F[-1]_1F[-2]_1F[-1]_2+2\,F[-2]_1F[-1]_1F[-1]_2+2\,F[-2]_1F[-2]_2\\ &-2\,F[-2]_1^2+F[-1]_1F[-2]_2F[-1]_2-3\,F[-1]_1F[-3]_2+3\,F[-1]_1F[-3]_1\Big).\end{aligned}$$

Applying again Lemmas 1.3.2 and 8.2.1, this time with $m=3$, and writing the symmetrizer $S^{(2)}$ explicitly, we get

$$\begin{aligned}\phi_{44} = \frac{1}{12}\, \mathrm{tr}_{1,2}\Big(\frac{1}{2}+\frac{P_{12}}{2}-\frac{Q_{12}}{N}\Big)\Big(&3\,F[-1]_1^2F[-1]_2\big(F[-1]_1+F[-1]_2\big)\\ &+(6N+9)\,F[-1]_1F[-1]_2\big(F[-2]_1+F[-2]_2\big)+(8N^2+5N)\,F[-1]_1F[-3]_2\\ &+(3N^2+5N)\,F[-2]_1F[-2]_2-6N\,F[-1]_1F[-2]_1F[-1]_2+2N\,F[-2]_1F[-1]_1F[-1]_2\\ &-2N\,F[-2]_1^2+N\,F[-1]_1F[-2]_2F[-1]_2+3N\,F[-1]_1F[-3]_1+6N(N+1)\,F[-4]_1\Big).\end{aligned}$$

Finally, calculating the traces in a way similar to the case of ϕ_{22} above, using (8.5) and taking into account the identity $\big(F[r]_1+F[r]_2\big)\,Q_{12}=0$, we come to the formula

$$\phi_{44} = \frac{1}{24}\Big(6\,\mathrm{tr}\,F[-1]^4 + 3\,\big(\mathrm{tr}\,F[-1]^2\big)^2 + (3N+15)\,\mathrm{tr}\,F[-1]^2F[-2] \\ + (N-1)\,\mathrm{tr}\,F[-1]F[-2]F[-1] + (2N+4)\,\mathrm{tr}\,F[-2]F[-1]^2 \\ + (5N^2+27N+22)\,\mathrm{tr}\,F[-1]F[-3] + (N^2+9N+14)\,\mathrm{tr}\,F[-2]^2\Big).$$

Observe that the following relations hold for the traces occurring in the formula,

$$\mathrm{tr}\,F[-1]^2F[-2] = \mathrm{tr}\,F[-2]F[-1]^2 = (N/2-1)\,\mathrm{tr}\,F[-2]^2$$

and

$$\mathrm{tr}\,F[-1]F[-2]F[-1] = -(N/2-1)\,\mathrm{tr}\,F[-2]^2 + (N-2)\,\mathrm{tr}\,F[-1]F[-3],$$

which are easy to verify. For instance, using (8.5), for the first element we can write

$$\mathrm{tr}\,F[-1]^2F[-2] = -\mathrm{tr}_{12}\,Q_{12}F[-1]_1F[-1]_2F[-2]_2 \\ = -\mathrm{tr}_{12}\,Q_{12}F[-1]_2F[-1]_1F[-2]_2 - \mathrm{tr}_{12}\,Q_{12}(P_{12}-Q_{12})F[-2]_2F[-2]_2 \\ + \mathrm{tr}_{12}\,Q_{12}F[-2]_2(P_{12}-Q_{12})F[-2]_2.$$

Observing that $Q_{12}F[-2]_2Q_{12} = 0$ and using the cyclic property of trace together with (1.71) we get

$$\mathrm{tr}\,F[-1]^2F[-2] = -\mathrm{tr}\,F[-1]^2F[-2] + (N-2)\,\mathrm{tr}\,F[-2]^2$$

as required. Therefore, the formula for ϕ_{44} simplifies to

$$\phi_{44} = \frac{1}{24}\Big(6\,\mathrm{tr}\,F[-1]^4 + 3\,\big(\mathrm{tr}\,F[-1]^2\big)^2 \\ + 6\,(N+2)^2\,\mathrm{tr}\,F[-1]F[-3] + (3N^2+15N-6)\,\mathrm{tr}\,F[-2]^2\Big).$$

Recall that the Feigin–Frenkel center $\mathfrak{z}(\widehat{\mathfrak{o}}_N)$ is invariant with respect to the translation operator T; see (6.7). In particular, the elements $T\phi_{22}$ and $T^2\phi_{22}$ are Segal–Sugawara vectors. We have

$$T\phi_{22} = \mathrm{tr}\,F[-1]F[-2] \qquad \text{and} \qquad T^2\phi_{22} = \mathrm{tr}\,F[-2]^2 + 2\,\mathrm{tr}\,F[-1]F[-3].$$

Therefore,

$$4\phi_{44} - 2\phi_{22}^2 - \frac{1}{2}\,(N+2)^2\,T^2\phi_{22} = \mathrm{tr}\,F[-1]^4 + (N/2-3)\,\mathrm{tr}\,F[-2]^2.$$

Hence, replacing ϕ_{44} by the element

$$\mathrm{tr}\,F[-1]^4 + (N/2-3)\,\mathrm{tr}\,F[-2]^2 \tag{8.25}$$

in the complete sets of Segal–Sugawara vectors provided by Theorem 8.1.9 we get new complete sets of Segal–Sugawara vectors.

EXAMPLE 8.2.2. With the conventions of Remark 8.1.1, the following are complete sets of Segal–Sugawara vectors:

$$
\begin{aligned}
&\text{for}\quad \mathfrak{o}_3: && \operatorname{tr} F[-1]^2\\
&\text{for}\quad \mathfrak{o}_4: && \operatorname{tr} F[-1]^2,\quad \operatorname{Pf} F[-1]\\
&\text{for}\quad \mathfrak{o}_5: && \operatorname{tr} F[-1]^2,\quad \operatorname{tr} F[-1]^4-\frac{1}{2}\operatorname{tr} F[-2]^2\\
&\text{for}\quad \mathfrak{o}_6: && \operatorname{tr} F[-1]^2,\quad \operatorname{tr} F[-1]^4,\quad \operatorname{Pf} F[-1].
\end{aligned}
$$

□

We will also give a direct proof that the element (8.25) is a Segal–Sugawara vector. Consider the algebra (8.20), where U now denotes the universal enveloping algebra $\mathrm{U}(\widehat{\mathfrak{o}}_N)$.

LEMMA 8.2.3. *For any $m \geqslant 1$ we have the relations*

$$F[0]_0 \operatorname{tr}_1 F[-1]_1^m = 0 \tag{8.26}$$

and

$$
\begin{aligned}
&F[1]_0 \operatorname{tr}_1 F[-1]_1^m \\
&\quad = \sum_{i=1}^{m} \operatorname{tr}_1 \Big(F[-1]_1^{i-1} \Phi_{01} F[-1]_1^{m-i} \Phi_{01} - \Phi_{01} F[-1]_1^{i-1} \Phi_{01} F[-1]_1^{m-i} \Big)
\end{aligned}
\tag{8.27}
$$

modulo the left ideal of U *generated by $\mathfrak{o}_N[t]$ and $K + N - 2$, where $\Phi = P - Q$.*

PROOF. Note the relation

$$F[0]_0 F[-1]_1^m - F[-1]_1^m F[0]_0 = \Phi_{01} F[-1]_1^m - F[-1]_1^m \Phi_{01} \tag{8.28}$$

which is implied by (8.5). By taking tr_1 and using the cyclic property of trace we get (8.26). Furthermore, calculating modulo the left ideal, we find

$$F[1]_0 \operatorname{tr}_1 F[-1]_1^m = \sum_{i=1}^{m} \operatorname{tr}_1 F[-1]_1^{i-1} \big(-\Phi_{01} F[0]_0 + F[0]_0 \Phi_{01} + \Phi_{01} K \big) F[-1]_1^{m-i}.$$

By (8.28) we have

$$F[0]_0 F[-1]_1^{m-i} = \Phi_{01} F[-1]_1^{m-i} - F[-1]_1^{m-i} \Phi_{01}. \tag{8.29}$$

Now use the following general property of the transposition (2.27) which is applied to the copy of $\operatorname{End} \mathbb{C}^N$ labelled by 1:

$$\operatorname{tr}_1 XY = \operatorname{tr}_1 X'Y'. \tag{8.30}$$

Taking $X = F[-1]_1^{i-1} F[0]_0$ and $Y = \Phi_{01} F[-1]_1^{m-i}$ we obtain

$$\operatorname{tr}_1 F[-1]_1^{i-1} F[0]_0 \Phi_{01} F[-1]_1^{m-i} = -\operatorname{tr}_1 \big(F[-1]^{i-1} \big)_1' F[0]_0 \big(F[-1]^{m-i} \big)_1' \Phi_{01},$$

where we also used Lemma 2.2.1 and the property $\Phi_{01}' = -\Phi_{01}$. Applying the transposition to both sides of (8.29) we get

$$F[0]_0\,\big(F[-1]^{m-i}\big)'_1 = \Phi_{01}\,\big(F[-1]^{m-i}\big)'_1 - \big(F[-1]^{m-i}\big)'_1\,\Phi_{01}.$$

Thus,

$$\begin{aligned}F[1]_0\,\mathrm{tr}_1\,F[-1]_1^m = \sum_{i=1}^m \mathrm{tr}_1\,\Big(&F[-1]_1^{i-1}\,\Phi_{01}\,F[-1]_1^{m-i}\,\Phi_{01} - F[-1]_1^{i-1}\Phi_{01}^2\,F[-1]_1^{m-i}\\ &- \big(F[-1]^{i-1}\big)'_1\,\Phi_{01}\,\big(F[-1]^{m-i}\big)'_1\,\Phi_{01} + \big(F[-1]^{i-1}\big)'_1\,\big(F[-1]^{m-i}\big)'_1\,\Phi_{01}^2\\ &+ K\,F[-1]_1^{i-1}\,\Phi_{01}\,F[-1]_1^{m-i}\Big).\end{aligned}$$

Finally, the application of (8.30) to the terms in the middle line yields the desired expression as in the right hand side of (8.27) together with the sum

$$\sum_{i=1}^m \mathrm{tr}_1\,F[-1]_1^{i-1}\big((\Phi_{01}^2)' - \Phi_{01}^2 + K\,\Phi_{01}\big)\,F[-1]_1^{m-i}.$$

However, this sum is zero since $(\Phi_{01}^2)' = \Phi_{01}^2 + (N-2)\,\Phi_{01}$ and $K + N - 2 = 0$. □

Note that $\mathrm{tr}\,F[-1] = 0$, and it is immediate from Lemma 8.2.3 that both $\mathrm{tr}\,F[-1]^2$ and $\mathrm{tr}\,F[-1]^3$ are Segal–Sugawara vectors for $\mathfrak{o}_N$. Furthermore, by (8.27),

$$\begin{aligned}F[1]_0\,\mathrm{tr}_1\,F[-1]_1^4 = \mathrm{tr}_1\,\Big(&F[-1]_1\,\Phi_{01}\,F[-1]_1^2\,\Phi_{01} + F[-1]_1^2\,\Phi_{01}\,F[-1]_1\,\Phi_{01}\\ &- \Phi_{01}\,F[-1]_1\,\Phi_{01}\,F[-1]_1^2 - \Phi_{01}\,F[-1]_1^2\,\Phi_{01}\,F[-1]_1\Big).\end{aligned}$$

By applying (8.30), we can write this as

$$\mathrm{tr}_1\,\Big(\big(F[-1]_1^2 + (F[-1]^2)'_1\big)\,\Phi_{01}\,F[-1]_1\,\Phi_{01} - \Phi_{01}\,F[-1]_1\,\Phi_{01}\,\big(F[-1]_1^2 + (F[-1]^2)'_1\big)\Big).$$

On the other hand,

$$\Phi_{01}\,F[-1]_1\,\Phi_{01} = F[-1]_0 + Q_{01}\,F[-1]_1 + F[-1]_1\,Q_{01}$$

so that another application of (8.30) brings the expression to the form

$$\begin{aligned}\mathrm{tr}_1\,\Big(&\big(F[-1]_1^2 + (F[-1]^2)'_1\big)\,\big(F[-1]_0 - F[-1]_1\,\Phi_{01}\big)\\ &- \big(F[-1]_0 - \Phi_{01}\,F[-1]_1\big)\,\big(F[-1]_1^2 + (F[-1]^2)'_1\big)\Big).\end{aligned}$$

Now observe that

$$\begin{aligned}(F[-1]^2)'_1 &= \mathrm{tr}_2\,F[-1]_1^2\,Q_{12} = -\mathrm{tr}_2\,F[-1]_1\,F[-1]_2\,Q_{12}\\ &= -\mathrm{tr}_2\,F[-1]_2\,F[-1]_1\,Q_{12} + \mathrm{tr}_2\,\big(\Phi_{12}\,F[-2]_1 - F[-2]_1\,\Phi_{12}\big)\,Q_{12}\end{aligned}$$

which gives

$$(F[-1]^2)' = F[-1]^2 - (N-2)\,F[-2].$$

Using this observation and the commutation relations (8.5), we come to the simplified expression

$$F[1]_0\,\mathrm{tr}_1\,F[-1]_1^4 = (N-6)\,\mathrm{tr}_1\,\big(F[-2]_1\,F[-1]_1\,\Phi_{01} - \Phi_{01}\,F[-1]_1\,F[-2]_1\big).$$

Furthermore, we also find with the use of (8.30) that

$$\operatorname{tr}_1 F[-2]_1^2 = -2\operatorname{tr}_1\big(F[-2]_1 F[-1]_1 \Phi_{01} - \Phi_{01} F[-1]_1 F[-2]_1\big)$$

thus concluding that

$$F[1]_0\Big(\operatorname{tr}_1 F[-1]_1^4 + (N/2-3)\operatorname{tr}_1 F[-2]_1^2\Big) = 0$$

and so (8.25) is a Segal–Sugawara vector.

8.3. Segal–Sugawara vectors in type C

We will keep the notation for the generators of the symplectic Lie algebra $\mathfrak{sp}_{2n}$ introduced in Chapter 5. That is, we can view $\mathfrak{sp}_{2n}$ is the subalgebra of $\mathfrak{gl}_{2n}$ spanned by the elements F_{ij} defined in (5.3). Consider the affine Kac–Moody algebra $\widehat{\mathfrak{sp}}_{2n} = \mathfrak{sp}_{2n}[t,t^{-1}] \oplus \mathbb{C}K$ as defined in (6.5) and set $F_{ij}[r] = F_{ij}t^r$ for any $r \in \mathbb{Z}$. The dual Coxeter number for the Lie algebra $\mathfrak{sp}_{2n}$ is given by

$$h^\vee = n+1 \tag{8.31}$$

and the normalized Killing form (6.4) can be written as

$$\langle X, Y\rangle = \operatorname{tr} XY, \qquad X, Y \in \mathfrak{sp}_{2n}.$$

By (5.4), the affine Kac–Moody algebra $\widehat{\mathfrak{sp}}_{2n}$ has the commutation relations

$$\begin{aligned}\big[F_{ij}[r], F_{kl}[s]\big] &= \delta_{kj}\, F_{il}[r+s] - \delta_{il}\, F_{kj}[r+s] \\ &- \varepsilon_i\varepsilon_j\big(\delta_{ki'}\, F_{j'l}[r+s] + \delta_{j'l}\, F_{ki'}[r+s]\big) + 2r\,\delta_{r,-s}\, K\big(\delta_{kj}\,\delta_{il} - \varepsilon_i\varepsilon_j\,\delta_{ki'}\,\delta_{j'l}\big),\end{aligned} \tag{8.32}$$

where the element K is central and the symmetry property $F_{ij}[r] + \varepsilon_i\varepsilon_j\, F_{j'i'}[r] = 0$ holds. As with the orthogonal case, for any $r \in \mathbb{Z}$ regard the matrix $F[r] = \big[F_{ij}[r]\big]$ as the element

$$F[r] = \sum_{i,j=1}^{2n} e_{ij} \otimes F_{ij}[r] \in \operatorname{End}\mathbb{C}^{2n} \otimes \mathrm{U}(\widehat{\mathfrak{sp}}_{2n}).$$

It has the symmetry property $F[r] + F[r]' = 0$; see (2.24). For each $a \in \{1,\dots,m\}$ introduce the element $F[r]_a$ of the algebra

$$\underbrace{\operatorname{End}\mathbb{C}^{2n} \otimes \dots \otimes \operatorname{End}\mathbb{C}^{2n}}_{m} \otimes \mathrm{U}(\widehat{\mathfrak{sp}}_{2n}) \tag{8.33}$$

by

$$F[r]_a = \sum_{i,j=1}^{2n} 1^{\otimes(a-1)} \otimes e_{ij} \otimes 1^{\otimes(m-a)} \otimes F_{ij}[r]. \tag{8.34}$$

As in (5.6), the defining relations of the algebra $\mathrm{U}(\widehat{\mathfrak{sp}}_{2n})$ can be written in the matrix form

$$\begin{aligned}&F[r]_1\, F[s]_2 - F[s]_2\, F[r]_1 \\ &\qquad = (P-Q)\, F[r+s]_2 - F[r+s]_2\,(P-Q) + 2r\,\delta_{r,-s}\,(P-Q)\,K,\end{aligned} \tag{8.35}$$

where both sides are elements of the algebra (8.33) with $m=2$.

The symmetrizer $s^{(m)}$ in the Brauer algebra $\mathcal{B}_m(\omega)$ was defined in Section 1.2. We keep the notation $S^{(m)}$ for the image of the symmetrizer $s^{(m)}$ under the action of the Brauer algebra $\mathcal{B}_m(-2n)$ in the tensor space defined by (1.73). We also regard $S^{(m)}$ as an element of the algebra (8.33), identifying it with $S^{(m)} \otimes 1$. We will keep using the constant $\gamma_m(-2n)$ defined in (2.40) so that

$$\gamma_m(-2n) = \frac{2n - m + 2}{2(n - m + 1)}.$$

The critical level of the vacuum module $V_\kappa(\mathfrak{sp}_{2n})$ corresponds to the value $\kappa = -h^\vee$. We will construct Segal–Sugawara vectors for $\mathfrak{sp}_{2n}$ in a way similar to the orthogonal case. However, an expression analogous to (8.15) will only make sense for the values of m bound by the condition $m \leqslant n + 1$. The reason is that the formulas for the symmetrizer $s^{(m)}$ given in Section 1.2 will contain zero denominators under the specialization $\omega = -2n$ for the values $m \geqslant n + 2$. These formulas do define the operator $S^{(m)}$ for $m = n+1$ but it turns out to be identically zero; see Proposition 2.3.2. We will go around this problem in a way similar to the construction of Casimir elements in Section 5.5. Namely, we will let the parameter m be fixed and let n vary. Then we will show that the symplectic analogue of (8.15) still defines Segal–Sugawara vectors, and the values of n can be 'analytically continued' to the region $n \geqslant (m - 1)/2$.

Suppose first that $n \geqslant m$ and introduce the elements $\phi_{m\,a}$ of the vacuum module $V_{-h^\vee}(\mathfrak{sp}_{2n}) \cong \mathrm{U}\left(t^{-1}\mathfrak{sp}_{2n}[t^{-1}]\right)$ by the expansion

$$\gamma_m(-2n)\,\mathrm{tr}_{1,\dots,m}\, S^{(m)} \mathcal{F}_1 \dots \mathcal{F}_m = \phi_{m\,0}\, \tau^m + \phi_{m\,1}\, \tau^{m-1} + \dots + \phi_{m\,m}, \tag{8.36}$$

where $\mathcal{F} = \tau + F[-1] = \left[\delta_{ij}\tau + F_{ij}[-1]\right]$ is the $2n \times 2n$ matrix with entries in the universal enveloping algebra of $\widehat{\mathfrak{sp}}_{2n} \oplus \mathbb{C}\tau$, and the commutation relations for τ are defined in (7.1).

Recall the Manin matrices of type C as introduced in Definition 5.6.1.

LEMMA 8.3.1. *The matrix $\mathcal{F}$ is a Manin matrix of type C. Moreover, if $n \geqslant m$ then for any $1 \leqslant a < b \leqslant m$ and any $r, s < 0$ we have the following relations in the algebra* (8.33):

$$S^{(m)}\big(F[r]_a\, F[s]_b - F[s]_b\, F[r]_a\big) = S^{(m)}\big(F[r+s]_a - F[r+s]_b\big) \tag{8.37}$$

and

$$\big(F[r]_a\, F[s]_b - F[s]_b\, F[r]_a\big) S^{(m)} = -\big(F[r+s]_a - F[r+s]_b\big) S^{(m)}. \tag{8.38}$$

PROOF. As in the proof of Lemma 8.1.5, relations (8.37) and (8.38) follow from (8.35) with the use of (1.31), (1.73) and the identities (8.18) which are valid in the symplectic case as well. Furthermore,

$$\mathcal{F}_1\mathcal{F}_2 - \mathcal{F}_2\mathcal{F}_1 = F[-1]_1\, F[-1]_2 - F[-1]_2\, F[-1]_1 - F[-2]_1 + F[-2]_2.$$

Hence, applying (8.37) with $m = 2$ we get

$$S^{(2)}\big(\mathcal{F}_1\mathcal{F}_2 - \mathcal{F}_2\mathcal{F}_1\big) = 0$$

so that $\mathcal{F}$ satisfies (5.51), as required. □

THEOREM 8.3.2. *Suppose that $n \geqslant m$. Then all elements $\phi_{m\,a}$ belong to the Feigin–Frenkel center $\mathfrak{z}(\widehat{\mathfrak{sp}}_{2n})$.*

PROOF. We will repeat the arguments used for the proof of Theorem 8.1.6 and indicate a few changes to be made. Consider the tensor product algebra

$$\underbrace{\mathrm{End}\,\mathbb{C}^N \otimes \dots \otimes \mathrm{End}\,\mathbb{C}^N}_{m+1} \otimes \mathrm{U} \tag{8.39}$$

with $m+1$ copies of $\operatorname{End}\mathbb{C}^N$ labeled by $0,1,\dots,m$, where U stands for the universal enveloping algebra of the Lie algebra $\widehat{\mathfrak{sp}}_{2n}\oplus\mathbb{C}\tau$. We need to verify the relations

$$F[0]_0\operatorname{tr}_{1,\dots,m}S^{(m)}\mathcal{F}_1\dots\mathcal{F}_m=0\quad\text{and}\quad F[1]_0\operatorname{tr}_{1,\dots,m}S^{(m)}\mathcal{F}_1\dots\mathcal{F}_m=0 \tag{8.40}$$

modulo the left ideal of U generated by $\mathfrak{sp}_{2n}[t]$ and $K+h^\vee$. The first relation follows by the same calculation as for $\mathfrak{o}_N$ which relies on a consequence of (8.35) given by

$$[F[0]_0,\mathcal{F}_a]=\Phi_{0\,a}\,\mathcal{F}_a-\mathcal{F}_a\,\Phi_{0\,a}, \tag{8.41}$$

where we set $\Phi=P-Q$. To prove the second relation in (8.40) use another consequence of (8.35),

$$[F[1]_0,\mathcal{F}_a]=F[0]_0+\Phi_{0\,a}\,F[0]_a-F[0]_a\,\Phi_{0\,a}+2\,\Phi_{0\,a}\,K,$$

so that

$$F[1]_0\operatorname{tr}_{1,\dots,m}S^{(m)}\mathcal{F}_1\dots\mathcal{F}_m=\sum_{a=1}^{m}\operatorname{tr}_{1,\dots,m}S^{(m)}\mathcal{F}_1\dots\mathcal{F}_{a-1}$$
$$\times\Big(F[0]_0+\Phi_{0\,a}\,F[0]_a-F[0]_a\,\Phi_{0\,a}+2\,\Phi_{0\,a}\,K\Big)\mathcal{F}_{a+1}\dots\mathcal{F}_m.$$

This expression is then transformed in the same way as for the orthogonal case, taking into account that now

$$S^{(m)}P_{a\,b}=P_{a\,b}\,S^{(m)}=-S^{(m)}$$

for $a<b$. Repeating the respective arguments, we come to a counterpart of (8.24):

$$F[1]_0\operatorname{tr}_{1,\dots,m}S^{(m)}\mathcal{F}_1\dots\mathcal{F}_m=-\sum_{a=1}^{m-1}2\,a\operatorname{tr}_{1,\dots,m}\Phi_{0\,a}\,S^{(m)}\mathcal{F}_1\dots\mathcal{F}_{m-1}$$
$$+\big(2mK+m(m-1)\big)\operatorname{tr}_{1,\dots,m}S^{(m)}\Phi_{0\,m}\,\mathcal{F}_1\dots\mathcal{F}_{m-1}. \tag{8.42}$$

The following is a symplectic analogue of Lemma 8.1.7

LEMMA 8.3.3. *If $m\leqslant n$ then for the m-th partial trace we have*

$$\operatorname{tr}_m S^{(m)}\Phi_{0\,m}=-\frac{n-m+1}{m\,(n-m+2)}\,S^{(m-1)}\sum_{a=1}^{m-1}\Phi_{0\,a}.$$

PROOF. The argument is the same as for Lemma 2.2.7, where the role of F_m is taken by $\Phi_{0\,m}$. □

By calculating the partial trace tr_m on the right hand side of (8.42) with the use of (5.43) and Lemma 8.3.3, we get

$$F[1]_0\operatorname{tr}_{1,\dots,m}S^{(m)}\mathcal{F}_1\dots\mathcal{F}_m=-\frac{n-m+1}{m\,(n-m+2)}$$
$$\times\Big((2n-m+3)\sum_{a=1}^{m-1}2\,a\operatorname{tr}_{1,\dots,m-1}\Phi_{0\,a}\,S^{(m-1)}\mathcal{F}_1\dots\mathcal{F}_{m-1}$$
$$+m\,(2K+m-1)\sum_{a=1}^{m-1}\operatorname{tr}_{1,\dots,m-1}\Phi_{0\,a}\,S^{(m-1)}\mathcal{F}_1\dots\mathcal{F}_{m-1}\Big).$$

Now prove a counterpart of Lemma 8.1.8.

LEMMA 8.3.4. *If $m \leqslant n$ then for any $1 \leqslant a < b \leqslant m$ we have*

$$\mathrm{tr}_{1,\dots,m}\, \Phi_{0\,a}\, S^{(m)} \mathcal{F}_1 \dots \mathcal{F}_m = \mathrm{tr}_{1,\dots,m}\, \Phi_{0\,b}\, S^{(m)} \mathcal{F}_1 \dots \mathcal{F}_m.$$

PROOF. By Lemma 8.3.1, the matrix $\mathcal{F}$ is a Manin matrix of type C so that by the property (5.55) we find that

$$\begin{aligned}\mathrm{tr}_{1,\dots,m}\, \Phi_{0\,a}\, S^{(m)} \mathcal{F}_1 \dots \mathcal{F}_m &= \mathrm{tr}_{1,\dots,m} \Phi_{0\,a}\, S^{(m)}\, \mathcal{F}_1 \dots \mathcal{F}_m\, P_{a\,b} \\ &= \mathrm{tr}_{1,\dots,m} P_{a\,b} \Phi_{0\,a}\, S^{(m)}\, \mathcal{F}_1 \dots \mathcal{F}_m\end{aligned}$$

where we used the cyclic property of trace. This expression equals

$$\mathrm{tr}_{1,\dots,m} \Phi_{0\,b}\, P_{a\,b}\, S^{(m)}\, \mathcal{F}_1 \dots \mathcal{F}_m = \mathrm{tr}_{1,\dots,m} \Phi_{0\,b}\, S^{(m)}\, \mathcal{F}_1 \dots \mathcal{F}_m,$$

as required. □

The proof of the theorem is now completed by the application of Lemma 8.3.4 with m replaced by $m-1$. We have

$$\begin{aligned}&F[1]_0\, \mathrm{tr}_{1,\dots,m}\, S^{(m)} \mathcal{F}_1 \dots \mathcal{F}_m \\ &\qquad = -2\,(K+n+1)\, \frac{n-m+1}{n-m+2} \sum_{a=1}^{m-1} \mathrm{tr}_{1,\dots,m-1}\, S^{(m-1)} \Phi_{0\,a}\, \mathcal{F}_1 \dots \mathcal{F}_{m-1}\end{aligned}$$

which is zero since $K+n+1=0$ at the critical level. □

Our next goal is to extend the definition of the elements $\phi_{m\,a}$ to the values $n \geqslant (m-1)/2$. Consider the symmetric algebra $\mathrm{S}\,(t^{-1}\mathfrak{sp}_{2n}[t^{-1}])$ as a module over $\mathfrak{sp}_{2n}$ defined in (6.18) with $r=0$, so that $\mathfrak{sp}_{2n}$ is identified with a subalgebra of $\mathfrak{sp}_{2n}[t]$ via the embedding $Y \mapsto Y[0]$. It follows from the results of Section 2.2 that for any positive integers $r_1, \dots, r_k$ the elements of $\mathrm{S}\,(t^{-1}\mathfrak{sp}_{2n}[t^{-1}])$ defined by

$$\mathrm{tr}\, F[-r_1] \dots F[-r_k] \tag{8.43}$$

are $\mathfrak{sp}_{2n}$-invariants. Furthermore, by a standard application of the first fundamental theorem of the invariant theory for the symplectic group, the algebra of invariants $\mathrm{S}\,(t^{-1}\mathfrak{sp}_{2n}[t^{-1}])^{\mathfrak{sp}_{2n}}$ is generated by elements of the form (8.43). We have the graded decomposition of this algebra into finite-dimensional subspaces,

$$\mathrm{S}\,(t^{-1}\mathfrak{sp}_{2n}[t^{-1}])^{\mathfrak{sp}_{2n}} = \bigoplus_{d \geqslant 0} \mathrm{S}^d_n, \tag{8.44}$$

where S^d_n denotes the subspace of homogeneous elements of degree d with respect to the grading on the symmetric algebra defined by $\deg F_{ij}[-r] = r$.

For all $n \geqslant 2$ identify $\mathfrak{sp}_{2n-2}$ with the subalgebra of $\mathfrak{sp}_{2n}$ spanned by all generators F_{ij} with $i, j \in \{1, \dots, n-1, (n-1)', \dots, 1'\}$. Introduce the projections $\pi_n : \mathfrak{sp}_{2n} \to \mathfrak{sp}_{2n-2}$ such that $\pi_n(F_{ij}) = 0$ for all $F_{ij} \notin \mathfrak{sp}_{2n-2}$. The projection π_n extends to the vector space $t^{-1}\mathfrak{sp}_{2n}[t^{-1}]$ where it acts on the coefficients of polynomials in t^{-1}. We will keep the same notation for the associated homomorphism of symmetric algebras

$$\pi_n : \mathrm{S}\,(t^{-1}\mathfrak{sp}_{2n}[t^{-1}]) \to \mathrm{S}\,(t^{-1}\mathfrak{sp}_{2n-2}[t^{-1}]).$$

Note that the images of the elements (8.43) under π_n take the same trace form and so belong to the algebra of $\mathfrak{sp}_{2n-2}$-invariants in $\mathrm{S}\,(t^{-1}\mathfrak{sp}_{2n-2}[t^{-1}])$.

LEMMA 8.3.5. *For each fixed value $d \geqslant 0$ there exists a nonnegative integer n_0 such that the dimension of the subspace S_n^d is independent of n for all $n \geqslant n_0$. These subspaces possess respective bases B_n^d formed by monomials in the elements (8.43) such that $\pi_n(B_n^d) = B_{n-1}^d$ for all $n > n_0$. Moreover, all subspaces S_n^d with $1 \leqslant n < n_0$ are spanned by the elements $(\pi_n \circ \pi_{n+1} \circ \dots \circ \pi_{n_0})(B_{n_0}^d)$.*

PROOF. Starting from $n = 1$, use the following procedure for consecutive values of n. Choose a basis of S_n^d formed by monomials in the elements (8.43). The same monomials regarded as elements of S_{n+1}^d (with $F[-r_i]$ understood as generator matrices for $\mathfrak{sp}_{2n+2}$) will be linearly independent, as follows by the application of the projection π_{n+1}. Hence, $\dim \mathrm{S}_n^d \leqslant \dim \mathrm{S}_{n+1}^d$. If the inequality is strict, then extend this family of linearly independent monomials of S_{n+1}^d to a basis by adjoining some other monomials in the elements of the form (8.43). Since the algebra $\mathrm{S}\left(t^{-1}\mathfrak{sp}_{2n}[t^{-1}]\right)^{\mathfrak{sp}_{2n}}$ is generated by the elements (8.43), the dimension of the subspace S_n^d has an upper bound independent of n. This ensures that the process stabilizers at a certain value $n = n_0$ thus producing a uniform basis for the spaces S_n^d for $n \geqslant n_0$, as required. □

The following is a symplectic analogue of Lemma 8.2.1.

LEMMA 8.3.6. *Suppose that $n \geqslant m$. Let X be a certain linear combination of products of elements $F[s]_b$ of the algebra (8.33) with $s < 0$ and $1 \leqslant b \leqslant m - 1$. Then for any $r < 0$ we have*

$$\mathrm{tr}_{1,\dots,m}\, S^{(m)} X\, F[r]_m = -\frac{n-m+1}{m(n-m+2)}\, \mathrm{tr}_{1,\dots,m-1}\, S^{(m-1)} X \sum_{a=1}^{m-1} F[r]_a.$$

PROOF. This is just an affine version of Lemma 2.2.7. The proof amounts to the calculation of the partial trace as in (2.55), where F_m should be replaced by $F[r]_m$. □

PROPOSITION 8.3.7. *The expression in (8.36) admits an equivalent form which represents a well-defined element of $\mathfrak{z}(\widehat{\mathfrak{sp}}_{2n})$ for all $n \geqslant (m-1)/2$.*

PROOF. Writing the product $\left(\tau + F[-1]_1\right) \dots \left(\tau + F[-1]_m\right)$ as a polynomial in τ, represent each coefficient ϕ_{ma} as a linear combination of expressions of the form

$$\frac{1}{n-m+1}\, \mathrm{tr}_{1,\dots,m}\, S^{(m)} F[r_1]_{a_1} \dots F[r_k]_{a_k}, \tag{8.45}$$

where the r_i are negative integers and $1 \leqslant a_1 < \dots < a_k \leqslant m$. We proceed by the same argument as in the proof of Proposition 5.5.3 with the use of (5.43) together with Lemmas 8.3.1 and 8.3.6. As a result, each expression (8.45) will be represented as a linear combination of expressions of the form

$$\frac{1}{n-h+1}\, \mathrm{tr}_{1,\dots,h}\, S^{(h)} F[s_1]_{b_1} \dots F[s_p]_{b_p}, \tag{8.46}$$

where $h = \lfloor m/2 \rfloor$, the s_i are negative integers and all indices $b_1, \dots, b_p$ belong to the set $\{1, \dots, h\}$. Such a linear combination defines an element of $\mathrm{U}\left(t^{-1}\mathfrak{sp}_{2n}[t^{-1}]\right)$ for all $n \geqslant h$ (which is equivalent to the condition $n \geqslant (m-1)/2$).

All possible expressions representing (8.36), which are obtained by the above procedure coincide as elements of the vacuum module for $n \geqslant m$ and belong to $\mathfrak{z}(\widehat{\mathfrak{sp}}_{2n})$ by Theorem 8.3.2. We need to show that this property extends to all values

$n \geqslant h$. We will regard $\mathrm{U}\left(t^{-1}\mathfrak{sp}_{2n}[t^{-1}]\right)$ as the adjoint $\mathfrak{sp}_{2n}$-module defined via the embedding $\mathfrak{sp}_{2n} \hookrightarrow \widehat{\mathfrak{sp}}_{2n}$ such that $Y \mapsto Y[0]$. Consider the grading on the universal enveloping algebra defined by the derivation D in (6.14) so that homogeneous elements of degree d are eigenvectors of D with the eigenvalue d. We have the graded decomposition of the algebra of $\mathfrak{sp}_{2n}$-invariants, analogous to (8.44),

$$\mathrm{U}\left(t^{-1}\mathfrak{sp}_{2n}[t^{-1}]\right)^{\mathfrak{sp}_{2n}} = \bigoplus_{d \geqslant 0} \mathrm{U}_n^d,$$

where U_n^d denotes the subspace of homogeneous elements of degree d. The canonical symmetrization map

$$\mathrm{S}\left(t^{-1}\mathfrak{sp}_{2n}[t^{-1}]\right) \to \mathrm{U}\left(t^{-1}\mathfrak{sp}_{2n}[t^{-1}]\right)$$

is an isomorphism of the graded $\mathfrak{sp}_{2n}$-modules. Hence, by Lemma 8.3.5 for any fixed $d \geqslant 0$ the subspaces U_n^d possess respective bases chosen in a uniform way for all $n \geqslant n_0$, where n_0 is a certain nonnegative integer depending on d. In particular, a basis $\{T_l\}_{l \in I}$ of U_n^d can be labeled by a set I which only depends on d and does not depend on $n \geqslant n_0$. Moreover, the set $\{T_l\}_{l \in I}$ spans U_n^d for all $n \geqslant 1$.

All expressions representing the coefficient $\phi_{m\,a}$ in (8.36) are homogeneous elements of degree a. The argument in the first part of the proof of Theorem 8.3.2 based on relation (8.41) can be easily extended to demonstrate that each component (8.46) with $s_1 + \cdots + s_p = -a$ belongs to U_n^a. Therefore, any such expression can be represented as a unique linear combination $\sum_{l \in I} c_l(n) T_l$ of the basis vectors of U_n^a, where the coefficients $c_l(n)$ are $\mathbb{C}$-valued rational functions in n with $n \geqslant n_0$ for a certain value n_0 depending on a. By Lemma 8.3.5, this representation extends to all values $n \geqslant h$. Since all possible expressions representing (8.36) coincide for all $n \geqslant m$, and the rational functions $c_l(n)$ are determined by infinitely many values of n, these expressions should coincide for all values $n \geqslant h$ for which they are defined.

Finally, observe that an element $\phi \in \mathrm{U}_n^a$ belongs to $\mathfrak{z}(\widehat{\mathfrak{sp}}_{2n})$ if and only if $F_{11}[1]\phi = 0$ in the vacuum module $V_{-h^\vee}(\mathfrak{sp}_{2n})$. Therefore, the condition that $\phi_{m\,a} \in \mathfrak{z}(\widehat{\mathfrak{sp}}_{2n})$ for $n \geqslant m$ implies a system of linear equations on the coefficients $c_l(n)$ arising from the property $F_{11}[1]\phi_{m\,a} = 0$ in $V_{-h^\vee}(\mathfrak{sp}_{2n})$. Since this system is satisfied by the coefficients $c_l(n)$ for infinitely many values of n, this shows that any expression representing $\phi_{m\,a}$ belongs to $\mathfrak{z}(\widehat{\mathfrak{sp}}_{2n})$ for each $n \geqslant h$. □

The above argument will be illustrated below in Section 8.4 by calculations of $\phi_{2\,2}$ and $\phi_{4\,4}$; cf. Section 8.2. Proposition 8.3.7 allows us to extend the definition of the elements $\phi_{m\,a}$ in (8.36) to all values $1 \leqslant m \leqslant 2n+1$. We will keep the same notation (8.36) for all these values with the understanding that it represents a well-defined element of $\mathfrak{z}(\widehat{\mathfrak{sp}}_{2n})$ obtained in the proof of Proposition 8.3.7. Theorem 8.3.2 thus extends to all values $1 \leqslant m \leqslant 2n+1$.

The next theorem implies Theorem 6.3.1 for type C.

THEOREM 8.3.8. *The family $\phi_{2\,2}, \phi_{4\,4}, \dots, \phi_{2n\,2n}$ is a complete set of Segal–Sugawara vectors for $\mathfrak{sp}_{2n}$.*

PROOF. The symbols of the elements of this family coincide with the images of certain elements of the symmetric algebra $\mathrm{S}(\mathfrak{sp}_{2n})$ under the embedding $\mathrm{S}(\mathfrak{sp}_{2n}) \hookrightarrow \mathrm{S}\left(t^{-1}\mathfrak{sp}_{2n}[t^{-1}]\right)$ taking an element $X \in \mathfrak{sp}_{2n}$ to $X[-1]$. Due to Theorem 6.3.3 and the above extension of Theorem 8.3.2, we only need to verify that the corresponding elements of $\mathrm{S}(\mathfrak{sp}_{2n})$ are algebraically independent generators of the algebra of invariants $\mathrm{S}(\mathfrak{sp}_{2n})^{\mathfrak{sp}_{2n}}$; see also the argument following the proof of Theorem 6.3.3. However, this was proved in Section 2.2; see Corollary 2.2.10. $\square$

The special role of the parameters satisfying the conditions $n+1 \leqslant m \leqslant 2n+1$ can be understood from the properties of the homomorphism

$$\mathcal{B}_m(-2n) \to \operatorname{End}_{\mathfrak{sp}_{2n}} \big(\mathbb{C}^{2n}\big)^{\otimes m} \tag{8.47}$$

defined by (1.73). The Brauer algebra $\mathcal{B}_m(-2n)$ is known to be semisimple for the values $m \leqslant 2n+1$; see [**2**], [**138**]. The homomorphism (8.47) is an isomorphism for $m \leqslant n$ but it has a nonzero kernel for the values $n+1 \leqslant m \leqslant 2n+1$. In particular, if $m = n+1$ then the kernel is one-dimensional and spanned by the symmetrizer $s^{(m)}$. To get explicit formulas for Segal–Sugawara vectors in this case, note the recurrence formula for the symmetrizer $S^{(m)}$; cf. the proof of Lemma 2.2.7:

$$S^{(m)} = \frac{1}{2m(n-m+2)}$$
$$\times \Big(2n - 2m + 4 - (2n-m+2) \sum_{a=1}^{m-1} P_{a\,m} - m \sum_{a=1}^{m-1} Q_{a\,m} - \sum_{a\neq b} \Phi_{a\,m}\Phi_{b\,m}\Big)\, S^{(m-1)},$$

where $\Phi_{a\,m} = P_{a\,m} - Q_{a\,m}$ for $a = 1,\dots,m-1$. Hence, applying the transposition (2.27) with respect to the m-th copy of $\operatorname{End}\mathbb{C}^{2n}$, we get

$$S^{(m)\prime} - S^{(m)} = -\frac{n-m+1}{m(n-m+2)}\, S^{(m-1)}\, Y^{(m)},$$

where

$$Y^{(m)} = -\sum_{a=1}^{m-1} \Phi_{a\,m}$$

is the image of the Jucys–Murphy element $y_m - (\omega-1)/2$ under the map (1.73); see (1.26). Therefore, assuming $n \geqslant m$ and using (8.30) for the m-th partial transposition, we obtain that the trace

$$\operatorname{tr}_{1,\dots,m} S^{(m)} \mathcal{F}_1 \dots \mathcal{F}_m$$

equals

$$\operatorname{tr}_{1,\dots,m} S^{(m)} \mathcal{F}_1 \dots \mathcal{F}_{m-1} \big(\tau - F[-1]_m\big)$$
$$- \frac{n-m+1}{m(n-m+2)} \operatorname{tr}_{1,\dots,m} S^{(m-1)}\, Y^{(m)} \mathcal{F}_1 \dots \mathcal{F}_{m-1} \big(\tau - F[-1]_m\big).$$

Since $Y^{(m)\prime} = -Y^{(m)}$, this implies

$$\operatorname{tr}_{1,\dots,m} S^{(m)} \mathcal{F}_1 \dots \mathcal{F}_m = \operatorname{tr}_{1,\dots,m} S^{(m)} \mathcal{F}_1 \dots \mathcal{F}_{m-1}\, \tau$$
$$+ \frac{n-m+1}{2m(n-m+2)} \operatorname{tr}_{1,\dots,m} S^{(m-1)}\, Y^{(m)} \mathcal{F}_1 \dots \mathcal{F}_m.$$

Together with (5.43) this gives

$$\frac{1}{n-m+1}\operatorname{tr}_{1,\dots,m} S^{(m)}\mathcal{F}_1\dots\mathcal{F}_m$$
$$= \frac{2n-m+3}{m(n-m+2)}\operatorname{tr}_{1,\dots,m} S^{(m-1)}\mathcal{F}_1\dots\mathcal{F}_{m-1}\tau$$
$$+\frac{1}{2m(n-m+2)}\operatorname{tr}_{1,\dots,m} S^{(m-1)}Y^{(m)}\mathcal{F}_1\dots\mathcal{F}_m$$

which can be taken as a definition of the left hand side for $n=m-1$. This definition is essentially equivalent to the one obtained in the proof of Proposition 8.3.7. We can now conclude that all coefficients of the polynomial in τ

$$\operatorname{tr}_{1,\dots,m} S^{(m-1)}Y^{(m)}\mathcal{F}_1\dots\mathcal{F}_m$$

are Segal–Sugawara vectors for $\mathfrak{sp}_{2n}$ for all $n\geqslant m-1$. The branching properties (1.30) of the primitive idempotents imply the relation

$$S^{(m-1)}=S^{(m)}+\mathcal{E}_{\mathcal{U}}+\mathcal{E}_{\mathcal{V}},$$

where $\mathcal{E}_{\mathcal{U}}$ and $\mathcal{E}_{\mathcal{V}}$ denote the images of the idempotents $e_{\mathcal{U}}$ and $e_{\mathcal{V}}$ under the map (1.73), which are associated with the respective updown tableaux

$$\mathcal{U}=\big((1),(2),\dots(m-1),(m-1,1)\big)\quad\text{and}\quad\mathcal{V}=\big((1),(2),\dots,(m-1),(m-2)\big).$$

Furthermore, by (1.27) we have

$$S^{(m)}Y^{(m)}=(m-1)\,S^{(m)},\qquad \mathcal{E}_{\mathcal{U}}Y^{(m)}=-\mathcal{E}_{\mathcal{U}}\quad\text{and}\quad\mathcal{E}_{\mathcal{V}}Y^{(m)}=(2n-m+3)\,\mathcal{E}_{\mathcal{V}}$$

and so

$$S^{(m-1)}\big(1+Y^{(m)}\big)=m\,S^{(m)}+(2n-m+4)\,\mathcal{E}_{\mathcal{V}}$$

and

$$S^{(m-1)}\big(2n-m+3-Y^{(m)}\big)=2(n-m+2)\,S^{(m)}+(2n-m+4)\,\mathcal{E}_{\mathcal{U}}.$$

This implies that all coefficients of the polynomials

$$\operatorname{tr}_{1,\dots,m}\mathcal{E}_{\mathcal{U}}\mathcal{F}_1\dots\mathcal{F}_m\qquad\text{and}\qquad\operatorname{tr}_{1,\dots,m}\mathcal{E}_{\mathcal{V}}\mathcal{F}_1\dots\mathcal{F}_m$$

are Segal–Sugawara vectors for $\mathfrak{sp}_{2n}$ for all $n\geqslant m-1$. It would be interesting to find analogous formulas for Segal–Sugawara vectors for the values $n\geqslant(m-1)/2$ together with their direct proofs.

To conclude this section we give equivalent expressions for the Segal–Sugawara vectors ϕ_{ma} in the context of skew Howe duality. In the notation of Section 2.3, introduce the element $\mathcal{F}^{(m)}\in\operatorname{End}\Lambda^m_{2n}\otimes\mathrm{U}\big(\widehat{\mathfrak{sp}}_{2n}\oplus\mathbb{C}\tau\big)$ by

$$\mathcal{F}^{(m)}:\zeta_{j_1}\wedge\dots\wedge\zeta_{j_m}\mapsto\sum_{i_1<\dots<i_m}\zeta_{i_1}\wedge\dots\wedge\zeta_{i_m}\otimes\mathcal{F}^{i_1\dots i_m}_{j_1\dots j_m},$$

where

$$\mathcal{F}^{i_1\dots i_m}_{j_1\dots j_m}=\frac{1}{m!}\sum_{\sigma,\pi\in\mathfrak{S}_m}\operatorname{sgn}\sigma\pi\cdot\mathcal{F}_{i_{\sigma(1)}j_{\pi(1)}}\dots\mathcal{F}_{i_{\sigma(m)}j_{\pi(m)}}.$$

We will regard the extremal projector p defined by the equivalent formulas (2.64) and (2.66) as an element of $\operatorname{End}\Lambda^m_{2n}$ via the action (2.69).

COROLLARY 8.3.9. *The Segal–Sugawara vectors* $\phi_{m\,a}$ *with* $m \leqslant n$ *can be found from the expansion*

$$\gamma_m(-2n)\,\mathrm{tr}\,p\,\mathcal{F}^{(m)} = \phi_{m\,0}\,\tau^m + \phi_{m\,1}\,\tau^{m-1} + \dots + \phi_{m\,m}$$

with the trace taken over the subspace of $\mathfrak{sl}_2$*-singular vectors in* Λ^m_{2n}.

PROOF. Using the cyclic property of trace and applying the formula (2.52) for the symmetrizer, write the left hand side of (8.36) in the form

$$\gamma_m(-2n)\,\mathrm{tr}_{1,\dots,m}\, S^{(m)}\mathcal{F}_1\dots\mathcal{F}_m = \gamma_m(-2n)\,\mathrm{tr}_{1,\dots,m}\, S^{(m)}\,H^{(m)}\mathcal{F}_1\dots\mathcal{F}_m\,H^{(m)}.$$

The product $H^{(m)}\mathcal{F}_1\dots\mathcal{F}_m H^{(m)}$ can be regarded as an operator on the vector space Λ^m_{2n} with coefficients in $\mathrm{U}\big(\widehat{\mathfrak{sp}}_{2n}\oplus\mathbb{C}\tau\big)$, due to the isomorphism (2.67). This operator coincides with $\mathcal{F}^{(m)}$. Finally, by Proposition 2.3.2, the operator $S^{(m)}$ acts on Λ^m_{2n} as the extremal projector p. □

8.4. Low degree invariants in trace form

As we pointed out in Section 8.2, the coefficients of the polynomials $\mathrm{tr}\,\mathcal{F}^m$ are not, in general, Segal–Sugawara vectors for $\mathfrak{sp}_{2n}$. In this section we construct examples of such vectors in trace form by using Theorem 8.3.8. We will calculate the Segal–Sugawara vectors $\phi_{2\,2}$ and $\phi_{4\,4}$ in a way similar to the orthogonal case; see Section 8.2. By (1.46) and (1.73) for $n \geqslant 2$ we get

$$\phi_{2\,2} = \frac{n}{n-1}\,\mathrm{tr}_{1,2}\Big(\frac{1}{2} - \frac{P_{12}}{2} - \frac{Q_{12}}{2n}\Big)\big(F[-1]_1F[-1]_2 + F[-2]_2\big).$$

The same argument as in the orthogonal case yields

$$\phi_{2\,2} = -\frac{1}{2}\,\mathrm{tr}\,F[-1]^2 = -\frac{1}{2}\sum_{i,j=1}^{N} F_{ij}[-1]F_{ji}[-1]$$

which is proportional to the canonical Segal–Sugawara vector (6.11) for all $n \geqslant 1$.

For $n \geqslant 4$ the Segal–Sugawara vector $\phi_{4\,4}$ is found as the constant term of the polynomial in τ defined by

$$\frac{n-1}{n-3}\,\mathrm{tr}_{1,2,3,4}\,S^{(4)}\big(\tau + F[-1]_1\big)\big(\tau + F[-1]_2\big)\big(\tau + F[-1]_3\big)F[-1]_4.$$

Applying Lemma 8.3.6 we can write this polynomial in the form

$$(8.48)\qquad -\frac{n-1}{4\,(n-2)}\,\mathrm{tr}_{1,2,3}\,S^{(3)}\big(\tau + F[-1]_1\big)\big(\tau + F[-1]_2\big)\big(\tau + F[-1]_3\big) \times \big(F[-1]_1 + F[-1]_2 + F[-1]_3\big).$$

Due to Lemma 8.3.1, $\tau + F[-1]$ is a Manin matrix of type C so that using relation (5.55) we get

$$\mathrm{tr}_{1,2,3}\,S^{(3)}\big(\tau + F[-1]_1\big)\big(\tau + F[-1]_2\big)\big(\tau + F[-1]_3\big)F[-1]_2 = \mathrm{tr}_{1,2,3}\,S^{(3)}\big(\tau + F[-1]_2\big)\big(\tau + F[-1]_1\big)\big(\tau + F[-1]_3\big)F[-1]_2.$$

Applying now the conjugation by P_{12} and using the cyclic property of trace we can write this expression as

$$\mathrm{tr}_{1,2,3}\,S^{(3)}\big(\tau + F[-1]_1\big)\big(\tau + F[-1]_2\big)\big(\tau + F[-1]_3\big)F[-1]_1.$$

A similar calculation applied to the third summand in (8.48) shows that ϕ_{44} equals the constant term of the polynomial

$$-\frac{3(n-1)}{4(n-2)}\,\mathrm{tr}_{1,2,3}\, S^{(3)}\,(\tau+F[-1]_1)(\tau+F[-1]_2)(\tau+F[-1]_3)F[-1]_1.$$

Expanding and using Lemma 8.3.1 we get

$$\begin{aligned}\phi_{44}=-\frac{3(n-1)}{4(n-2)}\,\mathrm{tr}_{1,2,3}\, S^{(3)}\Big(&F[-1]_1^2F[-1]_2F[-1]_3+F[-1]_1F[-1]_2F[-2]_3\\&+4F[-1]_1F[-2]_1F[-1]_2+F[-1]_1^2F[-2]_2+3F[-2]_1F[-2]_2\\&+2F[-1]_1F[-3]_2+6F[-1]_1F[-3]_1+6F[-4]_1\Big).\end{aligned}$$

Now use Lemma 8.3.6 and relation (5.43) for $m=3$ to bring this to the form

$$\begin{aligned}\phi_{44}=\frac{1}{8}\,\mathrm{tr}_{1,2}\Big(1-P_{12}-\frac{Q_{12}}{n}\Big)\Big(&F[-1]_1^2F[-1]_2(F[-1]_1+F[-1]_2)\\+F[-1]_1F[-1]_2(F[-1]_1+F[-1]_2)&-8n\,F[-1]_1F[-2]_1F[-1]_2-2n\,F[-1]_1^2F[-2]_2\\-6n\,F[-2]_1F[-2]_2-4n\,F[-1]_1F[-3]_2&-12n\,F[-1]_1F[-3]_1-12n\,F[-4]_1\Big).\end{aligned}$$

Calculating the traces with the use of (8.35) and the identity $\big(F[r]_1+F[r]_2\big)\,Q_{12}=0$, we obtain

$$\begin{aligned}\phi_{44}=\frac{1}{8}\Big(\big(\mathrm{tr}\,F[-1]^2\big)^2-2\,\mathrm{tr}\,F[-1]^4+(4n-1)\,\mathrm{tr}\,F[-1]^2F[-2]+(6n-6)\,\mathrm{tr}\,F[-2]^2\\+(8n-9)\,\mathrm{tr}\,F[-1]F[-2]F[-1]+(-24n^2+18n+10)\,\mathrm{tr}\,F[-1]F[-3]\Big).\end{aligned}$$

The following relations are analogous to the orthogonal case and verified in the same way:

$$\mathrm{tr}\,F[-1]^2F[-2]=(n+1)\,\mathrm{tr}\,F[-2]^2$$

and

$$\mathrm{tr}\,F[-1]F[-2]F[-1]=-(n+1)\,\mathrm{tr}\,F[-2]^2+(2n+2)\,\mathrm{tr}\,F[-1]F[-3].$$

Hence, the formula for ϕ_{44} simplifies to

$$\begin{aligned}\phi_{44}=\frac{1}{8}\Big(\big(\mathrm{tr}\,F[-1]^2\big)^2-2\,\mathrm{tr}\,F[-1]^4\\-8\,(n-1)^2\,\mathrm{tr}\,F[-1]F[-3]+(-4n^2+10n+2)\,\mathrm{tr}\,F[-2]^2\Big).\end{aligned}$$

This is a Segal–Sugawara vector for all $n\geqslant 2$. In fact, ϕ_{44} is well-defined for $n=1$ as well, but is equal to zero because $S^{(2)}=0$.

Since the Feigin–Frenkel center $\mathfrak{z}(\widehat{\mathfrak{sp}}_{2n})$ is invariant with respect to the translation operator T defined in (6.7), the elements $T\phi_{22}$ and $T^2\phi_{22}$ are Segal–Sugawara vectors. We have

$$T\phi_{22}=-\mathrm{tr}\,F[-1]F[-2]\qquad\text{and}\qquad T^2\phi_{22}=-\mathrm{tr}\,F[-2]^2-2\,\mathrm{tr}\,F[-1]F[-3].$$

Therefore, the element ϕ_{44} in the complete set of Segal–Sugawara vectors of Theorem 8.3.8 can be replaced with the element

$$-4\phi_{44}+2\phi_{22}^2+2\,(n-1)^2\,T^2\phi_{22}$$

which coincides with

$$\operatorname{tr} F[-1]^4 - (n+3) \operatorname{tr} F[-2]^2. \tag{8.49}$$

EXAMPLE 8.4.1. The following are complete sets of Segal–Sugawara vectors:

$$\text{for} \quad \mathfrak{sp}_2 : \qquad \operatorname{tr} F[-1]^2$$

$$\text{for} \quad \mathfrak{sp}_4 : \qquad \operatorname{tr} F[-1]^2, \quad \operatorname{tr} F[-1]^4 - 5 \operatorname{tr} F[-2]^2.$$

We will also give a direct proof that the element (8.49) is a Segal–Sugawara vector. In the next lemma we use the tensor product algebra (8.39).

LEMMA 8.4.2. *For any $m \geqslant 1$ we have the relations*

$$F[0]_0 \operatorname{tr}_1 F[-1]_1^m = 0$$

and

$$F[1]_0 \operatorname{tr}_1 F[-1]_1^m$$
$$= \sum_{i=1}^{m} \operatorname{tr}_1 \Big(F[-1]_1^{i-1} \Phi_{01} F[-1]_1^{m-i} \Phi_{01} - \Phi_{01} F[-1]_1^{i-1} \Phi_{01} F[-1]_1^{m-i} \Big)$$

modulo the left ideal of U *generated by $\mathfrak{sp}_{2n}[t]$ and $K + n + 1$, where $\Phi = P - Q$.*

PROOF. The argument is essentially the same as for Lemma 8.2.3; it uses the relation $(\Phi_{01}^2)' = \Phi_{01}^2 + 2\,(n+1)\,\Phi_{01}$ and the corresponding version of (8.30) which holds for the transposition (2.27) in the symplectic case. □

We have $\operatorname{tr} F[-1] = 0$, and Lemma 8.4.2 implies that both $\operatorname{tr} F[-1]^2$ and $\operatorname{tr} F[-1]^3$ are Segal–Sugawara vectors for $\mathfrak{sp}_{2n}$. Now we repeat the corresponding argument of Section 8.2 taking into account (1.75) and the respective symplectic analogues of the relations used in the argument:

$$\Phi_{01}\, F[-1]_1\, \Phi_{01} = F[-1]_0 - Q_{01}\, F[-1]_1 - F[-1]_1\, Q_{01}$$

and

$$(F[-1]^2)' = F[-1]^2 - 2\,(n+1)\, F[-2].$$

This leads to the formulas

$$F[1]_0 \operatorname{tr}_1 F[-1]_1^4 = -2\,(n+3) \operatorname{tr}_1 \big(F[-2]_1\, F[-1]_1\, \Phi_{01} - \Phi_{01}\, F[-1]_1\, F[-2]_1 \big)$$

and

$$F[1]_0 \operatorname{tr}_1 F[-2]_1^2 = -2 \operatorname{tr}_1 \big(F[-2]_1\, F[-1]_1\, \Phi_{01} - \Phi_{01}\, F[-1]_1\, F[-2]_1 \big)$$

which imply

$$F[1]_0 \Big(\operatorname{tr}_1 F[-1]_1^4 - (n+3) \operatorname{tr}_1 F[-2]_1^2 \Big) = 0$$

so that (8.49) is a Segal–Sugawara vector.

8.5. Sugawara operators in types B, C and D

Here we use the general results on Sugawara operators outlined in Section 6.6 to construct generators of the completed universal enveloping algebras at the critical level. As in Section 5.1, we will consider the three cases B, C and D simultaneously and use the notation $\mathfrak{g}_N$ for any of the Lie algebras $\mathfrak{o}_N$ or $\mathfrak{sp}_N$. The critical level $\kappa = -h^\vee$ is defined in (8.1) and (8.31); see also Remark 8.1.1. Introduce the Laurent series in z with coefficients in $\mathrm{U}_{-h^\vee}(\widehat{\mathfrak{g}}_N)$ by

$$F_{ij}(z) = \sum_{r\in\mathbb{Z}} F_{ij}[r]\, z^{-r-1}, \qquad i,j = 1,\dots,N, \tag{8.50}$$

and combine them into the matrix $F(z)$ so that

$$F(z) = \sum_{i,j=1}^{N} e_{ij} \otimes F_{ij}(z).$$

As in (8.4) and (8.34), for $a \in \{1,\dots,m\}$ we use the notation $F(z)_a$ for the corresponding elements of the algebra

$$\underbrace{\mathrm{End}\,\mathbb{C}^N \otimes \dots \otimes \mathrm{End}\,\mathbb{C}^N}_{m} \otimes \mathrm{U},$$

where U stands for the algebra of polynomial differential operators over z with coefficients in $\widetilde{\mathrm{U}}_{-h^\vee}(\widehat{\mathfrak{g}}_N)[[z,z^{-1}]]$.

We will keep using the rational function $\gamma_m(\omega)$ defined in (2.40). The variable ω takes value $\omega = N$ in the orthogonal case and $\omega = -N$ in the symplectic case. As suggested by (8.15) and (8.36) together with Proposition 8.3.7, define the Laurent series $\phi_{m\,a}(z)$ by the following expansion with the usual reading of normal ordering from right to left:

$$\begin{aligned} &: \gamma_m(\omega)\,\mathrm{tr}_{1,\dots,m}\, S^{(m)}\big(\partial_z + F(z)_1\big)\dots\big(\partial_z + F(z)_m\big) : \\ &\qquad\qquad = \phi_{m\,0}(z)\,\partial_z^m + \phi_{m\,1}(z)\,\partial_z^{m-1} + \dots + \phi_{m\,m}(z). \end{aligned}$$

Following (8.6), in the case $\mathfrak{g}_N = \mathfrak{o}_{2n}$ we also introduce the (noncommutative) *Pfaffian* as the Laurent series

$$\mathrm{Pf}\ F(z) = \frac{1}{2^n n!} \sum_{\sigma\in\mathfrak{S}_{2n}} \mathrm{sgn}\,\sigma\ : F_{\sigma(1)\,\sigma(2)'}(z)\dots F_{\sigma(2n-1)\,\sigma(2n)'}(z) : . \tag{8.51}$$

The results of Section 8.1 in the orthogonal case and Section 8.3 in the symplectic case together with Proposition 6.6.1 lead to the following.

THEOREM 8.5.1. *All coefficients of the Laurent series $\phi_{m\,a}(z)$, belong to the center of the algebra $\widetilde{\mathrm{U}}_{-h^\vee}(\widehat{\mathfrak{g}}_N)$. Moreover, the coefficients of the Laurent series $\mathrm{Pf}\ F(z)$ belong to the center of $\widetilde{\mathrm{U}}_{-h^\vee}(\widehat{\mathfrak{o}}_{2n})$. The coefficients of the Laurent series*

$$\phi_{2\,2}(z), \phi_{4\,4}(z), \dots, \phi_{2n\,2n}(z)$$

are topological generators of the center of $\widetilde{\mathrm{U}}_{-h^\vee}(\widehat{\mathfrak{g}}_N)$ in the cases $\mathfrak{g}_N = \mathfrak{o}_{2n+1}$ and $\mathfrak{g}_N = \mathfrak{sp}_{2n}$. Furthermore, the coefficients of the Laurent series

$$\phi_{2\,2}(z), \phi_{4\,4}(z), \dots, \phi_{2n-2\,2n-2}(z), \mathrm{Pf}\ F(z)$$

are topological generators of the center of $\widetilde{\mathrm{U}}_{-h^\vee}(\widehat{\mathfrak{o}}_{2n})$. □

EXAMPLE 8.5.2. By the calculations of Section 8.2 and Section 8.4 for the Segal–Sugawara vector ϕ_{22}, the Laurent series $\phi_{22}(z)$ is a scalar multiple of the series $:\operatorname{tr} F(z)^2:$. We have

$$:\operatorname{tr} F(z)^2: \; = \; : \sum_{i,j=1}^{N} F_{ij}(z) F_{ji}(z) := \sum_{i,j=1}^{N} \Big(F_{ij}(z)_+ F_{ji}(z) + F_{ji}(z) F_{ij}(z)_- \Big).$$

Hence for each $p \in \mathbb{Z}$ the coefficient of z^{-p-2} in this series is a central element of $\widetilde{\mathrm{U}}_{-h^\vee}(\widehat{\mathfrak{g}}_N)$ given by

$$\sum_{i,j=1}^{N} \Big(\sum_{r<0} F_{ij}[r] F_{ji}[p-r] + \sum_{r\geqslant 0} F_{ji}[p-r] F_{ij}[r] \Big).$$

EXAMPLE 8.5.3. In the case $\mathfrak{g}_N = \mathfrak{o}_4$ we have

$$\begin{aligned} \operatorname{Pf} F(z) = \frac{1}{2} \Big(&: F_{13}(z) F_{31}(z) : \; - \; : F_{12}(z) F_{21}(z) : \; + \; : F_{11}(z) F_{22}(z) : \\ &+ \; : F_{31}(z) F_{13}(z) : \; - \; : F_{21}(z) F_{12}(z) : \; + \; : F_{22}(z) F_{11}(z) : \Big). \end{aligned}$$

By Example 8.1.2, this can also be written as

$$\operatorname{Pf} F(z) = \; : F_{11}(z) F_{22}(z) : \; - \; : F_{12}(z) F_{21}(z) : \; + \; : F_{13}(z) F_{31}(z) : \; - \partial_z F_{22}(z).$$

Taking the coefficient of z^{-p-2} in this series for any $p \in \mathbb{Z}$, we get a central element of $\widetilde{\mathrm{U}}_{-h^\vee}(\widehat{\mathfrak{o}}_4)$ (assuming $h^\vee = 2$) given by

$$\begin{aligned} \sum_{r\in\mathbb{Z}} F_{11}[r] F_{22}[p-r] &+ \sum_{r<0} \big(F_{13}[r] F_{31}[p-r] - F_{12}[r] F_{21}[p-r] \big) \\ &+ \sum_{r\geqslant 0} \big(F_{31}[p-r] F_{13}[r] - F_{21}[p-r] F_{12}[r] \big) + (p+1) F_{22}[p]; \end{aligned}$$

see Remark 8.1.1. □

By the general construction of Section 6.5, some Casimir elements produced in Sections 5.4 and 5.5 can be obtained from the Segal–Sugawara vectors; cf. Section 7.1. Namely, apply the homomorphism ϱ defined in (6.30) to the coefficients of the polynomials (8.15) and (8.36). Multiplying the image by z^m from the right we get

$$\begin{aligned} &\gamma_m(\omega) \operatorname{tr}_{1,\dots,m} S^{(m)} \big(-\partial_z + F_1 z^{-1} \big) \dots \big(-\partial_z + F_m z^{-1} \big) z^m \\ &\qquad = \gamma_m(\omega) \operatorname{tr}_{1,\dots,m} S^{(m)} \big(u + F_1 \big) \dots \big(u + F_m - m + 1 \big), \end{aligned}$$

where $u = -\partial_z z$ and the value of ω is N or $-N$, respectively. Using Lemma 5.3.4, and taking $m = 2k$ with the specializations $u = k-1$ and $u = k$ we reproduce the respective Casimir elements (5.23) and (5.44). Equivalently, these Casimir elements can be obtained from the Sugawara operators constructed above in the same way as for type A as outlined at the end of Chapter 7.

8.6. Bibliographical notes

The exposition follows [**110**] with some modifications, especially in the symplectic case. If the Feigin–Frenkel theorem (Theorem 6.3.1) is assumed in this case, then the second part of Proposition 8.3.7 follows by a simpler argument as in the proof of Proposition 5.5.3. A noncommutative form of the property that the elements (8.43) generate the algebra of invariants was derived by Kumar [**97**].

CHAPTER 9

Commutative subalgebras of $\mathrm{U}(\mathfrak{g})$

We will now use the explicit generators of the algebra $\mathfrak{z}(\widehat{\mathfrak{g}})$ produced in Chapters 7 and 8 to construct commutative subalgebras $\mathcal{A}_\mu$ of the universal enveloping algebra $\mathrm{U}(\mathfrak{g})$ parameterized by elements $\mu \in \mathfrak{g}^*$. The corresponding subalgebras of the symmetric algebra $\mathrm{S}(\mathfrak{g})$ are Poisson commutative and known as the Mishchenko–Fomenko or shift of argument subalgebras. The construction thus provides an affirmative solution of Vinberg's quantization problem in terms of explicit generators. Furthermore, we obtain a direct proof of a theorem of Feigin, Frenkel and Toledano Laredo [**42**] describing the structure of $\mathcal{A}_\mu$ for regular elements μ in the case where $\mathfrak{g}$ is a classical Lie algebra. In a more general context, a connection of the center at the critical level with a commutative family of Gaudin Hamiltonians will be discussed in Chapter 14.

9.1. Mishchenko–Fomenko subalgebras

A *Poisson algebra* A is a commutative associative algebra equipped with a *Poisson bracket* which is a bilinear map

$$\{\ ,\ \} : A \times A \to A \tag{9.1}$$

satisfying the following properties: A is a Lie algebra with respect to this bracket, and the *Leibniz rule*

$$\{x, yz\} = \{x, y\}\, z + y\,\{x, z\} \tag{9.2}$$

holds for any three elements $x, y, z \in A$. In particular, the map (9.1) is *skew-symmetric*,

$$\{x, y\} = -\{y, x\},$$

and satisfies the *Jacobi identity*

$$\{x, \{y, z\}\} + \{y, \{z, x\}\} + \{z, \{x, y\}\} = 0$$

for all $x, y, z \in A$. The *Poisson center* of A is defined by

$$\mathrm{Z}(A) = \{P \in A \mid \{x, P\} = 0 \qquad \text{for all} \quad x \in A\}. \tag{9.3}$$

Clearly, $\mathrm{Z}(A)$ is a subalgebra of A.

Now let $\mathfrak{g}$ be an arbitrary finite-dimensional Lie algebra over $\mathbb{C}$. The symmetric algebra $\mathrm{S}(\mathfrak{g})$ is a Poisson algebra with the *Lie–Poisson bracket* such that for any two elements $x, y \in \mathfrak{g}$ the value $\{x, y\}$ is the commutator $[x, y]$. The bracket extends to all elements of $\mathrm{S}(\mathfrak{g})$ via the Leibniz rule (9.2). If $J_1, \dots, J_d$ is a basis of $\mathfrak{g}$ such that

$$[J_i, J_j] = \sum_{k=1}^{d} c_{ij}^{k}\, J_k \tag{9.4}$$

for structure constants c_{ij}^k, then

$$\{J_i, J_j\} = \sum_{k=1}^{d} c_{ij}^k \, J_k. \tag{9.5}$$

By the Leibniz rule, the Poisson bracket of two monomials is found explicitly by

$$\{x_1 \dots x_r, y_1 \dots y_s\} = \sum_{a=1}^{r} \sum_{b=1}^{s} x_1 \dots \widehat{x}_a \dots x_r \, y_1 \dots \widehat{y}_b \dots y_s \, \{x_a, y_b\} \tag{9.6}$$

with the hats indicating that the corresponding factors should be skipped. By the definition (9.3), the Poisson center of $\mathrm{S}(\mathfrak{g})$ coincides with the subalgebra $\mathrm{S}(\mathfrak{g})^{\mathfrak{g}}$ of $\mathfrak{g}$-invariants in $\mathrm{S}(\mathfrak{g})$.

Fix $\mu \in \mathfrak{g}^*$ and $t \in \mathbb{C}$. Consider another set of generators of the algebra $\mathrm{S}(\mathfrak{g})$ by 'shifting the arguments'. Namely, set $J_i' = J_i + t\,\mu_i$ for $i = 1, \dots, d$ with $\mu_i = \mu(J_i)$. Equip $\mathrm{S}(\mathfrak{g})$ with another Poisson bracket $\{\ ,\ \}_t$ which is defined by the same formulas as in (9.5) but for the shifted generators:

$$\{J_i', J_j'\}_t = \sum_{k=1}^{d} c_{ij}^k \, J_k'. \tag{9.7}$$

In terms of the basis elements of $\mathfrak{g}$ it takes the form

$$\{J_i, J_j\}_t = \sum_{k=1}^{d} c_{ij}^k \, (J_k + t\,\mu_k). \tag{9.8}$$

It is clear from (9.7), that the Poisson center Z_t of $\mathrm{S}(\mathfrak{g})$ with respect to the bracket $\{\ ,\ \}_t$ coincides with the subalgebra obtained from $\mathrm{S}(\mathfrak{g})^{\mathfrak{g}}$ by shifting the arguments of the invariants $J_i \mapsto J_i + t\,\mu_i$. That is,

$$\mathrm{Z}_t = \big\{ P(J + t\,\mu) \mid P \in \mathrm{S}(\mathfrak{g})^{\mathfrak{g}} \big\}, \tag{9.9}$$

where we regard elements of the symmetric algebra $\mathrm{S}(\mathfrak{g})$ as polynomials $P = P(J)$ in the variables $J = (J_1, \dots, J_d)$ and write

$$P(J + t\,\mu) = P\big(J_1 + t\,\mu_1, \dots, J_d + t\,\mu_d\big).$$

DEFINITION 9.1.1. Denote by $\overline{\mathcal{A}}_\mu$ the subalgebra of $\mathrm{S}(\mathfrak{g})$ generated by the Poisson centers Z_t for all $t \in \mathbb{C}$. This subalgebra $\overline{\mathcal{A}}_\mu$ of $\mathrm{S}(\mathfrak{g})$ is called the *Mishchenko–Fomenko subalgebra* or *shift of argument subalgebra*. □

PROPOSITION 9.1.2. *The subalgebra $\overline{\mathcal{A}}_\mu$ is Poisson commutative with respect to any bracket $\{\ ,\ \}_t$. That is, $\{x, y\}_t = 0$ for all $x, y \in \overline{\mathcal{A}}_\mu$ and any $t \in \mathbb{C}$.*

PROOF. Suppose that $t_1, t_2 \in \mathbb{C}$. Take an arbitrary element of Z_{t_1} and an arbitrary element of Z_{t_2}. By (9.9) such elements should have the respective forms $P(J + t_1\,\mu)$ and $Q(J + t_2\,\mu)$ for some $P, Q \in \mathrm{S}(\mathfrak{g})^{\mathfrak{g}}$. By (9.6) and (9.8), the expression

$$\big\{ P(J + t_1\,\mu), Q(J + t_2\,\mu) \big\}_t \tag{9.10}$$

is linear in t. If $t_1 \neq t_2$ then this expression vanishes at $t = t_1$ and $t = t_2$ because the elements belong to the Poisson centers of the respective brackets. This implies that (9.10) is zero for all t. On the other hand, for a fixed value of t, the expression (9.10) depends polynomially on t_1 and t_2. Since this polynomial vanishes for all $t_1 \neq t_2$ it must be identically zero. □

Taking a slightly different point of view, let $P = P(J_1, \dots, J_d)$ be an element of $\mathrm{S}(\mathfrak{g})^{\mathfrak{g}}$ of a certain degree m as a polynomial in the J_i. Let us now regard t as a variable. Make the substitution $J_i \mapsto J_i + t\mu_i$ and expand as a polynomial in t,

$$P(J_1 + t\mu_1, \dots, J_d + t\mu_d) = P_{(0)} + P_{(1)}t + \dots + P_{(m)}t^m, \tag{9.11}$$

to define elements $P_{(i)} \in \mathrm{S}(\mathfrak{g})$ associated with P and μ. The subalgebra $\overline{\mathcal{A}}_\mu$ is generated by all elements $P_{(i)}$ associated with all $\mathfrak{g}$-invariants $P \in \mathrm{S}(\mathfrak{g})^{\mathfrak{g}}$. By Proposition 9.1.2, $\overline{\mathcal{A}}_\mu$ is Poisson commutative with respect to the initial bracket $\{\ ,\ \}$ (which corresponds to $t = 0$).

When P is a homogeneous polynomial, we will also use an equivalent form of (9.11), where we make the substitution $J_i \mapsto J_i z^{-1} + \mu_i$ for a variable z, and expand as a polynomial in z^{-1},

$$P(J_1 z^{-1} + \mu_1, \dots, J_d z^{-1} + \mu_d) = P_{(0)} z^{-m} + \dots + P_{(m-1)} z^{-1} + P_{(m)}. \tag{9.12}$$

Now suppose that $\mathfrak{g}$ is a simple Lie algebra. The bilinear form (6.4) allows us to identify $\mathfrak{g}^*$ with $\mathfrak{g}$, as in the proof of Theorem 6.3.3. It will also be convenient to use the basis $J^1, \dots, J^d$ of $\mathfrak{g}$ which is dual to $J_1, \dots, J_d$ with respect to (6.4). For any element $\mu \in \mathfrak{g}^*$ set $\mu_i = \mu(J_i)$ and $\mu^i = \mu(J^i)$. Then μ is identified with the element of $\mathfrak{g}$ given by

$$\mu = \mu_1 J^1 + \dots + \mu_d J^d = \mu^1 J_1 + \dots + \mu^d J_d \tag{9.13}$$

and we have $\langle \mu, J_i \rangle = \mu_i$ and $\langle \mu, J^i \rangle = \mu^i$. We let n denote the rank of $\mathfrak{g}$. An element $\mu \in \mathfrak{g}^* \cong \mathfrak{g}$ is called *regular*, if the centralizer $\mathfrak{g}^\mu$ of μ in $\mathfrak{g}$ has minimal possible dimension; this minimal dimension coincides with n.

As we recalled in Chapter 2, the subalgebra of $\mathfrak{g}$-invariants in the symmetric algebra $\mathrm{S}(\mathfrak{g})$ admits a family $P_1, \dots, P_n$ of algebraically independent generators; see (2.2).

THEOREM 9.1.3. *Let μ be regular. If $P_1, \dots, P_n$ are algebraically independent generators of $\mathrm{S}(\mathfrak{g})^{\mathfrak{g}}$ of the respective degrees $d_1, \dots, d_n$, then the elements*

$$P_{k(i)}, \qquad k = 1, \dots, n, \quad i = 0, 1, \dots, d_k - 1, \tag{9.14}$$

are algebraically independent generators of $\overline{\mathcal{A}}_\mu$.

PROOF. It is clear that the $P_{k(i)}$ generate the algebra $\overline{\mathcal{A}}_\mu$. We only need to show that these polynomials are algebraically independent. It is enough to demonstrate that the differentials $dP_{k(i)}$ evaluated at a certain point are linearly independent.

We will regard the differential dP of any polynomial $P = P(J) \in \mathrm{S}(\mathfrak{g})$ as the element of $\mathrm{S}(\mathfrak{g}) \otimes \mathfrak{g}$ defined by

$$dP = \frac{\partial P}{\partial J_1} \otimes J_1 + \dots + \frac{\partial P}{\partial J_d} \otimes J_d.$$

It is straightforward to verify that it does not depend on the chosen basis. The vector space $\mathrm{S}(\mathfrak{g}) \otimes \mathfrak{g}$ will be understood as a Lie algebra with the bracket

$$[P \otimes X, Q \otimes Y] = PQ \otimes [X, Y], \qquad P, Q \in \mathrm{S}(\mathfrak{g}) \quad \text{and} \quad X, Y \in \mathfrak{g}.$$

Introduce the element $\mathcal{X}$ of this Lie algebra by

$$\mathcal{X} = \sum_{i=1}^{d} J^i \otimes J_i.$$

LEMMA 9.1.4. *If $P \in \mathrm{S}(\mathfrak{g})^{\mathfrak{g}}$ then in the Lie algebra $\mathrm{S}(\mathfrak{g}) \otimes \mathfrak{g}$ we have*

$$\big[\mathcal{X}, dP\big] = 0.$$

PROOF. By using the structure coefficients defined by (9.4) we can write

$$\Big[\sum_{i=1}^{d} J^i \otimes J_i, dP\Big] = \sum_{i,j=1}^{d} J^i \frac{\partial P}{\partial J_j} \otimes [J_i, J_j] = \sum_{i,j,k=1}^{d} c_{ij}^{k}\, J^i \frac{\partial P}{\partial J_j} \otimes J_k. \tag{9.15}$$

On the other hand, we have $c_{ij}^{k} = \langle [J_i, J_j], J^k \rangle$, and so $c_{ij}^{k} = \langle J_i, [J_j, J^k] \rangle$ since the form (6.4) is invariant. This implies the expansion

$$[J^k, J_j] = -\sum_{i=1}^{d} c_{ij}^{k}\, J^i.$$

The condition $P \in \mathrm{S}(\mathfrak{g})^{\mathfrak{g}}$ means that for any $k = 1, \dots, d$,

$$J^k \cdot P = \sum_{j=1}^{d} \frac{\partial P}{\partial J_j} [J^k, J_j] = -\sum_{i,j=1}^{d} c_{ij}^{k} \frac{\partial P}{\partial J_j} J^i = 0$$

and so (9.15) is equal to zero. □

The identity of Lemma 9.1.4 will hold if we replace the generators of $\mathrm{S}(\mathfrak{g})$ by $J_i \mapsto J_i + t\mu_i$ for all $i = 1, \dots, d$, where t is a variable. In terms of the dual generators this replacement is written as $J^i \mapsto J^i + t\mu^i$. That is,

$$\Big[\sum_{i=1}^{d} (J^i + t\mu^i) \otimes J_i, dP\big|_{J_i \mapsto J_i + t\mu_i}\Big] = 0.$$

Therefore, expanding $P(J + t\mu)$ as in (9.11) and taking the coefficients of all powers of t we get the relations

$$\begin{aligned} \big[\mathcal{X}, dP_{(0)}\big] &= 0, \\ \big[\mathcal{X}, dP_{(i+1)}\big] + \big[1 \otimes \mu, dP_{(i)}\big] &= 0, \qquad i = 0, 1, \dots, m-2, \\ \big[1 \otimes \mu, dP_{(m-1)}\big] &= 0. \end{aligned} \tag{9.16}$$

As a next step, we complete the proof of the theorem for the particular case where μ is regular nilpotent. Consider an $\mathfrak{sl}_2$-subalgebra of $\mathfrak{g}$ with the standard basis elements e, f, h satisfying the commutation relations (6.12) where $e = \mu$. Set $x = h/2$ and introduce the evaluation map

$$\mathrm{ev} : \mathrm{S}(\mathfrak{g}) \to \mathbb{C}, \qquad J_i \mapsto x_i \quad \text{for} \quad i = 1, \dots, d,$$

where x is written as $x = x_1 J^1 + \dots + x_d J^d$. We will use the same notation for its extension to the Lie algebra

$$\mathrm{ev} : \mathrm{S}(\mathfrak{g}) \otimes \mathfrak{g} \to \mathfrak{g} \tag{9.17}$$

which is the identity map on $\mathfrak{g}$. Clearly, $\mathrm{ev}(\mathcal{X}) = x$. Consider the respective evaluations of the differentials of the polynomials $P_{k(i)}$ and set

$$\mathcal{P}_{k(i)} = \mathrm{ev}\big(dP_{k(i)}\big), \qquad k = 1, \dots, n, \quad i = 0, 1, \dots, d_k - 1. \tag{9.18}$$

The maximum of all integers d_k is the *Coxeter number* $h_{\mathfrak{g}}$ of the Lie algebra $\mathfrak{g}$. For each $j = 0, 1, \dots, h_{\mathfrak{g}} - 1$ introduce the subspace of $\mathfrak{g}$ by

$$\mathcal{D}_j = \text{span of } \{\mathcal{P}_{k(i)} \mid k = 1, \dots, n, \quad i = 0, 1, \dots, j\},$$

assuming that $\mathcal{P}_{k(i)} = 0$ for $i \geqslant d_k$.

By the results of Kostant [**93**], we have the decomposition of $\mathfrak{g}$ regarded as an adjoint $\mathfrak{sl}_2$-module, into irreducible components,

$$\mathfrak{g} = \mathfrak{m}_1 \oplus \dots \oplus \mathfrak{m}_n,$$

where $\dim \mathfrak{m}_k = 2d_k - 1$. The highest vector of the $\mathfrak{sl}_2$-module $\mathfrak{m}_k$ is annihilated by e and has the x-weight equal to $d_k - 1$. Hence its weight decomposition with respect to the adjoint action of x has the form

$$\mathfrak{m}_k = \mathfrak{m}_k^{(d_k-1)} \oplus \mathfrak{m}_k^{(d_k-2)} \oplus \dots \oplus \mathfrak{m}_k^{(-d_k+1)}.$$

For each integer $-h_{\mathfrak{g}} + 1 \leqslant i \leqslant h_{\mathfrak{g}} - 1$ introduce the subspace of $\mathfrak{g}$ by

$$\mathfrak{g}^{(i)} = \mathfrak{m}_1^{(i)} \oplus \dots \oplus \mathfrak{m}_n^{(i)}. \tag{9.19}$$

LEMMA 9.1.5. *For each $j = 0, 1, \dots, h_{\mathfrak{g}} - 1$ we have*

$$\mathcal{D}_j = \mathfrak{g}^{(0)} \oplus \mathfrak{g}^{(1)} \oplus \dots \oplus \mathfrak{g}^{(j)}. \tag{9.20}$$

PROOF. By applying the evaluation map (9.17) to the relations (9.16) we get

$$\begin{aligned} &[x, \mathcal{P}_{k(0)}] = 0, \\ &[x, \mathcal{P}_{k(i+1)}] + [e, \mathcal{P}_{k(i)}] = 0, \qquad i = 0, 1, \dots, d_k - 2, \\ &[e, \mathcal{P}_{k(d_k-1)}] = 0 \end{aligned} \tag{9.21}$$

for each $k = 1, \dots, n$. We will be proving (9.20) by the induction on j. If $j = 0$ then the first relation in (9.21) shows that $\mathcal{D}_0 \subset \mathfrak{g}^{(0)}$. However, x is a regular element of $\mathfrak{g}$ as its centralizer coincides with the n-dimensional subspace $\mathfrak{g}^{(0)}$. Hence, by another theorem of Kostant [**94**], the differentials $dP_1, \dots, dP_n$ evaluated at the regular point x are linearly independent and so $\mathcal{D}_0 = \mathfrak{g}^{(0)}$.

Suppose that (9.20) folds for $j = 0, 1, \dots, m$. By (9.21) we have

$$\operatorname{ad} x \cdot \mathcal{D}_{m+1} = \operatorname{ad} e \cdot \mathcal{D}_m. \tag{9.22}$$

For each k the map $\operatorname{ad} e : \mathfrak{m}_k^{(i)} \to \mathfrak{m}_k^{(i+1)}$ is surjective for all $i \geqslant 0$ and so is the map $\operatorname{ad} e : \mathfrak{g}^{(i)} \to \mathfrak{g}^{(i+1)}$. Hence by the induction hypothesis,

$$\operatorname{ad} e \cdot \mathcal{D}_m = \mathfrak{g}^{(1)} \oplus \dots \oplus \mathfrak{g}^{(m+1)}. \tag{9.23}$$

Observe that for each $i \neq 0$ the restriction of the operator $\operatorname{ad} x$ on the subspace $\mathfrak{g}^{(i)}$ is invertible. Moreover, the kernel of this operator on $\mathfrak{g}$ coincides with $\mathfrak{g}^{(0)}$. Therefore, (9.22) and (9.23) imply that

$$\mathcal{D}_{m+1} \subset \mathfrak{g}^{(0)} \oplus \mathfrak{g}^{(1)} \oplus \dots \oplus \mathfrak{g}^{(m+1)}$$

and that the projection of $\mathcal{D}_{m+1}$ on the direct sum in (9.23) along $\mathfrak{g}^{(0)}$ is surjective. However, $\mathcal{D}_{m+1}$ contains $\mathcal{D}_0 = \mathfrak{g}^{(0)}$ and thus (9.20) holds for $j = m+1$, completing the proof of the lemma. □

Taking $j = h_{\mathfrak{g}} - 1$ in Lemma 9.1.5 we find that the span of all differentials (9.18) coincides with the direct sum

$$\mathfrak{g}^{(0)} \oplus \mathfrak{g}^{(1)} \oplus \cdots \oplus \mathfrak{g}^{(h_{\mathfrak{g}}-1)} \tag{9.24}$$

which is a Borel subalgebra of $\mathfrak{g}$. Therefore the differentials (9.18) are linearly independent since their total number $d_1 + \cdots + d_n$ equals the dimension of the Borel subalgebra. Thus we showed that the polynomials (9.14) are algebraically independent in the case where μ is regular nilpotent.

To complete the proof of the theorem in the general case, introduce the subset $S \subset \mathfrak{g}$ which consists of all elements μ satisfying the property that the corresponding polynomials (9.14) are algebraically independent. Then S is a nonempty Zariski open subset of $\mathfrak{g}$. It follows from the definition of the algebra $\overline{\mathcal{A}}_\mu$ via the expansions (9.11) that $\overline{\mathcal{A}}_\mu = \overline{\mathcal{A}}_{c\mu}$ for any nonzero $c \in \mathbb{C}$. Moreover, $\mathrm{Ad}_g(\overline{\mathcal{A}}_\mu) = \overline{\mathcal{A}}_{\mathrm{Ad}_g(\mu)}$ for any element $g \in G$ of the adjoint group G of $\mathfrak{g}$. Hence, S is a conic subset of $\mathfrak{g}$ which is invariant under the adjoint action of G. We will keep the $\mathfrak{sl}_2$-subalgebra of $\mathfrak{g}$ introduced in the previous argument. It is well-known by Kostant [**94**], that the adjoint orbit of any regular element of $\mathfrak{g}$ intersects the affine subspace $e + \mathfrak{g}^f$ (*Kostant's slice*) at one point. Here $\mathfrak{g}^f$ denotes the centralizer of f in $\mathfrak{g}$ which is a subspace of the Borel subalgebra opposite to (9.24),

$$\mathfrak{g}^f \subset \mathfrak{g}^{(0)} \oplus \mathfrak{g}^{(-1)} \oplus \cdots \oplus \mathfrak{g}^{(-h_{\mathfrak{g}}+1)}.$$

Thus, for the rest of the proof we may assume that μ is regular and $\mu \in e + \mathfrak{g}^f$. Suppose that $\mu \notin S$ and for any $a \in \mathbb{C}$ set

$$\chi_a = \exp a \cdot \mathrm{Ad}_{\exp(-ax)}(\mu).$$

All these elements do not belong to S. Note that if $y \in \mathfrak{g}^{(p)}$ then

$$\exp\big(\mathrm{ad}(-ax)\big)(y) = \exp(-ap)y.$$

Since $e \in \mathfrak{g}^{(1)}$, this implies that if $\exp a \to 0$ then $\chi_a \to e$. This makes a contradiction as the element e of the open set S is approximated by elements not belonging to S. So, $\mu \in S$ and the proof is complete. □

The total number of the algebraically independent generators of the algebra $\overline{\mathcal{A}}_\mu$ provided by Theorem 9.1.3 is $d_1 + \cdots + d_n$ which equals $(\dim \mathfrak{g} + n)/2$ and coincides with the dimension of a Borel subalgebra of $\mathfrak{g}$. The transcendence degree of any Poisson commutative subalgebra of $\mathrm{S}(\mathfrak{g})$ is known not to exceed this number.

REMARK 9.1.6. The Mishchenko–Fomenko subalgebra $\overline{\mathcal{A}}_\mu$ of $\mathrm{S}(\mathfrak{g})$ associated with a regular element $\mu \in \mathfrak{g}^*$ is maximal Poisson commutative; see [**132**] and [**147**]. □

EXAMPLE 9.1.7. Theorem 9.1.3 extends to the reductive Lie algebra $\mathfrak{g} = \mathfrak{gl}_N$ in an obvious way; we just need to include an extra invariant of degree one, as compared to the case of the simple Lie algebra $\mathfrak{sl}_N$. Any element $\mu \in \mathfrak{gl}_N^*$ can be regarded as the $N \times N$ matrix $\mu = [\mu_{ij}]$, where $\mu_{ij} = \mu(E_{ij})$. Equip the Lie algebra $\mathfrak{gl}_N$ with the nondegenerate invariant bilinear form[1]

$$\langle X, Y \rangle = \mathrm{tr}\, XY, \qquad X, Y \in \mathfrak{gl}_N.$$

In accordance with (9.13), we can identify μ with the element of $\mathfrak{gl}_N$ represented by the transposed matrix μ^t. Then μ is a *regular element* of $\mathfrak{gl}_N^*$ if and only if

[1]Although we use the same notation, this form is different from the Killing form (7.2).

the Jordan canonical form of the matrix $[\mu_{ij}]$ (or, equivalently, its transpose) does not contain two Jordan cells with the same eigenvalue. Furthermore, μ is *regular semisimple* if this matrix is diagonalizable and all eigenvalues are distinct. In the case where the canonical form is a single Jordan cell with the zero eigenvalue, μ is called *regular nilpotent.*

Using the invariants $C_1, \dots, C_N$ of $\mathrm{S}(\mathfrak{gl}_N)$ defined in (2.6), introduce elements $C_{m\,(i)}$ of the subalgebra $\overline{\mathcal{A}}_\mu$ by the expansion

$$\det\big(u + \mu + E\,z^{-1}\big) = u^N + \sum_{m=1}^{N} C_m(z)\,u^{N-m} \tag{9.25}$$

with

$$C_m(z) = C_{m\,(0)}\,z^{-m} + \dots + C_{m\,(m-1)}\,z^{-1} + C_{m\,(m)}.$$

If μ is regular, then the elements $C_{m\,(i)}$ with $m = 1, \dots, N$ and $i = 0, 1, \dots, m-1$ are algebraically independent generators of $\overline{\mathcal{A}}_\mu$.

Similarly, taking the invariants T_m defined in (2.11), set

$$\mathrm{tr}\,\big(\mu + E\,z^{-1}\big)^m = T_{m\,(0)}\,z^{-m} + \dots + T_{m\,(m-1)}\,z^{-1} + T_{m\,(m)}.$$

All elements $T_{m\,(i)}$ belong to the subalgebra $\overline{\mathcal{A}}_\mu$ of $\mathrm{S}(\mathfrak{gl}_N)$. Furthermore, if μ is regular, then the elements $T_{m\,(i)}$ with $m = 1, \dots, N$ and $i = 0, 1, \dots, m-1$ are algebraically independent generators of $\overline{\mathcal{A}}_\mu$. □

The invariants in types B, C and D constructed in Section 2.2 can be used to produce analogous families of algebraically independent generators of the subalgebra $\overline{\mathcal{A}}_\mu$ of $\mathrm{S}(\mathfrak{g}_N)$ for regular elements μ.

9.2. Vinberg's quantization problem

The universal enveloping algebra $\mathrm{U}(\mathfrak{g})$ can be regarded as a 'quantization' of the Poisson algebra $\mathrm{S}(\mathfrak{g})$. Namely, we can re-scale the basis elements of the Lie algebra $\mathfrak{g}$ by introducing the 'deformation parameter' h (taking nonzero complex values) so that the new basis elements are $\widetilde{J}_i = h\,J_i$ with $i = 1, \dots, d$. In the new basis the commutation relations (9.4) take the form

$$[\widetilde{J}_i, \widetilde{J}_j] = h \sum_{k=1}^{d} c_{ij}^k\,\widetilde{J}_k.$$

In the limit $h \to 0$ we get the symmetric algebra $\mathrm{S}(\mathfrak{g})$, and the Poisson bracket (9.5) is recovered by

$$\{\widetilde{J}_i, \widetilde{J}_j\} = \lim_{h\to 0} \frac{[\widetilde{J}_i, \widetilde{J}_j]}{h}.$$

An equivalent viewpoint which we took in Chapter 4 is to equip the universal enveloping algebra $\mathrm{U}(\mathfrak{g})$ with the canonical filtration so that the associated graded algebra $\mathrm{gr}\ \mathrm{U}(\mathfrak{g})$ is isomorphic to the symmetric algebra $\mathrm{S}(\mathfrak{g})$.

Given that the subalgebra $\overline{\mathcal{A}}_\mu$ of $\mathrm{S}(\mathfrak{g})$ is Poisson commutative, one could wonder whether it is possible to construct a commutative subalgebra $\mathcal{A}_\mu$ of $\mathrm{U}(\mathfrak{g})$ which 'quantizes' $\overline{\mathcal{A}}_\mu$ in the sense that $\mathrm{gr}\,\mathcal{A}_\mu = \overline{\mathcal{A}}_\mu$. This *quantization problem* was raised by Vinberg in [**149**], where, in particular, some commuting families of elements of $\mathrm{U}(\mathfrak{g})$ were produced. Theorem 9.2.4 below provides a positive solution of Vinberg's problem which uses the Feigin–Frenkel center $\mathfrak{z}(\widehat{\mathfrak{g}})$.

Suppose that z is a variable. For any $\mu \in \mathfrak{g}^*$ consider the following generalization of the evaluation homomorphism (6.30),

$$\varrho_\mu : \mathrm{U}\left(t^{-1}\mathfrak{g}[t^{-1}]\right) \to \mathrm{U}(\mathfrak{g}) \otimes \mathbb{C}[z^{-1}], \qquad X[r] \mapsto X\, z^r + \delta_{r,-1}\, \mu(X), \tag{9.26}$$

for any $X \in \mathfrak{g}$ and $r < 0$. If $S \in \mathfrak{z}(\widehat{\mathfrak{g}})$ is a homogeneous element of degree d with respect to the grading defined by the derivation (6.14), define the elements $S_{(a)} \in \mathrm{U}(\mathfrak{g})$ (depending on μ) by the expansion

$$\varrho_\mu(S) = S_{(0)}\, z^{-d} + \dots + S_{(d-1)}\, z^{-1} + S_{(d)}. \tag{9.27}$$

Since $\mathfrak{z}(\widehat{\mathfrak{g}})$ is invariant with respect to the translation operator T defined in (6.7), the property (6.31) extends to the homomorphism (9.26) so that

$$\varrho_\mu(TS) = -\partial_z\, \varrho_\mu(S). \tag{9.28}$$

Note that by Proposition 6.5.1 the element $S_{(0)}$ is central in $\mathrm{U}(\mathfrak{g})$.

Suppose that the variable z takes a particular nonzero value in $\mathbb{C}$. Then the formula (9.26) defines a homomorphism

$$\varrho_{\mu,z} : \mathrm{U}\left(t^{-1}\mathfrak{g}[t^{-1}]\right) \to \mathrm{U}(\mathfrak{g}), \qquad X[r] \mapsto X\, z^r + \delta_{r,-1}\, \mu(X), \tag{9.29}$$

for any $X \in \mathfrak{g}$ and $r < 0$. The Feigin–Frenkel center $\mathfrak{z}(\widehat{\mathfrak{g}})$ is a commutative subalgebra of $\mathrm{U}\left(t^{-1}\mathfrak{g}[t^{-1}]\right)$, and so its image under $\varrho_{\mu,z}$ is a commutative subalgebra of $\mathrm{U}(\mathfrak{g})$. We will see below in Lemma 9.2.2 that the image does not depend on the value of z.

DEFINITION 9.2.1. The commutative subalgebra $\mathcal{A}_\mu$ of $\mathrm{U}(\mathfrak{g})$ is defined as the image of $\mathfrak{z}(\widehat{\mathfrak{g}})$ under the homomorphism (9.29) with a nonzero $z \in \mathbb{C}$. □

LEMMA 9.2.2. *The subalgebra $\mathcal{A}_\mu$ does not depend on z.*

PROOF. Suppose that the algebra $\mathcal{A}_\mu$ is defined for a certain fixed nonzero value $z \in \mathbb{C}$. It will be sufficient to verify that for any homogeneous element $S \in \mathfrak{z}(\widehat{\mathfrak{g}})$ of degree d all coefficients $S_{(a)}$ of the polynomial (9.27) belong to $\mathcal{A}_\mu$. Introduce another polynomial

$$P(z) = S_{(0)} + S_{(1)}z + \dots + S_{(d)}z^d \in \mathrm{U}(\mathfrak{g}) \otimes \mathbb{C}[z].$$

Then $P(z) = z^d \varrho_\mu(S)$ and by (9.28) all derivatives of $P(z)$ over z can be expressed as $\mathbb{C}[z]$-linear combinations of the images $\varrho_\mu(T^r S)$. The claim now follows by evaluating z at the chosen fixed value. □

Suppose $S_1, \dots, S_n$ is a complete set of Segal–Sugawara vectors of the respective degrees $d_1, \dots, d_n$ as provided by Theorem 6.3.1. Introduce the corresponding polynomials (9.27) by

$$\varrho_\mu(S_k) = S_{k\,(0)}\, z^{-d_k} + \dots + S_{k\,(d_k-1)}\, z^{-1} + S_{k\,(d_k)}. \tag{9.30}$$

COROLLARY 9.2.3. *The algebra $\mathcal{A}_\mu$ is generated by the elements*

$$S_{k\,(i)}, \qquad k = 1, \dots, n, \quad i = 0, 1, \dots, d_k - 1.$$

PROOF. As the proof of Lemma 9.2.2 shows, the algebra $\mathcal{A}_\mu$ can be defined as the subalgebra of $\mathrm{U}(\mathfrak{g})$ generated by all coefficients $S_{(a)}$ of the polynomials (9.27) associated to all homogeneous elements $S \in \mathfrak{z}(\widehat{\mathfrak{g}})$. On the other hand, the algebra $\mathfrak{z}(\widehat{\mathfrak{g}})$ is generated by the elements $T^r S_k$; see (6.15). Hence the required property follows from (9.28). □

Due to Proposition 6.5.2 the algebra $\mathcal{A}_\mu$ contains the center $\mathrm{Z}(\mathfrak{g})$ of $\mathrm{U}(\mathfrak{g})$. In the particular case $\mu = 0$ it coincides with the center.

The following theorem provides a positive solution of Vinberg's quantization problem for an arbitrary simple Lie algebra $\mathfrak{g}$ and a regular element $\mu \in \mathfrak{g}^*$. We use the notation (9.30).

THEOREM 9.2.4. *Let $\mu \in \mathfrak{g}^*$ be regular. If $S_1, \dots, S_n$ is a complete set of Segal–Sugawara vectors of the respective degrees $d_1, \dots, d_n$ then the elements*

$$S_{k\,(i)}, \qquad k = 1, \dots, n, \quad i = 0, 1, \dots, d_k - 1,$$

are algebraically independent generators of $\mathcal{A}_\mu$. Moreover, $\mathrm{gr}\,\mathcal{A}_\mu = \overline{\mathcal{A}}_\mu$.

A general proof can be found in [**42**]. In the next sections we give a direct proof for the Lie algebras $\mathfrak{g}$ of classical types relying on the explicit constructions of complete sets of Segal–Sugawara vectors provided in Chapters 7 and 8.

REMARK 9.2.5. Under the assumptions of Theorem 9.2.4, the subalgebra $\mathcal{A}_\mu$ of $\mathrm{U}(\mathfrak{g})$ is maximal commutative. This follows from the corresponding property of the Poisson commutative subalgebra $\overline{\mathcal{A}}_\mu$ of $\mathrm{S}(\mathfrak{g})$; see Remark 9.1.6. □

9.3. Generators of commutative subalgebras of $\mathrm{U}(\mathfrak{gl}_N)$

We let $\mathfrak{g} = \mathfrak{gl}_N$ and use the tensor notation as in Section 7.1. Consider the tensor product algebra

$$\underbrace{\mathrm{End}\,\mathbb{C}^N \otimes \dots \otimes \mathrm{End}\,\mathbb{C}^N}_{m} \otimes \mathrm{U}, \tag{9.31}$$

where $\mathrm{U} = \mathrm{U}(\mathfrak{gl}_N) \otimes \mathbb{C}[z^{-1}, \partial_z]$ is the algebra of differential operators whose elements are finite sums of the form

$$\sum_{k,l \geqslant 0} x_{kl}\, z^{-k}\, \partial_z^l, \qquad x_{kl} \in \mathrm{U}(\mathfrak{gl}_N).$$

As in Example 9.1.7, take an element $\mu \in \mathfrak{gl}_N^*$ which will be regarded as the $N \times N$ matrix $\mu = [\mu_{ij}]$, where $\mu_{ij} = \mu(E_{ij})$. Introduce the matrix

$$M = -\partial_z + \mu + E z^{-1} = \big[-\delta_{ij}\,\partial_z + \mu_{ij} + E_{ij} z^{-1}\big] \tag{9.32}$$

with entries in U. It follows from Lemma 4.5.3 that M is a Manin matrix.

Define polynomials $\phi_{m\,a}(z)$, $\psi_{m\,a}(z)$ and $\theta_{m\,a}(z)$ in z^{-1} (depending on μ) with coefficients in $\mathrm{U}(\mathfrak{gl}_N)$ by the expansions

$$\mathrm{tr}_{1,\dots,m}\, A^{(m)} M_1 \dots M_m = \phi_{m\,0}(z)\,(-\partial_z)^m + \phi_{m\,1}(z)\,(-\partial_z)^{m-1} + \dots + \phi_{m\,m}(z),$$

$$\mathrm{tr}_{1,\dots,m}\, H^{(m)} M_1 \dots M_m = \psi_{m\,0}(z)\,(-\partial_z)^m + \psi_{m\,1}(z)\,(-\partial_z)^{m-1} + \dots + \psi_{m\,m}(z),$$

and

$$\mathrm{tr}\, M^m = \theta_{m\,0}(z)\,(-\partial_z)^m + \theta_{m\,1}(z)\,(-\partial_z)^{m-1} + \dots + \theta_{m\,m}(z);$$

cf. (7.7), (7.8) and (7.9). Furthermore, following (7.11) define the polynomials $\phi_a(z)$ by

$$\mathrm{cdet}\, M = (-\partial_z)^N + \phi_1(z)\,(-\partial_z)^{N-1} + \dots + \phi_N(z). \tag{9.33}$$

By the same argument as for the proof of (7.12) we get

$$\phi_{m\,a}(z) = \binom{N-a}{m-a}\,\phi_a(z), \qquad 0 \leqslant a \leqslant m \leqslant N, \tag{9.34}$$

and so $\phi_{mm}(z) = \phi_m(z)$ for all m. Introduce the coefficients of the polynomials by

$$\phi_m(z) = \phi_{m\,(0)} z^{-m} + \dots + \phi_{m\,(m-1)} z^{-1} + \phi_{m\,(m)},$$

$$\psi_{mm}(z) = \psi_{mm\,(0)} z^{-m} + \dots + \psi_{mm\,(m-1)} z^{-1} + \psi_{mm\,(m)},$$

and

$$\theta_{mm}(z) = \theta_{mm\,(0)} z^{-m} + \dots + \theta_{mm\,(m-1)} z^{-1} + \theta_{mm\,(m)}.$$

The following theorem will imply Theorem 9.2.4 in type A.

THEOREM 9.3.1. (i) *Given any $\mu \in \mathfrak{gl}_N^*$, all coefficients of the polynomials $\phi_m(z)$, $\psi_{ma}(z)$ and $\theta_{ma}(z)$ belong to the commutative subalgebra $\mathcal{A}_\mu$ of* $\mathrm{U}(\mathfrak{gl}_N)$. *Moreover, the elements of each of the families*

$$\phi_{m\,(k)}, \qquad \psi_{mm\,(k)} \qquad \text{and} \qquad \theta_{mm\,(k)} \tag{9.35}$$

with $m = 1, \dots, N$ and $k = 0, 1, \dots, m-1$, are generators of the algebra $\mathcal{A}_\mu$.

(ii) *If μ is regular, then each of the three families* (9.35) *is algebraically independent. Moreover,* $\mathrm{gr}\,\mathcal{A}_\mu = \overline{\mathcal{A}}_\mu$.

PROOF. Under the homomorphism ϱ_μ defined in (9.26) we have

$$\varrho_\mu : E_{ij}[-1] \mapsto E_{ij}\, z^{-1} + \mu_{ij}$$

so that $E[-1] \mapsto \mu + E z^{-1}$. Extend ϱ_μ to the homomorphism of the tensor product algebras

$$\varrho_\mu : \mathrm{U}\big(t^{-1}\mathfrak{gl}_N[t^{-1}]\big) \otimes \mathbb{C}[\tau] \to \mathrm{U}(\mathfrak{gl}_N) \otimes \mathbb{C}[z^{-1}, \partial_z] \tag{9.36}$$

by setting $\tau \mapsto -\partial_z$. The coefficients of the polynomials $\phi_m(z)$, $\psi_{ma}(z)$ and $\theta_{ma}(z)$ are the images under the homomorphism ϱ_μ of the respective elements of the Feigin–Frenkel center $\mathfrak{z}(\widehat{\mathfrak{gl}}_N)$ provided by Theorem 7.1.3. Hence they belong to $\mathcal{A}_\mu$. Together with Theorem 7.1.4 and Corollary 9.2.3 this completed the proof of part (i) of the theorem.

Now suppose that μ is regular and consider the symbols of the elements of the families (9.35) in the associated graded algebra $\mathrm{S}(\mathfrak{gl}_N) \cong \mathrm{gr}\ \mathrm{U}(\mathfrak{gl}_N)$. These symbols coincide with the respective coefficients of the polynomials in z^{-1} given by

$$\mathrm{tr}_{1,\dots,m}\, A^{(m)} \big(\mu_1 + E_1 z^{-1}\big) \dots \big(\mu_m + E_m z^{-1}\big),$$

$$\mathrm{tr}_{1,\dots,m}\, H^{(m)} \big(\mu_1 + E_1 z^{-1}\big) \dots \big(\mu_m + E_m z^{-1}\big),$$

and

$$\mathrm{tr}\,\big(\mu + E z^{-1}\big)^m$$

for $m = 1, \dots, N$; cf. Example 9.1.7. However, these coefficients belong to the Mishchenko–Fomenko subalgebra $\overline{\mathcal{A}}_\mu$ of $\mathrm{S}(\mathfrak{gl}_N)$ since they arise from generators of $\mathrm{S}(\mathfrak{gl}_N)^{\mathfrak{gl}_N}$ by the rule (9.12); see Corollary 2.1.6 and (2.12). Moreover, by Theorem 9.1.3 the coefficients are algebraically independent generators of $\overline{\mathcal{A}}_\mu$. Hence part (ii) of the theorem follows. □

Note that, in particular, the coefficients of the column-determinant $\mathrm{cdet}\, M$ are respective quantizations of the generators of the algebra $\overline{\mathcal{A}}_\mu$ provided by the 'shifted' characteristic polynomial $\det\big(u + \mu + E\, z^{-1}\big)$ defined in (9.25).

EXAMPLE 9.3.2. Using the complete sets of Segal–Sugawara vectors given in Example 7.1.6, we get the following algebraically independent generators of the algebra $\mathcal{A}_\mu$ for regular μ:

$$
\begin{aligned}
&\text{for } \mathfrak{gl}_2: \quad && \operatorname{tr} E, \quad \operatorname{tr} \mu E, \quad \operatorname{tr} E^2\\
&\text{for } \mathfrak{gl}_3 && \operatorname{tr} E, \quad \operatorname{tr} \mu E, \quad \operatorname{tr} \mu^2 E, \quad \operatorname{tr} E^2, \quad \operatorname{tr} \mu E^2, \quad \operatorname{tr} E^3\\
&\text{for } \mathfrak{gl}_4: && \operatorname{tr} E, \quad \operatorname{tr} \mu E, \quad \operatorname{tr} \mu^2 E, \quad \operatorname{tr} \mu^3 E, \quad \operatorname{tr} E^2, \quad \operatorname{tr} \mu E^2,\\
& && \qquad 2 \operatorname{tr} \mu^2 E^2 + \operatorname{tr} (\mu E)^2, \quad \operatorname{tr} E^3, \quad \operatorname{tr} \mu E^3, \quad \operatorname{tr} E^4.
\end{aligned}
$$

This follows by calculating the coefficients of the powers of z^{-1} in the polynomials of the form $\operatorname{tr} (\mu + E z^{-1})^m$. There are some relations between these coefficients which we have used to simplify the set of generators. One easily verifies that

$$\operatorname{tr} E\mu E = \operatorname{tr} \mu E^2 + \operatorname{tr} \mu \operatorname{tr} E - N \operatorname{tr} \mu E$$

and

$$\operatorname{tr} E\mu E^2 = \operatorname{tr} E^2 \mu E = \operatorname{tr} \mu E^3 + \operatorname{tr} \mu \operatorname{tr} E^2 - N \operatorname{tr} \mu E^2.$$

Indeed, using (4.10), for the first relation we have

$$
\begin{aligned}
\operatorname{tr} E\mu E &= \operatorname{tr}_{12} P_{12} E_1 \mu_1 E_2\\
&= \operatorname{tr}_{12} \mu_1 P_{12} E_1 E_2 = \operatorname{tr}_{12} \mu_1 P_{12} (E_2 E_1 + P_{12} E_2 - P_{12} E_1)\\
&= \operatorname{tr}_{12} \big(\mu_1 E_1 P_{12} E_1 + \mu_1 E_2 - \mu_1 E_1\big) = \operatorname{tr} \mu E^2 + \operatorname{tr} \mu \operatorname{tr} E - N \operatorname{tr} \mu E,
\end{aligned}
$$

as claimed. The second relation is proved in a similar way. □

The condition that the element $\mu \in \mathfrak{gl}_N^*$ is regular turns out to be necessary for the subalgebra $\mathcal{A}_\mu$ of $\mathrm{U}(\mathfrak{gl}_N)$ to have the maximum possible transcendence degree. This can be seen directly from the next theorem providing an algebraically independent family of generators of $\mathcal{A}_\mu$ for an arbitrary element $\mu \in \mathfrak{gl}_N^*$. It also shows that the subalgebra $\mathcal{A}_\mu$ is a quantization of the shift of argument subalgebra $\overline{\mathcal{A}}_\mu$ for any μ thus proving a conjecture from [**42**, Conjecture 1].

Consider the matrix M introduced in (9.32). We will work with the generators $\phi_{m\,(k)}$ of $\mathcal{A}_\mu$ obtained as the coefficients of the column-determinant

$$\operatorname{cdet} M = (-\partial_z)^N + \sum_{m=1}^{N} \sum_{k=0}^{m-1} \phi_{m\,(k)}\, z^{-m+k} (-\partial_z)^{N-m}$$

as defined in (9.33). To state the theorem, suppose that the distinct eigenvalues of μ are $\lambda_1, \dots, \lambda_r$ and the Jordan canonical form of μ is the direct sum of the respective Jordan blocks $J_{\alpha_j^{(i)}}(\lambda_i)$ of sizes $\alpha_1^{(i)} \geqslant \alpha_2^{(i)} \geqslant \dots \geqslant \alpha_{s_i}^{(i)} \geqslant 1$, where we use the notation

$$J_p(\lambda) = \begin{bmatrix} \lambda & 1 & 0 & \dots & 0\\ 0 & \lambda & 1 & \dots & 0\\ \vdots & \vdots & \vdots & \ddots & \vdots\\ 0 & 0 & 0 & \dots & 1\\ 0 & 0 & 0 & \dots & \lambda \end{bmatrix}$$

for the Jordan block of size p. We let $\alpha^{(i)}$ denote the corresponding Young diagram whose j-th row is $\alpha_j^{(i)}$ and let $|\alpha^{(i)}|$ be the number of boxes of $\alpha^{(i)}$. Given these data, introduce another Young diagram $\gamma = (\gamma_1, \gamma_2, \dots)$ by setting

$$\gamma_l = \sum_{i=1}^{r} \sum_{j\geqslant l+1} \alpha_j^{(i)}, \tag{9.37}$$

so that γ_l is the total number of boxes which are strictly below the l-th rows in all diagrams $\alpha^{(i)}$. Furthermore, associate the elements of the family $\phi_{m\,(k)}$ with the boxes of the diagram $\Gamma = (N, N-1, \dots, 1)$, as illustrated:

$$\Gamma \quad = \quad \begin{matrix} \phi_{N\,(N-1)} & \phi_{N-1\,(N-2)} & \cdots & \phi_{2\,(1)} & \phi_{1\,(0)} \\ \phi_{N\,(N-2)} & \phi_{N-1\,(N-3)} & \cdots & \phi_{2\,(0)} & \\ \cdots & \cdots & \cdots & & \\ \phi_{N\,(1)} & \phi_{N-1\,(0)} & & & \\ \phi_{N\,(0)} & & & & \end{matrix} \tag{9.38}$$

It is clear that the diagram γ is contained in Γ. The *skew diagram* Γ/γ is obtained by deleting from Γ all boxes which belong to γ.

THEOREM 9.3.3. *The elements $\phi_{m\,(k)}$ corresponding to the boxes of the skew diagram Γ/γ are algebraically independent generators of the subalgebra $\mathcal{A}_\mu$. Moreover, the subalgebra $\mathcal{A}_\mu$ is a quantization of $\overline{\mathcal{A}}_\mu$ so that* $\operatorname{gr} \mathcal{A}_\mu = \overline{\mathcal{A}}_\mu$.

PROOF. As we pointed out above, the matrix M defined in (9.32) is a Manin matrix. Therefore, applying relation (3.17) in Proposition 3.2.2 we get

$$\operatorname{tr}_{1,\dots,m} A^{(m)} M_1 \dots M_m = \sum_{I,\,|I|=m} M_I^I, \tag{9.39}$$

summed over the subsets $I = \{i_1, \dots, i_m\}$ with $i_1 < \cdots < i_m$, where we used the following column-minor notation. Given two subsets $B = \{b_1, \dots, b_k\}$ and $C = \{c_1, \dots, c_k\}$ of $\{1, \dots, N\}$, the corresponding *column-minor* of a $N \times N$ matrix $X = [x_{ij}]$ is defined by

$$X_C^B = \sum_{\tau\in\mathfrak{S}_k} \operatorname{sgn}\tau \cdot x_{b_{\tau(1)}c_1} \cdots x_{b_{\tau(k)}c_k}.$$

Due to (9.34) with $a = m$, the definition of the generators $\phi_{m\,(k)}$ of the algebra $\mathcal{A}_\mu$ can be written in terms of column-minors as

$$\phi_{m\,(0)} z^{-m} + \dots + \phi_{m\,(m-1)} z^{-1} + \phi_{m\,(m)} = \sum_{I,\,|I|=m} M_I^I\, 1,$$

assuming that the differential operator on the right hand side applies to the element 1 so that $\partial_z 1 = 0$. This implies the explicit formula

$$\phi_{m\,(k)} = z^{m-k} \sum_{I,\,|I|=m} \sum_{\substack{B,C\subset I\\ |B|=|C|=k}} \operatorname{sgn}\sigma \cdot \mu_C^B \left[-\partial_z + E z^{-1}\right]_{I\setminus C}^{I\setminus B} 1, \tag{9.40}$$

where σ denotes the permutation of the set I given by

$$\sigma = \begin{pmatrix} B, & I\setminus B \\ C, & I\setminus C \end{pmatrix} = \begin{pmatrix} b_1, \dots, b_k, i_1, \dots, \widehat{b}_1, \dots, \widehat{b}_k, \dots, i_m \\ c_1, \dots, c_k, i_1, \dots, \widehat{c}_1, \dots, \widehat{c}_k, \dots, i_m \end{pmatrix},$$

and we assume that $b_1 < \cdots < b_k$ and $c_1 < \cdots < c_k$ for the respective elements of the subsets B and C in I.

For each $l = 1, \dots, N$ introduce the polynomial in a variable t with coefficients in $\mathcal{A}_\mu$ by

$$\Phi_l(t, \mu) = \phi_{l\,(0)}(\mu)\, t^{N-l} + \phi_{l+1\,(1)}(\mu)\, t^{N-l-1} + \dots + \phi_{N\,(N-l)}(\mu),$$

where we indicated the dependence of μ of the elements $\phi_{m\,(k)}$. Note that the coefficients of $\Phi_l(t, \mu)$ are the entries of the l-th row of the diagram Γ; see (9.38).

LEMMA 9.3.4. *For any $a \in \mathbb{C}$ we have the relation*

$$\Phi_l(t, \mu + a\,1) = \Phi_l(t + a, \mu).$$

PROOF. We have

$$\operatorname{tr}_{1,\dots,m} A^{(m)}(a + M_1) \dots (a + M_m) = \sum_{p=0}^{m} a^p \sum_{i_1 < \dots < i_{m-p}} \operatorname{tr}_{1,\dots,m} A^{(m)} M_{i_1} \dots M_{i_{m-p}}.$$

Furthermore, $A^{(m)} = \operatorname{sgn} \sigma \cdot A^{(m)} P_\sigma$ for any $\sigma \in \mathfrak{S}_m$, where P_σ denotes the image of $\sigma \in \mathfrak{S}_m$ in the algebra (9.31) under the action of $\mathfrak{S}_m$ defined in (1.65). Hence, applying conjugations by appropriate elements P_σ and using the cyclic property of trace, we can write the expression as

$$\sum_{p=0}^{m} \binom{m}{p} a^p \operatorname{tr}_{1,\dots,m} A^{(m)} M_1 \dots M_{m-p}.$$

The partial trace of the anti-symmetrizer over the m-th copy of $\operatorname{End} \mathbb{C}^n$ is found by (3.26) which implies

$$\operatorname{tr}_{m-p+1,\dots,m} A^{(m)} = \frac{(N - m + p)!\,(m-p)!}{(N-m)!\,m!} A^{(m-p)}.$$

Hence,

$$\operatorname{tr}_{1,\dots,m} A^{(m)}(a + M_1) \dots (a + M_m)$$
$$= \sum_{p=0}^{m} \binom{N - m + p}{p} a^p \operatorname{tr}_{1,\dots,m-p} A^{(m-p)} M_1 \dots M_{m-p}.$$

Now equate the constant terms of the differential operators on both sides and take the coefficients of z^{-m+k} to get the relation

$$\phi_{m\,(k)}(\mu + a\,1) = \sum_{p=0}^{k} \binom{N - m + p}{p} a^p \phi_{m-p\,(k-p)}(\mu).$$

Therefore, for the polynomial $\Phi_l(t, \mu + a\,1)$ we find

$$\Phi_l(t, \mu + a\,1) = \sum_{k=0}^{N-l} \phi_{l+k\,(k)}(\mu + a\,1)\, t^{N-l-k}$$
$$= \sum_{k=0}^{N-l} \sum_{p=0}^{k} \binom{N - l - k + p}{p} a^p \phi_{l+k-p\,(k-p)}(\mu)\, t^{N-l-k}$$

which equals

$$\sum_{p=0}^{N-l} a^p \sum_{r=0}^{N-l-p} \binom{N-l-r}{p} \phi_{l+r\,(r)}(\mu)\, t^{N-l-p-r}$$

$$= \sum_{p=0}^{N-l} \frac{a^p}{p!} \left(\frac{d}{dt}\right)^p \Phi_l(t,\mu) = \Phi_l(t+a,\mu),$$

as claimed. □

LEMMA 9.3.5. *Suppose that μ has the form of a block-diagonal matrix*

$$\mu = \begin{bmatrix} J_\alpha(0) & \mathrm{O} \\ \mathrm{O} & \widetilde{\mu} \end{bmatrix}, \tag{9.41}$$

where $J_\alpha(0)$ is the nilpotent Jordan matrix associated with a certain Young diagram $\alpha = (\alpha_1, \alpha_2, \dots)$ and $\widetilde{\mu}$ is an arbitrary square matrix of size q such that $|\alpha| + q = N$. Then for any $l \geqslant 1$ we have

$$\phi_{l+k\,(k)} = 0 \qquad \textit{for all} \quad N - l - \delta_l + 1 \leqslant k \leqslant N - l,$$

where $\delta_l = \alpha_{l+1} + \alpha_{l+2} + \dots$ is the number of boxes of α below its row l.

PROOF. The generator $\phi_{l+k\,(k)}$ is found by (9.40) for $m = l + k$. The internal sum is a linear combination of $k \times k$ minors of the matrix μ satisfying the condition that the union $B \cup C$ of the row and column indices of each minor is a set of size not exceeding $k + l$. Hence the lemma will be implied by the claim that under the given condition on k, the minor μ^B_C can be nonzero only if the union of row and column indices is of the size at least $k+l+1$. Indeed, observe that if p is a positive integer, then any nonzero $p \times p$ minor of a nilpotent Jordan block has the property that the minimal possible size of the union of its row and column indices is $p + 1$. However, the condition $k \geqslant N - l - \delta_l + 1$ means that $k \geqslant \alpha_1 + \dots + \alpha_l - l + 1 + q$. Therefore, any nonzero $k \times k$ minor must involve at least $l + 1$ Jordan blocks and the claim follows. □

In the notation of the theorem, for each diagram $\alpha^{(i)}$ denote by $\delta_l^{(i)}$ the corresponding parameter δ_l introduced in Lemma 9.3.5, so that for the number γ_l defined in (9.37) we have

$$\gamma_l = \sum_{i=1}^{r} \delta_l^{(i)}.$$

LEMMA 9.3.6. *The polynomial $\Phi_l(t,\mu)$ admits the factorization*

$$\Phi_l(t,\mu) = (t+\lambda_1)^{\delta_l^{(1)}} \dots (t+\lambda_r)^{\delta_l^{(r)}}\, \widetilde{\Phi}_l(t,\mu)$$

for a certain polynomial $\widetilde{\Phi}_l(t,\mu)$ in t.

PROOF. Observe that by formulas (9.39), the elements $\phi_{m\,(k)}$ are unchanged under the simultaneous replacements $\mu \mapsto g\mu g^{-1}$ and $E \mapsto gEg^{-1}$ for any element $g \in \mathrm{GL}_N$. This implies that $\mathcal{A}_{g\mu g^{-1}}$ can be identified with the algebra $\mathcal{A}_\mu$ associated with the image of $\mathrm{U}(\mathfrak{gl}_N)$ under the automorphism sending E to gEg^{-1}. Therefore, the algebra $\mathcal{A}_\mu$ depends only on the coadjoint orbit of μ. This is a particular case of the general property; cf. the proof of Theorem 9.1.3.

For any $i \in \{1, \ldots, r\}$ the Jordan canonical form of $\mu - \lambda_i 1$ is a matrix of the form (9.41), where $\alpha = \alpha^{(i)}$. By Lemma 9.3.5, the polynomial $\Phi_l(t, \mu - \lambda_i 1)$ is divisible by $t^{\delta_l^{(i)}}$. By Lemma 9.3.4, the polynomial $\Phi_l(t, \mu) = \Phi_l(t + \lambda_i, \mu - \lambda_i 1)$ is then divisible by $(t + \lambda_i)^{\delta_l^{(i)}}$. □

Returning to the proof of the theorem, note that by Lemma 9.3.6 for any $l = 1, \ldots, N$ the generators $\phi_{l+k\,(k)}$ with $N - l - \gamma_l + 1 \leqslant k \leqslant N - l$ are linear combinations of those generators with $k = 0, 1, \ldots, N - l - \gamma_l$. Therefore, the elements $\phi_{l+k\,(k)}$ corresponding to the boxes of the skew diagram Γ/γ generate the algebra $\mathcal{A}_\mu$. It remains to verify that these generators are algebraically independent.

Consider the elements $\overline{\phi}_{m\,(k)} \in \mathrm{S}(\mathfrak{gl}_N)$ which are defined by

$$\overline{\phi}_{m\,(k)} = \sum_{I,\,|I|=m} \; \sum_{\substack{B,C \subset I \\ |B|=|C|=k}} \operatorname{sgn} \sigma \cdot \mu_C^B \, E_{I\setminus C}^{I\setminus B},$$

with the notation as in (9.40), where the entries of the matrix E are now regarded as elements of the symmetric algebra $\mathrm{S}(\mathfrak{gl}_N)$. Equivalently, the elements $\overline{\phi}_{m\,(k)}$ are found by

$$\begin{aligned} \operatorname{tr}_{1,\ldots,m} A^{(m)} \big(\mu_1 + E_1 z^{-1}\big) \ldots \big(\mu_m + E_m z^{-1}\big) \\ = \overline{\phi}_{m\,(0)} z^{-m} + \cdots + \overline{\phi}_{m\,(m-1)} z^{-1} + \overline{\phi}_{m\,(m)}. \end{aligned}$$

By Theorem 9.3.1(i), they generate the subalgebra $\overline{\mathcal{A}}_\mu$. The above arguments applied to these generators instead of the $\phi_{m\,(k)}$ show that the algebra $\overline{\mathcal{A}}_\mu$ is generated by the subset of the elements $\overline{\phi}_{m\,(k)}$ corresponding to the boxes of the skew diagram Γ/γ. Furthermore, we have the following.

LEMMA 9.3.7. *The generators $\overline{\phi}_{m\,(k)}$ of the subalgebra $\overline{\mathcal{A}}_\mu$ corresponding to the boxes of the skew diagram Γ/γ are algebraically independent.*

PROOF. Regarding the elements $\overline{\phi}_{m\,(k)}$ as polynomials in the variables E_{ij}, we will see that their differentials $d\,\overline{\phi}_{m\,(k)}$ are linearly independent at a certain point. Since these elements generate $\overline{\mathcal{A}}_\mu$, the linear span of the differentials $d\,\overline{\phi}_{m\,(k)}$ at any point coincides with the linear span of all differentials

$$d\,\overline{\mathcal{A}}_\mu = \text{span of } \{d\phi \mid \phi \in \overline{\mathcal{A}}_\mu\}.$$

The dimension of the space of differentials evaluated at a certain regular point can be found from *Bolsinov's criterion* [**11**, Theorem 3.2] . To state it for the Lie algebra $\mathfrak{gl}_N$, recall that the *index* $\operatorname{ind}\mathfrak{g}$ of an arbitrary Lie algebra $\mathfrak{g}$ is the minimal dimension

$$\operatorname{ind}\mathfrak{g} = \min_{\xi \in \mathfrak{g}^*} \dim \operatorname{Ann} \xi$$

of the annihilators $\operatorname{Ann}\xi$ for the coadjoint representation, where

$$\operatorname{Ann}\xi = \{X \in \mathfrak{g} \mid \operatorname{ad}_X^* \xi = 0\}.$$

By Bolsinov's criterion, the relation

$$\dim d\,\overline{\mathcal{A}}_\mu = \operatorname{rank}\mathfrak{gl}_N + \frac{1}{2}\big(\dim \mathfrak{gl}_N - \dim \operatorname{Ann}\mu\big)$$

for the differentials evaluated at a certain regular point holds if and only if

$$\operatorname{ind} \operatorname{Ann}\mu = \operatorname{rank}\mathfrak{gl}_N. \tag{9.42}$$

This equality does hold for any μ [**154**], and so, to show that the differentials $d\,\overline{\phi}_{m\,(k)}$ of the generators are linearly independent at a certain point, we only need to verify that the number of boxes of the skew diagram Γ/γ coincides with

$$\operatorname{rank}\mathfrak{gl}_N + \frac{1}{2}\big(\dim\mathfrak{gl}_N - \dim\operatorname{Ann}\mu\big) = N + \frac{1}{2}\big(N^2 - \dim\operatorname{Ann}\mu\big).$$

Since $|\Gamma| = N(N+1)/2$, the desired formula is equivalent to the relation

$$\dim\operatorname{Ann}\mu = 2\,|\gamma| + N. \tag{9.43}$$

The dimension of the annihilator $\operatorname{Ann}\mu$ coincides with that of the centralizer of μ^t in $\dim\mathfrak{gl}_N$; see Example 9.1.7. We have

$$\dim\operatorname{Ann}\mu = \sum_{i=1}^{r}\dim\operatorname{Ann}_{\,\mathfrak{gl}_{n_i}}\mu^{(i)},$$

where $\mu^{(i)}$ denotes the direct sum of all Jordan blocks of μ with the eigenvalue λ_i, and n_i is the size of $\mu^{(i)}$. Hence, by the definition of γ, the verification of (9.43) reduces to the case where μ has only one eigenvalue. Let $\alpha_1 \geqslant \cdots \geqslant \alpha_s$ be the respective sizes of the Jordan blocks of such a matrix μ. The dimension of the annihilator (or, equivalently, the centralizer of μ^t) is found by the well-known formula [**80**, Section 3.1],

$$\dim\operatorname{Ann}\mu = \alpha_1 + 3\alpha_2 + \cdots + (2s-1)\,\alpha_s,$$

while

$$|\gamma| = \alpha_2 + 2\alpha_3 + \cdots + (s-1)\,\alpha_s \qquad\text{and}\qquad N = \alpha_1 + \cdots + \alpha_s,$$

thus implying (9.43). □

Now consider the generators $\phi_{m\,(k)}$ of the algebra $\mathcal{A}_\mu$ associated with the boxes of the diagram Γ/γ. By Lemma 9.3.7, the corresponding elements $\overline{\phi}_{m\,(k)}$ are nonzero, so that the image of $\phi_{m\,(k)}$ in the $(m-k)$-th component of the graded algebra gr $\mathrm{U}(\mathfrak{gl}_N) \cong \mathrm{S}(\mathfrak{gl}_N)$ coincides with $\overline{\phi}_{m\,(k)}$. Moreover, the generators $\phi_{m\,(k)}$ corresponding to the boxes of the diagram Γ/γ are algebraically independent. This completes the proof of the theorem. □

REMARK 9.3.8. Theorem 9.3.3 is valid in the same form for some other families of generators of the algebra $\mathcal{A}_\mu$ in place of the elements $\phi_{m\,(k)}$; see [**53**]. □

EXAMPLE 9.3.9. Take $N = 6$ and let μ be a nilpotent matrix with the Jordan blocks of sizes $(2, 2, 1, 1)$. Then $\gamma = (4, 2, 1)$ and the skew diagram Γ/γ is

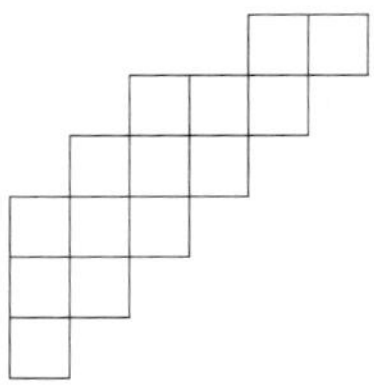

so that $\mathcal{A}_\mu$ has 14 algebraically independent generators. □

If the element μ is regular, then all Jordan blocks correspond to distinct eigenvalues so that each $\alpha^{(i)}$ is a singe row diagram. Therefore, $\gamma = \varnothing$ and all generators $\phi_{m\,(k)}$ associated with the boxes of Γ are algebraically independent. On the other hand, if μ is a scalar matrix (in particular, $\mu = 0$), then $\gamma = (N-1, N-2, \ldots, 1)$.

In this case, $\mathcal{A}_\mu$ is generated by $\phi_{1\,(0)},\dots,\phi_{N\,(0)}$ and it coincides with the center of $\mathrm{U}(\mathfrak{gl}_N)$; cf. Section 6.5.

9.4. Generators of commutative subalgebras of $\mathrm{U}(\mathfrak{o}_N)$ and $\mathrm{U}(\mathfrak{sp}_N)$

Now let $\mathfrak{g}=\mathfrak{g}_N$ be the orthogonal Lie algebra $\mathfrak{o}_N$ or the symplectic Lie algebra $\mathfrak{sp}_N$. We keep using the tensor product notation and consider the algebra

$$\underbrace{\operatorname{End}\mathbb{C}^N\otimes\dots\otimes\operatorname{End}\mathbb{C}^N}_{m}\otimes\mathrm{U},$$

where $\mathrm{U}=\mathrm{U}(\mathfrak{g}_N)\otimes\mathbb{C}[z^{-1},\partial_z]$ is the algebra of differential operators whose elements are finite sums of the form

$$\sum_{k,l\geqslant 0} x_{kl}\,z^{-k}\,\partial_z^{\,l},\qquad x_{kl}\in\mathrm{U}(\mathfrak{g}_N).$$

We will also use the rational function $\gamma_m(\omega)$ defined in (2.40) with the specializations $\omega=N$ and $\omega=-N$ in the orthogonal and symplectic case, respectively.

An arbitrary element $\mu\in\mathfrak{g}_N^*$ will be regarded as the $N\times N$ matrix $\mu=[\mu_{ij}]$, where $\mu_{ij}=\mu(F_{ij})$. The matrix μ has the property $\mu+\mu'=0$; see (2.24). Define the polynomials $\phi_{m\,a}(z)$ in z^{-1} (depending on μ) by the expansion

$$\begin{aligned}(9.44)\quad &\gamma_m(\omega)\,\mathrm{tr}_{1,\dots,m}\,S^{(m)}\big(-\partial_z+\mu_1+F_1\,z^{-1}\big)\dots\big(-\partial_z+\mu_m+F_m\,z^{-1}\big)\\ &\qquad=\phi_{m\,0}(z)\,(-\partial_z)^m+\phi_{m\,1}(z)\,(-\partial_z)^{m-1}+\dots+\phi_{m\,m}(z).\end{aligned}$$

In the symplectic case $\mathfrak{g}_N=\mathfrak{sp}_{2n}$ we adopt the same convention for (9.44) as for the elements (8.36); cf. the remark just before Theorem 8.3.8. That is, the left hand side is assumed to be written in an equivalent form which is well-defined for all values $1\leqslant m\leqslant 2n+1$ as provided by Proposition 8.3.7. In fact, in both the orthogonal and symplectic case the expression (9.44) coincides with the respective image of (8.15) or (8.36) under the homomorphism (9.26) with the extension to the differential operators by $\tau\mapsto-\partial_z$ as in (9.36).

In the case $\mathfrak{g}_N=\mathfrak{o}_{2n}$ we also define the (noncommutative) *Pfaffian* of the matrix $\mu+F\,z^{-1}$ by

$$\begin{aligned}&\mathrm{Pf}\,\big(\mu+F\,z^{-1}\big)\\ &\qquad=\frac{1}{2^n n!}\sum_{\sigma\in\mathfrak{S}_{2n}}\mathrm{sgn}\,\sigma\cdot\big(\mu+F\,z^{-1}\big)_{\sigma(1)\,\sigma(2)'}\cdots\big(\mu+F\,z^{-1}\big)_{\sigma(2n-1)\,\sigma(2n)'}.\end{aligned}$$

Introduce the coefficients of the polynomials by

$$\phi_{m\,m}(z)=\phi_{m\,m\,(0)}z^{-m}+\dots+\phi_{m\,m\,(m-1)}z^{-1}+\phi_{m\,m\,(m)}$$

and

$$\mathrm{Pf}\,\big(\mu+F\,z^{-1}\big)=\pi_{(0)}z^{-n}+\dots+\pi_{(n-1)}z^{-1}+\pi_{(n)}.$$

In particular, $\pi_{(0)}$ coincides with $\mathrm{Pf}\,F$.

The next theorem will imply Theorem 9.2.4 in types B,C and D.

THEOREM 9.4.1.

(i) *Given any $\mu\in\mathfrak{g}_N^*$, all coefficients of the polynomial $\phi_{m\,a}(z)$ belong to the commutative subalgebra $\mathcal{A}_\mu$ of $\mathrm{U}(\mathfrak{g}_N)$. Moreover, all coefficients of the polynomial $\mathrm{Pf}\,\big(\mu+F\,z^{-1}\big)$ belong to the subalgebra $\mathcal{A}_\mu$ of $\mathrm{U}(\mathfrak{o}_{2n})$.*

(ii) *In the case where μ is regular, the elements*

$$\phi_{2k\,2k\,(i)} \qquad \text{with} \quad k=1,\dots,n \quad \text{and} \quad i=0,1,\dots,2k-1$$

are algebraically independent generators of the algebra $\mathcal{A}_\mu$ in the cases B and C, while the elements

$$\phi_{2k\,2k\,(i)} \qquad \text{with} \quad k=1,\dots,n-1 \quad \text{and} \quad i=0,1,\dots,2k-1$$

together with $\pi_{(0)},\dots,\pi_{(n-1)}$ are algebraically independent generators of the algebra $\mathcal{A}_\mu$ in the case D. Moreover, in all three cases, $\operatorname{gr}\mathcal{A}_\mu=\overline{\mathcal{A}}_\mu$.

PROOF. The coefficients of the polynomials $\phi_{m\,a}(z)$ and $\operatorname{Pf}\big(\mu+F\,z^{-1}\big)$ are the images under the homomorphism ϱ_μ of the respective elements of the Feigin–Frenkel center $\mathfrak{z}(\widehat{\mathfrak{g}}_N)$ provided by Proposition 8.1.4, Theorem 8.1.6 and Proposition 8.3.7. Therefore, they belong to $\mathcal{A}_\mu$.

To prove part (ii), suppose that μ is regular and consider the symbols of the elements $\phi_{2k\,2k\,(i)}$ and $\pi_{(j)}$ in the associated graded algebra $\mathrm{S}(\mathfrak{g}_N)\cong\operatorname{gr}\mathrm{U}(\mathfrak{g}_N)$. These symbols coincide with the respective coefficients of the polynomials in z^{-1} given by

$$\gamma_{2k}(\omega)\,\operatorname{tr}_{1,\dots,2k}S^{(2k)}\big(\mu_1+F_1\,z^{-1}\big)\dots\big(\mu_{2k}+F_{2k}\,z^{-1}\big)$$

and $\operatorname{Pf}\big(\mu+F\,z^{-1}\big)$. The coefficients belong to the Mishchenko–Fomenko subalgebra $\overline{\mathcal{A}}_\mu$ of $\mathrm{S}(\mathfrak{g}_N)$ since they arise from generators of $\mathrm{S}(\mathfrak{g}_N)^{\mathfrak{g}_N}$ by the rule (9.12); see Section 2.2. Furthermore, taking into account Corollaries 2.2.5 and 2.2.10, we derive from Theorem 9.1.3 that the respective coefficients are algebraically independent generators of $\overline{\mathcal{A}}_\mu$. Hence part (ii) of the theorem follows. $\square$

EXAMPLE 9.4.2. By making use of the Segal–Sugawara vectors provided in Example 8.2.2, we get algebraically independent generators of the algebra $\mathcal{A}_\mu$ for regular μ (cf. Example 9.3.2):

for $\mathfrak{o}_3$: $\quad \operatorname{tr}\mu F,\quad \operatorname{tr}F^2$

for $\mathfrak{o}_4$: $\quad \operatorname{tr}\mu F,\quad \operatorname{tr}F^2,\quad \pi_{(0)},\quad \pi_{(1)}$

for $\mathfrak{o}_5$: $\quad \operatorname{tr}\mu F,\quad \operatorname{tr}F^2,\quad \operatorname{tr}\mu^3F,\quad 2\operatorname{tr}\mu^2F^2+\operatorname{tr}(\mu F)^2,\quad \operatorname{tr}\mu F^3,\quad \operatorname{tr}F^4$

for $\mathfrak{o}_6$: $\quad \operatorname{tr}\mu F,\quad \operatorname{tr}F^2,\quad \operatorname{tr}\mu^3F,\quad 2\operatorname{tr}\mu^2F^2+\operatorname{tr}(\mu F)^2,\quad \operatorname{tr}\mu F^3,\quad \operatorname{tr}F^4,$
$\quad \pi_{(0)},\quad \pi_{(1)},\quad \pi_{(2)}.$

As in Example 9.3.2, we calculate the coefficients of the powers of z^{-1} in the polynomials of the form $\operatorname{tr}(\mu+F\,z^{-1})^k$. We also use some relations (for $\mathfrak{o}_N$) between these coefficients which can be derived from (5.2) or (5.6):

$$\operatorname{tr}F\mu F^2=\operatorname{tr}F^2\mu F=\operatorname{tr}\mu\,F^3-(N-2)\operatorname{tr}\mu\,F^2,$$
$$2\operatorname{tr}\mu\,F^2=(N-2)\operatorname{tr}\mu\,F$$

and

$$\operatorname{tr}F\mu^2F=\operatorname{tr}\mu^2F^2,\qquad \operatorname{tr}\mu^2F=0.$$

EXAMPLE 9.4.3. Similarly, by Example 8.4.1 we get algebraically independent generators of the algebra $\mathcal{A}_\mu$ for regular μ in the symplectic case:

$$\text{for}\quad \mathfrak{sp}_2: \qquad \operatorname{tr}\mu F, \quad \operatorname{tr} F^2$$

$$\text{for}\quad \mathfrak{sp}_4: \qquad \operatorname{tr}\mu F, \quad \operatorname{tr} F^2, \quad \operatorname{tr}\mu^3 F, \quad 2\operatorname{tr}\mu^2 F^2 + \operatorname{tr}(\mu F)^2, \quad \operatorname{tr}\mu F^3, \quad \operatorname{tr} F^4.$$

9.5. Bibliographical notes

The Poisson commutative subalgebras $\overline{\mathcal{A}}_\mu$ of $\mathrm{S}(\mathfrak{g})$ were first introduced by Mishchenko and Fomenko [**108**] in connection with the Euler equations; see also Manakov [**105**]. Theorem 9.1.3 is contained in the original paper [**108**] in the case where μ is regular semisimple. The general regular case is due to Bolsinov [**11**], [**12**]. We followed the proof given by Feigin, Frenkel and Toledano Laredo [**42**]; see also Kostant [**95**]. The paper [**42**] also contains a general proof of Theorem 9.2.4. In the case of regular semisimple μ it was previously proved by Rybnikov [**139**].

Explicit formulas of Theorem 9.3.1 originate in the paper by Talalaev [**145**]; see also Chervov and Talalaev [**24**]. These families of generators are related to each other via the identities satisfied by Manin matrices; see the proof of Theorem 7.1.3. A different construction of a commutative subalgebra of $\mathrm{U}(\mathfrak{gl}_N)$ was provided by Tarasov [**146**] via a symmetrization map. In the case where $\mu \in \mathfrak{g}^*$ is regular semisimple, explicit constructions of maximal commutative subalgebras of $\mathrm{U}(\mathfrak{g})$ were given in an earlier work of Nazarov and Olshanski [**126**] with the use of the Yangian for $\mathfrak{gl}_N$ and the twisted Yangians associated with the orthogonal and symplectic Lie algebras. In the case of $\mathfrak{gl}_N$ their elements are found by an expansion of the quantum determinant which is essentially equivalent to (9.33). In types B, C and D their elements appear to be quite different from those of Theorem 9.4.1 which were provided in [**110**]. For an arbitrary simple Lie algebra $\mathfrak{g}$, the spectra of the subalgebra $\mathcal{A}_\mu$ of $\mathrm{U}(\mathfrak{g})$ in finite-dimensional irreducible representations were described by Feigin, Frenkel and Rybnikov [**41**].

In the proof of Theorem 9.3.3 we followed the arguments of [**53**] which rely on Bolsinov's work [**11**]; see also Bolsinov and Zhang [**13**]. Relation (9.42) is a particular case of the *Bolsinov–Elashvili conjecture* which originates in [**12**] and applies to any reductive Lie algebra; see also Panyushev [**130**]. The first published proof of (9.42) is due to Yakimova [**154**], which extends to all classical Lie algebras; see also Charbonnel and Moreau [**17**] for its proof covering all simple Lie algebras and more references.

CHAPTER 10

Yangian characters in type A

As we recalled in the Preface, the Harish–Chandra isomorphism (0.1) leads to a description of the structure of the center $\mathrm{Z}(\mathfrak{g})$ as an algebra of polynomials. We will now aim to prove an affine analogue of the isomorphism theorem for the classical Lie algebras. In its basic form, this analogue will involve the Feigin–Frenkel center $\mathfrak{z}(\widehat{\mathfrak{g}})$ which will play the role of $\mathrm{Z}(\mathfrak{g})$. The algebra of W-invariants in $\mathrm{U}(\mathfrak{h})$ will now be replaced by a *classical $\mathcal{W}$-algebra*; we will review these algebras in Chapter 12. Our arguments will rely on the explicit constructions of generators of $\mathfrak{z}(\widehat{\mathfrak{g}})$ given in Chapters 7 and 8. The calculation of the Harish–Chandra images as elements of the respective classical $\mathcal{W}$-algebra will be completed in Chapter 13 and will be based on the *Yangian characters*. We will derive the necessary character formulas in the current chapter for type A and in Chapter 11 for types B, C and D.

The *Yangian* $\mathrm{Y}(\mathfrak{g})$ is a Hopf algebra which is a deformation of the universal enveloping algebra $\mathrm{U}\left(\mathfrak{g}[t]\right)$ in the class of Hopf algebras. We will work with the *RTT-presentations* of the Yangians to calculate the characters of certain classes of representations. Essentially the same calculation will give the character formulas for representations of the *dual Yangian* $\mathrm{Y}^{+}(\mathfrak{g})$ which is a deformation of the universal enveloping algebra $\mathrm{U}\left(t^{-1}\mathfrak{g}[t^{-1}]\right)$. By taking appropriate classical limits we will recover the Harish-Chandra images of certain elements of $\mathrm{U}(\mathfrak{g}[t])$ and $\mathrm{U}\left(t^{-1}\mathfrak{g}[t^{-1}]\right)$ which will then be translated into the desired formulas for the corresponding Harish-Chandra images of the Segal–Sugawara vectors.

The Yangian approach can also be used to obtain an alternative construction of the Segal–Sugawara vectors of Section 7.1 which we discuss below in Section 10.5. Moreover, this approach can be extended to any simple Lie algebra $\mathfrak{g}$ provided that an appropriate version of the fusion procedure is developed. We begin by reviewing some well-known properties of the Yangian for $\mathfrak{gl}_N$ following [**109**].

10.1. Yangian for $\mathfrak{gl}_N$

The *Yangian* $\mathrm{Y}(\mathfrak{gl}_N)$ is a unital associative algebra with generators $t_{ij}^{(r)}$, where $1 \leqslant i, j \leqslant N$ and $r = 1, 2, \dots$ and the defining relations

$$[t_{ij}^{(r+1)}, t_{kl}^{(s)}] - [t_{ij}^{(r)}, t_{kl}^{(s+1)}] = t_{kj}^{(r)} t_{il}^{(s)} - t_{kj}^{(s)} t_{il}^{(r)}, \tag{10.1}$$

where $r, s = 0, 1, \dots$ and $t_{ij}^{(0)} = \delta_{ij}$ with the usual notation for the commutator $[a, b] = ab - ba$. In terms of the formal series

$$t_{ij}(u) = \delta_{ij} + \sum_{r=1}^{\infty} t_{ij}^{(r)} u^{-r} \in \mathrm{Y}(\mathfrak{gl}_N)[[u^{-1}]]$$

the defining relations are written in the form

$$(u - v)\, [t_{ij}(u), t_{kl}(v)] = t_{kj}(u)\, t_{il}(v) - t_{kj}(v)\, t_{il}(u). \tag{10.2}$$

By dividing both sides by $u-v$ we get an equivalent form of (10.1),

$$[t_{ij}^{(r)}, t_{kl}^{(s)}] = \sum_{a=1}^{\min\{r,s\}} \Big(t_{kj}^{(a-1)} t_{il}^{(r+s-a)} - t_{kj}^{(r+s-a)} t_{il}^{(a-1)} \Big). \tag{10.3}$$

Set

$$T(u) = \sum_{i,j=1}^{N} e_{ij} \otimes t_{ij}(u) \in \operatorname{End} \mathbb{C}^N \otimes \mathrm{Y}(\mathfrak{gl}_N)[[u^{-1}]] \tag{10.4}$$

and adopt the matrix notation of Sections 1.4 and 1.5, so that $T_a(u)$ for $a = 1, \dots, m$ will be understood as formal series in u^{-1} with coefficients in the tensor product algebra

$$\underbrace{\operatorname{End} \mathbb{C}^N \otimes \dots \otimes \operatorname{End} \mathbb{C}^N}_{m} \otimes \mathrm{Y}(\mathfrak{gl}_N).$$

The *Yang R-matrix* $R_{12}(u)$ is a rational function in a complex parameter u with values in the tensor product algebra $\operatorname{End} \mathbb{C}^N \otimes \operatorname{End} \mathbb{C}^N$ defined by

$$R_{12}(u) = 1 - P_{12}\, u^{-1}, \tag{10.5}$$

where P is the permutation operator (1.64). This function satisfies the *Yang–Baxter equation*

$$R_{12}(u)\, R_{13}(u+v)\, R_{23}(v) = R_{23}(v)\, R_{13}(u+v)\, R_{12}(u). \tag{10.6}$$

The defining relations for the algebra $\mathrm{Y}(\mathfrak{gl}_N)$ can be written in the matrix form as

$$R_{12}(u-v)\, T_1(u)\, T_2(v) = T_2(v)\, T_1(u)\, R_{12}(u-v). \tag{10.7}$$

The algebra $\mathrm{Y}(\mathfrak{gl}_N)$ possesses two natural ascending filtrations defined by

$$\deg t_{ij}^{(r)} = r \qquad \text{and} \qquad \deg' t_{ij}^{(r)} = r-1$$

for all $r \geqslant 1$. Denote by $\operatorname{gr} \mathrm{Y}(\mathfrak{gl}_N)$ and $\operatorname{gr}' \mathrm{Y}(\mathfrak{gl}_N)$ the corresponding graded algebras. We have the isomorphisms

$$\operatorname{gr} \mathrm{Y}(\mathfrak{gl}_N) \cong \mathrm{S}\left(\mathfrak{gl}_N[t]\right) \qquad \text{and} \qquad \operatorname{gr}' \mathrm{Y}(\mathfrak{gl}_N) \cong \mathrm{U}\left(\mathfrak{gl}_N[t]\right)$$

so that, in particular, the algebra $\operatorname{gr} \mathrm{Y}(\mathfrak{gl}_N)$ is commutative. The image $\bar{t}_{ij}^{(r)}$ of the generator $t_{ij}^{(r)}$ in the $(r-1)$-th component of the graded algebra $\operatorname{gr}' \mathrm{Y}(\mathfrak{gl}_N)$ corresponds to the element $E_{ij}[r-1]$ of $\mathrm{U}\left(\mathfrak{gl}_N[t]\right)$.

It follows easily from Proposition 4.2.2 that the mapping

$$\mathrm{ev} : T(u) \mapsto 1 + E\, u^{-1}, \tag{10.8}$$

defines a surjective homomorphism $\mathrm{Y}(\mathfrak{gl}_N) \to \mathrm{U}(\mathfrak{gl}_N)$ known as the *evaluation homomorphism*. Every representation V of $\mathfrak{gl}_N$ can thus be extended to a Yangian *evaluation module* via (10.8).

The universal enveloping algebra $\mathrm{U}(\mathfrak{gl}_N)$ will be identified with a subalgebra of the Yangian $\mathrm{Y}(\mathfrak{gl}_N)$ via the embedding $E_{ij} \mapsto t_{ij}^{(1)}$. Then $\mathrm{Y}(\mathfrak{gl}_N)$ can be regarded as a $\mathfrak{gl}_N$-module with the adjoint action, and the defining relations (10.3) give

$$[E_{ij}, t_{kl}^{(s)}] = \delta_{kj}\, t_{il}^{(s)} - \delta_{il}\, t_{kj}^{(s)}. \tag{10.9}$$

We have the following generalization of Theorem 4.5.1, where we keep the notation used in that theorem.

PROPOSITION 10.1.1. *For any $s \in \mathbb{C}[\mathfrak{S}_m]$ and variables $u_1, \ldots, u_m$, all coefficients of the series*

$$\operatorname{tr}_{1,\ldots,m} S\, T_1(u_1) \ldots T_m(u_m)$$

commute with $\mathfrak{gl}_N$.

PROOF. Consider the tensor product

$$\operatorname{End}\mathbb{C}^N \otimes \operatorname{End}(\mathbb{C}^N)^{\otimes m} \otimes \mathrm{Y}(\mathfrak{gl}_N)$$

with the additional copy of the endomorphism algebra $\operatorname{End}\mathbb{C}^N$ labeled by 0. Using (10.7) or writing (10.9) in a matrix form, we obtain the relations

$$[E_0, T_a(u_a)] = P_{0\,a} T_a(u_a) - T_a(u_a) P_{0\,a}$$

for $a = 1, \ldots, m$. Now the relation

$$\big[E_0, \operatorname{tr}_{1,\ldots,m} S\, T_1(u_1) \ldots T_m(u_m)\big] = 0,$$

which is a counterpart of (4.32), is verified by the same argument as in the proof of Theorem 4.5.1. □

Denote by $\mathrm{Y}(\mathfrak{gl}_N)^{\mathfrak{h}}$ the subalgebra of $\mathfrak{h}$-invariants under the action (10.9) so that

$$\mathrm{Y}(\mathfrak{gl}_N)^{\mathfrak{h}} = \{y \in \mathrm{Y}(\mathfrak{gl}_N) \mid [E_{ii}, y] = 0 \quad \text{for} \quad i = 1, \ldots, N\}.$$

Consider the left ideal I of the algebra $\mathrm{Y}(\mathfrak{gl}_N)$ generated by all elements $t_{ij}^{(r)}$ with the conditions $1 \leqslant i < j \leqslant N$ and $r \geqslant 1$. By the Poincaré–Birkhoff–Witt theorem for the Yangian [**109**, Section 1.4], the intersection $\mathrm{Y}(\mathfrak{gl}_N)^{\mathfrak{h}} \cap I$ is a two-sided ideal of $\mathrm{Y}(\mathfrak{gl}_N)^{\mathfrak{h}}$. Moreover, the quotient of $\mathrm{Y}(\mathfrak{gl}_N)^{\mathfrak{h}}$ by this ideal is isomorphic to the commutative algebra freely generated by the images of the elements $t_{ii}^{(r)}$ with $i = 1, \ldots, N$ and $r \geqslant 1$ in the quotient. We will use the notation $\lambda_i^{(r)}$ for this image of $t_{ii}^{(r)}$. Thus, we get an analogue of the Harish-Chandra homomorphism (4.6),

$$\mathrm{Y}(\mathfrak{gl}_N)^{\mathfrak{h}} \to \mathbb{C}[\lambda_i^{(r)} \mid i = 1, \ldots, N,\ r \geqslant 1]. \tag{10.10}$$

We combine the elements $\lambda_i^{(r)}$ into the formal series

$$\lambda_i(u) = 1 + \sum_{r=1}^{\infty} \lambda_i^{(r)} u^{-r}, \qquad i = 1, \ldots, N,$$

which will be understood as the images of the series $t_{ii}(u)$ under the homomorphism (10.10).

Using the notation of Section 4.9, consider a partition $\mu \vdash m$ of length $\ell(\mu) \leqslant N$. Suppose that $\mathcal{U}$ is a standard tableau of shape $\mu \vdash m$ and let $\mathcal{E}_{\mathcal{U}}$ be the image of the primitive idempotent $e_{\mathcal{U}}$ under the action of the symmetric group $\mathfrak{S}_m$ given by (1.65). The operator $\mathcal{E}_{\mathcal{U}}$ commutes with the natural action of $\mathfrak{gl}_N$ on the tensor product space $(\mathbb{C}^N)^{\otimes m}$. The image $\mathcal{E}_{\mathcal{U}}(\mathbb{C}^N)^{\otimes m}$ is an irreducible representation of $\mathfrak{gl}_N$ which is isomorphic to the highest weight representation $L(\mu)$, where μ is regarded as the N-tuple $(\mu_1, \ldots, \mu_N)$; see Section 4.4. Furthermore, the vector space $\mathcal{E}_{\mathcal{U}}(\mathbb{C}^N)^{\otimes m}$ is also a representation of the Yangian $\mathrm{Y}(\mathfrak{gl}_N)$, isomorphic to the evaluation module $L(\mu)$; see e.g. [**109**, Section 6.5].

The next theorem provides the Harish-Chandra image of a certain series $\mathbb{T}_\mu(u)$ associated with the partition μ. The coefficients of this series belong to $\mathrm{Y}(\mathfrak{gl}_N)^{\mathfrak{h}}$ by Proposition 10.1.1. We take as a definition that this image is the *Yangian character* of the evaluation module $L(\mu)$; see also Remark 10.1.5(ii) below.

As before, by $c_a = c_a(\mathcal{U})$ we denote the content $c(\alpha) = j-i$ of the box $\alpha = (i,j)$ in $\mathcal{U}$ occupied by $a \in \{1,\dots,m\}$.

THEOREM 10.1.2. *The series*

$$\mathbb{T}_\mu(u) = \mathrm{tr}_{1,\dots,m}\, \mathcal{E}_{\mathcal{U}}\, T_1(u+c_1)\dots T_m(u+c_m) \tag{10.11}$$

does not depend on the standard tableau $\mathcal{U}$ of shape μ. Moreover, the images of its coefficients under the Harish-Chandra homomorphism (10.10) *are found by*

$$\mathbb{T}_\mu(u) \mapsto \sum_{\mathrm{sh}(\mathcal{T})=\mu} \prod_{\alpha\in\mu} \lambda_{\mathcal{T}(\alpha)}(u+c(\alpha)),$$

summed over semistandard tableau $\mathcal{T}$ of shape μ with entries in $\{1,\dots,N\}$.

PROOF. To prove the first claim it suffices to show that the right hand side of (10.11) remains unchanged if $\mathcal{U}$ is replaced by the standard tableau $\mathcal{U}' = s_a\mathcal{U}$ for some $a \in \{1,\dots,m-1\}$ such that the entries a and $a+1$ of $\mathcal{U}$ do not belong to the same row or the same column. By (1.67) and the Yangian defining relations (10.7), we have

$$\begin{aligned}P_{a\,a+1}R_{a\,a+1}(c_a-c_{a+1})\,T_a(u+c_a)\,T_{a+1}(u+c_{a+1})&\\ = T_a(u+c_{a+1})\,T_{a+1}(u+c_a)&P_{a\,a+1}R_{a\,a+1}(c_a-c_{a+1}).\end{aligned}$$

Hence, representing the right hand side of (10.11) in the form

$$\begin{aligned}\mathrm{tr}_{1,\dots,m}\, \mathcal{E}_{\mathcal{U}} R_{a\,a+1}(c_a-c_{a+1})^{-1}P_{a\,a+1}P_{a\,a+1}R_{a\,a+1}(c_a-c_{a+1})&\\ \times T_1(u+c_1)\dots T_m(u+c_m)&\end{aligned}$$

and using the cyclic property of trace, we can write it as

$$\begin{aligned}\mathrm{tr}_{1,\dots,m}\, P_{a\,a+1}R_{a\,a+1}(c_a-c_{a+1})\,\mathcal{E}_{\mathcal{U}} R_{a\,a+1}(c_a-c_{a+1})^{-1}P_{a\,a+1}&\\ \times\, T_1(u+c_1)\dots T_a(u+c_{a+1})\,T_{a+1}(u+c_a)\dots T_m(u+c_m).&\end{aligned}$$

Using the notation $d = (c_{a+1}-c_a)^{-1}$ as in Lemma 1.1.1, we find

$$\begin{aligned}\mathcal{E}_{\mathcal{U}} R_{a\,a+1}(c_a-c_{a+1})^{-1}P_{a\,a+1}&\\ = \mathcal{E}_{\mathcal{U}}(P_{a\,a+1}-d)\,\frac{1}{1-d^2} &= \mathcal{E}_{\mathcal{U}}\, P_{a\,a+1}\, \mathcal{E}_{\mathcal{U}'}\,\frac{1}{1-d^2},\end{aligned} \tag{10.12}$$

where the last equality holds by (1.6).

LEMMA 10.1.3. *In the notation of the theorem, we have*

$$\mathcal{E}_{\mathcal{U}}\, T_1(u+c_1)\dots T_m(u+c_m) = \mathcal{E}_{\mathcal{U}}\, T_1(u+c_1)\dots T_m(u+c_m)\,\mathcal{E}_{\mathcal{U}}.$$

PROOF. Denote by $R(u_1,\dots,u_m)$ the image of the expression $\phi(u_1,\dots,u_m)$ defined in (1.23), under the action of the symmetric group in $(\mathbb{C}^N)^{\otimes m}$. Then

$$R(u_1,\dots,u_m) = \prod_{1\leqslant a<b\leqslant m} R_{a\,b}(u_a-u_b), \tag{10.13}$$

where the product is taken in the lexicographical order on the set of pairs (a,b). By applying (10.7) repeatedly, we obtain the identity

$$R(u_1,\dots,u_m)T_1(u_1)\dots T_m(u_m) = T_m(u_m)\dots T_1(u_1)R(u_1,\dots,u_m).$$

Due to the fusion procedure of Proposition 1.1.7, the consecutive evaluations of the variables $u_a = u + c_a$ for $a = 1, \dots, m$ imply the relation

$$\mathcal{E}_{\mathcal{U}}\, T_1(u+c_1)\dots T_m(u+c_m) = T_m(u+c_m)\dots T_1(u+c_1)\, \mathcal{E}_{\mathcal{U}}. \tag{10.14}$$

This completes the proof of the lemma since $\mathcal{E}_{\mathcal{U}}^2 = \mathcal{E}_{\mathcal{U}}$. □

Applying (10.12), the cyclic property of trace and Lemma 10.1.3 with $\mathcal{U}$ replaced by $\mathcal{U}'$, we bring the right hand side of (10.11) to the form

$$\frac{1}{1-d^2}\,\mathrm{tr}_{1,\dots,m}\, \mathcal{E}_{\mathcal{U}'} P_{a\, a+1}\, \mathcal{E}_{\mathcal{U}} P_{a\, a+1} \mathcal{E}_{\mathcal{U}'} \\ \times T_1(u+c_1)\dots T_a(u+c_{a+1})\, T_{a+1}(u+c_a)\dots T_m(u+c_m).$$

By (1.7), this coincides with the right hand side of (10.11), where $\mathcal{U}$ is replaced by $\mathcal{U}'$, as required.

To prove the second part of the theorem, set

$$\mathcal{E}_\mu(u) = \sum_{\mathrm{sh}(\mathcal{V})=\mu} \mathcal{E}_{\mathcal{V}}\, T_1(u+c_1)\dots T_m(u+c_m),$$

summed over the standard tableaux $\mathcal{V}$ of shape μ, where it is understood that the contents $c_a = c_a(\mathcal{V})$ depend on $\mathcal{V}$.

LEMMA 10.1.4. *For any $s \in \mathfrak{S}_m$ we have*

$$P_s\, \mathcal{E}_\mu(u) = \mathcal{E}_\mu(u)\, P_s.$$

PROOF. We will use the isomorphism (1.2) between the group algebra $\mathbb{C}[\mathfrak{S}_m]$ and the direct sum of matrix algebras. We will show that the element $\mathcal{E}_\mu(u)$ commutes with the image $\mathcal{E}_{\mathcal{U}\mathcal{U}'}$ of an arbitrary matrix unit $e_{\mathcal{U}\mathcal{U}'}$ under the action of the symmetric group in $(\mathbb{C}^N)^{\otimes m}$, where $\mathcal{U}$ and $\mathcal{U}'$ are standard tableaux of the same shape. By Lemma 10.1.3 we have

$$\mathcal{E}_\mu(u) = \sum_{\mathrm{sh}(\mathcal{V})=\mu} \mathcal{E}_{\mathcal{V}}\, T_1(u+c_1)\dots T_m(u+c_m)\, \mathcal{E}_{\mathcal{V}}.$$

Hence, if $\mathcal{U}$ and $\mathcal{U}'$ are of the shape ν and $\nu \neq \mu$, then

$$\mathcal{E}_{\mathcal{U}\mathcal{U}'}\, \mathcal{E}_\mu(u) = \mathcal{E}_\mu(u)\, \mathcal{E}_{\mathcal{U}\mathcal{U}'} = 0.$$

Now suppose that $\mathcal{U}$ and $\mathcal{U}'$ are of the shape μ. In the case where $\mathcal{U} = \mathcal{U}'$ we have

$$\mathcal{E}_{\mathcal{U}}\, \mathcal{E}_\mu(u) = \mathcal{E}_\mu(u)\, \mathcal{E}_{\mathcal{U}} = \mathcal{E}_{\mathcal{U}}\, T_1(u+c_1)\dots T_m(u+c_m)\, \mathcal{E}_{\mathcal{U}},$$

where $c_a = c_a(\mathcal{U})$ for all a. Finally, suppose $\mathcal{U}$ and $\mathcal{U}'$ are different tableaux of the shape μ. It suffices to consider the case, where $\mathcal{U}'$ is obtained from $\mathcal{U}$ by swapping the entries a and $a+1$ for some $a \in \{1, \dots, m-1\}$, which do not belong to the same row or the same column. That is, $\mathcal{U}' = s_a \mathcal{U}$. In this case, relation (1.8) implies

$$\mathcal{E}_{\mathcal{U}\mathcal{U}'} = \frac{1}{\sqrt{1-d^2}}\, \mathcal{E}_{\mathcal{U}}(P_{a\, a+1} - d)\, \mathcal{E}_{\mathcal{U}'} = \frac{1}{\sqrt{1-d^2}}\, \mathcal{E}_{\mathcal{U}}\, P_{a\, a+1}\, R_{a\, a+1}(c_{a+1} - c_a)\, \mathcal{E}_{\mathcal{U}'}.$$

Now, using (1.67), (10.7) and Lemma 10.1.3 we find

$$\begin{aligned}
&\mathcal{E}_\mu(u)\,\mathcal{E}_\mathcal{U} P_{a\,a+1}\,R_{a\,a+1}(c_{a+1}-c_a)\,\mathcal{E}_{\mathcal{U}'}\\
&\qquad=\mathcal{E}_\mathcal{U}\,T_1(u+c_1)\dots T_m(u+c_m)\,\mathcal{E}_\mathcal{U} P_{a\,a+1}\,R_{a\,a+1}(c_{a+1}-c_a)\,\mathcal{E}_{\mathcal{U}'}\\
&\qquad=\mathcal{E}_\mathcal{U}\,T_1(u+c_1)\dots T_m(u+c_m)\,P_{a\,a+1}\,R_{a\,a+1}(c_{a+1}-c_a)\,\mathcal{E}_{\mathcal{U}'}\\
&\qquad=\mathcal{E}_\mathcal{U}\,P_{a\,a+1}\,R_{a\,a+1}(c_{a+1}-c_a)\,T_1(u+c_1')\dots T_m(u+c_m')\,\mathcal{E}_{\mathcal{U}'},
\end{aligned}$$

where we denoted $c_a'=c_a(\mathcal{U}')$ for all a. Transform this expression further by using relation (1.6) which gives

$$\mathcal{E}_\mathcal{U}\,P_{a\,a+1}\,R_{a\,a+1}(c_{a+1}-c_a)=\mathcal{E}_\mathcal{U}\,P_{a\,a+1}\,R_{a\,a+1}(c_{a+1}-c_a)\,\mathcal{E}_{\mathcal{U}'}.$$

So, by Lemma 10.1.3, applied to $\mathcal{U}'$ instead of $\mathcal{U}$, the expression equals

$$\begin{aligned}
&\mathcal{E}_\mathcal{U}\,P_{a\,a+1}\,R_{a\,a+1}(c_{a+1}-c_a)\,\mathcal{E}_{\mathcal{U}'}\,T_1(u+c_1')\dots T_m(u+c_m')\,\mathcal{E}_{\mathcal{U}'}\\
&\qquad\qquad=\mathcal{E}_\mathcal{U}\,P_{a\,a+1}\,R_{a\,a+1}(c_{a+1}-c_a)\,\mathcal{E}_{\mathcal{U}'}\,\mathcal{E}_\mu(u),
\end{aligned}$$

thus completing the proof of the lemma. □

Write the series (10.11) in the form

$$\mathbb{T}_\mu(u)=\frac{1}{f_\mu}\,\mathrm{tr}_{1,\dots,m}\,\mathcal{E}_\mu(u),$$

where f_μ is the number of standard tableaux of shape μ. Introduce the matrix elements of $\mathcal{E}_\mu(u)$ by

$$\mathcal{E}_\mu(u)=\sum_{I,J} e_{i_1j_1}\otimes\dots\otimes e_{i_mj_m}\otimes\mathcal{E}_\mu(u)^{\,i_1\dots i_m}_{\,j_1\dots j_m},$$

summed over all m-tuples of indices $I=(i_1,\dots,i_m)$ and $J=(j_1,\dots,j_m)$ from $\{1,\dots,N\}$. Then

$$\mathrm{tr}_{1,\dots,m}\,\mathcal{E}_\mu(u)=\sum_I\mathcal{E}_\mu(u)^{\,i_1\dots i_m}_{\,i_1\dots i_m}.$$

By Lemma 10.1.4, for any $\sigma\in\mathfrak{S}_m$ the summands corresponding to the m-tuples $(i_1,\dots,i_m)$ and $(i_{\sigma(1)},\dots,i_{\sigma(m)})$ are equal. Hence we may restrict the summation to the weakly increasing sets of m-tuples. Taking into account the number of equal summands, we come to the expression

$$\mathrm{tr}_{1,\dots,m}\,\mathcal{E}_\mu(u)=\sum_{i_1\leqslant\dots\leqslant i_m}\frac{m!}{\alpha_1!\dots\alpha_N!}\,\mathcal{E}_\mu(u)^{\,i_1\dots i_m}_{\,i_1\dots i_m},\tag{10.15}$$

where α_i denotes the multiplicity of the index i among the m-tuple $(i_1,\dots,i_m)$. Let

$$\Phi_\mathcal{U}=\sum_{s\in\mathfrak{S}_m}\Phi_\mathcal{U}(s)\,P_{s^{-1}}$$

be the image of the diagonal matrix element $\varphi_\mathcal{U}$ defined in (1.5), under the action of the symmetric group in $(\mathbb{C}^N)^{\otimes m}$. Note that by the invariance of the inner product,

$$\Phi_\mathcal{U}(s)=(s\cdot v_\mathcal{U},v_\mathcal{U})=(v_\mathcal{U},s^{-1}\cdot v_\mathcal{U})=\Phi_\mathcal{U}(s^{-1}).$$

By (1.4),

$$\mathcal{E}_\mu(u)=\frac{f_\mu}{m!}\sum_{\mathrm{sh}(\mathcal{U})=\mu}\Phi_\mathcal{U}\,T_1(u+c_1)\dots T_m(u+c_m),\tag{10.16}$$

where $c_a = c_a(\mathcal{U})$ for $a = 1, \dots, m$, as before. Therefore,

$$\mathcal{E}_\mu(u)^{i_1 \dots i_m}_{i_1 \dots i_m} = \frac{f_\mu}{m!} \sum_{\mathrm{sh}(\mathcal{U})=\mu} \sum_{s \in \mathfrak{S}_m} \Phi_{\mathcal{U}}(s)\, t_{i_{s(1)}\, i_1}(u + c_1) \dots t_{i_{s(m)}\, i_m}(u + c_m).$$

The series (10.11) can now be written in the form

$$\mathbb{T}_\mu(u) = \sum_{i_1 \leqslant \dots \leqslant i_m} \frac{1}{\alpha_1! \dots \alpha_N!} \sum_{\mathrm{sh}(\mathcal{U})=\mu} \sum_{s \in \mathfrak{S}_m} \Phi_{\mathcal{U}}(s)\, t_{i_{s(1)}\, i_1}(u+c_1) \dots t_{i_{s(m)}\, i_m}(u+c_m).$$

This form of $\mathbb{T}_\mu(u)$ allows us to get a formula for its Harish-Chandra image which can be written as

$$\sum_{i_1 \leqslant \dots \leqslant i_m} \frac{1}{\alpha_1! \dots \alpha_N!} \sum_{\mathrm{sh}(\mathcal{U})=\mu} \sum_{s \in \mathfrak{S}_{(i)}} \Phi_{\mathcal{U}}(s)\, \lambda_{i_1}(u + c_1) \dots \lambda_{i_m}(u + c_m), \tag{10.17}$$

where $\mathfrak{S}_{(i)} \cong \mathfrak{S}_{\alpha_1} \times \dots \times \mathfrak{S}_{\alpha_N}$ denotes the subgroup of $\mathfrak{S}_m$ which consists of the permutations stabilizing the m-tuple $(i_1, \dots, i_m) = (1^{\alpha_1} 2^{\alpha_2} \dots N^{\alpha_N})$. Indeed, $t_{i_{s(m)}\, i_m}(u + c_m)$ is zero modulo the left ideal I of $\mathrm{Y}(\mathfrak{gl}_N)$ unless $i_{s(m)} = i_m$. In this case the permutation s preserves the subset of indices $\{r \mid i_r = i_m\}$. The image of $t_{i_m i_m}(u + c_m)$ under the homomorphism (10.10) is $\lambda_{i_m}(u + c_m)$, and the claim follows by applying the same argument to $t_{i_{s(m-1)}\, i_{m-1}}(u + c_{m-1})$ etc.

Given an m-tuple $(i_1, \dots, i_m)$ with $1 \leqslant i_1 \leqslant \dots \leqslant i_m \leqslant N$, for each standard tableau $\mathcal{U}$ of shape μ denote by $\mathcal{T} = i(\mathcal{U})$ the tableau obtained from $\mathcal{U}$ by replacing the entry r with the number i_r for $r = 1, \dots, m$. The entries of $\mathcal{T}$ then weakly increase along the rows and down the columns. Changing the order of summation in the expression (10.17), we can write it as

$$\sum_{i_1 \leqslant \dots \leqslant i_m} \frac{1}{\alpha_1! \dots \alpha_N!} \sum_{\mathrm{sh}(\mathcal{T})=\mu} \Psi_{\mathcal{T}} \prod_{\alpha \in \mu} \lambda_{\mathcal{T}(\alpha)}(u + c(\alpha)), \tag{10.18}$$

summed over the tableaux $\mathcal{T}$ with the entries $i_1, \dots, i_m$, where we have set

$$\Psi_{\mathcal{T}} = \sum_{\mathcal{U},\, i(\mathcal{U})=\mathcal{T}} \sum_{s \in \mathfrak{S}_{(i)}} \Phi_{\mathcal{U}}(s). \tag{10.19}$$

For a μ-tableau $\mathcal{T}$ with entries in $\{1, \dots, N\}$ such that the entries of $\mathcal{T}$ weakly increase along the rows and down the columns, consider the skew diagrams $\omega_1, \dots, \omega_N$, where ω_r is the union of the cells of $\mathcal{T}$ occupied by r. Then (10.19) can be rewritten as

$$\Psi_{\mathcal{T}} = \prod_{r=1}^{N} \sum_{s_r \in \mathfrak{S}_{\alpha_r}} \chi_{\omega_r}(s_r),$$

where χ_{ω_r} denotes the skew character of $\mathfrak{S}_{\alpha_r}$ associated with ω_r. However, the expression

$$\frac{1}{\alpha_r!} \sum_{s \in \mathfrak{S}_{\alpha_r}} \chi_{\omega_r}(s)$$

coincides with the standard inner product of χ_{ω_r} with the trivial character of $\mathfrak{S}_{\alpha_r}$. This equals the multiplicity of the trivial representation in the skew representation of $\mathfrak{S}_{\alpha_r}$ associated with ω_r. By the Pieri rule, the multiplicity is nonzero only if ω_r does not contain two cells in the same column, in which case the multiplicity is 1.

This implies that the summation in (10.18) can be restricted to those tableaux $\mathcal{T}$ whose columns strictly increase and so expression (10.18) equals

$$\sum_{\mathrm{sh}(\mathcal{T})=\mu} \prod_{\alpha\in\mu} \lambda_{\mathcal{T}(\alpha)}\big(u+c(\alpha)\big),$$

completing the proof of the theorem. □

REMARK 10.1.5. (i) Applying the evaluation homomorphism (10.8) to the series $\mathbb{T}_\mu(u)$ defined in (10.11) and multiplying by a polynomial in u we recover the Casimir element $\mathbb{S}_\mu(u)$ defined in (4.44):

$$\mathbb{S}_\mu(u) = (u+c_1)\dots(u+c_m)\ \mathrm{ev}\,\big(\mathbb{T}_\mu(u)\big).$$

Consider the Harish-Chandra homomorphism (4.6) for $\mathfrak{g} = \mathfrak{gl}_N$ with the standard Cartan subalgebra $\mathfrak{h}$ spanned by the elements $\lambda_i = E_{ii}$ with $i = 1,\dots,N$ and the subalgebra $\mathfrak{n}_+$ spanned by the elements E_{ij} with $1 \leqslant i < j \leqslant N$. We have the commutative diagram

$$\begin{array}{ccc}
\mathrm{Y}(\mathfrak{gl}_N)^{\mathfrak{h}} & \longrightarrow & \mathbb{C}[\lambda_1^{(r)},\dots,\lambda_N^{(r)} \mid r\geqslant 1] \\
{\scriptstyle \mathrm{ev}}\big\downarrow & & \big\downarrow \\
\mathrm{U}(\mathfrak{gl}_N)^{\mathfrak{h}} & \longrightarrow & \mathbb{C}[\lambda_1,\dots,\lambda_N],
\end{array}$$

where the horizontal arrows are the respective Harish-Chandra homomorphisms (10.10) and (4.6), while the right vertical arrow is the homomorphism defined by

$$\lambda_i^{(1)} \mapsto \lambda_i \qquad \text{and} \qquad \lambda_i^{(r)} \mapsto 0, \quad r \geqslant 2.$$

Using the diagram, we get the Harish-Chandra images of the polynomials $\mathbb{S}_\mu(u)$ from Theorem 10.1.2 and hence prove Proposition 4.9.1.

Note also that apart from the calculation of the central characters $\chi(\mathbb{S}_\mu)$, Theorem 10.1.2 also implies the formula for the character of the irreducible representation $L(\lambda)$ of GL_N. Namely, the character formula (4.27) is recovered by the specialization $T(u) \mapsto \mathbf{h}$.

(ii) The Yangian characters can be defined for all finite-dimensional representations of the Yangian as the Harish-Chandra images of the associated transfer matrices by analogy with the quantum affine algebras; see [**51**]. However, the proof that $\mathbb{T}_\mu(u)$ coincides with the *transfer matrix* associated with $L(\mu)$ requires additional arguments which use some properties of the universal R-matrix; cf. [**48**] and [**124**]. By the properties of the transfer matrices, the coefficients of all series $\mathbb{T}_\mu(u)$ belong to a commutative subalgebra of the Yangian. This fact can also be established directly with the use of the fusion formula of Proposition 1.1.7 for the idempotents $e_{\mathcal{U}}$ involved in the definition of the series. Another proof of commutativity of the subalgebra generated by the coefficients can be obtained by using a quantum vertex algebra structure associated with the double Yangian [**37**]; see Remark 10.4.2 below. □

We will point out two particular cases of Theorem 10.1.2 corresponding to the respective constructions of the Casimir elements for $\mathfrak{gl}_N$ in Propositions 4.6.1 and 4.7.1. They provide the characters for the Yangian representations in the exterior powers $\Lambda^m(\mathbb{C}^N)$ and symmetric powers $\mathrm{S}^m(\mathbb{C}^N)$ of the vector representation. Here μ is a column or row diagram with m boxes and $\mathcal{U}$ is the unique standard tableau of shape μ. The primitive idempotent $\mathcal{E}_{\mathcal{U}}$ in (10.11) coincides with the anti-symmetrizer $A^{(m)}$ or symmetrizer $H^{(m)}$, respectively.

COROLLARY 10.1.6. *For the images under the Harish-Chandra homomorphism* (10.10) *we have*

$$\mathrm{tr}_{1,\dots,m}\, A^{(m)}\, T_1(u)\dots T_m(u-m+1) \mapsto \sum_{1\leqslant i_1<\dots<i_m\leqslant N} \lambda_{i_1}(u)\cdots\lambda_{i_m}(u-m+1)$$

and

$$\mathrm{tr}_{1,\dots,m}\, H^{(m)}\, T_1(u)\dots T_m(u+m-1) \mapsto \sum_{1\leqslant i_1\leqslant\dots\leqslant i_m\leqslant N} \lambda_{i_1}(u)\cdots\lambda_{i_m}(u+m-1).$$

10.2. Dual Yangian for $\mathfrak{gl}_N$

By analogy with (10.3), define the *dual Yangian* $\mathrm{Y}^+(\mathfrak{gl}_N)$ as the unital associative algebra with generators $t_{ij}^{(-r)}$, where $1\leqslant i,j\leqslant N$ and $r=1,2,\dots$ subject to the defining relations

$$\begin{aligned}[t_{ij}^{(-r)},t_{kl}^{(-s)}] &= \delta_{kj}\, t_{il}^{(-r-s)} - \delta_{il}\, t_{kj}^{(-r-s)}\\ &\qquad + \sum_{a=1}^{\min\{r,s\}} \Big(t_{kj}^{(-r-s+a-1)} t_{il}^{(-a)} - t_{kj}^{(-a)} t_{il}^{(-r-s+a-1)}\Big).\end{aligned}$$

Combining the generators into the formal power series

$$t_{ij}^{+}(u) = \delta_{ij} - \sum_{r=1}^{\infty} t_{ij}^{(-r)}\, u^{r-1} \in \mathrm{Y}^+(\mathfrak{gl}_N)[[u]]$$

we can write the defining relations as

$$(u-v)\,[t_{ij}^+(u),t_{kl}^+(v)] = t_{kj}^+(u)\, t_{il}^+(v) - t_{kj}^+(v)\, t_{il}^+(u) \tag{10.20}$$

which thus take the same form as (10.2). Hence, they can also be written in the matrix form

$$R_{12}(u-v)\, T_1^+(u)\, T_2^+(v) = T_2^+(v)\, T_1^+(u)\, R_{12}(u-v) \tag{10.21}$$

as in (10.7), where we use the Yang R-matrix (10.5) and

$$T^+(u) = \sum_{i,j=1}^{N} e_{ij}\otimes t_{ij}^+(u) \in \mathrm{End}\,\mathbb{C}^N \otimes \mathrm{Y}^+(\mathfrak{gl}_N)[[u]]. \tag{10.22}$$

Similar to the matrix notation used above, $T_a^+(u)$ for $a=1,\dots,m$ denote formal series in u defined as in (1.61) with coefficients in the tensor product algebra

$$\underbrace{\mathrm{End}\,\mathbb{C}^N\otimes\dots\otimes\mathrm{End}\,\mathbb{C}^N}_{m}\otimes\mathrm{Y}^+(\mathfrak{gl}_N). \tag{10.23}$$

The algebra $\mathrm{Y}^+(\mathfrak{gl}_N)$ possesses a natural ascending filtration defined by setting $\deg' t_{ij}^{(-r)} = -r$ for all $r\geqslant 1$. We have the isomorphism for the associated graded algebra

$$\mathrm{gr}'\ \mathrm{Y}^+(\mathfrak{gl}_N) \cong \mathrm{U}\big(t^{-1}\mathfrak{gl}_N[t^{-1}]\big).$$

The image $\bar{t}_{ij}^{(-r)}$ of the generator $t_{ij}^{(-r)}$ in the $(-r)$-th component of the graded algebra $\mathrm{gr}'\ \mathrm{Y}^+(\mathfrak{gl}_N)$ corresponds to the element $E_{ij}[-r]$ of $\mathrm{U}\big(t^{-1}\mathfrak{gl}_N[t^{-1}]\big)$.

We will regard $\mathrm{Y}^+(\mathfrak{gl}_N)$ as a module over the Cartan subalgebra $\mathfrak{h}$ of $\mathfrak{gl}_N$, where each basis element E_{ii} of $\mathfrak{h}$ acts as a derivation and the action is defined on the generators by

$$E_{ii} \cdot t^+_{kl}(u) = \delta_{ki}\, t^+_{il}(u) - \delta_{il}\, t^+_{ki}(u).$$

Denote by $\mathrm{Y}^+(\mathfrak{gl}_N)^{\mathfrak{h}}$ the subalgebra of $\mathfrak{h}$-invariants under this action. Consider the left ideal J of the algebra $\mathrm{Y}^+(\mathfrak{gl}_N)$ generated by all elements $t^{(-r)}_{ij}$ with the conditions $N \geqslant i > j \geqslant 1$ and $r \geqslant 1$. As with the Yangian, the quotient of $\mathrm{Y}^+(\mathfrak{gl}_N)^{\mathfrak{h}}$ by the two-sided ideal $\mathrm{Y}^+(\mathfrak{gl}_N)^{\mathfrak{h}} \cap J$ is isomorphic to the commutative algebra freely generated by the images of the elements $t^{(-r)}_{ii}$ with $i = 1, \dots, N$ and $r \geqslant 1$ in the quotient. We will use the notation $\lambda^{(-r)}_i$ for this image of $t^{(-r)}_{ii}$. The corresponding analogue of the Harish-Chandra homomorphism (10.10) now takes the form

$$\mathrm{Y}^+(\mathfrak{gl}_N)^{\mathfrak{h}} \to \mathbb{C}[\lambda^{(-r)}_i \,|\, i = 1, \dots, N,\ r \geqslant 1]. \tag{10.24}$$

We combine the elements $\lambda^{(-r)}_i$ into the formal series

$$\lambda^+_i(u) = 1 - \sum_{r=1}^{\infty} \lambda^{(-r)}_i u^{r-1}, \qquad i = 1, \dots, N, \tag{10.25}$$

which will be understood as the images of the series $t^+_{ii}(u)$ under the homomorphism (10.24).

Our next goal is to prove a dual Yangian version of Theorem 10.1.2. Note, however, that the shift automorphism of the Yangian taking $T(u)$ to $T(u+c)$ for a given constant c, which is used in the definition of the series $\mathbb{T}_\mu(u)$ is not defined for the dual Yangian. To introduce counterparts of such series we need to extend the algebra $\mathrm{Y}^+(\mathfrak{gl}_N)$ by embedding it into the completed dual Yangian $\widehat{\mathrm{Y}}^+(\mathfrak{gl}_N)$ defined as follows. Consider the *descending* filtration

$$\mathrm{Y}^+(\mathfrak{gl}_N) = \mathrm{Y}^+_0 \supset \mathrm{Y}^+_1 \supset \mathrm{Y}^+_2 \supset \dots \tag{10.26}$$

on $\mathrm{Y}^+(\mathfrak{gl}_N)$ defined by setting the degree of the generator $t^{(-r)}_{ij}$ with $r \geqslant 1$ to be equal to r. That is, the subspace Y^+_k is spanned by all monomials in the generators of the total degree at least k. Then $\widehat{\mathrm{Y}}^+(\mathfrak{gl}_N)$ is the completion of $\mathrm{Y}^+(\mathfrak{gl}_N)$ with respect to this filtration. It is clear that for any $c \in \mathbb{C}$ the coefficients of u in the power series $t^+_{ij}(u+c)$ can be regarded as elements of $\widehat{\mathrm{Y}}^+(\mathfrak{gl}_N)$. The Harish-Chandra homomorphism (10.24) extends by continuity to the homomorphism from $\widehat{\mathrm{Y}}^+(\mathfrak{gl}_N)^{\mathfrak{h}}$ to the corresponding completed polynomial algebra $\widehat{\mathbb{C}}[\lambda^{(-r)}_i \,|\, i = 1, \dots, N,\ r \geqslant 1]$.

Working with the tensor product algebra (10.23) with $\mathrm{Y}^+(\mathfrak{gl}_N)$ replaced with $\widehat{\mathrm{Y}}^+(\mathfrak{gl}_N)$, we can now define the series $\mathbb{T}^+_\mu(u)$ by setting

$$\mathbb{T}^+_\mu(u) = \mathrm{tr}_{1,\dots,m}\, \mathcal{E}_{\mathcal{U}}\, T^+_1(u+c_1) \dots T^+_m(u+c_m), \tag{10.27}$$

using the same idempotents $\mathcal{E}_{\mathcal{U}}$ as in Theorem 10.1.2. Given a diagram μ, a *reverse tableau* of shape μ is obtained by filling in the boxes of the diagram with the numbers in a given set in such a way that the entries weakly decrease along each row from left to right and strictly decrease in each column from top to bottom.

THEOREM 10.2.1. *The series $\mathbb{T}^+_\mu(u)$ does not depend on the standard tableau $\mathcal{U}$ of shape μ. Moreover, the images of its coefficients under the Harish-Chandra homomorphism* (10.24) *are found by*

$$\mathbb{T}^+_\mu(u) \mapsto \sum_{\operatorname{sh}(\mathcal{T})=\mu} \prod_{\alpha\in\mu} \lambda^+_{\mathcal{T}(\alpha)}\big(u+c(\alpha)\big),$$

summed over reverse tableau $\mathcal{T}$ of shape μ with entries in $\{1,\dots,N\}$.

PROOF. The first part follows from the respective argument of the proof of Theorem 10.1.2 because it only relies on the matrix form of the defining relations which is the same for both the Yangian and dual Yangian. In particular, the relation (10.14) holds in the same form,

$$\mathcal{E}_{\mathcal{U}}\, T^+_1(u+c_1)\dots T^+_m(u+c_m) = T^+_m(u+c_m)\dots T^+_1(u+c_1)\,\mathcal{E}_{\mathcal{U}}. \tag{10.28}$$

A slight change is needed for the proof of the second part. Namely, using the matrices $T^+(u)$ in place of $T(u)$ throughout the arguments, we need the following counterpart of (10.15),

$$\operatorname{tr}_{1,\dots,m} \mathcal{E}^+_\mu(u) = \sum_{i_1\geqslant\dots\geqslant i_m} \frac{m!}{\alpha_1!\dots\alpha_N!}\, \mathcal{E}^+_\mu(u)^{i_1\dots i_m}_{i_1\dots i_m},$$

where α_i denotes the multiplicity of the index i among the m-tuple $(i_1,\dots,i_m)$ of weakly decreasing indices. By applying (10.28), we get the following counterpart of (10.16),

$$\mathcal{E}^+_\mu(u) = \frac{f_\mu}{m!} \sum_{\operatorname{sh}(\mathcal{U})=\mu} T^+_m(u+c_m)\dots T^+_1(u+c_1)\,\Phi_{\mathcal{U}},$$

where $c_a = c_a(\mathcal{U})$ for $a=1,\dots,m$, as before. Hence,

$$\mathcal{E}^+_\mu(u)^{i_1\dots i_m}_{i_1\dots i_m} = \frac{f_\mu}{m!} \sum_{\operatorname{sh}(\mathcal{U})=\mu} \sum_{s\in\mathfrak{S}_m} \Phi_{\mathcal{U}}(s)\, t^+_{i_m\, i_{s(m)}}(u+c_m)\dots t^+_{i_1\, i_{s(1)}}(u+c_1)$$

and so

$$\mathbb{T}^+_\mu(u) = \sum_{i_1\geqslant\dots\geqslant i_m} \frac{1}{\alpha_1!\dots\alpha_N!} \sum_{\operatorname{sh}(\mathcal{U})=\mu} \sum_{s\in\mathfrak{S}_m} \Phi_{\mathcal{U}}(s)\, t^+_{i_m\, i_{s(m)}}(u+c_m)\dots t^+_{i_1\, i_{s(1)}}(u+c_1).$$

Using the notation as in (10.17), for the Harish-Chandra image of $\mathbb{T}^+_\mu(u)$ we get

$$\sum_{i_1\geqslant\dots\geqslant i_m} \frac{1}{\alpha_1!\dots\alpha_N!} \sum_{\operatorname{sh}(\mathcal{U})=\mu} \sum_{s\in\mathfrak{S}_{(i)}} \Phi_{\mathcal{U}}(s)\, \lambda^+_{i_m}(u+c_m)\dots\lambda^+_{i_1}(u+c_1).$$

Repeating the concluding part of the argument with the reverse ordering of the indices we come to the desired expression for the Harish-Chandra image where the role of semistandard tableaux is played by the reverse tableaux. □

The following is the counterpart of Corollary 10.1.6 for the dual Yangian.

COROLLARY 10.2.2. *For the images under the Harish-Chandra homomorphism (10.24) we have*

$$\mathrm{tr}_{1,\dots,m} A^{(m)} T_1^+(u)\dots T_m^+(u-m+1) \mapsto \sum_{N\geqslant i_1>\dots>i_m\geqslant 1} \lambda_{i_1}^+(u)\cdots\lambda_{i_m}^+(u-m+1)$$

and

$$\mathrm{tr}_{1,\dots,m} H^{(m)} T_1^+(u)\dots T_m^+(u+m-1) \mapsto \sum_{N\geqslant i_1\geqslant\dots\geqslant i_m\geqslant 1} \lambda_{i_1}^+(u)\cdots\lambda_{i_m}^+(u+m-1).$$

10.3. Double Yangian for $\mathfrak{gl}_N$

The *double Yangian* $\mathrm{DY}(\mathfrak{gl}_N)$ for $\mathfrak{gl}_N$ is defined as the associative algebra generated by the central element C and elements $t_{ij}^{(r)}$ and $t_{ij}^{(-r)}$, where $1\leqslant i,j\leqslant N$ and $r=1,2,\dots$, subject to the defining relations written in terms of the generator matrices (10.4) and (10.22) as follows. They are given by (10.7), (10.21) together with the relation

$$\overline{R}_{12}(u-v+C/2)\,T_1(u)\,T_2^+(v)=T_2^+(v)\,T_1(u)\,\overline{R}_{12}(u-v-C/2), \tag{10.29}$$

where

$$\overline{R}_{12}(u)=g(u)\,R_{12}(u)=g(u)\,\big(1-P_{12}\,u^{-1}\big)$$

and

$$g(u)=1+\sum_{i=1}^{\infty} g_i\,u^{-i},\qquad g_i\in\mathbb{C}, \tag{10.30}$$

is a formal power series in u^{-1} whose coefficients are uniquely determined by the relation

$$g(u+N)=g(u)\,(1-u^{-2}). \tag{10.31}$$

Its first few terms are

$$g(u)=1+\frac{1}{N}\,u^{-1}+\frac{N^2+1}{2N^2}\,u^{-2}+\dots.$$

The relation (10.31) ensures that the R-matrix $\overline{R}_{12}(u)$ possesses the *crossing symmetry* properties

$$\big(\overline{R}_{12}(u)^{-1}\big)^{t_1}\,\overline{R}_{12}(u+N)^{t_1}=1\quad\text{and}\quad\big(\overline{R}_{12}(u)^{-1}\big)^{t_2}\,\overline{R}_{12}(u+N)^{t_2}=1, \tag{10.32}$$

where t_1 and t_2 are the standard matrix transpositions $t: e_{ij}\mapsto e_{ji}$ acting on the first and second copy of $\mathrm{End}\,\mathbb{C}^N$, respectively. Moreover, the following *unitarity property* holds

$$\overline{R}_{12}(u)\,\overline{R}_{12}(-u)=1. \tag{10.33}$$

Indeed, replacing u with $-u-N$ in (10.31) we get

$$g(-u)=g(-u-N)\,\big(1-(u+N)^{-2}\big)$$

and so

$$g(u)\,g(-u)\,(1-u^{-2})=g(u+N)\,g(-u-N)\,\big(1-(u+N)^{-2}\big).$$

So the series on the left hand side is invariant under the shift $u\mapsto u+N$ which is only possible when

$$g(u)\,g(-u)\,(1-u^{-2})=1$$

thus implying (10.33). The series $g(u)$ can be defined equivalently as the unique formal power series of the form (10.30) satisfying the relation

$$g(u)\,g(u+1)\dots g(u+N-1)=\big(1-u^{-1}\big)^{-1}. \tag{10.34}$$

To see the equivalence of the definitions, observe that by (10.31), the series $G(u)$ defined by the left hand side of (10.34) satisfies $G(u+1)=G(u)\,(1-u^{-2})$. However, $G(u)$ is uniquely determined by this relation and so coincides with the right hand side of (10.34).

Given any $c\in\mathbb{C}$ we will introduce the *double Yangian at the level* c as the quotient $\mathrm{DY}_c(\mathfrak{gl}_N)$ of $\mathrm{DY}(\mathfrak{gl}_N)$ by the ideal generated by $C-c$. As a vector space, this quotient is isomorphic to the tensor product

$$\mathrm{DY}_c(\mathfrak{gl}_N)\cong\mathrm{Y}^+(\mathfrak{gl}_N)\otimes\mathrm{Y}(\mathfrak{gl}_N) \tag{10.35}$$

of the Yangian and dual Yangian. This property relies on the Poincaré–Birkhoff–Witt theorem for the algebra $\mathrm{DY}(\mathfrak{gl}_N)$. For the level zero case $c=0$ the theorem can be proved by using the arguments of [**35**] and [**125**] and then extended to an arbitrary level with the use of level 1 representations constructed in [**72**]; see [**82**].

Consider an ascending filtration on the double Yangian $\mathrm{DY}(\mathfrak{gl}_N)$ defined by

$$\deg' t_{ij}^{(r)}=r-1\qquad\text{and}\qquad \deg' t_{ij}^{(-r)}=-r$$

for all $r\geqslant 1$; the degree of the central element C is defined to be equal to zero. Denote by $\mathrm{gr}'\,\mathrm{DY}(\mathfrak{gl}_N)$ the corresponding graded algebra. We will use the notation $\bar t_{ij}^{(r)}$ and $\bar t_{ij}^{(-r)}$ for the images of the generators in the respective components of the graded algebra and let $\overline{C}$ be the image of C in the zeroth component.

Recall the affine Kac–Moody algebra $\widehat{\mathfrak{gl}}_N=\mathfrak{gl}_N[t,t^{-1}]\oplus\mathbb{C}K$ defined by the commutation relations (7.3).

PROPOSITION 10.3.1. *The assignments*

$$E_{ij}[r-1]\mapsto \bar t_{ij}^{(r)},\qquad E_{ij}[-r]\mapsto \bar t_{ij}^{(-r)}\qquad\textit{and}\qquad K\mapsto\overline{C} \tag{10.36}$$

with $r\geqslant 1$ define an algebra isomorphism

$$\mathrm{U}(\widehat{\mathfrak{gl}}_N)\to\mathrm{gr}'\,\mathrm{DY}(\mathfrak{gl}_N).$$

PROOF. As we pointed out in the previous sections, there are isomorphisms

$$\mathrm{U}\big(\mathfrak{gl}_N[t]\big)\cong\mathrm{gr}'\,\mathrm{Y}(\mathfrak{gl}_N)\qquad\text{and}\qquad \mathrm{U}\big(t^{-1}\mathfrak{gl}_N[t^{-1}]\big)\cong\mathrm{gr}'\,\mathrm{Y}^+(\mathfrak{gl}_N).$$

They are defined by the assignments (10.36). We will now use the defining relations (10.29) to verify that the images of the generators $\bar t_{ij}^{(r)}$ and $\bar t_{kl}^{(-s)}$ with $r,s\geqslant 1$ satisfy the desired relations in $\mathrm{U}(\widehat{\mathfrak{gl}}_N)$. Introduce re-scaled generators of $\mathrm{DY}(\mathfrak{gl}_N)$ by setting

$$\widetilde t_{ij}^{(r)}=h^{r-1}\,t_{ij}^{(r)}\qquad\text{and}\qquad \widetilde t_{ij}^{(-r)}=h^{-r}\,t_{ij}^{(-r)}$$

for $r\geqslant 1$, where h is a complex-valued parameter. The relations satisfied by $\bar t_{ij}^{(r)}$ and $\bar t_{kl}^{(-s)}$ in the graded algebra $\mathrm{gr}'\,\mathrm{DY}(\mathfrak{gl}_N)$ will be recovered by calculating the relations between $\widetilde t_{ij}^{(r)}$ and $\widetilde t_{kl}^{(-s)}$ and then taking the limit as $h\to 0$. Set

$$\widetilde t_{ij}(u)=\sum_{r=1}^{\infty}\widetilde t_{ij}^{(r)}\,u^{-r}=\frac{1}{h}\Big(t_{ij}\Big(\frac{u}{h}\Big)-\delta_{ij}\Big)$$

and

$$\widetilde{t}^{+}_{kl}(v) = \sum_{s=1}^{\infty} \widetilde{t}^{\,(-s)}_{kl}\, v^{s-1} = \frac{1}{h}\Big(\delta_{kl} - t^{+}_{kl}\Big(\frac{v}{h}\Big)\Big).$$

Write (10.29) in terms of the generating series:

$$g\big(u-v+C/2\big)\Big(t_{ij}(u)\,t^{+}_{kl}(v) - \frac{1}{u-v+C/2}\,t_{kj}(u)\,t^{+}_{il}(v)\Big)$$
$$= g\big(u-v-C/2\big)\Big(t^{+}_{kl}(v)\,t_{ij}(u) - \frac{1}{u-v-C/2}\,t^{+}_{kj}(v)\,t_{il}(u)\Big). \tag{10.37}$$

Note the expansion into a power series in $(u-v)^{-1}$:

$$\frac{g\big(u-v-C/2\big)}{g\big(u-v+C/2\big)} = 1 + \frac{C}{N\,(u-v)^2} + \dots \tag{10.38}$$

Now replace u by u/h and v by v/h in (10.37) to get the corresponding relations between the series $\widetilde{t}_{ij}(u)$ and $\widetilde{t}^{+}_{kl}(v)$. We have

$$\big(\delta_{ij} + h\,\widetilde{t}_{ij}(u)\big)\big(\delta_{kl} - h\,\widetilde{t}^{+}_{kl}(v)\big) - \frac{h}{u-v+hC/2}\,\big(\delta_{kj} + h\,\widetilde{t}_{kj}(u)\big)\big(\delta_{il} - h\,\widetilde{t}^{+}_{il}(v)\big)$$
$$- \Big(\big(\delta_{kl} - h\,\widetilde{t}^{+}_{kl}(v)\big)\big(\delta_{ij} + h\,\widetilde{t}_{ij}(u)\big) - \frac{h}{u-v-hC/2}\,\big(\delta_{kj} - h\,\widetilde{t}^{+}_{kj}(v)\big)\big(\delta_{il} + h\,\widetilde{t}_{il}(u)\big)\Big)$$
$$\times \Big(1 + \frac{h^2 C}{N\,(u-v)^2} + \dots\Big) = 0.$$

As a power series in h, the left hand side is divisible by h^2. Hence, dividing by h^2 we get the relation modulo h,

$$\big[\widetilde{t}_{ij}(u), \widetilde{t}^{+}_{kl}(v)\big] \equiv \frac{1}{u-v}\Big(\delta_{kj}\,\big(\widetilde{t}_{il}(u) + \widetilde{t}^{+}_{il}(v)\big) - \delta_{il}\,\big(\widetilde{t}_{kj}(u) + \widetilde{t}^{+}_{kj}(v)\big)\Big)$$
$$+ \frac{C}{N\,(u-v)^2}\Big(N\,\delta_{kj}\,\delta_{il} - \delta_{ij}\,\delta_{kl}\Big).$$

Thus, taking the coefficients of $u^{-r}v^{s-1}$ with $r, s \geqslant 1$ on both sides, in the limit $h \to 0$ in the graded algebra we get

$$\big[\bar{t}^{\,(r)}_{ij}, \bar{t}^{\,(-s)}_{kl}\big] = \begin{cases} \delta_{kj}\,\bar{t}^{\,(r-s)}_{il} - \delta_{il}\,\bar{t}^{\,(r-s)}_{kj} + (r-1)\,\delta_{r,s+1}\,\overline{C}\Big(\delta_{kj}\,\delta_{il} - \dfrac{\delta_{ij}\,\delta_{kl}}{N}\Big) \\[2ex] \delta_{kj}\,\bar{t}^{\,(r-s-1)}_{il} - \delta_{il}\,\bar{t}^{\,(r-s-1)}_{kj} + (r-1)\,\delta_{r,s+1}\,\overline{C}\Big(\delta_{kj}\,\delta_{il} - \dfrac{\delta_{ij}\,\delta_{kl}}{N}\Big) \end{cases}$$

with the top line chosen for $r > s$ and the bottom line for $r \leqslant s$. Comparing with (7.3), we may conclude that the assignments (10.36) define a homomorphism $\mathrm{U}(\widehat{\mathfrak{gl}}_N) \to \mathrm{gr}'\,\mathrm{DY}(\mathfrak{gl}_N)$. It is clear that the homomorphism is surjective whereas its injectivity is implied by the Poincaré–Birkhoff–Witt theorem for the double Yangian. □

The *critical level* $c = -N$ will play a particular role in what follows. Under this specialization the fraction (10.38) admits a closed form,

$$\frac{g\big(u-v+N/2\big)}{g\big(u-v-N/2\big)} = 1 - \frac{1}{(u-v-N/2)^2}.$$

This implies the following form of the defining relations (10.37) at the critical level:

$$\big[t_{ij}(u), t^{+}_{kl}(v)\big] = -\frac{1}{u-v+N/2}\, t^{+}_{kj}(v)\, t_{il}(u) + \frac{1}{u-v-N/2}\, t^{+}_{il}(v)\, t_{kj}(u) - \frac{1}{(u-v)^2-N^2/4}\, t^{+}_{ij}(v)\, t_{kl}(u).$$

10.4. Invariants of the vacuum module over the double Yangian

By analogy with the vacuum module over $\widehat{\mathfrak{gl}}_N$ defined in (6.6) (with $\mathfrak{g} = \mathfrak{gl}_N$), introduce the *vacuum module* $\mathcal{V}_c(\mathfrak{gl}_N)$ *at the level* $c \in \mathbb{C}$ over $\mathrm{DY}(\mathfrak{gl}_N)$ as the quotient of the algebra $\mathrm{DY}_c(\mathfrak{gl}_N)$ by the left ideal generated by all elements $t_{ij}^{(r)}$ with $r \geqslant 1$:

$$\mathcal{V}_c(\mathfrak{gl}_N) = \mathrm{DY}_c(\mathfrak{gl}_N)/\mathrm{DY}_c(\mathfrak{gl}_N)\langle t_{ij}^{(r)} \,|\, r \geqslant 1\rangle.$$

We let $\mathbf{1}$ denote the image of 1 in the quotient. In particular, $\mathcal{V}_c(\mathfrak{gl}_N)$ has a structure of a module over the Yangian $\mathrm{Y}(\mathfrak{gl}_N)$ obtained by restriction. As a vector space, the vacuum module is isomorphic to the dual Yangian $\mathrm{Y}^{+}(\mathfrak{gl}_N)$ due to the decomposition (10.35). Therefore, $\mathcal{V}_c(\mathfrak{gl}_N)$ is equipped with the descending filtration (10.26). The completed vector space $\widehat{\mathcal{V}}_c(\mathfrak{gl}_N)$ is also a $\mathrm{Y}(\mathfrak{gl}_N)$-module.

Now assume that the level is critical, $c = -N$ and introduce the subspace of $\mathrm{Y}(\mathfrak{gl}_N)$-invariants by

$$\mathfrak{z}(\widehat{\mathcal{V}}_{\mathrm{cri}}) = \{v \in \widehat{\mathcal{V}}_{-N}(\mathfrak{gl}_N) \mid t_{ij}(u)\, v = \delta_{ij}\, v\},$$

so that any element of $\mathfrak{z}(\widehat{\mathcal{V}}_{\mathrm{cri}})$ is annihilated by all operators $t_{ij}^{(r)}$ with $r \geqslant 1$. As with the Feigin–Frenkel center $\mathfrak{z}(\widehat{\mathfrak{gl}}_N)$, the subspace $\mathfrak{z}(\widehat{\mathcal{V}}_{\mathrm{cri}})$ can be regarded as a subalgebra of the completed dual Yangian $\widehat{\mathrm{Y}}^{+}(\mathfrak{gl}_N)$. Moreover, this subalgebra turns out to be commutative; see [**82**].

Suppose that μ is a diagram with m boxes whose length does not exceed N. Consider the power series $\mathbb{T}^{+}_{\mu}(u)$ in u defined in (10.27).

THEOREM 10.4.1. *All coefficients of the series* $\mathbb{T}^{+}_{\mu}(u)\mathbf{1}$ *belong to the subalgebra of invariants* $\mathfrak{z}(\widehat{\mathcal{V}}_{cri})$ *of the vacuum module. In particular, the coefficients are pairwise commuting elements of the completed dual Yangian* $\widehat{\mathrm{Y}}^{+}(\mathfrak{gl}_N)$.

PROOF. Similar to the proof of Theorem 7.1.3, consider the tensor product space

$$\underbrace{\mathrm{End}\,\mathbb{C}^N \otimes \ldots \otimes \mathrm{End}\,\mathbb{C}^N}_{m+1} \otimes \widehat{\mathcal{V}}_{-N}(\mathfrak{gl}_N)$$

with the $m+1$ copies of $\mathrm{End}\,\mathbb{C}^N$ labeled by $0, 1, \ldots, m$. We need to verify the identity

$$T_0(z)\, \mathbb{T}^{+}_{\mu}(u)\mathbf{1} = \mathbb{T}^{+}_{\mu}(u)\mathbf{1}.$$

By the defining relations (10.29), for all $a = 1, \ldots, m$ we can write

$$T_0(z)\, T^{+}_a(u + c_a) = \overline{R}_{0\,a}(z-u-c_a-N/2)^{-1}\, T^{+}_a(u+c_a)\, T_0(z)\, \overline{R}_{0\,a}(z-u-c_a+N/2).$$

Hence, suppressing the arguments of the R-matrices we get

$$T_0(z)\,\mathrm{tr}_{1,\dots,m}\,\mathcal{E}_{\mathcal{U}}\,T_1^+(u+c_1)\dots T_m^+(u+c_m)\,\mathbf{1}$$

$$=\mathrm{tr}_{1,\dots,m}\,\mathcal{E}_{\mathcal{U}}\,\overline{R}_{01}^{\,-1}\dots\overline{R}_{0m}^{\,-1}\,T_1^+(u+c_1)\dots T_m^+(u+c_m)\,T_0(z)\,\overline{R}_{0m}\dots\overline{R}_{01}\,\mathbf{1}$$

$$=\mathrm{tr}_{1,\dots,m}\,\mathcal{E}_{\mathcal{U}}\,\overline{R}_{01}^{\,-1}\dots\overline{R}_{0m}^{\,-1}\,T_1^+(u+c_1)\dots T_m^+(u+c_m)\,\overline{R}_{0m}\dots\overline{R}_{01}\,\mathbf{1},$$

where the second equality holds since $T_0(z)$ acts as the identity operator on the subspace $\mathrm{End}\,(\mathbb{C}^N)^{\otimes(m+1)}\otimes\mathbf{1}$.

Note the following consequence of the Yang–Baxter equation (10.6),

$$R(u_1,\dots,u_m)\,R_{0m}(u_0-u_m)\dots R_{01}(u_0-u_1)$$
$$=R_{01}(u_0-u_1)\dots R_{0m}(u_0-u_m)\,R(u_1,\dots,u_m), \tag{10.39}$$

where u_0 is another variable and we use the notation (10.13). Relation (10.39) will remain valid if each factor $R_{0a}(u_0-u_a)$ is replaced with $\overline{R}_{0a}(u_0-u_a)$. Hence, by the fusion procedure of Proposition 1.1.7, the consecutive evaluations $u_a=c_a$ for $a=1,\dots,m$ imply

$$\mathcal{E}_{\mathcal{U}}\,\overline{R}_{0m}(u_0-c_m)\dots\overline{R}_{01}(u_0-c_1)=\overline{R}_{01}(u_0-c_1)\dots\overline{R}_{0m}(u_0-c_m)\,\mathcal{E}_{\mathcal{U}}.$$

By inverting the R-matrices we also get

$$\mathcal{E}_{\mathcal{U}}\,\overline{R}_{01}(u_0-c_1)^{-1}\dots\overline{R}_{0m}(u_0-c_m)^{-1}$$
$$=\overline{R}_{0m}(u_0-c_m)^{-1}\dots\overline{R}_{01}(u_0-c_1)^{-1}\,\mathcal{E}_{\mathcal{U}}.$$

Returning now to the calculation of $T_0(z)\,\mathbb{T}^+_\mu(u)\,\mathbf{1}$, recall that $\mathcal{E}_{\mathcal{U}}$ in an idempotent and use the cyclic property of trace to write

$$\mathrm{tr}_{1,\dots,m}\,\mathcal{E}_{\mathcal{U}}\,X\,Y=\mathrm{tr}_{1,\dots,m}\,X^{\circ}\,\mathcal{E}_{\mathcal{U}}\,Y=\mathrm{tr}_{1,\dots,m}\,X^{\circ}\,\mathcal{E}_{\mathcal{U}}^2\,Y$$

$$=\mathrm{tr}_{1,\dots,m}\,\mathcal{E}_{\mathcal{U}}\,X\,Y^{\circ}\,\mathcal{E}_{\mathcal{U}}=\mathrm{tr}_{1,\dots,m}\,X\,Y^{\circ}\,\mathcal{E}_{\mathcal{U}}^2=\mathrm{tr}_{1,\dots,m}\,X\,Y^{\circ}\,\mathcal{E}_{\mathcal{U}}=\mathrm{tr}_{1,\dots,m}\,X\,\mathcal{E}_{\mathcal{U}}\,Y,$$

where we set

$$X=\overline{R}_{01}^{\,-1}\dots\overline{R}_{0m}^{\,-1},\qquad Y=T_1^+(u+c_1)\dots T_m^+(u+c_m)\,\overline{R}_{0m}\dots\overline{R}_{01}$$

and used the notation X° and Y° for the same products written in the opposite order. Thus, we can write

$$T_0(z)\,\mathbb{T}^+_\mu(u)\,\mathbf{1}=\mathrm{tr}_{1,\dots,m}\,X\,\mathcal{E}_{\mathcal{U}}\,Y\,\mathbf{1}=\mathrm{tr}_{1,\dots,m}\,X^{t_1\dots t_m}\,\big(\mathcal{E}_{\mathcal{U}}\,Y\big)^{t_1\dots t_m}\,\mathbf{1}.$$

We have

$$\big(\mathcal{E}_{\mathcal{U}}\,Y\big)^{t_1\dots t_m}=\overline{R}_{0m}^{\,t_m}\dots\overline{R}_{01}^{\,t_1}\,\big(\mathcal{E}_{\mathcal{U}}\,T_1^+(u+c_1)\dots T_m^+(u+c_m)\big)^{t_1\dots t_m}$$

and

$$X^{t_1\dots t_m}=\Big(\overline{R}_{01}^{\,-1}\Big)^{t_1}\dots\Big(\overline{R}_{0m}^{\,-1}\Big)^{t_m}.$$

By the crossing symmetry (10.32), we have

$$\Big(\overline{R}_{0a}^{\,-1}\Big)^{t_a}\,\overline{R}_{0a}^{\,t_a}=1$$

for all $a=1,\dots,m$ and so

$$T_0(z)\,\mathbb{T}^+_\mu(u)\,\mathbf{1}=\mathrm{tr}_{1,\dots,m}\,\big(\mathcal{E}_{\mathcal{U}}\,T_1^+(u+c_1)\dots T_m^+(u+c_m)\big)^{t_1\dots t_m}\,\mathbf{1}=\mathbb{T}^+_\mu(u)\,\mathbf{1}$$

as required. □

REMARK 10.4.2. As pointed out above, it was derived in [**82**] that the coefficients of all series $\mathbb{T}^+_\mu(u)$ pairwise commute as they belong to a commutative subalgebra of $\mathfrak{z}(\widehat{\mathcal{V}}_{\text{cri}})$ of the completed dual Yangian $\widehat{\mathrm{Y}}^+(\mathfrak{gl}_N)$. Since the defining relations of both the Yangian and dual Yangian written in terms of the generating series are given by the identical formulas (10.2) and (10.20), we thus recover a well-known property that the coefficients of all series $\mathbb{T}_\mu(u)$ belong to a commutative subalgebra of the Yangian $\mathrm{Y}(\mathfrak{gl}_N)$. □

10.5. From Yangian invariants to Segal–Sugawara vectors

In this section we give another proof of Theorem 7.1.3. By Proposition 10.3.1, for any element $S \in \mathfrak{z}(\widehat{\mathcal{V}}_{\text{cri}})$ its image $\overline{S}$ in the graded algebra $\mathrm{gr}'\,\widehat{\mathrm{Y}}^+(\mathfrak{gl}_N)$ belongs to the Feigin–Frenkel center $\mathfrak{z}(\widehat{\mathfrak{gl}}_N)$. We will use Theorem 10.4.1 to construct an appropriate linear combination of elements of $\mathfrak{z}(\widehat{\mathcal{V}}_{\text{cri}})$ whose graded image will coincide with the elements given by (7.7). As explained in Section 7.1, the claim for the elements (7.8), (7.9) and (7.10) will then follow.

Extend the ascending filtration on the dual Yangian defined by $\deg' t_{ij}^{(-r)} = -r$ to the algebra of formal series $\mathrm{Y}^+(\mathfrak{gl}_N)[[u, \partial_u]]$ by setting $\deg u = 1$ and $\deg \partial_u = -1$ so that the associated graded algebra is isomorphic to $\mathrm{U}\left(t^{-1}\mathfrak{gl}_N[t^{-1}]\right)[[u, \partial_u]]$. Then the element

$$\operatorname{tr}_{1,\dots,m} A^{(m)} \big(1 - T_1^+(u)\, e^{-\partial_u}\big) \dots \big(1 - T_m^+(u)\, e^{-\partial_u}\big) \tag{10.40}$$

has degree $-m$ and its image in the graded algebra coincides with

$$\operatorname{tr}_{1,\dots,m} A^{(m)} \big(\partial_u + E(u)_{+1}\big) \dots \big(\partial_u + E(u)_{+m}\big), \tag{10.41}$$

where, as in Section 7.2,

$$E(u)_+ = \sum_{r=1}^{\infty} E[-r]\, u^{r-1}.$$

On the other hand, the element (10.40) equals

$$\operatorname{tr}_{1,\dots,m} A^{(m)} \sum_{k=0}^{m} \sum_{1 \leqslant i_1 < \dots < i_k \leqslant m} (-1)^k\, T_{i_1}^+(u) \dots T_{i_k}^+(u-k+1)\, e^{-k\,\partial_u}.$$

Transform the expression by applying conjugations by elements of $\mathfrak{S}_m$ and using the cyclic property of trace to bring it to the form

$$\operatorname{tr}_{1,\dots,m} A^{(m)} \sum_{k=0}^{m} (-1)^k \binom{m}{k} T_1^+(u) \dots T_k^+(u-k+1)\, e^{-k\,\partial_u}.$$

Calculating partial traces of the anti-symmetrizer with the use of (3.26) we can write this as

$$\sum_{k=0}^{m} (-1)^k \binom{N-k}{m-k} \operatorname{tr}_{1,\dots,k} A^{(k)}\, T_1^+(u) \dots T_k^+(u-k+1)\, e^{-k\,\partial_u}. \tag{10.42}$$

Apply Theorem 10.4.1 to the case of the one-column diagram $\mu = (1^k)$. There is a unique standard tableau $\mathcal{U}$ of shape μ so that $\mathcal{E}_{\mathcal{U}} = A^{(k)}$ and the contents are $c_i = -i + 1$ for $i = 1, \dots, k$. We find that all elements

$$\operatorname{tr}_{1,\dots,k} A^{(k)}\, T_1^+(u) \dots T_k^+(u-k+1)\, \mathbf{1}$$

belong to $\mathfrak{z}(\widehat{\mathcal{V}}_{\text{cri}})$. This proves that all coefficients of (10.41) are Segal–Sugawara vectors for $\mathfrak{gl}_N$. In particular, this holds for the coefficient of u^0 which coincides with (7.7) after the replacement of ∂_u by τ. The last step is essentially an application of the vacuum axiom for the affine vertex algebra. This completes the proof of Theorem 7.1.3.

The above construction of Segal–Sugawara vectors for $\mathfrak{gl}_N$ based on the use of the double Yangian can be extended, in principle, to any simple Lie algebra $\mathfrak{g}$. In accordance to Drinfeld [**32**], the corresponding Yangian $\mathrm{Y}(\mathfrak{g})$ admits an *RTT* presentation of the form (10.7). The double Yangian $\mathrm{DY}(\mathfrak{g})$ can then be defined by additional relations analogous to (10.21) and (10.29). Its associated graded algebra at the critical level should reproduce $\mathrm{U}_{-h^\vee}(\widehat{\mathfrak{g}})$. Hence, Segal–Sugawara vectors for $\mathfrak{g}$ could be obtained by taking the graded images of the invariants of the vacuum module over the double Yangian as in the above argument for $\mathfrak{gl}_N$.

The proof of Theorem 10.4.1 providing invariants of the vacuum module at the critical level relies on the existence of a fusion procedure. Namely, the primitive idempotents $\mathcal{E} = \mathcal{E}_{\mathcal{U}}$ are represented as consecutive evaluations of the product of the R-matrices

$$\mathcal{E} = \frac{1}{h(\mu)} R(u_1, \dots, u_m)\big|_{u_1=c_1}\big|_{u_2=c_2} \cdots \big|_{u_m=c_m} \tag{10.43}$$

as stated in Proposition 1.1.7, where

$$R(u_1, \dots, u_m) = \prod_{1 \leqslant a < b \leqslant m} R_{a\,b}(u_a - u_b),$$

with the product taken in the lexicographical order on the set of pairs (a, b). Therefore, to extend Theorem 10.4.1 to the double Yangian $\mathrm{DY}(\mathfrak{g})$, we need to have an idempotent $\mathcal{E}$ which admits a factorization of the form (10.43).

In type A such idempotents exist for any $m \geqslant 1$; they are associated with arbitrary standard tableaux with m boxes. The fusion procedures for the Brauer algebra indicate that in types B, C and D at least one idempotent of the form (10.43) does exist for each m; see Section 1.2. Namely, the factorization formula for the symmetrizer (1.46) will have the form (10.43) under the respective representations of the Brauer algebras (1.69) and (1.73) because the factors will coincide with the values of the R-matrices of types B, C and D; see (11.13) below. We thus get another proof of Theorems 8.1.6 and 8.3.2.

An extension of the fusion procedure to the exceptional types to produce idempotents of the form (10.43) (perhaps, for some values of m) would allow to apply this approach to construct Segal–Sugawara vectors. A different approach based on calculations performed with the use of computer programs was used in [**116**] to obtain explicit generators of the algebra $\mathfrak{z}(\widehat{\mathfrak{g}})$ for the Lie algebra $\mathfrak{g}$ of type G_2.

10.6. Screening operators

The Harish-Chandra images provided by Theorem 10.1.2 can be regarded as polynomials in the formal variables $\lambda_i(u + a)$, where $i = 1, \dots, N$ and $a \in \mathbb{C}$. It was conjectured in [**51**] and proved in [**48**], that the polynomials which arise as the characters of Yangian representations are precisely those which belong to the intersection of the kernels of the *screening operators*. Although the Harish-Chandra image of $\mathbb{T}_\mu(u)$ manifestly coincides with the q-character of the evaluation module under an appropriate identification of the parameters, we will show it belongs to the kernels of all screening operators thus illustrating the use of the result of [**48**].

We will suppress the variable u from the notation and introduce the algebra of polynomials

$$\mathcal{L} = \mathbb{C}[\lambda_i(a) \,|\, i = 1, \dots, N,\ a \in \mathbb{C}]$$

in the variables $\lambda_i(a)$. For every $i \in \{1, \dots, N-1\}$ consider the free left $\mathcal{L}$-module $\widetilde{\mathcal{L}}_i$ with the generators $\sigma_i(a)$, where a runs over $\mathbb{C}$ and denote by $\mathcal{L}_i$ its quotient by the relations

$$\lambda_i(a)\,\sigma_i(a) = \lambda_{i+1}(a)\,\sigma_i(a+1), \qquad a \in \mathbb{C}. \tag{10.44}$$

Define the linear operator $\widetilde{S}_i : \mathcal{L} \to \widetilde{\mathcal{L}}_i$ by the formula

$$\widetilde{S}_i : \lambda_j(a) \mapsto \begin{cases} \lambda_i(a)\,\sigma_i(a) & \text{for } \ j = i \\ -\lambda_{i+1}(a)\,\sigma_i(a+1) & \text{for } \ j = i+1 \\ 0 & \text{for } \ j \neq i, i+1 \end{cases} \tag{10.45}$$

and the Leibniz rule

$$\widetilde{S}_i(AB) = B\,\widetilde{S}_i(A) + A\,\widetilde{S}_i(B). \tag{10.46}$$

Now the *i-th screening operator*

$$S_i : \mathcal{L} \to \mathcal{L}_i$$

is defined as the composition of $\widetilde{S}_i$ and the projection $\widetilde{\mathcal{L}}_i \to \mathcal{L}_i$.

The subalgebra $\operatorname{Rep} \mathrm{Y}(\mathfrak{gl}_N)$ of Yangian characters in $\mathcal{L}$ can be defined as the intersection of the kernels of the screening operators:

$$\operatorname{Rep} \mathrm{Y}(\mathfrak{gl}_N) = \bigcap_{i=1}^{N-1} \ker S_i.$$

For any diagram μ of length $\ell(\mu) \leqslant N$ define the polynomial $\chi_{L(\mu)} \in \mathcal{L}$ by

$$\chi_{L(\mu)} = \sum_{\mathrm{sh}(\mathcal{T})=\mu} \prod_{\alpha \in \mu} \lambda_{\mathcal{T}(\alpha)}\big(c(\alpha)\big), \tag{10.47}$$

summed over all semistandard tableaux $\mathcal{T}$ of shape μ with entries in the set $\{1, \dots, N\}$; see Theorem 10.1.2.

PROPOSITION 10.6.1. *We have $\chi_{L(\mu)} \in \operatorname{Rep} \mathrm{Y}(\mathfrak{gl}_N)$.*

PROOF. We will verify that $\chi_{L(\mu)}$ is annihilated by all screening operators S_i for $i = 1, \dots, N-1$. By the definition (10.45), the operator S_i annihilates all variables except for $\lambda_i(a)$ and $\lambda_{i+1}(a)$. Since each tableau $\mathcal{U}$ is semistandard, its entries i and $i+1$ occupy a skew subdiagram which can have at most two boxes in the same column, as illustrated:

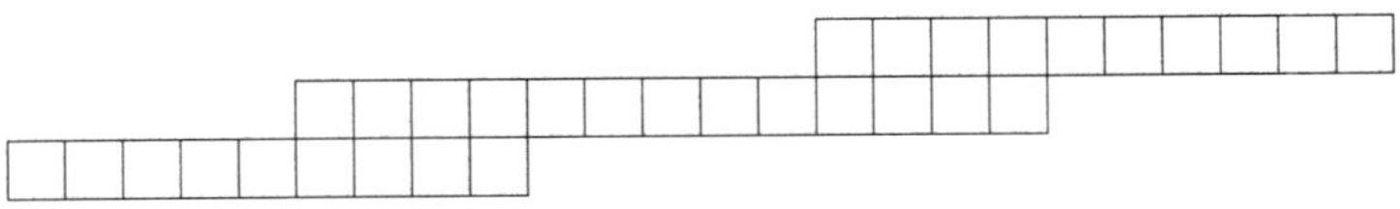

Therefore, changing the order of summation in (10.47), we can reduce the calculation to the case where the diagram μ is replaced by such a skew diagram and the corresponding tableaux take entries in the set $\{i, i+1\}$. Two such entries occurring in the same two-box column contribute the factor of the form $\lambda_i(c)\,\lambda_{i+1}(c-1)$ where c is the content of the upper box. However,

$$S_i : \lambda_i(c)\,\lambda_{i+1}(c-1) \mapsto \lambda_i(c)\,\lambda_{i+1}(c-1)\,\sigma_i(c) - \lambda_i(c)\,\lambda_{i+1}(c-1)\,\sigma_i(c) = 0.$$

Excluding the two-box columns from the diagram, we get a disjoint union of rows of boxes. Such a row of length k will contribute the sum

$$\sum_{p=0}^{k} \lambda_i(c)\,\lambda_i(c+1)\dots\lambda_i(c+p-1) \times \lambda_{i+1}(c+p)\,\lambda_{i+1}(c+p+1)\dots\lambda_{i+1}(c+k), \tag{10.48}$$

where c is now the content of the leftmost box of the row. By applying the screening operator S_i we get

$$\sum_{p=0}^{k} \lambda_i(c)\,\lambda_i(c+1)\dots\lambda_i(c+p-1)\,\lambda_{i+1}(c+p)\,\lambda_{i+1}(c+p+1)\dots\lambda_{i+1}(c+k)$$
$$\times \big(\sigma_i(c) + \sigma_i(c+1) + \dots + \sigma_i(c+p-1) - \sigma_i(c+p+1) - \dots - \sigma_i(c+k+1)\big).$$

By (10.44) we can transform the products by using the relations

$$\lambda_{i+1}(c+p)\dots\lambda_{i+1}(c+p+r)\,\sigma_i(c+p+r+1) = \lambda_i(c+p)\dots\lambda_i(c+p+r)\,\sigma_i(c+p).$$

This implies that the result of the application of S_i to the polynomial (10.48) will contain each monomial of the form

$$\lambda_i(c)\dots\lambda_i(c+q-1)\,\lambda_{i+1}(c+q)\dots\lambda_{i+1}(c+k)\,\sigma_i(c+s-1)$$

with $s = 1,\dots,q$ exactly twice with the opposite signs. The proof is completed by taking the Leibniz rule (10.46) into account. □

REMARK 10.6.2. By analogy with (4.34), introduce the shift operator e^{∂} such that

$$e^{\partial}\lambda_i(a) = \lambda_i(a+1)\,e^{\partial} \qquad \text{and} \qquad e^{\partial}\sigma_i(a) = \sigma_i(a+1)\,e^{\partial}.$$

Each screening operator S_i commutes with e^{∂}. The polynomials (10.48) can be combined into the generating series

$$\big(1 - \lambda_i(c)\,e^{\partial}\big)^{-1}\big(1 - \lambda_{i+1}(c)\,e^{\partial}\big)^{-1}.$$

Therefore, as an alternative to the last step of the proof of Proposition 10.6.1, we can verify that the inverse series

$$\big(1 - \lambda_{i+1}(c)\,e^{\partial}\big)\big(1 - \lambda_i(c)\,e^{\partial}\big)$$

is annihilated by S_i. By applying S_i we get

$$\lambda_{i+1}(c)\,\sigma_i(c+1)\,e^{\partial}\big(1 - \lambda_i(c)\,e^{\partial}\big) + \big(1 - \lambda_{i+1}(c)\,e^{\partial}\big)\big(-\lambda_i(c)\,\sigma_i(c)\,e^{\partial}\big)$$
$$= \lambda_{i+1}(c)\,\sigma_i(c+1)\,e^{\partial} - \lambda_i(c)\,\sigma_i(c)\,e^{\partial} = 0$$

by (10.44). Note also that the proof of Proposition 10.6.1 obviously extends to arbitrary skew diagrams μ which do not contain columns of length exceeding N. The corresponding polynomials (10.47) are the characters of the skew representations of the Yangian $\mathrm{Y}(\mathfrak{gl}_N)$ [**127**]; see also [**109**, Section 8.5]. □

We will now state the counterpart of Proposition 10.6.1 for the Harish-Chandra image of the series $\mathbb{T}^+_\mu(u)$ associated with the dual Yangian $\mathrm{Y}^+(\mathfrak{gl}_N)$ as provided by Theorem 10.2.1. Observe that the only formal difference with the corresponding polynomial of Theorem 10.1.2 is the use of reverse tableaux. However, there is an obvious bijection between the semistandard tableaux $\mathcal{U}$ and reverse tableaux $\mathcal{U}'$ of the same shape μ with the entries in the set $\{1,\dots,N\}$. The bijection replaces each entry i of $\mathcal{U}$ by $i' = N - i + 1$. Therefore the following definitions and results are essentially reformulations of those associated with the Yangian $\mathrm{Y}(\mathfrak{gl}_N)$.

Introduce the algebra of polynomials

$$\mathcal{L}^+ = \mathbb{C}[\lambda^+_i(a)\,|\,i = 1,\dots,N,\ a \in \mathbb{C}] \tag{10.49}$$

in the variables $\lambda^+_i(a)$. For every $i \in \{1,\dots,N-1\}$ consider the free left $\mathcal{L}^+$-module $\widetilde{\mathcal{L}}^+_i$ with the generators $\sigma^+_i(a)$, where a runs over $\mathbb{C}$ and denote by $\mathcal{L}^+_i$ its quotient by the relations

$$\lambda^+_i(a)\,\sigma^+_i(a+1) = \lambda^+_{i+1}(a)\,\sigma^+_i(a), \qquad a \in \mathbb{C}. \tag{10.50}$$

Define the linear operator $\widetilde{S}^+_i : \mathcal{L}^+ \to \widetilde{\mathcal{L}}^+_i$ by the formula

$$\widetilde{S}^+_i : \lambda^+_j(a) \mapsto \begin{cases} -\lambda^+_i(a)\,\sigma^+_i(a+1) & \text{for} \quad j = i \\ \lambda^+_{i+1}(a)\,\sigma^+_i(a) & \text{for} \quad j = i+1 \\ 0 & \text{for} \quad j \neq i, i+1 \end{cases} \tag{10.51}$$

and the Leibniz rule

$$\widetilde{S}^+_i(AB) = B\,\widetilde{S}^+_i(A) + A\,\widetilde{S}^+_i(B). \tag{10.52}$$

Now the *i-th screening operator*

$$S^+_i : \mathcal{L}^+ \to \mathcal{L}^+_i$$

is defined as the composition of $\widetilde{S}^+_i$ and the projection $\widetilde{\mathcal{L}}^+_i \to \mathcal{L}^+_i$.

The subalgebra $\mathrm{Rep}\ \mathrm{Y}^+(\mathfrak{gl}_N)$ of $\mathcal{L}^+$ is defined as the intersection of kernels of the screening operators:

$$\mathrm{Rep}\ \mathrm{Y}^+(\mathfrak{gl}_N) = \bigcap_{i=1}^{N-1} \ker S^+_i. \tag{10.53}$$

For any diagram μ of length $\ell(\mu) \leqslant N$ define the polynomial $\chi^+_{L(\mu)} \in \mathcal{L}^+$ by

$$\chi^+_{L(\mu)} = \sum_{\mathrm{sh}(\mathcal{T})=\mu} \prod_{\alpha\in\mu} \lambda^+_{\mathcal{T}(\alpha)}\big(c(\alpha)\big),$$

summed over all reverse tableaux $\mathcal{T}$ of shape μ with entries in the set $\{1,\dots,N\}$; see Theorem 10.2.1.

PROPOSITION 10.6.3. *We have* $\chi^+_{L(\mu)} \in \mathrm{Rep}\ \mathrm{Y}^+(\mathfrak{gl}_N)$.

PROOF. This follows by the same argument as for Proposition 10.6.1. □

10.7. Bibliographical notes

A detailed account of the properties of the Yangian $\mathrm{Y}(\mathfrak{gl}_N)$ and its representations together with references to original papers can be found in the books by Chari and Pressley [**18**] and the author [**109**]. The proof of Theorem 10.1.2 follows Okounkov's paper [**128**]. Although it deals with $\mathrm{U}(\mathfrak{gl}_N)$, the arguments extend to the Yangian without significant changes. Some other calculations of the character of the evaluation module can be found in Frenkel and Mukhin [**49**, Section 4.5] and Brundan and Kleshchev [**16**, Section 7.4].

Our definition of the double Yangian follows Iohara [**72**] (we put $h = -1$ in his notation); see also references therein. These algebras were used by Etingof and Kazhdan [**36**], [**37**] in the construction of quantum vertex algebras. Dual Yangians were studied in their previous work [**35**] and also by Nazarov [**125**] who introduced them for the queer Lie superalgebras. The results of Section 10.4 are based on [**45**] and follow [**82**].

The theory of characters originates in the work of Knight [**91**] in the Yangian context and Frenkel and Reshetikhin [**51**] in the context of quantum affine algebras (the latter are commonly known as the q-characters). The theory was further developed by Frenkel and Mukhin [**48**] where an algorithm for the calculation of the q-characters was proposed while conjectures for functional relations satisfied by the q-characters were proved by Hernandez [**66**] and Nakajima [**122**]. An extensive review of the role of the q-characters in classical and quantum integrable systems is given by Kuniba, Nakanishi and Suzuki [**99**].

CHAPTER 11

Yangian characters in types B, C and D

Our goal in this chapter is to prove some analogues of Theorem 10.1.2 for the Yangian characters in types B, C and D. We will do this for one particular idempotent in the Brauer algebra, the symmetrizer $s^{(m)}$ associated with the trivial representation; see Section 1.2. These particular Yangian characters will be used in Chapter 13 for the computation of the Harish-Chandra images of the generators of the Feigin–Frenkel center constructed in Chapter 8.

11.1. Yangian for $\mathfrak{g}_N$

Recall that the Lie subalgebra of $\mathfrak{gl}_N$ spanned by the elements F_{ij} defined in (2.23) is isomorphic to the orthogonal Lie algebra $\mathfrak{o}_N$ or the symplectic Lie algebra $\mathfrak{sp}_N$. We will keep the notation $\mathfrak{g}_N$ for $\mathfrak{o}_N$ (with $N = 2n$ or $N = 2n+1$) or $\mathfrak{sp}_N$ (with $N = 2n$). We let $\mathfrak{h}$ denote the Cartan subalgebra of $\mathfrak{g}_N$ spanned by the basis elements $F_{11}, \dots, F_{nn}$. Set

$$\kappa = \begin{cases} N/2 - 1 & \text{in the orthogonal case,} \\ N/2 + 1 & \text{in the symplectic case.} \end{cases}$$

In the notation of Section 1.5, the *R-matrix* $R_{12}(u)$ is a rational function in a complex parameter u with values in the tensor product algebra $\operatorname{End}\mathbb{C}^N \otimes \operatorname{End}\mathbb{C}^N$ defined by

$$R_{12}(u) = 1 - \frac{P_{12}}{u} + \frac{Q_{12}}{u - \kappa}. \tag{11.1}$$

It is well known by **[155]** that this function satisfies the Yang–Baxter equation (10.6).

The *extended Yangian* $\mathrm{X}(\mathfrak{g}_N)$ is an associative algebra with generators $t_{ij}^{(r)}$, where $1 \leqslant i, j \leqslant N$ and $r = 1, 2, \dots$, satisfying certain quadratic relations. Introduce the formal series

$$t_{ij}(u) = \delta_{ij} + \sum_{r=1}^{\infty} t_{ij}^{(r)}\, u^{-r} \in \mathrm{X}(\mathfrak{g}_N)[[u^{-1}]] \tag{11.2}$$

and set

$$T(u) = \sum_{i,j=1}^{N} e_{ij} \otimes t_{ij}(u) \in \operatorname{End}\mathbb{C}^N \otimes \mathrm{X}(\mathfrak{g}_N)[[u^{-1}]].$$

The defining relations for the algebra $\mathrm{X}(\mathfrak{g}_N)$ are written in the form

$$R_{12}(u - v)\, T_1(u)\, T_2(v) = T_2(v)\, T_1(u)\, R_{12}(u - v). \tag{11.3}$$

The *Yangian* $\mathrm{Y}(\mathfrak{g}_N)$ is the quotient of the algebra $\mathrm{X}(\mathfrak{g}_N)$ by the relation

$$T'(u + \kappa)\, T(u) = 1, \tag{11.4}$$

where we use the matrix notation of Section 1.4 and the prime denotes the matrix transposition defined in (2.24). In terms of the series (11.2) the defining relations (11.3) can be written as

$$[t_{ij}(u), t_{kl}(v)] = \frac{1}{u-v}\Big(t_{kj}(u)\, t_{il}(v) - t_{kj}(v)\, t_{il}(u)\Big)$$
$$- \frac{1}{u-v-\kappa}\Big(\delta_{ki'}\sum_{p=1}^{N}\theta_{ip}\, t_{pj}(u)\, t_{p'l}(v) - \delta_{lj'}\sum_{p=1}^{N}\theta_{jp}\, t_{kp'}(v)\, t_{ip}(u)\Big),$$

where we set $\theta_{ij} \equiv 1$ in the orthogonal case, and $\theta_{ij} = \varepsilon_i \varepsilon_j$ in the symplectic case. By taking the coefficients of $u^{-r}v^{-s}$ on both sides one can get the defining relations explicitly in terms of the $t_{ij}^{(r)}$ as in (10.3), although their general form will not be used below. Similarly, relation (11.4) reads as

$$\sum_{i=1}^{N}\theta_{ki}\, t_{i'k'}(u+\kappa)\, t_{il}(u) = \delta_{kl}.$$

We will identify the universal enveloping algebra $\mathrm{U}(\mathfrak{g}_N)$ with a subalgebra of the Yangian $\mathrm{Y}(\mathfrak{g}_N)$ via the embedding $F_{ij} \mapsto t_{ij}^{(1)}$. Then $\mathrm{Y}(\mathfrak{g}_N)$ can be regarded as a $\mathfrak{g}_N$-module with the adjoint action determined by the relations

$$[F_{ij}, t_{kl}^{(s)}] = \delta_{kj}\, t_{il}^{(s)} - \delta_{il}\, t_{kj}^{(s)} - \theta_{ij}\big(\delta_{ki'}\, t_{j'l}^{(s)} - \delta_{j'l}\, t_{ki'}^{(s)}\big). \tag{11.5}$$

In the following proposition we use the notation of Theorem 5.3.1.

PROPOSITION 11.1.1. *For any $s \in \mathcal{B}_m(\omega)$ and variables $u_1, \dots, u_m$, all coefficients of the series*

$$\mathrm{tr}_{1,\dots,m}\, S\, T_1(u_1)\dots T_m(u_m)$$

commute with $\mathfrak{g}_N$.

PROOF. Take the tensor product

$$\mathrm{End}\,\mathbb{C}^N \otimes \mathrm{End}\,(\mathbb{C}^N)^{\otimes m} \otimes \mathrm{Y}(\mathfrak{g}_N)$$

with the additional copy of the endomorphism algebra $\mathrm{End}\,\mathbb{C}^N$ labeled by 0. A matrix form of the relations (11.5) reads

$$[F_0, T_a(u_a)] = (P_{0\,a} - Q_{0\,a})T_a(u_a) - T_a(u_a)(P_{0\,a} - Q_{0\,a})$$

for $a = 1, \dots, m$. The relation

$$\big[E_0, \mathrm{tr}_{1,\dots,m}\, S\, T_1(u_1)\dots T_m(u_m)\big] = 0$$

is a counterpart of (5.20) and it follows by the same argument as in the proof of Theorem 5.3.1. □

Denote by $\mathrm{Y}(\mathfrak{g}_N)^{\mathfrak{h}}$ the subalgebra of $\mathfrak{h}$-invariants under the action (11.5),

$$\mathrm{Y}(\mathfrak{g}_N)^{\mathfrak{h}} = \{y \in \mathrm{Y}(\mathfrak{g}_N) \mid [F_{ii}, y] = 0 \quad \text{for} \quad i = 1, \dots, n\}.$$

Consider the left ideal I of the algebra $\mathrm{Y}(\mathfrak{g}_N)$ generated by all elements $t_{ij}^{(r)}$ with the conditions $1 \leqslant i < j \leqslant N$ and $r \geqslant 1$. It follows from the Poincaré–Birkhoff–Witt theorem for the Yangian [**7**, Section 3] that the intersection $\mathrm{Y}(\mathfrak{g}_N)^{\mathfrak{h}} \cap I$ is a two-sided ideal of $\mathrm{Y}(\mathfrak{g}_N)^{\mathfrak{h}}$. Moreover, the quotient of $\mathrm{Y}(\mathfrak{g}_N)^{\mathfrak{h}}$ by this ideal is isomorphic to the commutative algebra freely generated by the images of the elements $t_{ii}^{(r)}$ with $i = 1, \dots, n$ and $r \geqslant 1$ in the quotient. We will use the notation $\lambda_i^{(r)}$ for this image of $t_{ii}^{(r)}$ and extend this notation to all values $i = 1, \dots, N$. Thus, we get an analogue of the Harish-Chandra homomorphisms (4.6) and (10.10),

$$\mathrm{Y}(\mathfrak{g}_N)^{\mathfrak{h}} \to \mathbb{C}[\lambda_i^{(r)} \,|\, i = 1, \dots, n, \ r \geqslant 1]. \tag{11.6}$$

We combine the elements $\lambda_i^{(r)}$ into the formal series

$$\lambda_i(u) = 1 + \sum_{r=1}^{\infty} \lambda_i^{(r)} \, u^{-r}, \qquad i = 1, \dots, N, \tag{11.7}$$

which can be understood as the image of the series $t_{ii}(u)$ under the homomorphism (11.6). Evaluating the coefficients of each series $\lambda_i(u)$ in $\mathbb{C}$, we get the N-tuple of numerical series

$$\lambda(u) = \big(\lambda_1(u), \dots, \lambda_N(u)\big) \tag{11.8}$$

which can be regarded as the *highest weight* of the corresponding *Verma module* $M(\lambda(u))$ over the Yangian $\mathrm{Y}(\mathfrak{g}_N)$. By the Yangian representation theory, the Verma module is nonzero if and only if the N-tuple (11.8) satisfies certain conditions which are given in [**7**, Propositions 5.2 and 5.14]. These conditions imply that the Harish-Chandra images (11.7) must satisfy the relations

$$\lambda_i(u + \kappa - i) \, \lambda_{i'}(u) = \lambda_{i+1}(u + \kappa - i) \, \lambda_{(i+1)'}(u), \tag{11.9}$$

for $i = 0, 1, \dots, n-1$ if $\mathfrak{g}_N = \mathfrak{o}_{2n}$ or $\mathfrak{sp}_{2n}$, and for $i = 0, 1, \dots, n$ if $\mathfrak{g}_N = \mathfrak{o}_{2n+1}$, where $\lambda_0(u) = \lambda_{0'}(u) := 1$.

Note that the Harish-Chandra homomorphism can be defined in a similar way for the extended Yangian $\mathrm{X}(\mathfrak{g}_N)$. Then relations (11.9) will hold in the same form for all values $i \geqslant 1$. The relations for $i = 0$ are excluded for the extended Yangian because they are consequences of (11.4).

Recall the element $S^{(m)}$ introduced in Section 2.2. We will identify it with the element $S^{(m)} \otimes 1$ of the algebra

$$\underbrace{\mathrm{End}\, \mathbb{C}^N \otimes \dots \otimes \mathrm{End}\, \mathbb{C}^N}_{m} \otimes \mathrm{Y}(\mathfrak{g}_N). \tag{11.10}$$

By Proposition 1.2.8, it can be given by the following multiplicative formula in the orthogonal and symplectic case, respectively:

$$S^{(m)} = \frac{1}{m!} \prod_{1 \leqslant a < b \leqslant m} \Big(1 + \frac{P_{ab}}{b-a} - \frac{Q_{ab}}{N/2 + b - a - 1}\Big) \tag{11.11}$$

and

$$S^{(m)} = \frac{1}{m!} \prod_{1 \leqslant a < b \leqslant m} \Big(1 - \frac{P_{ab}}{b-a} - \frac{Q_{ab}}{n - b + a + 1}\Big), \tag{11.12}$$

where the products are taken in the lexicographic order on the pairs (a, b) and the condition $m \leqslant n + 1$ is assumed in (11.12). Note that in both cases the products involve evaluated R-matrices (11.1),

$$S^{(m)} = \frac{1}{m!} \prod_{1 \leqslant a < b \leqslant m} R_{ab}(u_a - u_b), \tag{11.13}$$

where $u_a = u + a - 1$ and $u_a = u - a + 1$ for $a = 1, \dots, m$ in the orthogonal and symplectic case, respectively.

We will now consider types B, C and D separately. In types B and D introduce the formal series $\mathbb{T}^{(m)}(u)$ with coefficients in the Yangian $\mathrm{Y}(\mathfrak{o}_N)$ by the formula

$$\mathbb{T}^{(m)}(u) = \mathrm{tr}_{1,\dots,m}\, S^{(m)} T_1(u)\, T_2(u+1)\dots T_m(u+m-1) \tag{11.14}$$

with the trace taken over all m copies of $\mathrm{End}\,\mathbb{C}^N$ in (11.10), where $\mathfrak{g}_N = \mathfrak{o}_N$. By Proposition 11.1.1, all coefficients of this series belong to $\mathrm{Y}(\mathfrak{o}_N)^{\mathfrak{h}}$.

Type B_n. Take $N = 2n+1$ in the definition (11.14).

THEOREM 11.1.2. *The image of the series $\mathbb{T}^{(m)}(u)$ under the homomorphism (11.6) is found by*

$$\mathbb{T}^{(m)}(u) \mapsto \sum_{1\leqslant i_1\leqslant\dots\leqslant i_m\leqslant N} \lambda_{i_1}(u)\, \lambda_{i_2}(u+1)\dots\lambda_{i_m}(u+m-1)$$

with the condition that $n+1$ occurs among the summation indices $i_1,\dots,i_m$ at most once.

PROOF. By (1.31) the symmetrizer $S^{(m)}$ has the properties

$$S^{(m)}\, Q_{ab} = Q_{ab}\, S^{(m)} = 0 \qquad \text{and} \qquad S^{(m)}\, P_{ab} = P_{ab}\, S^{(m)} = S^{(m)}. \tag{11.15}$$

The subspace of *harmonic tensors* in $(\mathbb{C}^N)^{\otimes m}$ is spanned by the tensors v with the property $Q_{ab}\, v = 0$ for all $1 \leqslant a < b \leqslant m$. It is clear from the equivalent formula (2.43) for the operator $S^{(m)}$ that it projects $(\mathbb{C}^N)^{\otimes m}$ to a subspace of symmetric harmonic tensors which we will denote by $\mathcal{H}_m$. As we recalled in Section 5.2, the vector space $\mathcal{H}_m = S^{(m)}(\mathbb{C}^N)^{\otimes m}$ carries an irreducible representation of the Lie algebra $\mathfrak{o}_N$ isomorphic to $L(m,0,\dots,0)$. The trace formula (1.52) implies that

$$\dim L(m,0,\dots,0) = \frac{N+2m-2}{N+m-2}\binom{N+m-2}{m} \tag{11.16}$$

which is also a particular case of (5.14).

So the trace in (11.14) can be calculated over $\mathcal{H}_m$. We will introduce a special basis of this subspace. To this end, identify the image of the symmetrizer $H^{(m)}$ with the space of homogeneous polynomials of degree m in variables $z_1,\dots,z_N$ via the isomorphism (2.62) so that

$$z_{i_1}\dots z_{i_m} = H^{(m)}(e_{i_1}\otimes\dots\otimes e_{i_m}). \tag{11.17}$$

By Proposition 2.3.1, $\mathcal{H}_m$ is then identified with the subspace of $\mathfrak{sl}_2$-singular vectors in $\mathcal{P}_N^m$ under the action (2.63) whose elements are homogeneous *harmonic polynomials* of degree m. They belong to the kernel of the *Laplace operator*

$$\sum_{i=1}^n \partial_i\, \partial_{i'} + \frac{1}{2}\,\partial_{n+1}^2,$$

where ∂_i denotes the partial derivative over z_i.

The basis vectors of $\mathcal{H}_m$ will be parameterized by the N-tuples of nonnegative integers $(k_1,\dots,k_n,\delta,l_n,\dots,l_1)$, where the values of δ are restricted to $\delta\in\{0,1\}$ and the sum of all entries is m. Given such a tuple, the corresponding harmonic polynomial is defined by

$$\sum_{a_1,\dots,a_n} \frac{(-2)^{a_1+\dots+a_n}(a_1+\dots+a_n)!\, z_{n+1}^{2a_1+\dots+2a_n+\delta}}{a_1!\dots a_n!\,(2a_1+\dots+2a_n+\delta)!} \times \prod_{i=1}^{n} \frac{z_i^{k_i-a_i}\, z_{i'}^{l_i-a_i}}{(k_i-a_i)!\,(l_i-a_i)!}, \tag{11.18}$$

summed over nonnegative integers a_i satisfying $a_i \leqslant \min\{k_i, l_i\}$. Each polynomial contains a unique monomial (which we call the *leading monomial*) where the variable z_{n+1} occurs with the power not exceeding 1. It is straightforward to see that these polynomials are indeed all harmonic and linearly independent. Furthermore, the number of the polynomials equals

$$\binom{N+m-2}{m} + \binom{N+m-3}{m-1}$$

which coincides with the dimension $\dim L(m,0,\dots,0)$ in (11.16). Hence the polynomials (11.18) form a basis of the subspace $\mathcal{H}_m$.

Similar to the proof of (10.14), relations (11.3) and (11.13) imply that the product occurring in (11.14) can be written as

$$S^{(m)}T_1(u)\dots T_m(u+m-1) = T_m(u+m-1)\dots T_1(u)\, S^{(m)}. \tag{11.19}$$

This relation shows that the product on each side can be regarded as an operator on $(\mathbb{C}^N)^{\otimes m}$ with coefficients in the algebra $\mathrm{Y}(\mathfrak{o}_N)[[u^{-1}]]$ such that the subspace $\mathcal{H}_m$ is invariant under this operator. Now fix a basis vector $v \in \mathcal{H}_m$ of the form (11.18). Denote the operator on the right hand side of (11.19) by A and consider the coefficient of v in the expansion of Av as a linear combination of the basis vectors. Use (11.17) to write the vector v as a linear combination of the tensors $e_{j_1} \otimes \dots \otimes e_{j_m}$. We have $S^{(m)}v = v$, while the matrix elements of the remaining product are found from the expansion

$$T_m(u+m-1)\dots T_1(u)(e_{j_1}\otimes\dots\otimes e_{j_m}) = \sum_{i_1,\dots,i_m} t_{i_m j_m}(u+m-1)\dots t_{i_1 j_1}(u)(e_{i_1}\otimes\dots\otimes e_{i_m}).$$

The coefficient of v in the expansion of Av is uniquely determined by the coefficient of the tensor $e_{i_1}\otimes\dots\otimes e_{i_m}$ with $i_1 \leqslant \dots \leqslant i_m$ which corresponds to the leading monomial of v under the isomorphism (11.17) so that

$$e_{i_1}\otimes\dots\otimes e_{i_m} = \underbrace{e_1\otimes\dots\otimes e_1}_{k_1} \otimes \underbrace{e_2\otimes\dots\otimes e_2}_{k_2} \otimes\dots\otimes \underbrace{e_{1'}\otimes\dots\otimes e_{1'}}_{l_1}. \tag{11.20}$$

It follows from the formula (11.18) that if a tensor of the form $e_{j_1}\otimes\dots\otimes e_{j_m}$ corresponds to a non-leading monomial occurring in the expansion of v, then the matrix element $t_{i_m j_m}(u+m-1)\dots t_{i_1 j_1}(u)$ vanishes under the homomorphism (11.6). Indeed, the multiplicity of the index 1 in the multiset $\{j_1,\dots,j_m\}$ equals $k_1 - a_1$ for a nonnegative integer a_1, whereas the coefficients of $t_{1 j_c}(u+c-1)$ belong to the left ideal I if $1 < j_c$. Hence, if we suppose that the matrix element does not vanish under the homomorphism (11.6), then we must have $a_1 = 0$. Similarly, the multiplicity of the index 2 in the multiset $\{j_1,\dots,j_m\}$ must coincide with k_2, etc., which forces all parameters $a_1,\dots,a_n$ to be equal to zero.

Thus, a nonzero contribution to the image of the diagonal matrix element of the operator A corresponding to v under the homomorphism (11.6) only comes from the term $t_{i_m i_m}(u+m-1)\dots t_{i_1 i_1}(u)$. Taking the sum over all basis vectors (11.18) yields the desired formula for the image of the element (11.14). □

REMARK 11.1.3. As with the Yangian $\mathrm{Y}(\mathfrak{gl}_N)$, we take the Harish-Chandra image provided by Theorem 11.1.2 as the *definition* of the Yangian character of the representation of $\mathrm{Y}(\mathfrak{o}_{2n+1})$ on $L(m,0,\dots,0)$ which extends the action of $\mathfrak{o}_{2n+1}$. The existence of such an extension is explained by the fact that the projection $S^{(m)}$ is a product of evaluated R-matrices (11.13). This Yangian representation is known as a *Kirillov–Reshetikhin module* and the character provided by Theorem 11.1.2 coincides with the *q-character* under an appropriate identification of the parameters; see reviews covering all classical types in [**99**, Section 7] and [**121**, Section 2]. In particular, the relations (11.9) coincide with those for the q-characters, as the $\lambda_i(u)$ correspond to the 'single box variables'. Note also that all coefficients of all series $\mathbb{T}^{(m)}(u)$ defined in (11.14) pairwise commute as it can be interpreted as a *transfer matrix* [**98**]; cf. Remark 10.1.5(ii). A direct proof of the commutativity property is given in [**110**]. □

Type D_n. Consider the series $\mathbb{T}^{(m)}(u)$ given by (11.14), where the parameter N now takes the even value $2n$.

THEOREM 11.1.4. *The image of the series $\mathbb{T}^{(m)}(u)$ under the homomorphism (11.6) is given by*

$$\mathbb{T}^{(m)}(u)\mapsto \sum_{1\leqslant i_1\leqslant\dots\leqslant i_m\leqslant N}\lambda_{i_1}(u)\,\lambda_{i_2}(u+1)\dots\lambda_{i_m}(u+m-1)$$

with the condition that n and n' do not occur simultaneously among the summation indices $i_1,\dots,i_m$.

PROOF. As in the proof of Theorem 11.1.2, we use the properties (11.15) of the symmetrizer $S^{(m)}$. Following the argument of that proof we identify the image $S^{(m)}(\mathbb{C}^N)^{\otimes m}$ with the space $\mathcal{H}_m$ of homogeneous *harmonic polynomials* of degree m in variables $z_1,\dots,z_N$ by (11.17). This time the harmonic polynomials are annihilated by the *Laplace operator* of the form

$$\sum_{i=1}^{n}\partial_i\,\partial_{i'}.$$

The vector space $\mathcal{H}_m$ carries an irreducible representation of $\mathfrak{o}_N$ isomorphic to $L(m,0,\dots,0)$ whose dimension is found by the formula (11.16) with $N=2n$. The basis vectors of $\mathcal{H}_m$ will be parameterized by the N-tuples $(k_1,\dots,k_n,l_n,\dots,l_1)$, where the k_i and l_i are arbitrary nonnegative integers, the sum of all entries is m and at least one of k_n and l_n is zero. Given such a tuple, the corresponding harmonic polynomial is now defined by

$$\sum_{a_1,\dots,a_{n-1}}\frac{(-1)^{a_1+\dots+a_{n-1}}(a_1+\dots+a_{n-1})!\,z_n^{a_1+\dots+a_{n-1}+k_n}\,z_{n'}^{a_1+\dots+a_{n-1}+l_n}}{a_1!\dots a_{n-1}!\,(a_1+\dots+a_{n-1}-k_n)!\,(a_1+\dots+a_{n-1}+l_n)!}\times\prod_{i=1}^{n-1}\frac{z_i^{k_i-a_i}\,z_{i'}^{l_i-a_i}}{(k_i-a_i)!\,(l_i-a_i)!},\tag{11.21}$$

summed over nonnegative integers $a_1, \dots, a_{n-1}$ satisfying $a_i \leqslant \min\{k_i, l_i\}$. A unique *leading monomial* corresponds to the values $a_1 = \dots = a_{n-1} = 0$. It is easily verified that these polynomials are indeed all harmonic and linearly independent. The number of the polynomials equals

$$2\binom{N+m-2}{m} - \binom{N+m-3}{m}$$

which coincides with the dimension $\dim L(m, 0, \dots, 0)$ in (11.16). Hence the polynomials (11.21) form a basis of $\mathcal{H}_m$.

The argument is now completed in the same way as for Theorem 11.1.2 by considering the diagonal matrix elements of the operator A on right hand side of (11.19) for the basis vectors (11.21). These matrix elements are determined by the coefficients of the tensors $e_{i_1} \otimes \dots \otimes e_{i_m}$ with $i_1 \leqslant \dots \leqslant i_m$ corresponding to the leading monomials of v under the isomorphism (11.17). The tensors are given by the same formula (11.20) with $N = 2n$ and the condition that at least one of k_n and l_n is zero. A nonzero contribution to the image of the diagonal matrix element of the operator A corresponding to a given basis vector v under the homomorphism (11.6) only comes from the term $t_{i_m i_m}(u+m-1) \dots t_{i_1 i_1}(u)$ with the condition that the indices $i_1, \dots, i_m$ correspond to a leading monomial. This yields the desired formula for the image of the element (11.14). □

As with type B_n, the coefficients of the series $\mathbb{T}^{(m)}(u)$ belong to a commutative subalgebra of the Yangian $\mathrm{Y}(\mathfrak{o}_{2n})$; cf. Remark 11.1.3.

Type C_n. Now let $\mathfrak{g}_N = \mathfrak{sp}_N$ with $N = 2n$. The symmetrizer $S^{(m)}$ is defined in (11.12) with $m \leqslant n+1$. Introduce the formal series

$$\mathbb{T}^{(m)}(u) = \mathrm{tr}_{1,\dots,m}\, S^{(m)} T_1(u)\, T_2(u-1) \dots T_m(u-m+1), \tag{11.22}$$

with the trace taken over all m copies of $\mathrm{End}\,\mathbb{C}^{2n}$ in (11.10). By Proposition 11.1.1, all coefficients of $\mathbb{T}^{(m)}(u)$ belong to $\mathrm{Y}(\mathfrak{sp}_{2n})^{\mathfrak{h}}$.

THEOREM 11.1.5. *The image of the series $\mathbb{T}^{(m)}(u)$ with $m \leqslant n$ under the homomorphism* (11.6) *is given by*

$$\mathbb{T}^{(m)}(u) \mapsto \sum_{1 \leqslant i_1 < \dots < i_m \leqslant 2n} \lambda_{i_1}(u)\, \lambda_{i_2}(u-1) \dots \lambda_{i_m}(u-m+1) \tag{11.23}$$

with the condition that if for any i both i and i' occur among the summation indices as $i = i_r$ and $i' = i_s$ for some $1 \leqslant r < s \leqslant m$, then $s - r \leqslant n - i$.

PROOF. Due to (1.31) and the formulas (1.73) for the action of the Brauer algebra, we have the properties

$$S^{(m)} Q_{ab} = Q_{ab} S^{(m)} = 0 \qquad \text{and} \qquad S^{(m)} P_{ab} = P_{ab} S^{(m)} = -S^{(m)}. \tag{11.24}$$

As with the orthogonal case, the subspace of *harmonic tensors* in $(\mathbb{C}^N)^{\otimes m}$ is spanned by the tensors v with the property $Q_{ab}\, v = 0$ for all $1 \leqslant a < b \leqslant m$. By the formula (2.52) for the operator $S^{(m)}$, it projects $(\mathbb{C}^{2n})^{\otimes m}$ to a subspace of skew-symmetric harmonic tensors which we denote by $\mathcal{H}_m$. The vector space $\mathcal{H}_m = S^{(m)} (\mathbb{C}^{2n})^{\otimes m}$ carries an irreducible representation of the Lie algebra $\mathfrak{sp}_{2n}$

isomorphic to $L(1,\dots,1,0,\dots,0)$ (with m copies of 1); see Section 5.2. Taking into account formula (1.52) and the trace diagram (1.77), we find

$$\dim L(1,\dots,1,0,\dots,0)=\frac{2n-2m+2}{2n-m+2}\binom{2n+1}{m}, \tag{11.25}$$

which is also a particular case of (5.15).

Thus, the trace in (11.22) can be calculated over the subspace $\mathcal{H}_m$. We introduce a special basis of this subspace by identifying the image of the anti-symmetrizer $H^{(m)}$ with the space of homogeneous polynomials of degree m in the anti-commuting variables $\zeta_1,\dots,\zeta_{2n}$ via the isomorphism (2.68) so that

$$\zeta_{i_1}\wedge\dots\wedge\zeta_{i_m}=H^{(m)}(e_{i_1}\otimes\ldots\otimes e_{i_m}).$$

The subspace $\mathcal{H}_m$ is then identified with the subspace of homogeneous *harmonic polynomials* of degree m; they belong to the kernel of the *skew Laplace operator*

$$\sum_{i=1}^{n}\partial_i\wedge\partial_{i'}, \tag{11.26}$$

where ∂_i denotes the (left) partial derivative over ζ_i.

The basis vectors of $\mathcal{H}_m$ will be parameterized by the subsets $\{i_1,\dots,i_m\}$ of the set $\{1,\dots,2n\}$ satisfying the condition as stated in the theorem, assuming that the elements $i_1,\dots,i_m$ are written in the increasing order. We will call such subsets *admissible* and extend this definition to all values $1\leqslant m\leqslant n+1$.

LEMMA 11.1.6. *For $m\leqslant n+1$ the number of admissible subsets $\{i_1,\dots,i_m\}$ of the set $\{1,\dots,2n\}$ equals*

$$\binom{2n}{m}-\binom{2n}{m-2}.$$

In particular, this number is zero for $m=n+1$ and it coincides with the dimension $\dim L(1,\dots,1,0,\dots,0)$ in (11.25) for $m\leqslant n$.

PROOF. We will construct a bijection π between the set of non-admissible subsets $\{i_1,\dots,i_m\}$ and the set of all subsets $\{j_1,\dots,j_{m-2}\}$ of $\{1,\dots,2n\}$ which will imply the desired formula. Our bijection will involve some multisets so it will be convenient to introduce some related notation. Let $\{i_1,\dots,i_m\}$ be a multiset with elements in $\{1,\dots,2n\}$ with $i_1\leqslant i_2\leqslant\dots\leqslant i_m$. If (q,q') is a pair of indices such that $q=i_r$ and $q'=i_s$ for some $1\leqslant r<s\leqslant m$ (as before, $q'=2n-q+1$), we will say that the *distance* between these entries q and q' in the given multiset is $s-r-1$.

By definition, a non-admissible subset $\{i_1,\dots,i_m\}$ contains a pair of indices (q,q') (which we will call a *breaking pair*) such that the distance between q and q' is at least $n-q$. By the *maximal breaking pair* of a non-admissible subset we will mean the breaking pair (p,p') with the maximal possible value of p. Note that $1\leqslant p\leqslant n$. Observe that if (p,p') is the maximal breaking pair of a non-admissible subset, then the distance between p and p' is exactly $n-p$. Indeed, if we suppose that the distance exceeds $n-p$, then there exists a pair (t,t') with $p<t\leqslant n$ which belongs to the subset. Take such a pair with the minimal possible value of t. Then the distance between t and t' will exceed

$$n-p-(t-p-1)-2=n-t-1$$

so that (t,t') is another breaking pair which contradicts the maximality of (p,p').

Now we will construct the desired bijection as a composition of certain transformations π_q of multisets. The transformations π_q are parameterized by elements $q \in \{1,\dots,n-1\}$ and defined on the multisets $\{i_1,\dots,i_m\}$ as follows. Consider all pairs (q,q') occurring in a given multiset $\{i_1,\dots,i_m\}$ such that the distance between q and q' is $n-q$. Note that such pairs are disjoint. The transformation π_q replaces each of these pairs by the pair $\big(q+1,(q+1)'\big)$ so that $q\mapsto q+1$ and $q'\mapsto (q+1)'$. If no pair (q,q') with that distance condition occurs, then π_q is the identity map.

Given a non-admissible subset $\{i_1,\dots,i_m\}$, let (p,p') be its maximal breaking pair so that the distance between p and p' is $n-p$. One verifies that the application of the compositions of the transformations π_q yields a multiset of the following form

$$\pi_{n-1}\circ\dots\circ\pi_{p+1}\circ\pi_p : \{i_1,\dots,i_m\}\mapsto\{j_1,\dots,j_a,n,n',j_{a+1},\dots,j_{m-2}\},$$

where

$$j_1<\dots<j_a\leqslant n<n'\leqslant j_{a+1}<\dots<j_{m-2} \tag{11.27}$$

for some $a\in\{0,\dots,m-2\}$. Indeed, it follows from the definition of the transformations that each multiset obtained by an application of π_q at any step has a special form: only the entries $q+1$ and $(q+1)'$ can occur with multiplicities exceeding 1. Moreover, the difference between these two multiplicities can only be $-1,0$ or 1. In addition, at the final step, the multiplicities of n and n' can only be 1 or 2.

The desired map π is now defined by the rule

$$\pi : \{i_1,\dots,i_m\}\mapsto\{j_1,\dots,j_a,j_{a+1},\dots,j_{m-2}\}.$$

To show that π is a bijection, we construct the inverse map. Starting with an arbitrary subset $\{j_1,\dots,j_{m-2}\}$ of $\{1,\dots,2n\}$, adjoin the entries n and n' to make it into an m-multiset J satisfying (11.27). Define the map τ by setting

$$\tau\big(\{j_1,\dots,j_{m-2}\}\big)=\big(\tau_{p+1}\circ\tau_{p+2}\circ\dots\circ\tau_n\big)(J),$$

where the transformations τ_q and the parameter p are defined as follows. Consider all pairs (q,q') occurring in a given multiset $\{i_1,\dots,i_m\}$ such that the distance between q and q' is $n-q+1$. The transformation τ_q replaces each of these pairs by the pair $\big(q-1,(q-1)'\big)$. If no pair (q,q') with this distance condition occurs then τ_q is the identity map. The parameter $1\leqslant p\leqslant n$ is determined by the property that this is the maximal value of $q\in\{1,\dots,n\}$ such that

$$\big(\tau_q\circ\tau_{q+1}\circ\dots\circ\tau_n\big)(J)=\big(\tau_{q+1}\circ\tau_{q+2}\circ\dots\circ\tau_n\big)(J),$$

where the right hand side is understood as equal to J if $q=n$. It is straightforward to verify that $\tau\big(\{j_1,\dots,j_{m-2}\}\big)$ is indeed a non-admissible subset of $\{1,\dots,2n\}$ and that the compositions $\pi\circ\tau$ and $\tau\circ\pi$ are the identity maps. $\square$

EXAMPLE 11.1.7. To illustrate the bijection used in the proof of Lemma 11.1.6, take $n=m=7$ and consider the non-admissible set $\{3,4,5,6',4',3',1'\}$. Its unique breaking pair is $(3,3')$ which is therefore maximal. The transformations act by

$$\pi_3 : \{3,4,5,6',4',3',1'\}\mapsto\{4,4,5,6',4',4',1'\},$$
$$\pi_4 : \{4,4,5,6',4',4',1'\}\mapsto\{5,5,5,6',5',5',1'\},$$
$$\pi_5 : \{5,5,5,6',5',5',1'\}\mapsto\{5,6,6,6',6',6',1'\},$$
$$\pi_6 : \{5,6,6,6',6',6',1'\}\mapsto\{5,7,7,7',7',6',1'\}.$$

Hence, for the image with respect to the map π we have

$$\pi : \{3,4,5,6',4',3',1'\} \mapsto \{5,7,7',6',1'\}.$$

Clearly, the transformations $\tau_7, \tau_6, \tau_5, \tau_4$ are respectively inverse to $\pi_6, \pi_5, \pi_4, \pi_3$ and τ_3 acts as the identity map on $\{3,4,5,6',4',3',1'\}$. □

To construct a basis of $\mathcal{H}_m$, consider monomials of the form

$$\zeta_{a_1} \wedge \zeta_{a_1'} \wedge \cdots \wedge \zeta_{a_k} \wedge \zeta_{a_k'} \wedge \zeta_{b_1} \wedge \cdots \wedge \zeta_{b_l} \tag{11.28}$$

with $1 \leqslant a_1 < \cdots < a_k \leqslant n$ and $1 \leqslant b_1 < \cdots < b_l \leqslant 2n$, associated with subsets $\{a_1, a_1', \ldots, a_k, a_k', b_1, \ldots, b_l\}$ of $\{1, \ldots, 2n\}$ of cardinality $m = 2k + l$, where $b_i \neq b_j'$ for all i and j. When the elements b_i are fixed we will label the monomial (11.28) by the k-tuple $(a_1, \ldots, a_k)$. We will order the k-tuples and the corresponding monomials lexicographically.

Now let the subset $\{a_1, a_1', \ldots, a_k, a_k', b_1, \ldots, b_l\}$ be admissible. The corresponding monomial (11.28) will also be called admissible. Take $i \in \{1, \ldots, k\}$. Let s be the number of the elements b_j of the subset satisfying $a_i < b_j < a_i'$. By the admissibility condition applied to a_i and a_i', we have the inequality $2(k-i) + s < n - a_i$. Therefore, there exist elements $c_i, \ldots, c_k$ satisfying $a_i < c_i < \cdots < c_k \leqslant n$ so that none of c_j or c_j' with $j = i, \ldots, k$ belongs to the subset $\{a_1, a_1', \ldots, a_k, a_k', b_1, \ldots, b_l\}$. We will choose such elements in an unique way by imposing an additional maximality condition. Namely, choose the maximal possible element c_k for $i = k$, then choose the maximal possible element c_{k-1} for $i = k - 1$, etc., to get a family of elements $c_1 < \cdots < c_k$ uniquely determined by the admissible subset. In particular, $c_i > a_i$ for all i.

Our condition on the parameters b_i implies that the monomial $\zeta_{b_1} \wedge \cdots \wedge \zeta_{b_l}$ is annihilated by the skew Laplace operator (11.26). We denote this monomial by y and set $x_a = \zeta_a \wedge \zeta_{a'}$ for $a = 1, \ldots, n$. The vector

$$\sum_{p=0}^{k} (-1)^p \sum_{d_1 < \cdots < d_p} x_{a_1} \wedge \cdots \wedge \widehat{x}_{a_{d_1}} \wedge \cdots \wedge \widehat{x}_{a_{d_p}} \wedge \cdots \wedge x_{a_k} \wedge x_{c_{d_1}} \wedge \cdots \wedge x_{c_{d_p}} \wedge y,$$

where the hats indicate the factors to be skipped, is easily seen to belong to the kernel of the operator (11.26) so it is an element of the subspace $\mathcal{H}_m$. Furthermore, these vectors parameterized by all admissible subsets form a basis of $\mathcal{H}_m$. Indeed, the vectors are linearly independent because the linear combination defining each vector is uniquely determined by the admissible monomial $x_{a_1} \wedge \cdots \wedge x_{a_k} \wedge y$ which precedes all the other monomials occurring in the linear combination with respect to the lexicographic order. The number of admissible monomials coincides with the dimension of $\mathcal{H}_m$ by Lemma 11.1.6.

Note that apart from the minimal admissible monomial $x_{a_1} \wedge \cdots \wedge x_{a_k} \wedge y$, the linear combination defining a basis vector may contain some other admissible monomials. By eliminating such additional admissible monomials with the use of the induction on the lexicographic order, we can produce another basis of the space $\mathcal{H}_m$ parameterized by all admissible subsets with the property that each basis vector is given by a linear combination of monomials of the same form as above, but containing a unique admissible monomial.

By relations (11.3) and (11.13) we can write the product occurring in (11.22) as

$$S^{(m)} T_1(u) \ldots T_m(u - m + 1) = T_m(u - m + 1) \ldots T_1(u)\, S^{(m)} \tag{11.29}$$

and complete the argument exactly as in the proof of Theorem 11.1.2. Indeed, relations (11.24) and (11.29) show that the product on each side can be regarded as an operator on $(\mathbb{C}^{2n})^{\otimes m}$ with coefficients in the algebra $\mathrm{Y}(\mathfrak{sp}_{2n})[[u^{-1}]]$ such that the subspace $\mathcal{H}_m$ is invariant under this operator. Denote the operator on the right hand side of (11.29) by A and let v denote the basis vector of $\mathcal{H}_m$ corresponding to an admissible subset $\{i_1, \dots, i_m\}$ with $i_1 < \dots < i_m$. The properties of the basis vectors imply that a nonzero contribution to the image of the diagonal matrix element of the operator A corresponding to v under the homomorphism (11.6) only comes from the term $t_{i_m i_m}(u - m + 1) \dots t_{i_1 i_1}(u)$. □

REMARK 11.1.8. As with the orthogonal case (see Remark 11.1.3), the Harish-Chandra image provided by Theorem 11.1.5 is *by definition* the Yangian character of the representation of $\mathrm{Y}(\mathfrak{sp}_{2n})$ on the m-th fundamental module $L(1, \dots, 1, 0, \dots, 0)$ (with m copies of 1) of the Lie algebra $\mathfrak{sp}_{2n}$. This character coincides with the *q-character* written in terms of different variables; see **[50]** and **[98]**. Note also that the coefficients of all series $\mathbb{T}^{(m)}(u)$ with $m \leqslant n$ defined in (11.22) pairwise commute in $\mathrm{Y}(\mathfrak{sp}_{2n})$; see **[98]** and **[110]**. □

We will be using an equivalent formula for the expression (11.23). Introduce new parameters $\varkappa_i(u)$ for $i = 1, \dots, 2n+2$ by

$$\varkappa_i(u) = \lambda_i(u), \qquad \varkappa_{2n-i+3}(u) = \lambda_{2n-i+1}(u) \qquad \text{for} \quad i = 1, \dots, n,$$

and

$$\varkappa_{n+2}(u) = -\varkappa_{n+1}(u), \tag{11.30}$$

where $\varkappa_{n+1}(u)$ is the formal series in u^{-1} with constant term 1 uniquely determined by the condition

$$\varkappa_{n+1}(u)\, \varkappa_{n+1}(u-1) = \lambda_n(u)\, \lambda_{n'}(u-1). \tag{11.31}$$

COROLLARY 11.1.9. *The image of the series* (11.22) *with* $m \leqslant n$ *under the homomorphism* (11.6) *can be written as*

$$\mathbb{T}^{(m)}(u) \mapsto \sum_{1 \leqslant i_1 < \dots < i_m \leqslant 2n+2} \varkappa_{i_1}(u)\, \varkappa_{i_2}(u-1) \dots \varkappa_{i_m}(u-m+1). \tag{11.32}$$

Moreover, if $m = n+1$ *then the sum in* (11.32) *is zero.*

PROOF. Note that $\mathbb{T}^{(n+1)}(u) = 0$ since $S^{(n+1)} = 0$ by Proposition 2.3.2. Therefore, Theorem 11.1.5 extends trivially to the case $m = n+1$ since the set of admissible subsets is empty for $m = n+1$ by Lemma 11.1.6, implying that the sum in (11.23) is zero. We will show that the sum in (11.32) coincides with the sum in (11.23) for all values $m \leqslant n+1$.

Note first that the terms in (11.32) corresponding to the sets of summation indices $\{i_1, \dots, i_m\}$ which contain exactly one of $n+1$ or $n+2$, cancel pairwise because of the condition (11.30). If none of $n+1$ or $n+2$ occurs in the sets of summation indices, then the corresponding partial sum will be found as

$$\sum_{1 \leqslant i_1 < \dots < i_m \leqslant 2n} \lambda_{i_1}(u)\, \lambda_{i_2}(u-1) \dots \lambda_{i_m}(u-m+1). \tag{11.33}$$

The remaining partial sum corresponding to the sets of summations indices containing both $n+1$ and $n+2$ can be written in terms of the $\lambda_i(u)$ by taking into account

(11.30) and (11.31). Therefore, the sum in (11.32) is obtained by subtracting the expression

$$\sum_{a=0}^{m-2} \sum_{1\leqslant j_1<\dots<j_a\leqslant n<n'\leqslant j_{a+1}<\dots<j_{m-2}\leqslant 2n} \lambda_{j_1}(u)\dots\lambda_{j_a}(u-a+1) \\ \times \lambda_n(u-a)\,\lambda_{n'}(u-a-1)\,\lambda_{j_{a+1}}(u-a-2)\dots\lambda_{j_{m-2}}(u-m+1) \tag{11.34}$$

from (11.33). Hence, the proof will be completed if we show that (11.34) coincides with the sum

$$\sum_{1\leqslant i_1<\dots<i_m\leqslant 2n} \lambda_{i_1}(u)\,\lambda_{i_2}(u-1)\dots\lambda_{i_m}(u-m+1), \tag{11.35}$$

taken over all non-admissible subsets $\{i_1,\dots,i_m\}$. However, a bijection between the sets of summation indices in (11.34) and (11.35) was constructed in the proof of Lemma 11.1.6. Moreover, this bijection turns out to be weight-preserving in the sense that the corresponding terms in the sums coincide. Indeed, this follows from relations (11.9) which in the symplectic case imply

$$\lambda_q(v)\,\lambda_{q'}(v-n+q-1)=\lambda_{q+1}(v)\,\lambda_{(q+1)'}(v-n+q-1) \tag{11.36}$$

for $q=1,\dots,n-1$. So the products in (11.35) are preserved by the respective transformations π_q used in the construction of the bijection π since the relation (11.36) is invoked precisely in the case where the distance between the entries q and q' is $n-q$. □

REMARK 11.1.10. Since the relation (11.4) was not used in the proofs, the Harish-Chandra images provided by Theorems 11.1.2, 11.1.4 and 11.1.5 take the same form if the Yangian $\mathrm{Y}(\mathfrak{g}_N)$ is replaced by the extended Yangian $\mathrm{X}(\mathfrak{g}_N)$. □

11.2. Dual Yangian for $\mathfrak{g}_N$

Define the *extended dual Yangian* $\mathrm{X}^+(\mathfrak{g}_N)$ as the associative algebra with generators $t_{ij}^{(-r)}$, where $1\leqslant i,j\leqslant N$ and $r=1,2,\dots$ subject to the defining relations written in a matrix form as follows. Combine the generators into the formal power series

$$t^+_{ij}(u)=\delta_{ij}-\sum_{r=1}^{\infty} t_{ij}^{(-r)}\,u^{r-1}\in \mathrm{X}^+(\mathfrak{g}_N)[[u]]$$

and introduce the matrix

$$T^+(u)=\sum_{i,j=1}^{N} e_{ij}\otimes t^+_{ij}(u)\in \operatorname{End}\mathbb{C}^N\otimes \mathrm{X}^+(\mathfrak{g}_N)[[u]].$$

The defining relations are

$$R_{12}(u-v)\,T^+_1(u)\,T^+_2(v)=T^+_2(v)\,T^+_1(u)\,R_{12}(u-v), \tag{11.37}$$

where we use the R-matrix (11.1). As in (1.61), $T^+_a(u)$ for $a=1,\dots,m$ denotes the corresponding element of the tensor product algebra

$$\underbrace{\operatorname{End}\mathbb{C}^N\otimes\dots\otimes\operatorname{End}\mathbb{C}^N}_{m}\otimes \mathrm{X}^+(\mathfrak{g}_N).$$

Similar to (10.26), define the *descending* filtration on $\mathrm{X}^+(\mathfrak{g}_N)$ by setting the degree of the generator $t_{ij}^{(-r)}$ with $r \geqslant 1$ to be equal to r. We let $\widehat{\mathrm{X}}^+(\mathfrak{g}_N)$ denote the completion of $\mathrm{X}^+(\mathfrak{g}_N)$ with respect to this filtration.

The *dual Yangian* $\widehat{\mathrm{Y}}^+(\mathfrak{g}_N)$ is now defined as the quotient of the algebra $\widehat{\mathrm{X}}^+(\mathfrak{g}_N)$ by the relations

$$T^{+\prime}(u+\kappa)\,T^+(u) = 1,$$

where the prime denotes the matrix transposition defined in (2.24).

Now consider the *ascending* filtration on $\widehat{\mathrm{Y}}^+(\mathfrak{g}_N)$ defined by $\deg' t_{ij}^{(-r)} = -r$ for all $r \geqslant 1$. A version of the Poincaré–Birkhoff–Witt theorem can be proved for $\widehat{\mathrm{Y}}^+(\mathfrak{g}_N)$ by using the approach of [**7**] which implies the isomorphism for the associated graded algebra

$$\mathrm{gr}'\,\widehat{\mathrm{Y}}^+(\mathfrak{g}_N) \cong \mathrm{U}\left(t^{-1}\mathfrak{g}_N[t^{-1}]\right).$$

The image $\bar{t}_{ij}^{(-r)}$ of the generator $t_{ij}^{(-r)}$ in the $(-r)$-th component of the graded algebra $\mathrm{gr}'\,\widehat{\mathrm{Y}}^+(\mathfrak{g}_N)$ corresponds to the element $F_{ij}[-r]$ of $\mathrm{U}\left(t^{-1}\mathfrak{g}_N[t^{-1}]\right)$.

We will regard $\widehat{\mathrm{Y}}^+(\mathfrak{g}_N)$ as a module over the Cartan subalgebra $\mathfrak{h}$ of $\mathfrak{g}_N$, where each basis element F_{ii} of $\mathfrak{h}$ with $i = 1,\dots,n$ acts as a derivation and the action is defined on the generators by

$$F_{ii}\cdot t_{kl}^+(u) = \delta_{ki}\,t_{il}^+(u) - \delta_{il}\,t_{ki}^+(u) - \delta_{ki'}\,t_{i'l}^+(u) + \delta_{il'}\,t_{ki'}^+(u).$$

Denote by $\widehat{\mathrm{Y}}^+(\mathfrak{g}_N)^{\mathfrak{h}}$ the subalgebra of $\mathfrak{h}$-invariants under this action. Consider the left ideal J of the algebra $\widehat{\mathrm{Y}}^+(\mathfrak{g}_N)$ generated by all elements $t_{ij}^{(-r)}$ with the conditions $N \geqslant i > j \geqslant 1$ and $r \geqslant 1$. The quotient of $\widehat{\mathrm{Y}}^+(\mathfrak{g}_N)^{\mathfrak{h}}$ by the two-sided ideal $\widehat{\mathrm{Y}}^+(\mathfrak{g}_N)^{\mathfrak{h}} \cap J$ is isomorphic to the commutative algebra freely generated by the images of the elements $t_{ii}^{(-r)}$ with $i = 1,\dots,n$ and $r \geqslant 1$ in the quotient. We will use the notation $\lambda_i^{(-r)}$ for this image of $t_{ii}^{(-r)}$ and extend this notation to all values $i = 1,\dots,N$. An analogue of the Harish-Chandra homomorphism (11.6) now takes the form

$$\widehat{\mathrm{Y}}^+(\mathfrak{g}_N)^{\mathfrak{h}} \to \widehat{\mathbb{C}}[\lambda_i^{(-r)}\,|\, i = 1,\dots,n,\ r \geqslant 1], \tag{11.38}$$

where $\widehat{\mathbb{C}}[\lambda_i^{(-r)}]$ is the completion of the algebra of polynomials with respect to the gradation defined by setting the degree of $\lambda_i^{(-r)}$ to be equal to r. We combine the elements $\lambda_i^{(-r)}$ into the formal series

$$\lambda_i^+(u) = 1 - \sum_{r=1}^{\infty} \lambda_i^{(-r)} u^{r-1}, \qquad i = 1,\dots,N,$$

which will be understood as the images of the series $t_{ii}^+(u)$ under the homomorphism (11.38). We will work with the tensor product algebra

$$\underbrace{\mathrm{End}\,\mathbb{C}^N \otimes \dots \otimes \mathrm{End}\,\mathbb{C}^N}_{m} \otimes \widehat{\mathrm{Y}}^+(\mathfrak{g}_N) \tag{11.39}$$

to state the dual Yangian analogues of the results of Section 11.1. These analogues follow in uniform way in all three cases as we will point out below.

Type B_n. Consider the element $S^{(m)}$ defined by (11.11) and introduce the formal series $\mathbb{T}^{+\,(m)}(u)$ with coefficients in the dual Yangian $\widehat{\mathrm{Y}}^{+}(\mathfrak{o}_N)$ with odd $N = 2n+1$ by the formula

$$\mathbb{T}^{+\,(m)}(u) = \mathrm{tr}_{1,\dots,m}\, S^{(m)} T_1^+(u)\, T_2^+(u+1) \dots T_m^+(u+m-1) \tag{11.40}$$

with the trace taken over all m copies of $\mathrm{End}\,\mathbb{C}^N$ in (11.39).

THEOREM 11.2.1. *The image of the series $\mathbb{T}^{+\,(m)}(u)$ under the homomorphism (11.38) is found by*

$$\mathbb{T}^{+\,(m)}(u) \mapsto \sum_{N \geqslant i_1 \geqslant \dots \geqslant i_m \geqslant 1} \lambda_{i_1}^+(u)\, \lambda_{i_2}^+(u+1) \dots \lambda_{i_m}^+(u+m-1)$$

with the condition that $n+1$ occurs among the summation indices $i_1, \dots, i_m$ at most once.

Type D_n. Define the series $\mathbb{T}^{+\,(m)}(u)$ by the same formula (11.40), where the parameter N now takes an even value $2n$.

THEOREM 11.2.2. *The image of the series $\mathbb{T}^{+\,(m)}(u)$ under the homomorphism (11.38) is given by*

$$\mathbb{T}^{+\,(m)}(u) \mapsto \sum_{N \geqslant i_1 \geqslant \dots \geqslant i_m \geqslant 1} \lambda_{i_1}^+(u)\, \lambda_{i_2}^+(u+1) \dots \lambda_{i_m}^+(u+m-1)$$

with the condition that n and n' do not occur simultaneously among the summation indices $i_1, \dots, i_m$.

Type C_n. Set

$$\mathbb{T}^{+\,(m)}(u) = \mathrm{tr}_{1,\dots,m}\, S^{(m)} T_1^+(u)\, T_2^+(u-1) \dots T_m^+(u-m+1), \tag{11.41}$$

with the trace taken over all m copies of $\mathrm{End}\,\mathbb{C}^N$ in (11.39) with $\mathfrak{g}_N = \mathfrak{sp}_N$ and $S^{(m)}$ defined in (11.12) with $m \leqslant n$.

THEOREM 11.2.3. *The image of the series $\mathbb{T}^{+\,(m)}(u)$ with $m \leqslant n$ under the homomorphism (11.38) is given by*

$$\mathbb{T}^{+\,(m)}(u) \mapsto \sum_{2n \geqslant i_1 > \dots > i_m \geqslant 1} \lambda_{i_1}^+(u)\, \lambda_{i_2}^+(u-1) \dots \lambda_{i_m}^+(u-m+1)$$

with the condition that if for any i both i and i' occur among the summation indices as $i' = i_r$ and $i = i_s$ for some $1 \leqslant r < s \leqslant m$, then $s - r \leqslant n - i$.

To state the dual Yangian counterpart of Corollary 11.1.9, we would like to introduce new parameters $\varkappa_i^+(u)$ for $i = 1, \dots, 2n+2$ by the relations

$$\varkappa_i^+(u) = \lambda_i^+(u), \qquad \varkappa_{2n-i+3}^+(u) = \lambda_{2n-i+1}^+(u) \qquad \text{for} \quad i = 1, \dots, n,$$

together with $\varkappa_{n+1}^+(u)$ and $\varkappa_{n+2}^+(u)$ which are supposed to be defined by the conditions

$$\varkappa_{n+1}^+(u) + \varkappa_{n+2}^+(u) = 0 \quad \text{and} \quad \varkappa_{n+2}^+(u)\, \varkappa_{n+1}^+(u-1) = -\lambda_{n'}^+(u)\, \lambda_n^+(u-1). \tag{11.42}$$

However, unlike $\varkappa_{n+1}(u)$ and $\varkappa_{n+2}(u)$ which are well-defined series in u^{-1}, we have to consider $\varkappa_{n+1}^+(u)$ and $\varkappa_{n+2}^+(u)$ as formal parameters satisfying relations (11.42) and the extensions of these relations for any replacements $u \mapsto u + a$ with $a \in \mathbb{C}$.

COROLLARY 11.2.4. *The image of the series* (11.41) *with* $m \leqslant n$ *under the homomorphism* (11.38) *can be written as*

$$\mathbb{T}^{+(m)}(u) \mapsto \sum_{2n+2\geqslant i_1>\dots>i_m\geqslant 1} \varkappa^+_{i_1}(u)\,\varkappa^+_{i_2}(u-1)\dots\varkappa^+_{i_m}(u-m+1). \tag{11.43}$$

Moreover, the sum in (11.43) *is zero for* $m = n+1$. □

Note that the formal parameters $\varkappa^+_{n+1}(u)$ and $\varkappa^+_{n+2}(u)$ are eliminated from the formula for the image in (11.43) as in the proof of Corollary 11.1.9. Namely, due to (11.42), the terms corresponding to two sets of summation indices which differ only at one index with $i_k = n+2$ and $i_k = n+1$ for some $k \in \{1,\dots,m\}$ cancel with each other, whereas the term corresponding to a set of indices with $i_k = n+2$ and $i_{k+1} = n+1$ is expressed in terms of the $\lambda^+_i(u+a)$.

Theorems 11.2.1, 11.2.2 and 11.2.3 follow from their respective counterparts which we stated in Section 11.1. Namely, the mapping

$$\theta : T(u) \mapsto T^t(-u), \tag{11.44}$$

where $t : e_{ij} \mapsto e_{ji}$ is the standard transposition, defines an automorphism of the extended Yangian $\mathrm{X}(\mathfrak{g}_N)$. This follows by the application of the transposition t_1t_2 to both sides of (11.3) with the use of Lemma 2.2.1 and the following properties of the R-matrix (11.1):

$$R^{t_1t_2}(u) = R(u) \qquad \text{and} \qquad R(u)\,R(-u) = 1 - u^{-2}.$$

Hence, applying the automorphism (11.44), we obtain the image of $\theta\big(\mathbb{T}^{(m)}(u)\big)$ under the Harish-Chandra-type homomorphism, where instead of I we use the left ideal of the extended Yangian generated by all elements $t^{(r)}_{ij}$ with the conditions $N \geqslant i > j \geqslant 1$ and $r \geqslant 1$; see Remark 11.1.10. Since the defining relations take the same matrix form for the extended Yangian and extended dual Yangian, we may conclude that in the orthogonal case the image of the series

$$\mathrm{tr}_{1,\dots,m}\, S^{(m)} T^{+\,t}_1(-u)\, T^{+\,t}_2(-u-1)\dots T^{+\,t}_m(-u-m+1) \tag{11.45}$$

under the homomorphism (11.38) equals

$$\sum_{1\leqslant i_1\leqslant\dots\leqslant i_m\leqslant N} \lambda^+_{i_1}(-u)\,\lambda^+_{i_2}(-u-1)\dots\lambda^+_{i_m}(-u-m+1)$$

with the respective additional conditions on the summation as in Theorems 11.1.2 and 11.1.4. Now apply the transposition $t_1t_2\dots t_m$ to the expression under the trace in (11.45) and observe that it leaves $S^{(m)}$ unchanged. The latter property holds since the application of this transposition to the right hand sides of (11.11) and (11.12) will result in the products written in the opposite order. However, the factors are the values of the R-matrix (11.1) so that the products coincide with $S^{(m)}$ by the Yang–Baxter equation (10.6). Thus, (11.45) equals

$$\begin{aligned}
&\mathrm{tr}_{1,\dots,m}\, T^+_1(-u)\dots T^+_m(-u-m+1)\, S^{(m)}\\
&= \mathrm{tr}_{1,\dots,m}\, T^+_m(-u)\dots T^+_1(-u-m+1)\, S^{(m)}\\
&= \mathrm{tr}_{1,\dots,m}\, S^{(m)}\, T^+_1(-u-m+1)\dots T^+_m(-u) = \mathbb{T}^{+(m)}(-u-m+1),
\end{aligned}$$

where we used the conjugation by the permutation $P_{1\,m}P_{2\,m-1}\cdots$ and repeated application of (11.37). Theorems 11.2.1 and 11.2.2 now follow by the replacement of u with $-u-m+1$.

Theorem 11.2.3 and Corollary 11.2.4 are derived by the same argument from Theorem 11.1.5 and Corollary 11.1.9 with the use of (11.12).

REMARK 11.2.5. The coefficients of all series $\mathbb{T}^{+\,(m)}(u)$ pairwise commute as elements of the dual Yangian. This can be shown with the use of the quantum vertex algebra structure associated with the corresponding double Yangian; see [**37**]. Moreover, this also implies that the coefficients of all series $\mathbb{T}^{(m)}(u)$ pairwise commute as elements of the Yangian $\mathrm{Y}(\mathfrak{g}_N)$; cf. Remark 10.4.2. □

11.3. Screening operators

We will show that the Harish-Chandra images of the series $\mathbb{T}^{(m)}(u)$ belong to the intersection of the kernels of the screening operators and so they coincide with the characters of Yangian representations; cf. Section 10.6.

Suppressing the variable u from the notation $\lambda_i(u+a)$, consider the algebra of polynomials in variables $\lambda_i(a)$ with $i=1,\dots,N$ and $a\in\mathbb{C}$ and denote by $\mathcal{L}=\mathcal{L}(\mathfrak{g}_N)$ its quotient by the relations

$$\lambda_i(a+\kappa-i)\,\lambda_{i'}(a)=\lambda_{i+1}(a+\kappa-i)\,\lambda_{(i+1)'}(a),\qquad a\in\mathbb{C},\tag{11.46}$$

for $i=0,1,\dots,n-1$ if $\mathfrak{g}_N=\mathfrak{o}_{2n}$ or $\mathfrak{sp}_{2n}$, and for $i=0,1,\dots,n$ if $\mathfrak{g}_N=\mathfrak{o}_{2n+1}$, where $\lambda_0(a)=\lambda_{0'}(a)=1$; cf. (11.9).

For $i=1,\dots,n$ consider the free left $\mathcal{L}$-module $\widetilde{\mathcal{L}}_i$ with the generators $\sigma_i(a)$, where a runs over $\mathbb{C}$ and denote by $\mathcal{L}_i$ its quotient by the relations

$$\lambda_i(a)\,\sigma_i(a)=\lambda_{i+1}(a)\,\sigma_i(a+1),\qquad i=1,\dots,n-1,\quad a\in\mathbb{C},$$

together with

$$\begin{aligned}\lambda_n(a)\,\sigma_n(a)&=\lambda_{n+1}(a)\,\sigma_n(a+1/2)&&\text{for}\quad \mathfrak{g}_N=\mathfrak{o}_{2n+1},\\ \lambda_n(a)\,\sigma_n(a)&=\lambda_{n+1}(a)\,\sigma_n(a+2)&&\text{for}\quad \mathfrak{g}_N=\mathfrak{sp}_{2n},\\ \lambda_{n-1}(a)\,\sigma_n(a)&=\lambda_{n+1}(a)\,\sigma_n(a+1)&&\text{for}\quad \mathfrak{g}_N=\mathfrak{o}_{2n}.\end{aligned}\tag{11.47}$$

For every $i\in\{1,\dots,n\}$ define a linear operator $\widetilde{S}_i:\mathcal{L}\to\widetilde{\mathcal{L}}_i$ satisfying the Leibniz rule (10.46). For $i=1,\dots,n-1$ set

$$\widetilde{S}_i:\lambda_j(a)\mapsto\begin{cases}\lambda_i(a)\,\sigma_i(a)&\text{for}\quad j=i\\ -\lambda_{i+1}(a)\,\sigma_i(a+1)&\text{for}\quad j=i+1\\ -\lambda_{i'}(a)\,\sigma_i(a+\kappa-i+1)&\text{for}\quad j=i'\\ \lambda_{(i+1)'}(a)\,\sigma_i(a+\kappa-i)&\text{for}\quad j=(i+1)'\\ 0&\text{for}\quad j\neq i,i',i+1,(i+1)'.\end{cases}$$

The action of $\widetilde{S}_n$ depends on the type and is given as follows.

Case $\mathfrak{g}_N=\mathfrak{o}_{2n+1}$. $\widetilde{S}_n:\lambda_j(a)\mapsto 0$ if $j<n$ or $j>n'$ and

$$\begin{aligned}\lambda_n(a)&\mapsto\lambda_n(a)\,\big(\sigma_n(a)+\sigma_n(a-1/2)\big)\\ \lambda_{n+1}(a)&\mapsto\lambda_{n+1}(a)\,\big(\sigma_n(a-1/2)-\sigma_n(a+1/2)\big)\\ \lambda_{n'}(a)&\mapsto-\lambda_{n'}(a)\,\big(\sigma_n(a)+\sigma_n(a+1/2)\big).\end{aligned}$$

Case $\mathfrak{g}_N = \mathfrak{sp}_{2n}$. $\widetilde{S}_n : \lambda_j(a) \mapsto 0$ if $j < n$ or $j > n'$ and

$$\lambda_n(a) \mapsto \lambda_n(a)\,\sigma_n(a)$$
$$\lambda_{n'}(a) \mapsto -\lambda_{n'}(a)\,\sigma_n(a+2).$$

Case $\mathfrak{g}_N = \mathfrak{o}_{2n}$. $\widetilde{S}_n : \lambda_j(a) \mapsto 0$ if $j < n-1$ or $j > (n-1)'$ and

$$\lambda_{n-1}(a) \mapsto \lambda_{n-1}(a)\,\sigma_n(a)$$
$$\lambda_n(a) \mapsto \lambda_n(a)\,\sigma_n(a)$$
$$\lambda_{n'}(a) \mapsto -\lambda_{n'}(a)\,\sigma_n(a+1)$$
$$\lambda_{(n-1)'}(a) \mapsto -\lambda_{(n-1)'}(a)\,\sigma_n(a+1).$$

Relations (11.46) are easily seen to be preserved by the action of the $\widetilde{S}_i$ so that the operators on $\mathcal{L}$ are well-defined. The *i-th screening operator*

$$S_i : \mathcal{L} \to \mathcal{L}_i$$

is now defined as the composition of $\widetilde{S}_i$ and the projection $\widetilde{\mathcal{L}}_i \to \mathcal{L}_i$. Define the subalgebra Rep $\mathrm{Y}(\mathfrak{g}_N)$ of Yangian characters in $\mathcal{L}$ as the intersection of the kernels of the screening operators:

$$\mathrm{Rep}\ \mathrm{Y}(\mathfrak{g}_N) = \bigcap_{i=1}^{n} \ker S_i.$$

Following Theorem 11.1.2, for $\mathfrak{g}_N = \mathfrak{o}_{2n+1}$ define the polynomial $\chi_m \in \mathcal{L}$ by

$$\chi_m = \sum_{1 \leqslant i_1 \leqslant \dots \leqslant i_m \leqslant N} \lambda_{i_1}(0)\,\lambda_{i_2}(1) \dots \lambda_{i_m}(m-1)$$

with the condition that $n+1$ occurs among the summation indices $i_1, \dots, i_m$ at most once. Similarly, taking Theorem 11.1.4 into account, for $\mathfrak{g}_N = \mathfrak{o}_{2n}$ define the polynomial $\chi_m \in \mathcal{L}$ by

$$\chi_m = \sum_{1 \leqslant i_1 \leqslant \dots \leqslant i_m \leqslant N} \lambda_{i_1}(0)\,\lambda_{i_2}(1) \dots \lambda_{i_m}(m-1)$$

with the condition that n and n' do not occur simultaneously among the summation indices $i_1, \dots, i_m$. Following Theorem 11.1.5, for $\mathfrak{g}_N = \mathfrak{sp}_{2n}$ define the polynomial $\chi_m \in \mathcal{L}$ by

$$\chi_m = \sum_{1 \leqslant i_1 < \dots < i_m \leqslant 2n} \lambda_{i_1}(0)\,\lambda_{i_2}(-1) \dots \lambda_{i_m}(-m+1)$$

with the condition that if for any i both i and i' occur among the summation indices as $i = i_r$ and $i' = i_s$ for some $1 \leqslant r < s \leqslant m$, then $s - r \leqslant n - i$. Also, using the notation of Corollary 11.1.9, we have

$$\chi_m = \sum_{1 \leqslant i_1 < \dots < i_m \leqslant 2n+2} \varkappa_{i_1}(0)\,\varkappa_{i_2}(-1) \dots \varkappa_{i_m}(-m+1).$$

This formula defines χ_m for all values $m = 1, 2, \dots, 2n+2$.

PROPOSITION 11.3.1. *In all three cases we have* $\chi_m \in \mathrm{Rep}\ \mathrm{Y}(\mathfrak{g}_N)$.

PROOF. A calculation quite analogous to the one used in the proof of Proposition 10.6.1 (see also Remark 10.6.2) shows that χ_m belongs to the kernel of the screening operator S_i for each $i = 1, \dots, n-1$. To verify this property for S_n we will consider the three cases separately.

Case $\mathfrak{g}_N = \mathfrak{o}_{2n+1}$. Using the notation of Remark 10.6.2, combine the polynomials χ_m into the generating series

$$1 + \sum_{m=1}^{\infty} \chi_m \, e^{m\,\partial} = \big(1 - \lambda_1(0)\, e^{\partial}\big)^{-1} \dots \big(1 - \lambda_n(0)\, e^{\partial}\big)^{-1}$$
$$\times \big(1 + \lambda_{n+1}(0)\, e^{\partial}\big)\big(1 - \lambda_{n'}(0)\, e^{\partial}\big)^{-1} \dots \big(1 - \lambda_{1'}(0)\, e^{\partial}\big)^{-1}.$$

It suffices to verify the desired property for the inverse series so that we need to see that the product

$$\big(1 - \lambda_{n'}(0)\, e^{\partial}\big)\big(1 + \lambda_{n+1}(0)\, e^{\partial}\big)^{-1}\big(1 - \lambda_n(0)\, e^{\partial}\big) \tag{11.48}$$

is annihilated by S_n. We have

$$S_n : \big(1 + \lambda_{n+1}(0)\, e^{\partial}\big)^{-1}$$
$$\mapsto \sigma_n(-1/2)\, \big(1 + \lambda_{n+1}(0)\, e^{\partial}\big)^{-1} - \big(1 + \lambda_{n+1}(0)\, e^{\partial}\big)^{-1} \sigma_n(-1/2).$$

Hence the image of (11.48) under S_n is

$$\lambda_{n'}(0)\big(\sigma_n(0) + \sigma_n(1/2)\big)\, e^{\partial}\big(1 + \lambda_{n+1}(0)\, e^{\partial}\big)^{-1}\big(1 - \lambda_n(0)\, e^{\partial}\big)$$
$$+ \big(1 - \lambda_{n'}(0)\, e^{\partial}\big)\Big(\sigma_n(-1/2)\, \big(1 + \lambda_{n+1}(0)\, e^{\partial}\big)^{-1}$$
$$- \big(1 + \lambda_{n+1}(0)\, e^{\partial}\big)^{-1}\sigma_n(-1/2)\Big)\big(1 - \lambda_n(0)\, e^{\partial}\big)$$
$$- \big(1 - \lambda_{n'}(0)\, e^{\partial}\big)\big(1 + \lambda_{n+1}(0)\, e^{\partial}\big)^{-1}\, \lambda_n(0)\big(\sigma_n(0) + \sigma_n(-1/2)\big)\, e^{\partial}$$

which simplifies to

$$\big(\lambda_{n'}(0)\, \sigma_n(0)\, e^{\partial} + \sigma_n(-1/2)\big)\, \big(1 + \lambda_{n+1}(0)\, e^{\partial}\big)^{-1}\big(1 - \lambda_n(0)\, e^{\partial}\big)$$
$$- \big(1 - \lambda_{n'}(0)\, e^{\partial}\big)\big(1 + \lambda_{n+1}(0)\, e^{\partial}\big)^{-1}\, \big(\lambda_n(0)\, \sigma_n(0)\, e^{\partial} + \sigma_n(-1/2)\big).$$

Using (11.46) for $i = n$ and (11.47) write

$$\lambda_{n'}(0)\, \sigma_n(0) = \lambda_{n+1}(0)\, \sigma_n(-1/2) \quad \text{and} \quad \lambda_n(0)\, \sigma_n(0) = \lambda_{n+1}(0)\, \sigma_n(1/2) \tag{11.49}$$

to simplify the image further to

$$\sigma_n(-1/2)\, \big(1 - \lambda_n(0)\, e^{\partial}\big) - \big(1 - \lambda_{n'}(0)\, e^{\partial}\big)\, \sigma_n(-1/2).$$

This is zero due to the relation $\lambda_n(0)\, \sigma_n(-1/2) = \lambda_{n'}(0)\, \sigma_n(1/2)$ implied by (11.49).

Case $\mathfrak{g}_N = \mathfrak{o}_{2n}$. We have the generating series

$$1 + \sum_{m=1}^{\infty} \chi_m \, e^{m\,\partial} = \big(1 - \lambda_1(0)\, e^{\partial}\big)^{-1} \dots \big(1 - \lambda_{n-1}(0)\, e^{\partial}\big)^{-1}$$

$$\times\Big(\big(1-\lambda_n(0)\, e^{\partial}\big)^{-1} + \big(1-\lambda_{n'}(0)\, e^{\partial}\big)^{-1} - 1\Big)\big(1-\lambda_{(n-1)'}(0)\, e^{\partial}\big)^{-1} \dots \big(1-\lambda_{1'}(0)\, e^{\partial}\big)^{-1}$$

which can also be written as

$$\big(1 - \lambda_1(0)\, e^{\partial}\big)^{-1} \dots \big(1 - \lambda_n(0)\, e^{\partial}\big)^{-1}$$
$$\times \big(1 - \lambda_n(0)\, e^{\partial}\, \lambda_{n'}(0)\, e^{\partial}\big)\big(1 - \lambda_{n'}(0)\, e^{\partial}\big)^{-1} \dots \big(1 - \lambda_{1'}(0)\, e^{\partial}\big)^{-1}.$$

By taking the inverse series, we come to verifying that the product

$$\big(1 - \lambda_{(n-1)'}(0)\, e^{\partial}\big)\big(1 - \lambda_{n'}(0)\, e^{\partial}\big)$$
$$\times \big(1 - \lambda_n(0)\, e^{\partial}\, \lambda_{n'}(0)\, e^{\partial}\big)^{-1}\big(1 - \lambda_n(0)\, e^{\partial}\big)\big(1 - \lambda_{n-1}(0)\, e^{\partial}\big)$$

is annihilated by S_n. This follows easily by using

$$S_n : \big(1 - \lambda_n(0)\, e^{\partial}\, \lambda_{n'}(0)\, e^{\partial}\big)^{-1}$$
$$\mapsto \sigma_n(0)\, \big(1 - \lambda_n(0)\, e^{\partial}\, \lambda_{n'}(0)\, e^{\partial}\big)^{-1} - \big(1 - \lambda_n(0)\, e^{\partial}\, \lambda_{n'}(0)\, e^{\partial}\big)^{-1} \sigma_n(0)$$

and the relations

$$\lambda_{(n-1)'}(0)\, \sigma_n(1) = \lambda_n(0)\, \sigma_n(0) \qquad \text{and} \qquad \lambda_{n-1}(0)\, \sigma_n(0) = \lambda_{n'}(0)\, \sigma_n(1).$$

Case $\mathfrak{g}_N = \mathfrak{sp}_{2n}$. We have

$$1 + \sum_{m=1}^{2n+2} \chi_m \, e^{-m\,\partial} = \big(1 + \varkappa_1(0)\, e^{-\partial}\big) \dots \big(1 + \varkappa_{2n+2}(0)\, e^{-\partial}\big).$$

It suffices to verify that the product

$$\big(1 + \varkappa_n(0)\, e^{-\partial}\big)\big(1 + \varkappa_{n+1}(0)\, e^{-\partial}\big)\big(1 + \varkappa_{n+2}(0)\, e^{-\partial}\big)\big(1 + \varkappa_{n+3}(0)\, e^{-\partial}\big)$$
$$= \big(1 + \lambda_n(0)\, e^{-\partial}\big)\big(1 - \lambda_n(0)\, e^{-\partial}\, \lambda_{n'}(0)\, e^{-\partial}\big)\big(1 + \lambda_{n'}(0)\, e^{-\partial}\big)$$

is annihilated by S_n. This follows by an easy calculation with the use of the relation

$$\lambda_n(0)\, \sigma_n(0) = \lambda_{n'}(0)\, \sigma_n(2)$$

thus completing the proof. □

To state the counterpart of Proposition 11.3.1 for the dual Yangian (see Proposition 11.3.2 below), note that the Harish-Chandra images calculated in Sections 11.1 and 11.2 can be obtained from each other by the replacements

$$\lambda_i(u) \longleftrightarrow \lambda^+_{i'}(u), \qquad i = 1, \dots, N.$$

Therefore, the above definitions and properties of the screening operators associated with the Yangian $\mathrm{Y}(\mathfrak{g}_N)$ are easily carried over to the dual Yangian by replacing the variables as follows:

$$\lambda_i(a) \mapsto \lambda^+_{i'}(a), \qquad i = 1, \dots, N,$$

and

$$\sigma_i(a) \mapsto \sigma^+_i(a - \kappa + i), \qquad i = 1, \dots, n,$$

for all $a \in \mathbb{C}$. Now we consider the algebra of polynomials in variables $\lambda_i^+(a)$ with $i = 1, \dots, N$ and $a \in \mathbb{C}$ and denote by $\mathcal{L}^+ = \mathcal{L}^+(\mathfrak{g}_N)$ its quotient by the relations

$$\lambda_i^+(a)\,\lambda_{i'}^+(a+\kappa-i) = \lambda_{i+1}^+(a)\,\lambda_{(i+1)'}^+(a+\kappa-i), \qquad a \in \mathbb{C}, \tag{11.50}$$

for $i = 0, 1, \dots, n-1$ if $\mathfrak{g}_N = \mathfrak{o}_{2n}$ or $\mathfrak{sp}_{2n}$, and for $i = 0, 1, \dots, n$ if $\mathfrak{g}_N = \mathfrak{o}_{2n+1}$, where $\lambda_0^+(a) = \lambda_{0'}^+(a) = 1$; cf. (11.46).

For $i = 1, \dots, n$ consider the free left $\mathcal{L}^+$-module $\widetilde{\mathcal{L}}_i^+$ with the generators $\sigma_i^+(a)$, where a runs over $\mathbb{C}$ and denote by $\mathcal{L}_i^+$ its quotient by the relations

$$\lambda_i^+(a)\,\sigma_i^+(a+1) = \lambda_{i+1}^+(a)\,\sigma_i^+(a), \qquad i = 1, \dots, n-1, \quad a \in \mathbb{C}, \tag{11.51}$$

together with[1]

$$\begin{aligned} \lambda_n^+(a)\,\sigma_n^+(a+1/2) &= \lambda_{n+1}^+(a)\,\sigma_n^+(a) && \text{for} \quad \mathfrak{g}_N = \mathfrak{o}_{2n+1}, \\ \lambda_n^+(a)\,\sigma_n^+(a+1) &= \lambda_{n+1}^+(a)\,\sigma_n^+(a-1) && \text{for} \quad \mathfrak{g}_N = \mathfrak{sp}_{2n}, \\ \lambda_{n-1}^+(a)\,\sigma_n^+(a+2) &= \lambda_{n+1}^+(a)\,\sigma_n^+(a+1) && \text{for} \quad \mathfrak{g}_N = \mathfrak{o}_{2n}. \end{aligned} \tag{11.52}$$

For every $i \in \{1, \dots, n\}$ define a linear operator $\widetilde{S}_i^+ : \mathcal{L}^+ \to \widetilde{\mathcal{L}}_i^+$ satisfying the Leibniz rule (10.52). For $i = 1, \dots, n-1$ set

$$\widetilde{S}_i^+ : \lambda_j^+(a) \mapsto \begin{cases} -\lambda_i^+(a)\,\sigma_i^+(a+1) & \text{for} \quad j = i \\ \lambda_{i+1}^+(a)\,\sigma_i^+(a) & \text{for} \quad j = i+1 \\ \lambda_{i'}^+(a)\,\sigma_i^+(a-\kappa+i) & \text{for} \quad j = i' \\ -\lambda_{(i+1)'}^+(a)\,\sigma_i^+(a-\kappa+i+1) & \text{for} \quad j = (i+1)' \\ 0 & \text{for} \quad j \neq i, i', i+1, (i+1)'. \end{cases}$$

The action of $\widetilde{S}_n^+$ is defined separately for each type.

Case $\mathfrak{g}_N = \mathfrak{o}_{2n+1}$. $\widetilde{S}_n^+ : \lambda_j^+(a) \mapsto 0$ if $j < n$ or $j > n'$ and

$$\begin{aligned} \lambda_n^+(a) &\mapsto -\lambda_n^+(a)\,\big(\sigma_n^+(a+1/2) + \sigma_n^+(a+1)\big) \\ \lambda_{n+1}^+(a) &\mapsto \lambda_{n+1}^+(a)\,\big(\sigma_n^+(a) - \sigma_n^+(a+1)\big) \\ \lambda_{n'}^+(a) &\mapsto \lambda_{n'}^+(a)\,\big(\sigma_n^+(a) + \sigma_n^+(a+1/2)\big). \end{aligned}$$

Case $\mathfrak{g}_N = \mathfrak{sp}_{2n}$. $\widetilde{S}_n^+ : \lambda_j^+(a) \mapsto 0$ if $j < n$ or $j > n'$ and

$$\begin{aligned} \lambda_n^+(a) &\mapsto -\lambda_n^+(a)\,\sigma_n^+(a+1) \\ \lambda_{n'}^+(a) &\mapsto \lambda_{n'}^+(a)\,\sigma_n^+(a-1). \end{aligned}$$

Case $\mathfrak{g}_N = \mathfrak{o}_{2n}$. $\widetilde{S}_n^+ : \lambda_j^+(a) \mapsto 0$ if $j < n-1$ or $j > (n-1)'$ and

$$\begin{aligned} \lambda_{n-1}^+(a) &\mapsto -\lambda_{n-1}^+(a)\,\sigma_n^+(a+2) \\ \lambda_n^+(a) &\mapsto -\lambda_n^+(a)\,\sigma_n^+(a+2) \\ \lambda_{n'}^+(a) &\mapsto \lambda_{n'}^+(a)\,\sigma_n^+(a+1) \\ \lambda_{(n-1)'}^+(a) &\mapsto \lambda_{(n-1)'}^+(a)\,\sigma_n^+(a+1). \end{aligned}$$

[1] One could replace $\sigma_n^+(a) \mapsto \sigma_n^+(a \pm 1)$ for types C_n and D_n, respectively, to make the conditions analogous to (11.47). This would not affect the properties which we discuss below.

One easily verifies that the action of the operators $\widetilde{S}_i^+$ on $\mathcal{L}^+$ is well-defined. The *i-th screening operator*

$$S_i^+ : \mathcal{L}^+ \to \mathcal{L}_i^+$$

is now defined as the composition of $\widetilde{S}_i^+$ and the projection $\widetilde{\mathcal{L}}_i^+ \to \mathcal{L}_i^+$. Define the subalgebra $\operatorname{Rep}\widehat{\mathrm{Y}}^+(\mathfrak{g}_N)$ of Yangian characters in $\mathcal{L}^+$ as the intersection of the kernels of the screening operators:

$$\operatorname{Rep}\widehat{\mathrm{Y}}^+(\mathfrak{g}_N) = \bigcap_{i=1}^{n} \ker S_i^+. \tag{11.53}$$

Following Theorem 11.2.1, for $\mathfrak{g}_N = \mathfrak{o}_{2n+1}$ define the polynomial $\chi_m^+ \in \mathcal{L}^+$ by

$$\chi_m^+ = \sum_{N \geqslant i_1 \geqslant \dots \geqslant i_m \geqslant 1} \lambda_{i_1}^+(0)\, \lambda_{i_2}^+(1) \dots \lambda_{i_m}^+(m-1)$$

with the condition that $n+1$ occurs among the summation indices $i_1, \dots, i_m$ at most once. Taking Theorem 11.2.2 into account, for $\mathfrak{g}_N = \mathfrak{o}_{2n}$ define the polynomial $\chi_m^+ \in \mathcal{L}^+$ by

$$\chi_m^+ = \sum_{N \geqslant i_1 \geqslant \dots \geqslant i_m \geqslant 1} \lambda_{i_1}^+(0)\, \lambda_{i_2}^+(1) \dots \lambda_{i_m}^+(m-1)$$

with the condition that n and n' do not occur simultaneously among the summation indices $i_1, \dots, i_m$. Following Theorem 11.2.3, for $\mathfrak{g}_N = \mathfrak{sp}_{2n}$ define the polynomial $\chi_m^+ \in \mathcal{L}$ by

$$\chi_m^+ = \sum_{2n \geqslant i_1 > \dots > i_m \geqslant 1} \lambda_{i_1}^+(0)\, \lambda_{i_2}^+(-1) \dots \lambda_{i_m}^+(-m+1)$$

with the condition that if for any i both i and i' occur among the summation indices as $i' = i_r$ and $i = i_s$ for some $1 \leqslant r < s \leqslant m$, then $s - r \leqslant n - i$. By Corollary 11.2.4, we also have

$$\chi_m^+ = \sum_{2n+2 \geqslant i_1 > \dots > i_m \geqslant 1} \varkappa_{i_1}^+(0)\, \varkappa_{i_2}^+(-1) \dots \varkappa_{i_m}^+(-m+1).$$

This allows us to extend the definition of χ_m^+ to all values $m = 1, 2, \dots, 2n+2$.

PROPOSITION 11.3.2. *In all three cases we have* $\chi_m^+ \in \operatorname{Rep}\widehat{\mathrm{Y}}^+(\mathfrak{g}_N)$. □

11.4. Bibliographical notes

The *RTT* presentation of the Yangian $\mathrm{Y}(\mathfrak{g}_N)$ and its representations were studied in the papers by Arnaudon *et al.* [**6**], [**7**] and Guay *et al.* [**61**], [**62**]. Explicit isomorphisms between various presentations of $\mathrm{Y}(\mathfrak{g}_N)$ were constructed recently in [**63**] and [**83**]. In Section 11.1 we followed [**111**]. The proofs of Lemma 11.1.6 and Corollary 11.1.9 are taken from the paper by Kuniba, Okado, Suzuki and Yamada [**100**]. The screening operators originate in the paper by Frenkel and Reshetikhin [**51**]; see also Frenkel and Mukhin [**48**]. We followed [**113**] for explicit formulas in the Yangian context.

CHAPTER 12

Classical $\mathcal{W}$-algebras

Classical $\mathcal{W}$-algebras $\mathcal{W}(\mathfrak{g}, f)$ are commutative algebras equipped with Poisson brackets and derivations. The word 'classical' refers to the fact that they are 'classical limits' of a certain family of vertex algebras known as the *affine $\mathcal{W}$-algebras* so that these limits possess the structures of *Poisson vertex algebras*. The algebras $\mathcal{W}(\mathfrak{g}, f)$ are associated with finite-dimensional simple Lie algebras $\mathfrak{g}$ over $\mathbb{C}$ and nilpotent elements $f \in \mathfrak{g}$. For the applications below, we will only be interested in the case where f is a regular (principal) nilpotent and therefore suppress f from the notation by denoting the algebra by $\mathcal{W}(\mathfrak{g})$.

Recall that the center $\mathrm{Z}(\mathfrak{g})$ of the universal enveloping algebra is isomorphic to the subalgebra $\mathrm{U}(\mathfrak{h})^W$ of W-invariants in $\mathrm{U}(\mathfrak{h})$; see (4.7). An affine version of the Harish-Chandra isomorphism between the center $\mathfrak{z}(\widehat{\mathfrak{g}})$ of the affine vertex algebra at the critical level (as defined in Chapter 6) and the classical $\mathcal{W}$-algebra $\mathcal{W}({}^L\mathfrak{g})$ will be discussed in Chapter 13 below (by ${}^L\mathfrak{g}$ we denote the Langlands dual Lie algebra of $\mathfrak{g}$). This chapter is meant to review some properties of the algebras $\mathcal{W}(\mathfrak{g})$ and construct their generators. We begin with a description of the classical $\mathcal{W}$-algebras in the framework of Poisson vertex algebras following De Sole, Kac and Valeri [**26**], [**27**], [**28**] and Kac [**88**] (see Sections 12.1 and 12.2 below), then give their definition via screening operators and show the equivalence of the definitions via the Chevalley isomorphism (Section 12.3).

12.1. Poisson vertex algebras

We will use the term *differential algebra* for a unital commutative associative algebra $\mathcal{V}$ equipped with a derivation ∂. Recall also Lie conformal algebras as introduced in Definition 6.1.5.

Definition 12.1.1. A *Poisson vertex algebra* is a differential algebra $\mathcal{V} = (\mathcal{V}, \partial)$ endowed with a $\mathbb{C}$-linear map

$$\mathcal{V} \otimes \mathcal{V} \to \mathbb{C}[\lambda] \otimes \mathcal{V}, \qquad a \otimes b \mapsto \{a_\lambda b\},$$

called the *λ-bracket*, such that $(\mathcal{V}, \partial, \{_\lambda\})$ is a Lie conformal algebra and the *Leibniz rule*

$$\{a_\lambda bc\} = \{a_\lambda b\}c + b\{a_\lambda c\} \tag{12.1}$$

holds for all $a, b, c \in \mathcal{V}$. □

Suppose that $\mathfrak{g}$ is a simple Lie algebra and let $J_1, \dots, J_d$ be a basis of $\mathfrak{g}$. Consider the differential algebra $\mathcal{V}(\mathfrak{g})$ which is defined as the algebra of polynomials

$$\mathcal{V}(\mathfrak{g}) = \mathbb{C}[J_1^{(r)}, \dots, J_d^{(r)} \mid r = 0, 1, 2, \dots] \qquad \text{with} \quad J_i^{(0)} = J_i, \tag{12.2}$$

equipped with the derivation ∂ defined by $\partial(J_i^{(r)}) = J_i^{(r+1)}$ for all $i = 1, \dots, d$ and $r \geqslant 0$. Fix a non-degenerate symmetric invariant bilinear form $\langle \, , \rangle$ on $\mathfrak{g}$ and introduce the λ-bracket on $\mathcal{V}(\mathfrak{g})$ by setting

$$\{X_\lambda Y\} = [X, Y] + \lambda \langle X, Y \rangle \qquad \text{for} \quad X, Y \in \mathfrak{g},$$

and extending to $\mathcal{V}(\mathfrak{g})$ by sesquilinearity $\big(a, b \in \mathcal{V}(\mathfrak{g})\big)$:

$$\{\partial a_\lambda b\} = -\lambda \{a_\lambda b\}, \qquad \{a_\lambda \partial b\} = (\lambda + \partial)\{a_\lambda b\},$$

skewsymmetry

$$\{a_\lambda b\} = -\{b_{-\lambda-\partial}\, a\},$$

and the Leibniz rule (12.1); cf. Definition 6.1.5. One verifies that this λ-bracket equips the differential algebra $(\mathcal{V}(\mathfrak{g}), \partial)$ with a Poisson vertex algebra structure and $\mathcal{V}(\mathfrak{g})$ is known as the *affine Poisson vertex algebra* (associated with $\mathfrak{g}$).

REMARK 12.1.2. The affine Poisson vertex algebra $\mathcal{V}(\mathfrak{g})$ can be shown to be obtained from the affine vertex algebra $V_\kappa(\mathfrak{g})$ defined in Section 6.2 via a *quasiclassical limit*; see [**88**]. □

Introduce the standard *Chevalley generators* e_i, h_i, f_i with $i = 1, \dots, n$ of the simple Lie algebra $\mathfrak{g}$ of rank n. The linear span $\mathfrak{h}$ of the generators $h_1, \dots, h_n$ is a Cartan subalgebra of $\mathfrak{g}$, while the e_i and f_i generate the respective nilpotent subalgebras $\mathfrak{n}_+$ and $\mathfrak{n}_-$. Let $A = [a_{ij}]$ be the Cartan matrix of $\mathfrak{g}$ so that the defining relations of $\mathfrak{g}$ take the form

$$[e_i, f_j] = \delta_{ij} h_i, \qquad [h_i, h_j] = 0,$$
$$[h_i, e_j] = a_{ij}\, e_j, \qquad [h_i, f_j] = -a_{ij}\, f_j,$$

together with the Serre relations

$$(\operatorname{ad} e_i)^{1-a_{ij}}\, e_j = 0, \qquad (\operatorname{ad} f_i)^{1-a_{ij}}\, f_j = 0, \qquad i \neq j.$$

There exists a diagonal matrix

$$D = \operatorname{diag}[\epsilon_1, \dots, \epsilon_n] \tag{12.3}$$

with positive rational entries such that the matrix $B = D^{-1}A$ is symmetric. Normalize the symmetric invariant bilinear form on $\mathfrak{g}$ so that

$$\langle e_i, f_j \rangle = \delta_{ij}\, \epsilon_i. \tag{12.4}$$

The element $f \in \mathfrak{g}$ defined by

$$f = f_1 + \dots + f_n$$

is a regular nilpotent element of $\mathfrak{g}$. Introduce an $\mathfrak{sl}_2$-subalgebra of $\mathfrak{g}$ with the standard basis elements e, f, h satisfying the commutation relations (6.12) and set $x = h/2$. Consider the $\operatorname{ad} x$-eigenspace decomposition of $\mathfrak{g}$,

$$\mathfrak{g} = \bigoplus_{i \in \mathbb{Z}} \mathfrak{g}^{(i)}, \qquad \mathfrak{g}^{(i)} = \{Y \in \mathfrak{g} \mid [x, Y] = iY\}. \tag{12.5}$$

As with (9.19), the subspace $\mathfrak{g}^{(i)}$ is zero unless i is within the region determined by the Coxeter number: $-h_{\mathfrak{g}} < i < h_{\mathfrak{g}}$. Note that $f \in \mathfrak{g}^{(-1)}$, $e \in \mathfrak{g}^{(1)}$ and $h \in \mathfrak{g}^{(0)}$. Moreover, the Cartan subalgebra $\mathfrak{h}$ coincides with $\mathfrak{g}^{(0)}$ and we have a triangular decomposition $\mathfrak{g} = \mathfrak{n}_- \oplus \mathfrak{h} \oplus \mathfrak{n}_+$ with

$$\mathfrak{n}_- = \bigoplus_{i<0} \mathfrak{g}^{(i)} \qquad \text{and} \qquad \mathfrak{n}_+ = \bigoplus_{i>0} \mathfrak{g}^{(i)}.$$

Set $\mathfrak{p} = \mathfrak{n}_- \oplus \mathfrak{h}$ and introduce the projection map $\pi_\mathfrak{p} : \mathfrak{g} \to \mathfrak{p}$ with the kernel $\mathfrak{n}_+$.

We will extend the notation (12.2) to an arbitrary subspace $\mathfrak{a} \subset \mathfrak{g}$ so that $\mathcal{V}(\mathfrak{a})$ will denote the corresponding differential subalgebra of $\mathcal{V}(\mathfrak{g})$. Elements of $\mathcal{V}(\mathfrak{a})$ are polynomials in the variables $Y^{(r)}$ with $Y \in \mathfrak{a}$ and $r \geqslant 0$. Define the differential algebra homomorphism

$$\rho : \mathcal{V}(\mathfrak{g}) \to \mathcal{V}(\mathfrak{p})$$

by setting

$$\rho(X) = \pi_\mathfrak{p}(X) + \langle f, X \rangle, \qquad X \in \mathfrak{g}. \tag{12.6}$$

DEFINITION 12.1.3. The *classical $\mathcal{W}$-algebra* $\mathcal{W}(\mathfrak{g})$ is defined by

$$\mathcal{W}(\mathfrak{g}) = \{P \in \mathcal{V}(\mathfrak{p}) \mid \rho\{X_\lambda P\} = 0 \quad \text{for all} \quad X \in \mathfrak{n}_+\}. \qquad \square$$

THEOREM 12.1.4. *The subspace $\mathcal{W}(\mathfrak{g}) \subset \mathcal{V}(\mathfrak{p})$ is a differential subalgebra of $\mathcal{V}(\mathfrak{p})$. Moreover, $\mathcal{W}(\mathfrak{g})$ is a Poisson vertex algebra equipped with the λ-bracket*

$$\{a_\lambda b\}_\rho = \rho\{a_\lambda b\}, \qquad a, b \in \mathcal{W}(\mathfrak{g}). \qquad \square$$

To describe the structure of the classical $\mathcal{W}$-algebra $\mathcal{W}(\mathfrak{g})$, choose a subspace $U \subset \mathfrak{g}$ complementary to $[f, \mathfrak{g}]$,

$$\mathfrak{g} = [f, \mathfrak{g}] \oplus U, \tag{12.7}$$

which is compatible with the grading (12.5) in the sense that U is the direct sum of the subspaces $U \cap \mathfrak{g}^{(i)}$. Since

$$\mathfrak{g} = [f, \mathfrak{g}] \oplus \mathfrak{g}^e, \tag{12.8}$$

a possible universal choice is $U = \mathfrak{g}^e$, the centralizer of e in $\mathfrak{g}$. Observe that the map $\operatorname{ad} f : \mathfrak{g}^{(i)} \to \mathfrak{g}^{(i-1)}$ is surjective for $i \leqslant 0$ so that we have the inclusion $\mathfrak{n}_- \subset [f, \mathfrak{g}]$. Therefore, for any choice of U we have $U \subset \mathfrak{h} \oplus \mathfrak{n}_+$. Since the form (12.4) is invariant and nondegenerate, the orthogonal complement to $[f, \mathfrak{g}]$ in $\mathfrak{g}$ coincides with the centralizer $\mathfrak{g}^f$ of f in $\mathfrak{g}$. Hence, by taking the orthogonal complements in the decomposition (12.7) we get

$$\mathfrak{g} = \mathfrak{g}^f \oplus U_\mathfrak{g}^\perp.$$

Note that the centralizer $\mathfrak{g}^f$ is contained in $\mathfrak{p}$ so that this implies the decomposition

$$\mathfrak{p} = \mathfrak{g}^f \oplus U^\perp,$$

where we set

$$U^\perp = \mathfrak{p} \cap U_\mathfrak{g}^\perp = \{v \in \mathfrak{p} \mid \langle v, U \rangle = 0\}. \tag{12.9}$$

This induces another direct sum decomposition,

$$\mathcal{V}(\mathfrak{p}) = \mathcal{V}(\mathfrak{g}^f) \oplus \langle U^\perp \rangle, \tag{12.10}$$

where $\langle U^\perp \rangle$ denotes the differential algebra ideal of $\mathcal{V}(\mathfrak{p})$ generated by $U^\perp$. We let $\pi = \pi_{\mathfrak{g}^f}$ denote the projection

$$\pi : \mathcal{V}(\mathfrak{p}) \to \mathcal{V}(\mathfrak{g}^f)$$

to the first summand in (12.10) with the kernel $\langle U^\perp \rangle$.

THEOREM 12.1.5. *The restriction of π to the classical $\mathcal{W}$-algebra $\mathcal{W}(\mathfrak{g}) \subset \mathcal{V}(\mathfrak{p})$ defines a differential algebra isomorphism*

$$\pi : \mathcal{W}(\mathfrak{g}) \to \mathcal{V}(\mathfrak{g}^f). \tag{12.11} \qquad \square$$

Theorem 12.1.5 implies that for any element $q \in \mathfrak{g}^f$ there exists a unique element $w(q) \in \mathcal{W}(\mathfrak{g})$ of the form $w(q) = q + r$ with $r \in \langle U^\perp \rangle$.

12.2. Generators of $\mathcal{W}(\mathfrak{g})$

We will need some general determinant formulas to be used for all classical types. Let A be a square matrix of the form

$$A = \begin{bmatrix} a_{11} & a_{12} & 0 & 0 & \dots & 0 \\ a_{21} & a_{22} & a_{23} & 0 & \dots & 0 \\ a_{31} & a_{32} & a_{33} & a_{34} & \dots & 0 \\ \dots & \dots & \dots & \dots & \dots & \dots \\ a_{N-1\,1} & a_{N-1\,2} & a_{N-1\,3} & \dots & \dots & a_{N-1\,N} \\ a_{N1} & a_{N2} & a_{N3} & \dots & \dots & a_{NN} \end{bmatrix} \tag{12.12}$$

with entries in a ring. If the entries $a_{12}, a_{23}, \dots, a_{N-1\,N}$ belong to the center of the ring then one easily verifies that the column-determinant of A coincides with its row-determinant and we will set

$$\det A = \sum_{\sigma \in \mathfrak{S}_N} \operatorname{sgn} \sigma \cdot a_{\sigma(1)\,1} \dots a_{\sigma(N)\,N} = \sum_{\sigma \in \mathfrak{S}_N} \operatorname{sgn} \sigma \cdot a_{1\,\sigma(1)} \dots a_{N\,\sigma(N)}. \tag{12.13}$$

In particular, if the ring contains the identity and the matrix A has the form

$$A = \begin{bmatrix} a_{11} & 1 & 0 & 0 & \dots & 0 \\ a_{21} & a_{22} & 1 & 0 & \dots & 0 \\ a_{31} & a_{32} & a_{33} & 1 & \dots & 0 \\ \dots & \dots & \dots & \dots & \dots & \dots \\ a_{N-1\,1} & a_{N-1\,2} & a_{N-1\,3} & \dots & \dots & 1 \\ a_{N1} & a_{N2} & a_{N3} & \dots & \dots & a_{NN} \end{bmatrix} \tag{12.14}$$

then

$$\det A = \sum_{k=0}^{N-1} \sum_{0=i_0<i_1<\dots<i_{k+1}=N} (-1)^{N-k-1} a_{i_1\,i_0+1} a_{i_2\,i_1+1} \dots a_{i_{k+1}\,i_k+1}. \tag{12.15}$$

Returning to more general matrices (12.12), denote by D_i (respectively, $\overline{D}_i$) the determinant of the $i \times i$ submatrix of A corresponding to the first (respectively, last) i rows and columns. We suppose that $D_0 = \overline{D}_0 = 1$.

LEMMA 12.2.1. *Fix $p \in \{0, 1, \dots, N\}$. Then for the determinant of the matrix* (12.12) *we have*

$$\det A = D_p\,\overline{D}_{N-p} + \sum_{j=1}^{p} \sum_{i=p+1}^{N} (-1)^{j+i} D_{j-1}\,\widetilde{a}_{ij}\,\overline{D}_{N-i},$$

where

$$\widetilde{a}_{ij} = a_{ij}\,a_{j\,j+1}\,a_{j+1\,j+2} \dots a_{i-1\,i} \qquad \text{for} \quad i > j.$$

PROOF. The formula follows easily from the definition of the determinant (12.13). In the applications which we consider below, the central elements a_{12}, $a_{23}, \dots, a_{N-1\,N}$ turn out to be invertible. Then the lemma is reduced to the particular case of matrices (12.14) and it is immediate from the explicit formula (12.15). □

Type A_{N-1}. It will be convenient to work with the reductive Lie algebra $\mathfrak{g} = \mathfrak{gl}_N$ instead of the simple Lie algebra $\mathfrak{sl}_N$ in type A. We will use its standard basis elements E_{ij}, $i, j = 1, \dots, N$. The elements $E_{11}, \dots, E_{NN}$ span a Cartan subalgebra of $\mathfrak{gl}_N$ which we denote by $\mathfrak{h}$. The respective subsets of basis elements E_{ij} with $i < j$ and $i > j$ span the nilpotent subalgebras $\mathfrak{n}_+$ and $\mathfrak{n}_-$. The subalgebra $\mathfrak{p} = \mathfrak{n}_- \oplus \mathfrak{h}$ is then spanned by the elements E_{ij} with $i \geqslant j$. Take the principal nilpotent element f in the form

$$f = E_{21} + E_{32} + \dots + E_{N\,N-1}$$

and complete it to the $\mathfrak{sl}_2$-triple $\{e, f, h\}$ by

$$e = \sum_{i=1}^{N-1} i(N-i)\, E_{i\,i+1} \qquad \text{and} \qquad h = \sum_{i=1}^{N} (N-2i+1)\, E_{i\,i}.$$

The classical $\mathcal{W}$-algebra $\mathcal{W}(\mathfrak{gl}_N)$ is introduced by Definition 12.1.3, and Theorems 12.1.4 and 12.1.5 hold for the reductive Lie algebra $\mathfrak{gl}_N$ in the same form.

We will be working with the algebra of differential operators $\mathcal{V}(\mathfrak{p}) \otimes \mathbb{C}[\partial]$, where the commutation relations are given by

$$\partial\, E_{ij}^{(r)} - E_{ij}^{(r)}\, \partial = E_{ij}^{(r+1)}.$$

In other words, ∂ will be regarded as a generator of this algebra rather than the derivation on $\mathcal{V}(\mathfrak{p})$. For any element $g \in \mathcal{V}(\mathfrak{p})$ and any nonnegative integer r the element $g^{(r)} = \partial^r(g)$ coincides with the constant term of the differential operator $\partial^r g$ so that

$$g^{(r)} = \partial^r g\, 1, \tag{12.16}$$

assuming that $\partial\, 1 = 0$.

As in Example 9.1.7, the nondegenerate invariant symmetric bilinear form on $\mathfrak{gl}_N$ is defined by

$$\langle X, Y \rangle = \operatorname{tr} XY, \qquad X, Y \in \mathfrak{gl}_N,$$

where X and Y are understood as $N \times N$ matrices over $\mathbb{C}$.

Consider the determinant (12.13) of the matrix with entries in $\mathcal{V}(\mathfrak{p}) \otimes \mathbb{C}[\partial]$,

$$\det \begin{bmatrix} \partial + E_{11} & 1 & 0 & 0 & \dots & 0 \\ E_{21} & \partial + E_{22} & 1 & 0 & \dots & 0 \\ E_{31} & E_{32} & \partial + E_{33} & 1 & \dots & 0 \\ \dots & \dots & \dots & \dots & \dots & \dots \\ E_{N-1\,1} & E_{N-1\,2} & E_{N-1\,3} & \dots & \dots & 1 \\ E_{N1} & E_{N2} & E_{N3} & \dots & \dots & \partial + E_{NN} \end{bmatrix} \tag{12.17}$$

and write it as a differential operator

$$\partial^N + w_1\, \partial^{N-1} + \dots + w_N, \qquad w_i \in \mathcal{V}(\mathfrak{p}).$$

THEOREM 12.2.2. *All elements $w_1, \dots, w_N$ belong to the classical $\mathcal{W}$-algebra $\mathcal{W}(\mathfrak{gl}_N)$. Moreover, the elements $w_1^{(r)}, \dots, w_N^{(r)}$ with $r = 0, 1, \dots$ are algebraically independent and generate the algebra $\mathcal{W}(\mathfrak{gl}_N)$.*

PROOF. Denote the determinant in (12.17) by D_N. For each $1 \leqslant k < N$ we will identify the Lie algebra $\mathfrak{gl}_k$ with the natural subalgebra of $\mathfrak{gl}_N$ spanned by the basis elements E_{ij} with $1 \leqslant i, j \leqslant k$ so that the determinant D_k will be regarded as an element of the algebra $\mathcal{V}(\mathfrak{p}) \otimes \mathbb{C}[\partial]$. We also set $D_0 = 1$. We will be proving the first part of the theorem by induction on N. The induction base is trivial and for any $1 \leqslant i < N-1$ the induction hypothesis implies

$$\rho\{E_{i\,i+1\ \lambda} D_k\} = 0 \tag{12.18}$$

for all $k = i+1, \dots, N$. Moreover, relation (12.18) clearly holds for the values $k = 1, \dots, i-1$ as well, while

$$\rho\{E_{i\,i+1\ \lambda} D_i\} = -\rho\big(D^+_{i-1}\, E_{i\,i+1}\big) = -D^+_{i-1},$$

where for each polynomial $P = P(\partial) \in \mathcal{V}(\mathfrak{p}) \otimes \mathbb{C}[\partial]$ we denote by P^+ the polynomial $P(\partial+\lambda)$ regarded as an element of $\mathbb{C}[\lambda] \otimes \mathcal{V}(\mathfrak{p}) \otimes \mathbb{C}[\partial]$. By expanding the determinant along the last row, we get the relation

$$\begin{aligned} D_N = D_{N-1}\,(\partial + E_{NN}) - D_{N-2}\, E_{N\,N-1} + D_{N-3}\, E_{N\,N-2} \\ + \dots + (-1)^{N-2}\, D_1\, E_{N\,2} + (-1)^{N-1}\, D_0\, E_{N\,1}. \end{aligned} \tag{12.19}$$

Hence, using the properties of the λ-bracket we get

$$\rho\{E_{i\,i+1\ \lambda} D_N\} = (-1)^{N-i+1}\, D^+_{i-1}\, E_{N\,i+1} + (-1)^{N-i}\, D^+_{i-1}\, E_{N\,i+1} = 0$$

for all $i = 1, \dots, N-2$. Furthermore,

$$\begin{aligned} \rho\{E_{N-1\,N\ \lambda} D_N\} = D^+_{N-1} - D^+_{N-2}\,(\partial + E_{NN}) - D^+_{N-2}\,(E_{N-1\,N-1} - E_{N\,N} + \lambda) \\ + D^+_{N-3}\, E_{N-1\,N-2} + \dots + (-1)^{N-2}\, D^+_1\, E_{N-1\,2} + (-1)^{N-1}\, D^+_0\, E_{N-1\,1} \end{aligned}$$

which is zero due to relation (12.19) applied to the determinant D^+_{N-1} instead of D_N. Since $\rho\{E_{i\,i+1\ \lambda} D_N\} = 0$ for all $i = 1, \dots, N-1$, we may conclude that $\rho\{X\,{}_\lambda D_N\} = 0$ for all $X \in \mathfrak{n}_+$ so that all elements $w_1, \dots, w_N$ belong to the subalgebra $\mathcal{W}(\mathfrak{gl}_N)$ of $\mathcal{V}(\mathfrak{p})$.

To verify the second claim, apply Theorem 12.1.5. We can take

$$U = \text{span of} \quad \{E_{11}, E_{12}, \dots, E_{1N}\}$$

so that

$$U^\perp = \text{span of} \quad \{E_{ij} \mid N \geqslant i \geqslant j > 1\}.$$

To calculate the images of the elements $w_1, \dots, w_N$ under the isomorphism (12.11) write them first as differential polynomials in the variables $E_{11}, E_{21}, \dots, E_{N1}$ modulo the differential algebra ideal of $\mathcal{V}(\mathfrak{p})$ generated by $U^\perp$. This amounts to replacing all the elements E_{ij} with $N \geqslant i \geqslant j > 1$ in the determinant (12.17) by zero. By (12.15) the reduced determinant equals

$$(\partial + E_{11})\,\partial^{N-1} - E_{21}\,\partial^{N-1} + \dots + (-1)^{N-1} E_{N1}$$

so that $w_m \equiv (-1)^{m-1} E_{m\,1} \mod \langle U^\perp \rangle$. Hence, for the images with respect to the isomorphism (12.11) we get

$$\pi : w_m \mapsto (-1)^{m-1}\big(E_{m\,1} + E_{m+1\,2} + \dots + E_{N\,N-m+1}\big)$$

for $m = 1, \dots, N$. Since the images $\pi(w_1), \dots, \pi(w_N)$ form a basis of the centralizer $\mathfrak{gl}_N^f$, the second claim follows. $\square$

Type B_n. We will use the same presentation of the classical Lie algebras in types B, C and D as in Sections 2.2 and 5.1. The elements $F_{11}, \dots, F_{nn}$ span a Cartan subalgebra of $\mathfrak{o}_N$ for $N = 2n$ or $N = 2n+1$, which we denote by $\mathfrak{h}$. The respective subsets of elements F_{ij} with $i < j$ and $i > j$ span the nilpotent subalgebras $\mathfrak{n}_+$ and $\mathfrak{n}_-$. The subalgebra $\mathfrak{p} = \mathfrak{n}_- \oplus \mathfrak{h}$ is then spanned by the elements F_{ij} with $i \geqslant j$.

We will be working with the algebra of differential operators $\mathcal{V}(\mathfrak{p}) \otimes \mathbb{C}[\partial]$, where the commutation relations are given by

$$\partial\, F_{ij}^{(r)} - F_{ij}^{(r)}\, \partial = F_{ij}^{(r+1)}. \tag{12.20}$$

For any element $g \in \mathcal{V}(\mathfrak{p})$ and any nonnegative integer r the element $g^{(r)}$ coincides with the constant term of the differential operator $\partial^r g$ as in (12.16).

Take the principal nilpotent element $f \in \mathfrak{o}_{2n+1}$ in the form

$$f = F_{21} + F_{32} + \dots + F_{n+1\,n}.$$

The $\mathfrak{sl}_2$-triple is now formed by the elements $\{e, f, h\}$ with

$$e = \sum_{i=1}^{n} i\,(2n-i+1)\, F_{i\,i+1} \qquad \text{and} \qquad h = 2\sum_{i=1}^{n} (n-i+1)\, F_{ii}. \tag{12.21}$$

The nondegenerate invariant symmetric bilinear form on $\mathfrak{o}_{2n+1}$ is now defined by

$$\langle X, Y\rangle = \frac{1}{2}\,\mathrm{tr}\, XY, \qquad X, Y \in \mathfrak{o}_{2n+1},$$

where X and Y are understood as matrices over $\mathbb{C}$ which are skew-symmetric with respect to the antidiagonal.

Consider the determinant (12.13) of the matrix with entries in $\mathcal{V}(\mathfrak{p}) \otimes \mathbb{C}[\partial]$,

$$\det \begin{bmatrix} \partial + F_{11} & 1 & \dots & 0 & 0 & 0 & \dots & 0 \\ F_{21} & \partial + F_{22} & \dots & 0 & 0 & 0 & \dots & 0 \\ \dots & \dots & \ddots & \dots & \dots & \dots & \dots & \dots \\ F_{n1} & F_{n2} & \dots & \partial + F_{nn} & 1 & 0 & \dots & 0 \\ F_{n+1\,1} & F_{n+1\,2} & \dots & F_{n+1\,n} & \partial & -1 & \dots & 0 \\ F_{n'1} & F_{n'2} & \dots & 0 & F_{n'\,n+1} & \partial + F_{n'n'} & \dots & 0 \\ \dots & \dots & \dots & \dots & \dots & \dots & \ddots & \dots \\ F_{2'1} & 0 & \dots & \dots & F_{2'\,n+1} & F_{2'n'} & \dots & -1 \\ 0 & F_{1'2} & \dots & \dots & F_{1'\,n+1} & F_{1'n'} & \dots & \partial + F_{1'1'} \end{bmatrix}$$

which has the form

$$\partial^{2n+1} + w_2\, \partial^{2n-1} + w_3\, \partial^{2n-2} + \dots + w_{2n+1}, \qquad w_i \in \mathcal{V}(\mathfrak{p}). \tag{12.22}$$

THEOREM 12.2.3. *All elements $w_2, w_3, \dots, w_{2n+1}$ belong to the classical $\mathcal{W}$-algebra $\mathcal{W}(\mathfrak{o}_{2n+1})$. Moreover, the elements $w_2^{(r)}, w_4^{(r)}, \dots, w_{2n}^{(r)}$ with $r = 0, 1, \dots$ are algebraically independent and generate the algebra $\mathcal{W}(\mathfrak{o}_{2n+1})$.*

PROOF. The argument is similar to the proof of Theorem 12.2.2. Denote the determinant by D and let D_i (respectively, $\overline{D}_i$) denote the $i \times i$ minor corresponding to the first (respectively, last) i rows and columns. We suppose that $D_0 = \overline{D}_0 = 1$.

Lemma 12.2.1 implies the expansion

$$D = D_n\,\partial\,\overline{D}_n + \sum_{j,k=1}^{n+1} (-1)^{n-j+1}\, D_{j-1}\, F_{k'\,j}\, \overline{D}_{k-1}. \tag{12.23}$$

To prove the first part of the theorem, note that the elements F_{ij} with $1 \leqslant i, j \leqslant n$ span a subalgebra of $\mathfrak{o}_{2n+1}$ isomorphic to the Lie algebra $\mathfrak{gl}_n$. Hence, by Theorem 12.2.2, if $1 \leqslant i \leqslant n-1$ then

$$\rho\{F_{i\,i+1\;\lambda}\, D_k\} = 0 \qquad \text{and} \qquad \rho\{F_{i\,i+1\;\lambda}\, \overline{D}_k\} = 0$$

for all $1 \leqslant k \leqslant n$ with $k \neq i$. Moreover,

$$\rho\{F_{i\,i+1\;\lambda}\, D_i\} = -D^+_{i-1} \qquad \text{and} \qquad \rho\{F_{i\,i+1\;\lambda}\, \overline{D}_i\} = \overline{D}_{i-1},$$

where, as before, $P^+ = P(\partial + \lambda)$ for any polynomial $P = P(\partial) \in \mathcal{V}(\mathfrak{p}) \otimes \mathbb{C}[\partial]$. Hence, for any $k \in \{1, \dots, n+1\}$ and $k \neq i, i+1$ we have

$$\rho\big\{F_{i\,i+1\;\lambda}\, (D_i\, F_{k'\,i+1}\, \overline{D}_{k-1} - D_{i-1}\, F_{k'\,i}\, \overline{D}_{k-1})\big\} = 0.$$

Similarly, for any $j \in \{1, \dots, n+1\}$ and $j \neq i, i+1$ we have

$$\rho\big\{F_{i\,i+1\;\lambda}\, (D_{j-1}\, F_{i'\,j}\, \overline{D}_{i-1} + D_{j-1}\, F_{(i+1)'\,j}\, \overline{D}_i)\big\} = 0$$

and

$$\rho\big\{F_{i\,i+1\;\lambda}\, (D_{i-1}\, F_{(i+1)'\,i}\, \overline{D}_i - D_i\, F_{i'\,i+1}\, \overline{D}_{i-1})\big\} = 0.$$

These relations imply $\rho\{F_{i\,i+1\;\lambda}\, D\} = 0$. Finally, performing similar calculations we get

$$\begin{aligned}\rho\{F_{n\,n+1\;\lambda}\, D\} = {} & D^+_n\, (\partial + \lambda)\, \overline{D}_{n-1} + \sum_{k=1}^{n-1} D^+_n\, F_{k'\,n'}\, \overline{D}_{k-1} - D^+_n\, (F_{n\,n} + \lambda)\, \overline{D}_{n-1} \\ & - D^+_{n-1}\, \partial\, \overline{D}_n + \sum_{j=1}^{n-1} (-1)^{n-j+1}\, D^+_{j-1}\, F_{n\,j}\, \overline{D}_n - D^+_{n-1}\, (F_{n\,n} + \lambda)\, \overline{D}_n.\end{aligned}$$

Applying Lemma 12.2.1 to the determinants D^+_n and $\overline{D}_n$, we get the relations

$$D^+_n = D^+_{n-1}\, (\partial + \lambda + F_{n\,n}) + \sum_{j=1}^{n-1} (-1)^{n-j}\, D^+_{j-1}\, F_{n\,j} \tag{12.24}$$

and

$$\overline{D}_n = (\partial + F_{n'\,n'})\, \overline{D}_{n-1} + \sum_{k=1}^{n-1} F_{k'\,n'}\, \overline{D}_{k-1} \tag{12.25}$$

which imply that $\rho\{F_{n\,n+1\;\lambda}\, D\} = 0$. This shows that all elements $w_2, \dots, w_{2n+1}$ belong to the subalgebra $\mathcal{W}(\mathfrak{o}_{2n+1})$ of $\mathcal{V}(\mathfrak{p})$.

For the proof of the second part, apply Theorem 12.1.5. The odd powers $e, e^3, \dots, e^{2n-1}$ of the matrix e given in (12.21) form a basis of the centralizer $\mathfrak{o}^e_{2n+1}$. Therefore, using (12.8) we obtain that (12.7) holds for the subspace

$$U = \text{span of} \quad \{F_{12}, F_{14}, \dots, F_{1\,2n}\}$$

so that

$$U^\perp = \text{span of} \quad \{F_{ij} \ \text{with} \ 1' \geqslant i \geqslant j > 1 \quad \text{and} \quad F_{2k-1\,1} \ \text{with} \ k = 1, \dots, n\}.$$

Now calculate the images of the elements $w_2, w_4, \dots, w_{2n}$ under the isomorphism (12.11). As a first step, write each element w_{2m} as a differential polynomial in the variables $F_{21}, F_{41}, \dots, F_{2n\,1}$ modulo the differential ideal $\langle U^{\perp}\rangle$. Formula (12.15) and expansion (12.23) imply that modulo $\langle U^{\perp}\rangle$ the determinant D equals

$$D \equiv \big(\partial^n - F_{21}\,\partial^{n-2} + \dots + (-1)^{n-1}F_{n1}\big)\,\partial\,\big(\partial^n - \partial^{n-2}F_{21} - \dots - F_{n1}\big)$$
$$+ \sum_{k=2}^{n+1}(-1)^n F_{k'1}\big(\partial^{k-1} - \partial^{k-3}F_{21} - \dots - F_{k-1\,1}\big)$$
$$+ \sum_{j=2}^{n+1}(-1)^{n-j+1}\big(\partial^{j-1} - F_{21}\,\partial^{j-3} + \dots + (-1)^{j-2}F_{j-1\,1}\big)F_{1'j},$$

where we keep all variables F_{p1} with $p \geqslant 2$. It is clear from this expression, that if $2m \leqslant n$ then modulo the differential ideal $\langle U^{\perp}\rangle$, the element w_{2m} equals $-2F_{2m\,1}$ plus a linear combination of the elements $F_{q1}^{(s)}$ with $s+q = 2m$ and $s > 0$, and the products $F_{p1}\,F_{q1}^{(t)}$ with $p+q+t = 2m$, where all p, q, s, t are even. Similarly, if $2m \geqslant n+1$, then w_{2m} equals $2(-1)^n F_{2m\,1}$ plus a linear combination of the same form as above.

On the other hand, the odd powers $f, f^3, \dots, f^{2n-1}$ of the matrix f form a basis of the centralizer $\mathfrak{o}_{2n+1}^f$. Observe that f^{2m-1} coincides with $\pm F_{2m\,1}$ modulo $U^{\perp}$. Therefore, we may conclude that the images of the elements $w_2^{(r)}, w_4^{(r)}, \dots, w_{2n}^{(r)}$ with $r = 0, 1, \dots$ under the isomorphism (12.11) are algebraically independent generators of the algebra $\mathcal{V}(\mathfrak{o}_{2n+1}^f)$ thus completing the proof. □

Type D_n. We use the same notation for the generators of the Lie algebra $\mathfrak{o}_{2n}$ as for type B_n above. We will work with the algebra of pseudo-differential operators $\mathcal{V}(\mathfrak{p}) \otimes \mathbb{C}((\partial^{-1}))$, where the relations are given by (12.20) and

$$\partial^{-1}\,F_{ij}^{(r)} = \sum_{s=0}^{\infty}(-1)^s\,F_{ij}^{(r+s)}\,\partial^{-s-1}.$$

Take the principal nilpotent element $f \in \mathfrak{o}_{2n}$ in the form

$$f = F_{21} + F_{32} + \dots + F_{n\,n-1} + F_{n'\,n-1}.$$

The $\mathfrak{sl}_2$-triple is formed by the elements $\{e, f, h\}$ with

$$e = \sum_{i=1}^{n-2} i\,(2n-i-1)\,F_{i\,i+1} + \frac{n^2-n}{2}\,\big(F_{n-1\,n} + F_{n-1\,n'}\big) \quad \text{and} \tag{12.26}$$
$$h = 2\sum_{i=1}^{n-1}(n-i)\,F_{i\,i}.$$

The nondegenerate invariant symmetric bilinear form on $\mathfrak{o}_{2n}$ is defined by

$$\langle X, Y\rangle = \frac{1}{2}\,\mathrm{tr}\,XY, \qquad X, Y \in \mathfrak{o}_{2n},$$

where X and Y are understood as matrices over $\mathbb{C}$ which are skew-symmetric with respect to the antidiagonal.

Consider the following $(2n+1)\times(2n+1)$ matrix with entries in $\mathcal{V}(\mathfrak{p})\otimes\mathbb{C}((\partial^{-1}))$,

$$\begin{bmatrix} \partial+F_{11} & 1 & \dots & 0 & 0 & 0 & \dots & 0 \\ F_{21} & \partial+F_{22} & \dots & 0 & 0 & 0 & \dots & 0 \\ \dots & \dots & \ddots & \dots & \dots & \dots & \dots & \dots \\ F_{n1}-F_{n'1} & F_{n2}-F_{n'2} & \dots & \partial+F_{nn} & 0 & -2\partial & \dots & 0 \\ 0 & 0 & \dots & 0 & \partial^{-1} & 0 & \dots & 0 \\ F_{n'1} & F_{n'2} & \dots & 0 & 0 & \partial+F_{n'n'} & \dots & 0 \\ \dots & \dots & \dots & \dots & \dots & \dots & \ddots & \dots \\ F_{2'1} & 0 & \dots & \dots & 0 & F_{2'n'}-F_{2'n} & \dots & -1 \\ 0 & F_{1'2} & \dots & \dots & 0 & F_{1'n'}-F_{1'n} & \dots & \partial+F_{1'1'} \end{bmatrix}$$

where all entries in the row and column $n+1$ are zero, except for the $(n+1, n+1)$ entry which equals ∂^{-1}. The (n, j) entries are $F_{nj} - F_{n'j}$ for $j = 1, \dots, n-1$, the (n, n) entry is $\partial + F_{nn}$ and the $(n, n+2)$ entry is -2∂, while the remaining entries in row n are zero. The remaining nonzero entries in column $n+2$ are $F_{k'n'} - F_{k'n}$ for $k = 1, 2, \dots, n-1$ which occur in the respective rows $2n - k + 2$, and $\partial + F_{n'n'}$ which occurs in row $n + 2$.

One easily verifies that the column-determinant and row-determinant of this matrix coincide, so that the determinant (12.13) is well-defined and we denote it by D. Applying the simultaneous column expansion along the first n columns and using Lemma 12.2.1, one derives that it can be written in the form

$$D = D_n\,\partial^{-1}\,\overline{D}_n + 2\sum_{j,k=1}^{n}(-1)^{n-j}\,D_{j-1}\,F_{k'j}\,\overline{D}_{k-1}, \tag{12.27}$$

where D_i (respectively, $\overline{D}_i$) denotes the $i \times i$ minor corresponding to the first (respectively, last) i rows and columns. We suppose that $D_0 = \overline{D}_0 = 1$. Write

$$D_n = \partial^n + y_1\,\partial^{n-1} + y_2\,\partial^{n-2} + \dots + y_n,$$

$$\overline{D}_n = \partial^n + \partial^{n-1}\,\bar{y}_1 + \partial^{n-2}\,\bar{y}_2 + \dots + \bar{y}_n,$$

for certain uniquely determined elements $y_i, \bar{y}_i \in \mathcal{V}(\mathfrak{p})$.

LEMMA 12.2.4. *We have* $\bar{y}_i = (-1)^i\,y_i$ *for all* $i = 1, \dots, n$.

PROOF. Replace ∂ by $-\partial$ in the minor $\overline{D}_n$ and multiply each row by -1. The lemma can then be equivalently stated as the identity

$$(-1)^n\,\overline{D}_n\Big|_{\partial\mapsto-\partial} = \partial^n + \partial^{n-1}\,y_1 + \partial^{n-2}\,y_2 + \dots + y_n.$$

The left hand side is the determinant

$$\det\begin{bmatrix} \partial+F_{nn} & 1 & 0 & 0 & \dots & 0 \\ F_{n\,n-1}-F_{n'\,n-1} & \partial+F_{n-1\,n-1} & 1 & 0 & \dots & 0 \\ \dots & \dots & \dots & \dots & \dots & \dots \\ F_{n2}-F_{n'2} & F_{n-1\,2} & \dots & \dots & \partial+F_{22} & 1 \\ F_{n1}-F_{n'1} & F_{n-1\,1} & \dots & \dots & F_{21} & \partial+F_{11} \end{bmatrix}.$$

The claim now follows from the observation that this determinant coincides with the image of D_n under the anti-automorphism of the algebra $\mathcal{V}(\mathfrak{p}) \otimes \mathbb{C}[\partial]$ which is the identity on the generators F_{ij} and ∂. □

By Lemma 12.2.4, the pseudo-differential operator D can be written as

$$D = \partial^{2n-1} + w_2\,\partial^{2n-3} + w_3\,\partial^{2n-4} + \cdots + w_{2n-1} + (-1)^n\, y_n\,\partial^{-1}\, y_n \tag{12.28}$$

for certain elements $w_i \in \mathcal{V}(\mathfrak{p})$.

THEOREM 12.2.5. *The coefficients $w_2, w_3, \ldots, w_{2n-1}$ and y_n belong to the classical $\mathcal{W}$-algebra $\mathcal{W}(\mathfrak{o}_{2n})$. Moreover, the elements $w_2^{(r)}, w_4^{(r)}, \ldots, w_{2n-2}^{(r)}, y_n^{(r)}$ with $r = 0, 1, \ldots$ are algebraically independent and generate the algebra $\mathcal{W}(\mathfrak{o}_{2n})$.*

PROOF. The relation $\rho\{F_{i\,i+1\,\lambda}\, D\} = 0$ for $1 \leqslant i \leqslant n-1$ follows by the same calculations as in the proof of Theorem 12.2.3. Furthermore, let $\sigma = (n\, n')$ be the permutation of the set of indices $\{1, \ldots, 2n\}$ which swaps n and $n' = n+1$ and leaves all other indices fixed. The mapping

$$\varsigma : F_{ij} \mapsto F_{\sigma(i)\,\sigma(j)} \tag{12.29}$$

defines an involutive automorphism of the Lie algebra $\mathfrak{o}_{2n}$. It also extends to an involutive automorphism of the Poisson vertex algebra $\mathcal{V}(\mathfrak{o}_{2n})$. We claim that all coefficients of the pseudo-differential operator D are ς-invariant. Indeed, let us apply the following operations on the rows and columns of the given matrix. Replace row $n+2$ by the sum of rows n and $n+2$. Then replace column n by the sum of columns n and $n+2$. Finally, multiply row n and column $n+2$ by -1. As a result, we get the image of the matrix with respect to the involution (12.29). On the other hand, the determinant D remains unchanged. This proves the relation $\rho\{F_{n-1\,n'\,\lambda}\, D\} = 0$. This shows that all coefficients of the operator D belong to the subalgebra $\mathcal{W}(\mathfrak{o}_{2n})$.

Note that the minor D_n can be written in the form

$$\begin{aligned} D_n &= D_{n-1}\,(\partial + F_{n\,n}) - D_{n-2}\,(F_{n\,n-1} - F_{n'\,n-1}) \\ &\quad + \cdots + (-1)^{n-2}\, D_1\,(F_{n\,2} - F_{n'\,2}) + (-1)^{n-1}\, D_0\,(F_{n\,1} - F_{n'\,1}). \end{aligned} \tag{12.30}$$

Repeating the calculations used in the proof of Theorem 12.2.2, we find that $\rho\{F_{i\,i+1\,\lambda}\, D_n\} = 0$ for $i = 1, \ldots, n-1$. Furthermore,

$$\begin{aligned} \rho\{F_{n-1\,n'\,\lambda} D_n\} &= -D^+_{n-1} - D^+_{n-2}\,(\partial + F_{n\,n}) + D^+_{n-2}\,(F_{n-1\,n-1} + F_{n\,n} + \lambda) \\ &\quad - D^+_{n-3}\, F_{n-1\,n-2} + \cdots + (-1)^{n-1}\, D^+_1\, F_{n-1\,2} + (-1)^n\, D^+_0\, F_{n-1\,1}. \end{aligned}$$

Applying relation (12.19) to the determinant D^+_{n-1} we get

$$\rho\{F_{n-1\,n'\,\lambda} D_n\} = -2\, D^+_{n-2}\,\partial.$$

This implies $\rho\{F_{n-1\,n'\,\lambda}\, y_n\} = 0$ so that the constant term y_n of the differential operator D_n belongs to $\mathcal{W}(\mathfrak{o}_{2n})$.

The second part of the theorem will follow from Theorem 12.1.5. The odd powers $e, e^3, \ldots, e^{2n-3}$ of the matrix e given in (12.26) together with the element $F_{1n} - F_{1n'}$ form a basis of the centralizer $\mathfrak{o}^e_{2n}$. Hence (12.8) implies that (12.7) holds for the subspace U defined by

$$U = \text{span of } \{F_{12}, F_{14}, \ldots, F_{1\,n}, F_{1\,n+1}, F_{1\,n+3}, \ldots F_{1\,2n-1}\},$$

if n is even, and by

$$U = \text{span of } \{F_{12}, F_{14}, \ldots, F_{1\,n-1}, F_{1n} - F_{1n+1}, F_{1\,n+2}, F_{1\,n+4}, \ldots, F_{1\,2n-1}\},$$

if n is odd. The complementary subspace $U^{\perp}$ to $\mathfrak{o}^f_{2n}$ in $\mathfrak{p}$ is then given by (12.9).

As in the proof of Theorem 12.2.3, expansion (12.27) implies that modulo $\langle U^{\perp}\rangle$ the determinant D equals

$$D \equiv \big(\partial^n - F_{21}\,\partial^{n-2} + \dots + (-1)^{n-1}(F_{n1} - F_{n'1})\big) \\ \times \partial^{-1}\big(\partial^n - \partial^{n-2}F_{21} - \dots - (F_{n1} - F_{n'1})\big) \\ + 2\sum_{k=2}^{n}(-1)^{n-1}F_{k'1}\big(\partial^{k-1} - \partial^{k-3}F_{21} - \dots - F_{k-1\,1}\big) \\ + 2\sum_{j=2}^{n}(-1)^{n-j}\big(\partial^{j-1} - F_{21}\,\partial^{j-3} + \dots + (-1)^{j-2}F_{j-1\,1}\big)F_{1'j},$$

where we keep all variables F_{p1} with $p \geqslant 2$. Moreover, due to (12.30) we also have $y_n \equiv (-1)^{n-1}(F_{n1} - F_{n'1})$. Therefore, if $2m \leqslant n-1$ then modulo the differential ideal $\langle U^{\perp}\rangle$, the element w_{2m} equals $-2F_{2m\,1}$ plus a linear combination of the elements $F_{q1}^{(s)}$ with $s+q = 2m$ and $s > 0$, and the products $F_{p1}F_{q1}^{(t)}$ with $p+q+t = 2m$, where all p, q, s, t are even.

Next we look at the cases of even and odd n separately. If n is even, then modulo the differential ideal $\langle U^{\perp}\rangle$, the coefficient w_n equals $-2(F_{n1} + F_{n'1})$ plus a linear combination of the same form as above with $2m = n$. If $n+2 \leqslant 2m \leqslant 2n-2$ then w_{2m} equals $-4F_{2m+1\,1}$ plus a similar linear combination. If n is odd and $n+1 \leqslant 2m \leqslant 2n-2$ then modulo the differential ideal $\langle U^{\perp}\rangle$, the coefficient w_{2m} equals $4F_{2m+1\,1}$ plus a linear combination of the same kind.

The odd powers $f, f^3, \dots, f^{2n-3}$ of the matrix f together with the element $F_{n1} - F_{n'1}$ form a basis of the centralizer $\mathfrak{o}_{2n}^f$. Observe that modulo $U^{\perp}$, the odd power f^{2m-1} coincides with $F_{2m\,1}$ if $2m < n$, and with $2(-1)^n F_{2m+1\,1}$ if $2m > n$. Furthermore, if n is even, then f^{n-1} coincides with $F_{n\,1} + F_{n'1}$ modulo $U^{\perp}$. Therefore, we may conclude that the images of the elements $w_2^{(r)}, w_4^{(r)}, \dots, w_{2n-2}^{(r)}$ and $y_n^{(r)}$ with $r = 0, 1, \dots$ under the isomorphism (12.11) are algebraically independent generators of the algebra $\mathcal{V}(\mathfrak{o}_{2n}^f)$ which completes the proof. □

Type C_n. We will use the same presentation of the Lie algebra $\mathfrak{sp}_{2n}$ as in Sections 2.2 and 5.1. The elements $F_{11}, \dots, F_{nn}$ span a Cartan subalgebra of $\mathfrak{sp}_{2n}$ which we denote by $\mathfrak{h}$. The respective subsets of elements F_{ij} with $i < j$ and $i > j$ span the nilpotent subalgebras $\mathfrak{n}_+$ and $\mathfrak{n}_-$. The subalgebra $\mathfrak{p} = \mathfrak{n}_- \oplus \mathfrak{h}$ is then spanned by the elements F_{ij} with $i \geqslant j$.

We will work with the algebra of differential operators $\mathcal{V}(\mathfrak{p}) \otimes \mathbb{C}[\partial]$, where the commutation relations are given by

$$\partial\, F_{ij}^{(r)} - F_{ij}^{(r)}\,\partial = F_{ij}^{(r+1)}.$$

For any element $g \in \mathcal{V}(\mathfrak{p})$ and any nonnegative integer r the element $g^{(r)}$ coincides with the constant term of the differential operator $\partial^r g$ as in (12.16).

Take the principal nilpotent element $f \in \mathfrak{sp}_{2n}$ in the form

$$f = F_{21} + F_{32} + \dots + F_{n\,n-1} + \frac{1}{2}\,F_{n'\,n}.$$

The $\mathfrak{sl}_2$-triple is formed by the elements $\{e, f, h\}$ with

$$e = \sum_{i=1}^{n-1} i\,(2n-i)\,F_{i\,i+1} - \frac{n^2}{2}\,F_{n\,n'} \qquad \text{and} \qquad h = \sum_{i=1}^{n}(2n-2i+1)\,F_{i\,i}.$$

The invariant symmetric bilinear form on $\mathfrak{sp}_{2n}$ is defined by[1]

$$\langle X, Y\rangle = \frac{1}{2}\,\mathrm{tr}\, XY, \qquad X, Y \in \mathfrak{sp}_{2n},$$

where X and Y are understood as $2n \times 2n$ symplectic matrices over $\mathbb{C}$.

Consider the determinant (12.13) of the matrix with entries in $\mathcal{V}(\mathfrak{p}) \otimes \mathbb{C}[\partial]$,

$$\det \begin{bmatrix} \partial + F_{11} & 1 & \dots & 0 & 0 & 0 & \dots & 0 \\ F_{21} & \partial + F_{22} & \dots & 0 & 0 & 0 & \dots & 0 \\ & & \ddots & & & & & \\ \dots & \dots & & \dots & \dots & \dots & \dots & \dots \\ F_{n1} & F_{n2} & \dots & \partial + F_{nn} & 1 & 0 & \dots & 0 \\ F_{n'1} & F_{n'2} & \dots & F_{n'n} & \partial + F_{n'n'} & -1 & \dots & 0 \\ \dots & \dots & \dots & \dots & \dots & \dots & \dots & \dots \\ F_{2'1} & F_{2'2} & \dots & F_{2'n} & F_{2'n'} & \dots & \dots & -1 \\ F_{1'1} & F_{1'2} & \dots & F_{1'n} & F_{1'n'} & \dots & \dots & \partial + F_{1'1'} \end{bmatrix}$$

which has the form

$$\partial^{2n} + w_2\,\partial^{2n-2} + w_3\,\partial^{2n-3} + \dots + w_{2n}, \qquad w_i \in \mathcal{V}(\mathfrak{p}). \tag{12.31}$$

THEOREM 12.2.6. *All coefficients $w_2, w_3, \dots, w_{2n}$ belong to the classical $\mathcal{W}$-algebra $\mathcal{W}(\mathfrak{sp}_{2n})$. Moreover, the elements $w_2^{(r)}, w_4^{(r)}, \dots, w_{2n}^{(r)}$ with $r = 0, 1, \dots$ are algebraically independent and generate the algebra $\mathcal{W}(\mathfrak{sp}_{2n})$.*

PROOF. Denote the determinant by D and let D_i (respectively, $\overline{D}_i$) denote the $i \times i$ minor corresponding to the first (respectively, last) i rows and columns. We suppose that $D_0 = \overline{D}_0 = 1$. Lemma 12.2.1 implies the expansion

$$D = D_n\,\overline{D}_n + \sum_{j,k=1}^{n} (-1)^{n-j+1}\, D_{j-1}\, F_{k'j}\, \overline{D}_{k-1}. \tag{12.32}$$

If $1 \leqslant i \leqslant n-1$ then the relation $\rho\{F_{i\,i+1\;\lambda}\, D\} = 0$ follows by the same calculation as in the proof of Theorem 12.2.3. Furthermore,

$$\begin{aligned} \rho\{F_{n\,n'\;\lambda}\, D\} = &-2\,D^{+}_{n-1}\,\overline{D}_n + 2\,D^{+}_{n}\,\overline{D}_{n-1} + 2\sum_{k=1}^{n-1} D^{+}_{n-1}\,F_{k'n'}\,\overline{D}_{k-1} \\ &- 2\,D^{+}_{n-1}\,(2F_{nn} + \lambda)\,\overline{D}_{n-1} + 2\sum_{j=1}^{n-1} (-1)^{n-j+1}\, D^{+}_{j-1}\, F_{nj}\,\overline{D}_{n-1}. \end{aligned}$$

This is zero since the relations (12.24) and (12.25) are valid for the case of $\mathfrak{sp}_{2n}$ as well. Thus, all elements $w_2, w_3, \dots, w_{2n}$ belong to the subalgebra $\mathcal{W}(\mathfrak{sp}_{2n})$ of $\mathcal{V}(\mathfrak{p})$.

Now apply Theorem 12.1.5. The odd powers $e, e^3, \dots, e^{2n-1}$ of the matrix e form a basis of the centralizer $\mathfrak{sp}^{e}_{2n}$. Therefore, using (12.8) we obtain that (12.7) holds for the subspace

$$U = \text{span of} \quad \{F_{12}, F_{14}, \dots, F_{1\,2n}\}$$

so that

$$U^{\perp} = \text{span of} \quad \{F_{ij} \ \text{ with } \ 1' \geqslant i \geqslant j > 1 \quad \text{and} \quad F_{2k-1\,1} \ \text{ with } \ k = 1, \dots, n\}.$$

[1] Note the additional factor 1/2 as compared to the normalized Killing form of Section 8.3.

Represent each of the coefficients w_{2m} as a differential polynomial in the variables $F_{21}, F_{41}, \dots, F_{2n\,1}$ modulo the differential ideal $\langle U^{\perp}\rangle$. Formula (12.15) and expansion (12.32) imply that modulo $\langle U^{\perp}\rangle$ the determinant D equals

$$D \equiv \big(\partial^n - F_{21}\,\partial^{n-2} + \dots + (-1)^{n-1}F_{n1}\big)\big(\partial^n - \partial^{n-2}\,F_{21} - \dots - F_{n1}\big)$$
$$+ \sum_{k=1}^{n}(-1)^n F_{k'1}\,\big(\partial^{k-1} - \partial^{k-3}\,F_{21} - \dots - F_{k-1\,1}\big)$$
$$+ \sum_{j=2}^{n}(-1)^{n-j+1}\big(\partial^{j-1} - F_{21}\,\partial^{j-3} + \dots + (-1)^{j-2}F_{j-1\,1}\big)F_{1'j},$$

where we keep all variables F_{p1} with $p \geqslant 2$. Hence, if $2m \leqslant n$ then modulo the differential ideal $\langle U^{\perp}\rangle$, the element w_{2m} equals $-2F_{2m\,1}$ plus a linear combination of the elements $F_{q1}^{(s)}$ with $s+q = 2m$ and $s > 0$, and the products $F_{p1}F_{q1}^{(t)}$ with $p+q+t = 2m$, where all p, q, s, t are even. Similarly, if $n+1 \leqslant 2m$, then w_{2m} equals $2(-1)^n F_{2m\,1}$ or $(-1)^n F_{2n\,1}$ depending on whether $m < 2n$ or $m = 2n$, plus a linear combination of the same form.

As with type B, the odd powers $f, f^3, \dots, f^{2n-1}$ of the matrix f form a basis of the centralizer $\mathfrak{sp}_{2n}^f$. The matrix f^{2m-1} coincides with $\pm F_{2m\,1}$ modulo $U^{\perp}$. Therefore, the images of the elements $w_2^{(r)}, w_4^{(r)}, \dots, w_{2n}^{(r)}$ with $r = 0, 1, \dots$ under the isomorphism (12.11) are algebraically independent generators of the algebra $\mathcal{V}(\mathfrak{sp}_{2n}^f)$ thus completing the proof. □

12.3. Chevalley projection

Return to the general settings related to an arbitrary simple Lie algebra $\mathfrak{g}$ as in Section 12.1. The differential algebra $\mathcal{V}(\mathfrak{h})$ is the algebra of polynomials in the variables $h_j^{(r)} = \partial^r(h_j)$ with $r = 0, 1, \dots$ and $j = 1, \dots, n$. Consider the homomorphism of differential algebras

$$\phi : \mathcal{V}(\mathfrak{p}) \to \mathcal{V}(\mathfrak{h}) \qquad (12.33)$$

which is defined on the generators as the projection $\mathfrak{p} \to \mathfrak{h}$ with the kernel $\mathfrak{n}_-$. Recall that due to Theorem 12.1.4 the classical $\mathcal{W}$-algebra $\mathcal{W}(\mathfrak{g})$ is a differential subalgebra of $\mathcal{V}(\mathfrak{p})$.

THEOREM 12.3.1. *The restriction of the homomorphism ϕ to the subalgebra $\mathcal{W}(\mathfrak{g})$ is injective. Hence, the differential algebra $\mathcal{W}(\mathfrak{g})$ is isomorphic to the subalgebra $\widetilde{\mathcal{W}}(\mathfrak{g}) = \phi(\mathcal{W}(\mathfrak{g}))$ of $\mathcal{V}(\mathfrak{h})$.*

PROOF (FOR CLASSICAL TYPES). We will show that the images of the algebraically independent generators of $\mathcal{W}(\mathfrak{g})$ constructed in Section 12.2 are algebraically independent elements of $\mathcal{V}(\mathfrak{h})$. We will employ the following general lemma.

LEMMA 12.3.2. *Suppose that polynomials $v_1, \dots, v_n$ in the n variables $h_1, \dots, h_n$ are algebraically independent. Then all derivatives $v_1^{(r)}, \dots, v_n^{(r)}$ with $r = 0, 1, \dots$ are algebraically independent elements of $\mathcal{V}(\mathfrak{h})$.*

PROOF. It is sufficient to demonstrate that for any nonnegative integer p the elements $v_1^{(r)}, \dots, v_n^{(r)}$ with $r = 0, 1, \dots, p$ are algebraically independent polynomials in the variables $h_1^{(r)}, \dots, h_n^{(r)}$ with $r = 0, 1, \dots, p$. We will use the Jacobian criterion for algebraic independence [**71**, Proposition 3.10]. By the criterion, the polynomials $v_1^{(r)}, \dots, v_n^{(r)}$ are algebraically independent if and only if the Jacobian is nonzero; that is,

$$\det \left[\frac{\partial v_i^{(r)}}{\partial h_j^{(s)}} \right] \neq 0. \tag{12.34}$$

Let us order the rows and columns of the Jacobian in accordance with the ordering on the variables and polynomials defined by $h_i^{(r)} \prec h_j^{(s)}$ and $v_i^{(r)} \prec v_j^{(s)}$ if $r < s$ and by setting $h_1^{(r)} \prec \dots \prec h_n^{(r)}$ and $v_1^{(r)} \prec \dots \prec v_n^{(r)}$ for each r. Furthermore, observe that

$$\frac{\partial v_i^{(r)}}{\partial h_j^{(r)}} = \frac{\partial v_i}{\partial h_j}$$

for each $r = 0, 1, \dots, p$. This implies that the Jacobian in (12.34) has a block-diagonal form with $p+1$ identical diagonal blocks which coincide with the Jacobian

$$\det \left[\frac{\partial v_i}{\partial h_j} \right].$$

By the assumption, this Jacobian is nonzero which implies (12.34). □

Now consider the classical types separately.

Case A_{N-1}. The generators $w_1, \dots, w_N$ of the classical $\mathcal{W}$-algebra $\mathcal{W}(\mathfrak{gl}_N)$ were produced in Theorem 12.2.2. Their images under the homomorphism (12.33) are the elements $\widetilde{w}_m \in \mathcal{V}(\mathfrak{h})$ found from the relation

$$(\partial + E_{11}) \dots (\partial + E_{NN}) = \partial^N + \widetilde{w}_1 \, \partial^{N-1} + \dots + \widetilde{w}_N.$$

We want to show that all polynomials $\widetilde{w}_1^{(r)}, \dots, \widetilde{w}_N^{(r)}$ with $r = 0, 1, \dots$ are algebraically independent. It suffices to do this for their top degree components with respect to the grading on $\mathcal{V}(\mathfrak{h})$ defined by setting $\deg E_{ii}^{(r)} = 1$ for all i and r. However, the top degree component of $\widetilde{w}_m$ coincides with the m-th elementary symmetric polynomial (2.10) in the variables $E_{11}, \dots, E_{NN}$. Since the elementary symmetric polynomials are algebraically independent, the desired property of the polynomials $\widetilde{w}_1^{(r)}, \dots, \widetilde{w}_N^{(r)}$ follows from Lemma 12.3.2.

Case B_n. The images $\widetilde{w}_m \in \mathcal{V}(\mathfrak{h})$ of the coefficients w_m of the differential operator (12.22) under the homomorphism (12.33) are found from the relation

$$(\partial + F_{11}) \dots (\partial + F_{nn}) \, \partial \, (\partial - F_{nn}) \dots (\partial - F_{11})$$
$$= \partial^{2n+1} + \widetilde{w}_2 \, \partial^{2n-1} + \widetilde{w}_3 \, \partial^{2n-2} + \dots + \widetilde{w}_{2n+1}.$$

As with type A above, define the grading on $\mathcal{V}(\mathfrak{h})$ by setting $\deg F_{ii}^{(r)} = 1$. We find that for $k = 1, \dots, n$ the top degree component of $(-1)^k \widetilde{w}_{2k}$ coincides with the k-th elementary symmetric polynomial in the variables $F_{11}^2, \dots, F_{nn}^2$. Hence, the polynomials $\widetilde{w}_2^{(r)}, \widetilde{w}_4^{(r)}, \dots, \widetilde{w}_{2n}^{(r)}$ with $r = 0, 1, \dots$ are algebraically independent by Lemma 12.3.2.

Case D_n. The images of the coefficients $w_2, w_3, \dots, w_{2n-1}, y_n$ of the operator D given in (12.28) under the homomorphism (12.33) are the respective elements $\widetilde{w}_2, \widetilde{w}_3, \dots, \widetilde{w}_{2n-1}, \widetilde{y}_n \in \mathcal{V}(\mathfrak{h})$ found from the relation

$$(\partial + F_{11}) \dots (\partial + F_{nn})\, \partial^{-1} (\partial - F_{nn}) \dots (\partial - F_{11}) \\ = \partial^{2n-1} + \widetilde{w}_2\, \partial^{2n-3} + \widetilde{w}_3\, \partial^{2n-4} + \dots + \widetilde{w}_{2n-1} + (-1)^n\, \widetilde{y}_n\, \partial^{-1}\, \widetilde{y}_n.$$

In particular,

$$\widetilde{y}_n = (\partial + F_{11}) \dots (\partial + F_{nn})\, 1.$$

Similar to type B_n above, the top degree component of $(-1)^k \widetilde{w}_{2k}$ coincides with the k-th elementary symmetric polynomial in the variables $F_{11}^2, \dots, F_{nn}^2$ for $k = 1, \dots, n-1$. Furthermore, the top degree component of $\widetilde{y}_n$ is $F_{11} \dots F_{nn}$. These components are well-known to be algebraically independent so the proof is completed by an application of Lemma 12.3.2.

Case C_n. The images $\widetilde{w}_m \in \mathcal{V}(\mathfrak{h})$ of the coefficients w_m of the differential operator (12.31) under the homomorphism (12.33) are found from the relation

$$(\partial + F_{11}) \dots (\partial + F_{nn}) (\partial - F_{nn}) \dots (\partial - F_{11}) \\ = \partial^{2n} + \widetilde{w}_2\, \partial^{2n-2} + \widetilde{w}_3\, \partial^{2n-3} + \dots + \widetilde{w}_{2n}.$$

As with type B_n, the top degree component of $(-1)^k \widetilde{w}_{2k}$ coincides with the k-th elementary symmetric polynomial in the variables $F_{11}^2, \dots, F_{nn}^2$ for all $k = 1, \dots, n$. By Lemma 12.3.2 the polynomials $\widetilde{w}_2^{(r)}, \widetilde{w}_4^{(r)}, \dots, \widetilde{w}_{2n}^{(r)}$ with $r = 0, 1, \dots$ are algebraically independent. □

The injective homomorphism $\phi : \mathcal{W}(\mathfrak{g}) \hookrightarrow \mathcal{V}(\mathfrak{h})$ provided by Theorem 12.3.1 is known as the *Miura transformation*. Its image $\widetilde{\mathcal{W}}(\mathfrak{g}) = \phi(\mathcal{W}(\mathfrak{g}))$ can be regarded as an alternative presentation of the classical $\mathcal{W}$-algebra. It is this presentation which we will need to describe the Harish-Chandra images of the Segal–Sugawara vectors in Chapter 13. An alternative description of the subalgebra $\widetilde{\mathcal{W}}(\mathfrak{g}) \subset \mathcal{V}(\mathfrak{h})$ can be given by in terms of screening operators as we discuss in the next section.

12.4. Screening operators

For each $i = 1, \dots, n$ introduce the *screening operator*

$$V_i^{\circ} : \mathcal{V}(\mathfrak{h}) \to \mathcal{V}(\mathfrak{h})$$

by the formula

$$V_i^{\circ} = \sum_{r=0}^{\infty} V_{i\,r}^{\circ} \sum_{j=1}^{n} a_{ji} \frac{\partial}{\partial h_j^{(r)}}, \tag{12.35}$$

where $A = [a_{ij}]$ is the Cartan matrix of the simple Lie algebra $\mathfrak{g}$ and the coefficients $V_{i\,r}^{\circ}$ are elements of $\mathcal{V}(\mathfrak{h})$ found by the relation

$$\sum_{r=0}^{\infty} \frac{V_{i\,r}^{\circ}\, z^r}{r!} = \exp\Big(- \sum_{m=1}^{\infty} \frac{h_i^{(m-1)}\, z^m}{\epsilon_i\, m!} \Big). \tag{12.36}$$

The positive rational numbers ϵ_i are given in (12.3). Their particular values for the classical types will be chosen below.

PROPOSITION 12.4.1. *The image $\phi(P)$ of any element $P \in \mathcal{W}(\mathfrak{g})$ under the homomorphism* (12.33) *is annihilated by all screening operators,*

$$V_i^\circ\, \phi(P) = 0, \qquad i = 1, \dots, n.$$

Hence,

$$\widetilde{\mathcal{W}}(\mathfrak{g}) \subset \bigcap_{i=1}^{n} \ker V_i^\circ. \tag{12.37}$$

PROOF. It will be convenient to work with an equivalent affine version of the differential algebra $\mathcal{V}(\mathfrak{g})$. Recall that the affine Kac–Moody algebra $\widehat{\mathfrak{g}}$ is defined as the central extension (6.5). Consider the quotient $\mathrm{S}(\widehat{\mathfrak{g}})/\,\mathrm{I}$ of the symmetric algebra $\mathrm{S}(\widehat{\mathfrak{g}})$ by its ideal I generated by the subspace $\mathfrak{g}[t]$ and the element $K - 1$. This quotient can be identified with the symmetric algebra $\mathrm{S}\left(t^{-1}\mathfrak{g}[t^{-1}]\right)$, as a vector space. We will identify the differential algebras $\mathcal{V}(\mathfrak{g}) \cong \mathrm{S}\left(t^{-1}\mathfrak{g}[t^{-1}]\right)$ via the isomorphism

$$X^{(r)} \mapsto r!\, X[-r-1], \qquad X \in \mathfrak{g}, \quad r \geqslant 0, \tag{12.38}$$

so that the derivation ∂ will correspond to the derivation T of $\mathrm{S}\left(t^{-1}\mathfrak{g}[t^{-1}]\right)$ defined in (6.20). Similarly, we will identify the differential algebras $\mathcal{V}(\mathfrak{p}) \cong \mathrm{S}\left(t^{-1}\mathfrak{p}[t^{-1}]\right)$.

By Definition 12.1.3, if an element $P \in \mathcal{V}(\mathfrak{p})$ belongs to the subalgebra $\mathcal{W}(\mathfrak{g})$, then $\rho\{e_i\,{}_\lambda\, P\} = 0$ for all $i = 1, \dots, n$. Now observe that, regarding P as an element of the $\mathfrak{g}[t]$-module $\mathrm{S}(\widehat{\mathfrak{g}})/\,\mathrm{I} \cong \mathrm{S}\left(t^{-1}\mathfrak{g}[t^{-1}]\right)$, we can write[2]

$$\{e_i\,{}_\lambda\, P\} = \sum_{r=0}^{\infty} \frac{\lambda^r}{r!}\, e_i[r]\, P.$$

We have the following relations in $\widehat{\mathfrak{g}}$,

$$\big[e_i[r], f_i[-s-1]\big] = h_i[r-s-1] + r\,\delta_{r,\,s+1}\,\epsilon_i\, K,$$

$$\big[e_i[r], h_j[-s-1]\big] = -a_{ji}\, e_i[r-s-1].$$

Moreover, for each positive root $\alpha \neq \alpha_i$ we also have

$$\big[e_i[r], e_{-\alpha}[-s-1]\big] = c_i(\alpha)\, e_{-\alpha+\alpha_i}[r-s-1]$$

for a certain constant $c_i(\alpha)$, if $\alpha - \alpha_i$ is a root; otherwise the commutator is zero. Hence, recalling the definition (12.6) of the homomorphism ρ, we can conclude that the condition that P belongs to the subalgebra $\mathcal{W}(\mathfrak{g})$ implies the relations

$$\widehat{e}_i[r]\, P = 0 \qquad \text{for all} \quad i = 1, \dots, n \quad \text{and} \quad r \geqslant 0, \tag{12.39}$$

where $\widehat{e}_i[r]$ is the operator on $\mathrm{S}\left(t^{-1}\mathfrak{p}[t^{-1}]\right)$ given by

$$\widehat{e}_i[r] = \sum_{s=r}^{\infty} h_i[r-s-1]\, \frac{\partial}{\partial f_i[-s-1]} + \epsilon_i\, r\, \frac{\partial}{\partial f_i[-r]} - \sum_{j=1}^{n} a_{ji}\, \frac{\partial}{\partial h_j[-r-1]}$$

$$+ \sum_{\alpha\in\Delta^+,\, \alpha\neq\alpha_i}\ \sum_{s=r}^{\infty} c_i(\alpha)\, e_{-\alpha+\alpha_i}[r-s-1]\, \frac{\partial}{\partial\, e_{-\alpha}[-s-1]},$$

and $e_{-\alpha+\alpha_i}$ is understood as being equal to zero, if $\alpha - \alpha_i$ is not a root. Denote the generating function in z introduced in (12.36) by $V_i^\circ(z)$ and replace $h_i^{(m-1)}/(m-1)!$

[2]This module structure differs from the one described by (6.18) since the central element K now takes the value 1 in the quotient.

with $h_i[-m]$ for $m \geqslant 1$ in accordance with (12.38). We have the relation for its derivative,

$$V_i^{\circ\prime}(z) = V_i^\circ(z)\Big(-\sum_{m=1}^{\infty}\frac{h_i[-m]\,z^{m-1}}{\epsilon_i}\Big).$$

Taking the coefficient of z^{p-1} with $p \geqslant 1$ we get the relations

$$\epsilon_i\, p\,\frac{V_{i\,p}^\circ}{p!} + \sum_{r=0}^{p-1}\frac{V_{i\,r}^\circ}{r!}\,h_i[r-p] = 0. \tag{12.40}$$

By (12.39) the element P has the property

$$\sum_{r=0}^{\infty}\frac{V_{i\,r}^\circ}{r!}\,\widehat{e}_i[r]\,P = 0. \tag{12.41}$$

Note that by (12.40) all the differentiations $\partial/\partial f_i[-s-1]$ with $s \geqslant 0$ will cancel in the expansion of the left hand side of (12.41). Moreover, the elements of the form $e_{-\alpha+\alpha_i}[r-s-1]$ occurring in the expansion of $\widehat{e}_i[r]$ will vanish under the projection (12.33). Therefore, (12.41) implies that the image $\phi(P)$ with respect to this projection satisfies the relation

$$\sum_{r=0}^{\infty}\frac{V_{i\,r}^\circ}{r!}\sum_{j=1}^{n}a_{ji}\,\frac{\partial}{\partial h_j[-r-1]}\,\phi(P) = 0$$

which is equivalent to

$$\sum_{r=0}^{\infty}V_{i\,r}^\circ\sum_{j=1}^{n}a_{ji}\,\frac{\partial}{\partial h_j^{(r)}}\,\phi(P) = 0,$$

that is, $V_i^\circ\,\phi(P) = 0$, as claimed. □

The inclusion (12.37) is in fact an equality which can be proved by comparing the Hilbert–Poincaré series of both subalgebras of $\mathcal{V}(\mathfrak{h})$; see [**46**, Chapter 8]. We omit this computation and state this result without a proof.

THEOREM 12.4.2. *The restriction of the homomorphism ϕ to $\mathcal{W}(\mathfrak{g})$ yields an isomorphism*

$$\phi : \mathcal{W}(\mathfrak{g}) \to \widetilde{\mathcal{W}}(\mathfrak{g}),$$

where $\widetilde{\mathcal{W}}(\mathfrak{g})$ is the subalgebra of $\mathcal{V}(\mathfrak{h})$ which consists of the elements annihilated by all screening operators V_i°,

$$\widetilde{\mathcal{W}}(\mathfrak{g}) = \bigcap_{i=1}^{n}\ker V_i^\circ.$$

□

In the rest of this section we will give a description of the generators of the classical $\mathcal{W}$-algebras and screening operators by regarding $\mathcal{W}(\mathfrak{g})$ as a subalgebra of the symmetric algebra $\mathrm{S}\left(t^{-1}\mathfrak{h}[t^{-1}]\right)$ via the isomorphism (12.38).

Type A_{N-1}. For $i = 1, \ldots, N$ and all $r \geqslant 0$ introduce new variables

$$\mu_{i'}[-r-1] = \frac{1}{r!}\, E_{ii}^{(r)}, \qquad i' = N - i + 1.$$

Hence

$$h_i^{(r)} = r!\,\big(\mu_{i'}[-r-1] - \mu_{(i+1)'}[-r-1]\big), \qquad i = 1, \ldots, N-1,$$

so that for the partial derivatives over new variables we have

$$\frac{\partial}{\partial \mu_{i'}[-r-1]} = -r!\,\frac{\partial}{\partial h_{i-1}^{(r)}} + r!\,\frac{\partial}{\partial h_i^{(r)}}$$

with out-of-range terms omitted. The Cartan matrix is of the size $(N-1)\times(N-1)$,

$$A = \begin{bmatrix} 2 & -1 & 0 & \ldots & 0 & 0 \\ -1 & 2 & -1 & \ldots & 0 & 0 \\ 0 & -1 & 2 & \ldots & 0 & 0 \\ \ldots & \ldots & \ldots & \ddots & \ldots & \ldots \\ 0 & 0 & 0 & \ldots & 2 & -1 \\ 0 & 0 & 0 & \ldots & -1 & 2 \end{bmatrix}$$

and the diagonal matrix D given in (12.3) is the identity matrix. Therefore, the sum in (12.35) equals

$$\begin{aligned} r! \sum_{j=1}^{N-1} a_{ji}\,\frac{\partial}{\partial h_j^{(r)}} &= r!\,\Big(-\frac{\partial}{\partial h_{i-1}^{(r)}} + 2\,\frac{\partial}{\partial h_i^{(r)}} - \frac{\partial}{\partial h_{i+1}^{(r)}}\Big) \\ &= \frac{\partial}{\partial \mu_{i'}[-r-1]} - \frac{\partial}{\partial \mu_{(i+1)'}[-r-1]} \end{aligned}$$

so that the screening operator (12.35) takes the form $V_i^{\circ} = -V_{(i+1)'}$, where

$$V_i = \sum_{r=0}^{\infty} V_{i\,r}\,\Big(\frac{\partial}{\partial \mu_i[-r-1]} - \frac{\partial}{\partial \mu_{i+1}[-r-1]}\Big), \qquad i = 1, \ldots, N-1, \tag{12.42}$$

and the coefficients $V_{i\,r}$ are found from the expansion

$$\sum_{r=0}^{\infty} V_{i\,r}\, z^r = \exp \sum_{m=1}^{\infty} \frac{\mu_i[-m] - \mu_{i+1}[-m]}{m}\, z^m. \tag{12.43}$$

In the new variables the elements $\widetilde{w}_1, \ldots, \widetilde{w}_N$ of $\widetilde{\mathcal{W}}(\mathfrak{gl}_N)$ become the respective coefficients $\mathcal{E}_1, \ldots, \mathcal{E}_N$ in the expansion in $\mathrm{S}\big(t^{-1}\mathfrak{h}[t^{-1}]\big) \otimes \mathbb{C}[\tau]$,

$$\big(\tau + \mu_N[-1]\big) \ldots \big(\tau + \mu_1[-1]\big) = \tau^N + \mathcal{E}_1\, \tau^{N-1} + \cdots + \mathcal{E}_N, \tag{12.44}$$

where for the element τ we have the relations

$$\big[\tau, \mu_i[r]\big] = -r\, \mu_i[r-1], \tag{12.45}$$

implied by (7.1). We will also need the operator $T = \operatorname{ad} \tau$ which is the derivation of the algebra $\mathrm{S}\big(t^{-1}\mathfrak{h}[t^{-1}]\big)$ so that $T\,1 = 0$ and it is defined on the generators by

$$T\, \mu_i[r] = -r\, \mu_i[r-1]. \tag{12.46}$$

To write a more explicit formula for $\mathcal{E}_m$, we need to recall standard noncommutative versions of symmetric functions. Suppose that $x_1, \ldots, x_N$ are variables which can be thought of as elements of a not necessarily commutative associative algebra.

The (generally, noncommutative) *complete* and *elementary symmetric functions* in the variables $x_1, \dots, x_N$ are defined by the respective formulas

$$h_m(x_1, \dots, x_N) = \sum_{i_1 \leqslant \dots \leqslant i_m} x_{i_1} \dots x_{i_m}, \tag{12.47}$$

$$e_m(x_1, \dots, x_N) = \sum_{i_1 > \dots > i_m} x_{i_1} \dots x_{i_m}, \tag{12.48}$$

for $m \geqslant 1$, and $h_0(x_1, \dots, x_N) = e_0(x_1, \dots, x_N) = 1$. If the variables do commute, the notation agrees with (2.10) and (2.22).

LEMMA 12.4.3. *The complete and elementary symmetric functions are related by the formulas*

$$h_m = \sum_{k=1}^{m} (-1)^{m-k} \sum_{a_1 + \dots + a_k = m} e_{a_1} \dots e_{a_k}, \tag{12.49}$$

$$e_m = \sum_{k=1}^{m} (-1)^{m-k} \sum_{a_1 + \dots + a_k = m} h_{a_1} \dots h_{a_k}, \tag{12.50}$$

with the second sums taken over positive integers a_i, where $h_a = h_a(x_1, \dots, x_N)$ and $e_a = e_a(x_1, \dots, x_N)$.

PROOF. Writing the definitions (12.47) and (12.48) in terms of generating functions we get

$$\sum_{m=0}^{\infty} h_m q^m = (1 - q x_1)^{-1} \dots (1 - q x_N)^{-1},$$

$$\sum_{m=0}^{N} e_m q^m = (1 + q x_N) \dots (1 + q x_1).$$

This implies that

$$\sum_{m=0}^{\infty} h_m q^m = \Big(1 + \sum_{a=1}^{N} e_a (-q)^a\Big)^{-1} = 1 + \sum_{k=1}^{\infty} (-1)^k \Big(\sum_{a=1}^{N} e_a (-q)^a\Big)^k.$$

Therefore (12.49) follows by taking the coefficients of q^m on both sides. By swapping h_m and e_m we get (12.50) by the same argument. □

Now return to the expansion (12.44). Let us specialize the variables by setting $x_i = \tau + \mu_i[-1]$ for $i = 1, \dots, N$ and write the elementary symmetric function as a polynomial in τ.

PROPOSITION 12.4.4. *For $m = 1, \dots, N$ we have*

$$e_m\big(\tau + \mu_1[-1], \dots, \tau + \mu_N[-1]\big) = \sum_{k=0}^{m} \binom{N-k}{m-k} \mathcal{E}_k \tau^{m-k}. \tag{12.51}$$

In particular, $\mathcal{E}_m$ coincides with the constant term of the polynomial (12.51),

$$\mathcal{E}_m = e_m\big(\tau + \mu_1[-1], \dots, \tau + \mu_N[-1]\big) 1, \tag{12.52}$$

assuming $\tau 1 = 0$.

PROOF. Note that relations (12.45) will remain valid under the replacement $\tau \mapsto u + \tau$ for a (commutative) variable u. Hence (12.44) implies

$$\big(u + \tau + \mu_N[-1]\big) \dots \big(u + \tau + \mu_1[-1]\big) = \sum_{k=0}^{N} \mathcal{E}_k \, (u + \tau)^{N-k}.$$

Since the left hand side equals

$$\sum_{m=0}^{N} e_m \big(\tau + \mu_1[-1], \dots, \tau + \mu_N[-1]\big) \, u^{N-m},$$

the desired relation follows by taking the coefficients of u^{N-m} on both sides. □

EXAMPLE 12.4.5. We have

$$\mathcal{E}_1 = \mu_1[-1] + \dots + \mu_N[-1],$$

$$\mathcal{E}_2 = \sum_{i>j} \mu_i[-1] \, \mu_j[-1] + \sum_{j=1}^{N} (N - j) \, \mu_j[-2].$$

□

By analogy with (12.52), for all $m \geqslant 0$ introduce elements $\mathcal{H}_m \in \mathrm{S}\left(t^{-1}\mathfrak{h}[t^{-1}]\right)$ by taking the constant terms of the complete symmetric functions,

$$\mathcal{H}_m = h_m \big(\tau + \mu_1[-1], \dots, \tau + \mu_N[-1]\big) 1.$$

EXAMPLE 12.4.6. We have

$$\mathcal{H}_1 = \mu_1[-1] + \dots + \mu_N[-1],$$

$$\mathcal{H}_2 = \sum_{i \leqslant j} \mu_i[-1] \, \mu_j[-1] + \sum_{j=1}^{N} j \, \mu_j[-2].$$

□

The following is a counterpart of Proposition 12.4.4.

PROPOSITION 12.4.7. *For all $m \geqslant 1$ we have*

$$h_m \big(\tau + \mu_1[-1], \dots, \tau + \mu_N[-1]\big) = \sum_{k=0}^{m} \binom{N+m-1}{m-k} \mathcal{H}_k \, \tau^{m-k}.$$

PROOF. If u is a variable, commuting with all the x_i, then the following identity holds for the noncommutative complete symmetric functions (12.47),

$$h_m(u + x_1, \dots, u + x_N) = \sum_{k=0}^{m} \binom{N+m-1}{m-k} h_k(x_1, \dots, x_N) \, u^{m-k}. \tag{12.53}$$

Indeed, calculating the generating function for the sequence on the right hand side we get

$$\sum_{m=0}^{\infty} \sum_{k=0}^{m} \binom{N+m-1}{m-k} h_k(x_1, \dots, x_N) \, u^{m-k} q^m$$

$$= \sum_{k=0}^{\infty} h_k(x_1, \dots, x_N) q^k \sum_{m=k}^{\infty} \binom{N+m-1}{m-k} u^{m-k} q^{m-k}.$$

This equals

$$\sum_{k=0}^{\infty} h_k(x_1, \dots, x_N) \frac{q^k}{(1 - q u)^{N+k}} = (1 - q u - q x_1)^{-1} \dots (1 - q u - q x_N)^{-1}$$

and so coincides with the generating function of the sequence on the left hand side of (12.53).

Now specialize the variables by $x_i = \tau + \mu_i[-1]$ for $i = 1, \dots, N$ and introduce the coefficients $\mathcal{H}_k^{(m)}$ by

$$h_m\big(\tau + \mu_1[-1], \dots, \tau + \mu_N[-1]\big) = \sum_{k=0}^{m} \mathcal{H}_k^{(m)}\, \tau^{m-k}$$

so that $\mathcal{H}_k^{(k)} = \mathcal{H}_k$. The relation (12.53) gives

$$\sum_{k=0}^{m} \mathcal{H}_k^{(m)}\, (u+\tau)^{m-k} = \sum_{k=0}^{m} \binom{N+m-1}{m-k} \sum_{p=0}^{k} \mathcal{H}_p^{(k)}\, \tau^{k-p}\, u^{m-k}.$$

By taking the coefficients of $u^{m-k}\tau^0$ we get

$$\mathcal{H}_k^{(m)} = \binom{N+m-1}{m-k} \mathcal{H}_k,$$

as required. □

COROLLARY 12.4.8. *Each of the families*

$$T^r \mathcal{E}_m \qquad \text{and} \qquad T^r \mathcal{H}_m$$

with $m = 1, \dots, N$ and $r = 0, 1, \dots$ is algebraically independent and generates the classical $\mathcal{W}$-algebra $\widetilde{\mathcal{W}}(\mathfrak{gl}_N) \subset \mathrm{S}\big(t^{-1}\mathfrak{h}[t^{-1}]\big)$.

PROOF. The claim for the first family follows from the type A part of the proof of Theorem 12.3.1. Furthermore, relations (12.49) and (12.51) imply that each element $\mathcal{H}_m$ is a differential polynomial in the elements $\mathcal{E}_1, \dots, \mathcal{E}_N$ and so belongs to $\widetilde{\mathcal{W}}(\mathfrak{gl}_N)$. The algebraic independence of the family $T^r \mathcal{H}_m$ is deduced from Lemma 12.3.2 by taking into account the algebraic independence of the (commutative) complete symmetric polynomials $h_1, \dots, h_N$ in N variables. □

Note that the classical $\mathcal{W}$-algebra $\widetilde{\mathcal{W}}(\mathfrak{sl}_N)$ associated with the special linear Lie algebra $\mathfrak{sl}_N$ can be obtained as the quotient of $\widetilde{\mathcal{W}}(\mathfrak{gl}_N)$ by the relation $\mathcal{E}_1 = 0$.

Although this follows from Proposition 12.4.1, it is also possible to verify directly that all elements $\mathcal{E}_m$ are annihilated by the screening operators V_i defined in (12.42). This is implied by the following relations for the operators on $\mathrm{S}\big(t^{-1}\mathfrak{h}[t^{-1}]\big)$.

LEMMA 12.4.9. *For $i = 1, \dots, N-1$ we have*

$$V_i\, T = \big(T + \mu_i[-1] - \mu_{i+1}[-1]\big)\, V_i.$$

PROOF. Since both V_i and T are derivations of $\mathrm{S}\big(t^{-1}\mathfrak{h}[t^{-1}]\big)$ so is their commutator. Therefore, it suffices to check the relation $[V_i, T] = \big(\mu_i[-1] - \mu_{i+1}[-1]\big)\, V_i$ on the generators $\mu_j[-r]$. This holds trivially for $j \neq i, i+1$ so we take $j = i$ (the case $j = i+1$ will only differ by an additional sign). Using the notation (12.43) we get

$$V_i\, T\, \mu_i[-r] = r\, V_i\, \mu_i[-r-1] = r\, V_{i\,r}.$$

On the other hand,

$$\big(T + \mu_i[-1] - \mu_{i+1}[-1]\big)\, V_i\, \mu_i[-r] = \big(T + \mu_i[-1] - \mu_{i+1}[-1]\big)\, V_{i\,r-1}.$$

Hence, denoting the generating function (12.43) by $V_i(z)$, we come to verifying the relation

$$\partial_z V_i(z) = \big(T + \mu_i[-1] - \mu_{i+1}[-1]\big)\, V_i(z).$$

It does hold since both sides are equal to

$$V_i(z) \sum_{m=1}^{\infty} \big(\mu_i[-m] - \mu_{i+1}[-m]\big)\, z^{m-1},$$

thus completing the proof. □

As a consequence of Lemma 12.4.9 and the property $T = \operatorname{ad}\tau$, we obtain the corresponding relations for the operators on $\mathrm{S}\big(t^{-1}\mathfrak{h}[t^{-1}]\big) \otimes \mathbb{C}[\tau]$,

$$V_i\, \tau = \big(\tau + \mu_i[-1] - \mu_{i+1}[-1]\big)\, V_i, \qquad i = 1, \dots, N-1, \tag{12.54}$$

where τ is regarded as the operator of left multiplication by τ and V_i acts as the identity operator on $\mathbb{C}[\tau]$. For each i the relation

$$V_i\, \big(\tau + \mu_N[-1]\big) \dots \big(\tau + \mu_1[-1]\big) = 0$$

then follows easily. It reduces to the particular case $N = 2$ where we have

$$\begin{aligned}
&V_1\, \big(\tau + \mu_2[-1]\big)\big(\tau + \mu_1[-1]\big)\\
&\qquad = \Big(\big(\tau + \mu_1[-1] - \mu_2[-1]\big)\, V_1 + \mu_2[-1]\, V_1 - 1\Big)\big(\tau + \mu_1[-1]\big)\\
&\qquad = \big(\tau + \mu_1[-1]\big)\, V_1\, \big(\tau + \mu_1[-1]\big) - \big(\tau + \mu_1[-1]\big) = 0.
\end{aligned}$$

Type B_n. For $i = 1, \dots, n$ and all $r \geqslant 0$ introduce new variables

$$\mu_i[-r-1] = -\frac{1}{r!}\, F_{ii}^{(r)}.$$

Hence we have the relations $h_i^{(r)} = r!\,\big(-\mu_i[-r-1] + \mu_{i+1}[-r-1]\big)$ for $i = 1, \dots, n-1$ and $h_n^{(r)} = -2r!\, \mu_n[-r-1]$. The Cartan matrix of type B_n has the form

$$A = \begin{bmatrix} 2 & -1 & 0 & \dots & 0 & 0 \\ -1 & 2 & -1 & \dots & 0 & 0 \\ 0 & -1 & 2 & \dots & 0 & 0 \\ & & & \ddots & & \\ \dots & \dots & \dots & & \dots & \dots \\ 0 & 0 & 0 & \dots & 2 & -1 \\ 0 & 0 & 0 & \dots & -2 & 2 \end{bmatrix}$$

so that the entries of the diagonal matrix $D = \operatorname{diag}[\epsilon_1, \dots, \epsilon_n]$ in (12.3) are found by

$$\epsilon_1 = \dots = \epsilon_{n-1} = 1 \qquad \text{and} \qquad \epsilon_n = 2.$$

Therefore,

$$r! \sum_{j=1}^{n} a_{ji}\, \frac{\partial}{\partial h_j^{(r)}} = -\frac{\partial}{\partial \mu_i[-r-1]} + \frac{\partial}{\partial \mu_{i+1}[-r-1]}, \qquad i = 1, \dots, n-1,$$

and

$$r! \sum_{j=1}^{n} a_{jn}\, \frac{\partial}{\partial h_j^{(r)}} = -\frac{\partial}{\partial \mu_n[-r-1]}.$$

This implies that the screening operators (12.35) take the form $V_i^\circ = -V_i$, where

$$V_i = \sum_{r=0}^{\infty} V_{i\,r} \Big(\frac{\partial}{\partial \mu_i[-r-1]} - \frac{\partial}{\partial \mu_{i+1}[-r-1]} \Big),$$

for $i = 1, \dots, n-1$, and

$$V_n = \sum_{r=0}^{\infty} V_{n\,r} \frac{\partial}{\partial \mu_n[-r-1]}; \tag{12.55}$$

the coefficients $V_{i\,r}$ are found from the expansions

$$\sum_{r=0}^{\infty} V_{i\,r}\, z^r = \exp \sum_{m=1}^{\infty} \frac{\mu_i[-m] - \mu_{i+1}[-m]}{m} z^m, \qquad i = 1, \dots, n-1$$

and

$$\sum_{r=0}^{\infty} V_{n\,r}\, z^r = \exp \sum_{m=1}^{\infty} \frac{\mu_n[-m]}{m} z^m.$$

In the new variables the elements $\widetilde{w}_2, \dots, \widetilde{w}_{2n+1}$ of $\widetilde{\mathcal{W}}(\mathfrak{o}_{2n+1})$ become the respective coefficients $\mathcal{E}_2, \dots, \mathcal{E}_{2n+1}$ in the expansion in $\mathrm{S}\big(t^{-1}\mathfrak{h}[t^{-1}]\big) \otimes \mathbb{C}[\tau]$,

$$\begin{aligned}\big(\tau - \mu_1[-1]\big) \dots \big(\tau - \mu_n[-1]\big)\, \tau\, \big(\tau + \mu_n[-1]\big) \dots \big(\tau + \mu_1[-1]\big)\\ = \tau^{2n+1} + \mathcal{E}_2\, \tau^{2n-1} + \mathcal{E}_3\, \tau^{2n-2} + \dots + \mathcal{E}_{2n+1}.\end{aligned}$$

By (12.52) we have

$$\mathcal{E}_m = e_m\big(\tau + \mu_1[-1], \dots, \tau + \mu_n[-1], \tau, \tau - \mu_n[-1], \dots, \tau - \mu_1[-1]\big)\, 1.$$

The relation

$$V_i \big(\tau - \mu_1[-1]\big) \dots \big(\tau - \mu_n[-1]\big)\, \tau\, \big(\tau + \mu_n[-1]\big) \dots \big(\tau + \mu_1[-1]\big) = 0 \tag{12.56}$$

can be verified directly for $i = 1, \dots, n-1$ in the same way as for $\mathfrak{gl}_N$ with the use of (12.54). Furthermore, we also have

$$V_n\, \tau = \big(\tau - \mu_n[-1]\big)\, V_n,$$

which is deduced in the same way as (12.54). Therefore,

$$\begin{aligned}V_n\, \big(\tau - \mu_n[-1]\big)\, \tau\, \big(\tau + \mu_n[-1]\big) &= \big(\tau V_n - 1\big)\, \tau\, \big(\tau + \mu_n[-1]\big)\\ &= \tau\, \big(\tau + \mu_n[-1]\big)\, \big(\tau + 2\mu_n[-1]\big)\, V_n,\end{aligned}$$

thus implying (12.56) for $i = n$.

As with type A, relations (12.49) and (12.51) imply that for all $m \geqslant 1$ the elements

$$\mathcal{H}_m = h_m\big(\tau + \mu_1[-1], \dots, \tau + \mu_n[-1], \tau, \tau - \mu_n[-1], \dots, \tau - \mu_1[-1]\big)\, 1$$

belong to $\widetilde{\mathcal{W}}(\mathfrak{o}_{2n+1})$. The same argument as for Corollary 12.4.8 together with the type B part of the proof of Theorem 12.3.1 implies the following.

COROLLARY 12.4.10. *Each of the families*

$$T^r\, \mathcal{E}_m \qquad \text{and} \qquad T^r\, \mathcal{H}_m$$

with $m = 2, 4, \dots, 2n$ *and* $r = 0, 1, \dots$ *is algebraically independent and generates the classical* $\mathcal{W}$*-algebra* $\widetilde{\mathcal{W}}(\mathfrak{o}_{2n+1}) \subset \mathrm{S}\big(t^{-1}\mathfrak{h}[t^{-1}]\big)$. □

Type D_n. By using the determinant $D = D(\partial)$ given in (12.28) define the elements $e_m \in \mathcal{V}(\mathfrak{p}) \otimes \mathbb{C}[\partial]$ as the coefficients of the formal power series in q,

$$\sum_{m=0}^{\infty} e_m\, q^m = q^{2n-1} D(\partial + q^{-1}).$$

Denote this series by $e(q)$. Furthermore, define elements $h_m \in \mathcal{V}(\mathfrak{p}) \otimes \mathbb{C}[\partial]$ as the coefficients of the series

$$h(q) = \sum_{m=0}^{\infty} h_m\, q^m, \qquad h(q) = e(-q)^{-1}.$$

By Theorem 12.2.5, all coefficients of the differential operators e_m and h_m belong to $\mathcal{W}(\mathfrak{o}_{2n})$. Their images under the homomorphism (12.33) are described as follows. Set $a_{i\,i} = \partial + F_{i\,i}$. We have

$$\phi : e(q) \mapsto (1 + q\, a_{1\,1}) \dots (1 + q\, a_{n\,n})\, (1 + q\,\partial)^{-1}\, (1 + q\, a_{n'\,n'}) \dots (1 + q\, a_{1'\,1'})$$

and hence

$$\phi : h(q) \mapsto (1 - q\, a_{1'\,1'})^{-1} \dots (1 - q\, a_{n'\,n'})^{-1}\, (1 - q\,\partial)\, (1 - q\, a_{n\,n})^{-1} \dots (1 - q\, a_{1\,1})^{-1}.$$

Observe that since $a_{n\,n} + a_{n'\,n'} = 2\partial$, we have the relation

$$(1 - q\, a_{n'\,n'})^{-1}\, (1 - q\,\partial)\, (1 - q\, a_{n\,n})^{-1} = \frac{1}{2}\Big((1 - q\, a_{n\,n})^{-1} + (1 - q\, a_{n'\,n'})^{-1} \Big).$$

Therefore, the images of the elements h_m can be written explicitly as

$$\phi(h_m) = \frac{1}{2} \sum_{k_{1'} + \dots + k_1 = m} a_{1'\,1'}^{k_{1'}} \dots a_{n'\,n'}^{k_{n'}}\, a_{n-1\,n-1}^{k_{n-1}} \dots a_{1\,1}^{k_1} + \frac{1}{2} \sum_{k_{1'} + \dots + k_1 = m} a_{1'\,1'}^{k_{1'}} \dots a_{(n-1)'\,(n-1)'}^{k_{(n-1)'}}\, a_{n\,n}^{k_n} \dots a_{1\,1}^{k_1}. \tag{12.57}$$

To embed the classical $\mathcal{W}$-algebra into the symmetric algebra $\mathrm{S}\left(t^{-1}\mathfrak{h}[t^{-1}]\right)$, for all $i = 1, \dots, n$ and $r \geqslant 0$ introduce new variables

$$\mu_i[-r-1] = -\frac{1}{r!}\, F_{i\,i}^{(r)}.$$

Hence

$$h_i^{(r)} = r!\, \big(-\mu_i[-r-1] + \mu_{i+1}[-r-1] \big), \qquad i = 1, \dots, n-1,$$

and

$$h_n^{(r)} = -r!\, \big(\mu_{n-1}[-r-1] + \mu_n[-r-1] \big).$$

The Cartan matrix has the form

$$A = \begin{bmatrix} 2 & -1 & 0 & \dots & 0 & 0 & 0 \\ -1 & 2 & -1 & \dots & 0 & 0 & 0 \\ 0 & -1 & 2 & \dots & 0 & 0 & 0 \\ \dots & \dots & \dots & \ddots & \dots & \dots & \dots \\ 0 & 0 & 0 & \dots & 2 & -1 & -1 \\ 0 & 0 & 0 & \dots & -1 & 2 & 0 \\ 0 & 0 & 0 & \dots & -1 & 0 & 2 \end{bmatrix}$$

so that

$$r!\sum_{j=1}^{n} a_{ji}\,\frac{\partial}{\partial h_j^{(r)}} = -\frac{\partial}{\partial \mu_i[-r-1]} + \frac{\partial}{\partial \mu_{i+1}[-r-1]}, \qquad i=1,\dots,n-1,$$

and

$$r!\sum_{j=1}^{n} a_{jn}\,\frac{\partial}{\partial h_j^{(r)}} = -\frac{\partial}{\partial \mu_{n-1}[-r-1]} - \frac{\partial}{\partial \mu_n[-r-1]}.$$

The diagonal matrix D in (12.3) is the identity matrix. Therefore, the screening operators (12.35) take the form $V_i^\circ = -V_i$, where

$$V_i = \sum_{r=0}^{\infty} V_{i\,r}\Big(\frac{\partial}{\partial \mu_i[-r-1]} - \frac{\partial}{\partial \mu_{i+1}[-r-1]}\Big),$$

for $i=1,\dots,n-1$, and

$$V_n = \sum_{r=0}^{\infty} V_{n\,r}\Big(\frac{\partial}{\partial \mu_{n-1}[-r-1]} + \frac{\partial}{\partial \mu_n[-r-1]}\Big); \tag{12.58}$$

the coefficients $V_{i\,r}$ are found from the expansions

$$\sum_{r=0}^{\infty} V_{i\,r}\, z^r = \exp \sum_{m=1}^{\infty} \frac{\mu_i[-m]-\mu_{i+1}[-m]}{m}\, z^m, \qquad i=1,\dots,n-1$$

and

$$\sum_{r=0}^{\infty} V_{n\,r}\, z^r = \exp \sum_{m=1}^{\infty} \frac{\mu_{n-1}[-m]+\mu_n[-m]}{m}\, z^m.$$

The elements $\widetilde{w}_2, \widetilde{w}_3, \dots, \widetilde{w}_{2n-1}$ and $\widetilde{y}_n$ of the algebra $\widetilde{\mathcal{W}}(\mathfrak{o}_{2n})$ become the respective coefficients $\mathcal{E}_2, \mathcal{E}_3, \dots, \mathcal{E}_{2n-1}$ and $\mathcal{E}_n^\circ$ in the expansion of the pseudo-differential operator

$$\begin{aligned}\big(\tau-\mu_1[-1]\big)\dots\big(\tau-\mu_n[-1]\big)\,\tau^{-1}\,\big(\tau+\mu_n[-1]\big)\dots\big(\tau+\mu_1[-1]\big)\\ = \tau^{\,2n-1}+\mathcal{E}_2\,\tau^{\,2n-3}+\mathcal{E}_3\,\tau^{\,2n-4}+\dots+\mathcal{E}_{2n-1}+(-1)^n\,\mathcal{E}_n^\circ\,\tau^{-1}\,\mathcal{E}_n^\circ.\end{aligned}$$

In particular,

$$\mathcal{E}_n^\circ = \big(\tau-\mu_1[-1]\big)\dots\big(\tau-\mu_n[-1]\big)\,1. \tag{12.59}$$

The identity

$$V_i\,\big(\tau-\mu_1[-1]\big)\dots\big(\tau-\mu_n[-1]\big)\,\tau^{-1}\,\big(\tau+\mu_n[-1]\big)\dots\big(\tau+\mu_1[-1]\big)=0 \tag{12.60}$$

is verified with the use of (12.54) and the additional relations

$$V_i\,\tau^{-1} = \big(\tau+\mu_i[-1]-\mu_{i+1}[-1]\big)^{-1}\,V_i, \qquad i=1,\dots,n-1,$$

and

$$V_n\,\tau^{-1} = \big(\tau+\mu_{n-1}[-1]+\mu_n[-1]\big)^{-1}\,V_n. \tag{12.61}$$

To perform the calculation for the case $i = n$ in (12.60), we may assume that $n = 2$. We have

$$\begin{aligned} V_2 \left(\tau - \mu_1[-1]\right)\left(\tau - \mu_2[-1]\right) \tau^{-1} \left(\tau + \mu_2[-1]\right)\left(\tau + \mu_1[-1]\right) \\ = \Big(\left(\tau - \mu_2[-1]\right) V_2 - 1\Big)\left(\tau - \mu_2[-1]\right) \tau^{-1} \left(\tau + \mu_2[-1]\right)\left(\tau + \mu_1[-1]\right) \\ = \Big(\left(\tau - \mu_2[-1]\right)\left(\tau + \mu_1[-1]\right) V_2 - 2\tau\Big) \tau^{-1} \left(\tau + \mu_2[-1]\right)\left(\tau + \mu_1[-1]\right). \end{aligned}$$

Furthermore, applying the operator V_2 we find

$$\begin{aligned} V_2 \left(\tau + \mu_2[-1]\right)\left(\tau + \mu_1[-1]\right) &= \Big(\left(\tau + \mu_1[-1] + 2\mu_2[-1]\right) V_2 + 1\Big)\left(\tau + \mu_1[-1]\right) \\ &= 2\left(\tau + \mu_1[-1] + \mu_2[-1]\right) \end{aligned}$$

and so by (12.61),

$$V_2 \, \tau^{-1} \left(\tau + \mu_2[-1]\right)\left(\tau + \mu_1[-1]\right) = 2$$

thus completing the calculation. The relations

$$V_i \left(\tau - \mu_1[-1]\right) \dots \left(\tau - \mu_n[-1]\right) 1 = 0, \qquad i = 1, \dots, n,$$

are verified with the use of (12.54).

Using the notation (12.47), we can write the constant terms of the elements (12.57) in the new variables in the form

$$\begin{aligned} (12.62) \quad \mathcal{H}_m = \tfrac{1}{2}\, h_m\big(\tau + \mu_1[-1], \dots, \tau + \mu_{n-1}[-1], \tau - \mu_n[-1], \dots \tau - \mu_1[-1]\big) 1 \\ + \tfrac{1}{2}\, h_m\big(\tau + \mu_1[-1], \dots, \tau + \mu_n[-1], \tau - \mu_{n-1}[-1], \dots \tau - \mu_1[-1]\big) 1. \end{aligned}$$

To summarize, we have the following corollary of the type D part of the proof of Theorem 12.3.1.

COROLLARY 12.4.11. *Each of the families*

$$T^r \mathcal{E}_2,\, T^r \mathcal{E}_4, \dots, T^r \mathcal{E}_{2n-2},\, T^r \mathcal{E}^\circ_n$$

and

$$T^r \mathcal{H}_2,\, T^r \mathcal{H}_4, \dots, T^r \mathcal{H}_{2n-2},\, T^r \mathcal{E}^\circ_n$$

with r running over the set of nonnegative integers is algebraically independent and generates the classical $\mathcal{W}$-algebra $\widetilde{\mathcal{W}}(\mathfrak{o}_{2n}) \subset \mathrm{S}\left(t^{-1}\mathfrak{h}[t^{-1}]\right)$. □

Type C_n. To embed the classical $\mathcal{W}$-algebra into $\mathrm{S}\left(t^{-1}\mathfrak{h}[t^{-1}]\right)$, for $i = 1, \dots, n$ and all $r \geqslant 0$ introduce new variables

$$\mu_i[-r-1] = -\frac{1}{r!}\, F^{(r)}_{ii}.$$

We have the relations $h^{(r)}_i = r!\left(-\mu_i[-r-1] + \mu_{i+1}[-r-1]\right)$ for $i = 1, \dots, n-1$ and $h^{(r)}_n = -r!\, \mu_n[-r-1]$. The Cartan matrix has the form

$$A = \begin{bmatrix} 2 & -1 & 0 & \dots & 0 & 0 \\ -1 & 2 & -1 & \dots & 0 & 0 \\ 0 & -1 & 2 & \dots & 0 & 0 \\ & & & \ddots & & \\ \dots & \dots & \dots & & \dots & \dots \\ 0 & 0 & 0 & \dots & 2 & -2 \\ 0 & 0 & 0 & \dots & -1 & 2 \end{bmatrix}$$

so that the entries of the diagonal matrix $D = \mathrm{diag}\,[\epsilon_1, \dots, \epsilon_n]$ are found by

$$\epsilon_1 = \dots = \epsilon_{n-1} = 1 \qquad \text{and} \qquad \epsilon_n = 1/2.$$

Hence,

$$r! \sum_{j=1}^{n} a_{ji} \frac{\partial}{\partial h_j^{(r)}} = -\frac{\partial}{\partial \mu_i[-r-1]} + \frac{\partial}{\partial \mu_{i+1}[-r-1]}, \qquad i = 1, \dots, n-1,$$

and

$$r! \sum_{j=1}^{n} a_{jn} \frac{\partial}{\partial h_j^{(r)}} = -\frac{2\partial}{\partial \mu_n[-r-1]}.$$

Therefore, for $i = 1, \dots, n-1$ the screening operators (12.35) are $V_i^{\circ} = -V_i$, where

$$V_i = \sum_{r=0}^{\infty} V_{i\,r} \Big(\frac{\partial}{\partial \mu_i[-r-1]} - \frac{\partial}{\partial \mu_{i+1}[-r-1]} \Big),$$

and $V_n^{\circ} = -2V_n$ with

$$V_n = \sum_{r=0}^{\infty} V_{n\,r} \frac{\partial}{\partial \mu_n[-r-1]}; \tag{12.63}$$

the coefficients $V_{i\,r}$ are found from the expansions

$$\sum_{r=0}^{\infty} V_{i\,r}\, z^r = \exp \sum_{m=1}^{\infty} \frac{\mu_i[-m] - \mu_{i+1}[-m]}{m}\, z^m, \qquad i = 1, \dots, n-1$$

and

$$\sum_{r=0}^{\infty} V_{n\,r}\, z^r = \exp \sum_{m=1}^{\infty} \frac{2\mu_n[-m]}{m}\, z^m.$$

The elements $\widetilde{w}_2, \dots, \widetilde{w}_{2n}$ of the algebra $\widetilde{\mathcal{W}}(\mathfrak{sp}_{2n})$ become the respective coefficients $\mathcal{E}_2, \dots, \mathcal{E}_{2n}$ in the expansion in $\mathrm{S}\,(t^{-1}\mathfrak{h}[t^{-1}]) \otimes \mathbb{C}[\tau]$,

$$\begin{aligned}(\tau - \mu_1[-1]) \dots (\tau - \mu_n[-1])\, (\tau + \mu_n[-1]) \dots (\tau + \mu_1[-1]) \\ = \tau^{2n} + \mathcal{E}_2\, \tau^{2n-2} + \mathcal{E}_3\, \tau^{2n-3} + \dots + \mathcal{E}_{2n}.\end{aligned}$$

By (12.52) we have

$$\mathcal{E}_m = e_m\big(\tau + \mu_1[-1], \dots, \tau + \mu_n[-1], \tau - \mu_n[-1], \dots, \tau - \mu_1[-1]\big)1.$$

The relation

$$V_i\, (\tau - \mu_1[-1]) \dots (\tau - \mu_n[-1])\, (\tau + \mu_n[-1]) \dots (\tau + \mu_1[-1]) = 0 \tag{12.64}$$

is verified for $i = 1, \dots, n-1$ in the same way as for $\mathfrak{gl}_N$ with the use of (12.54). In the case $i = n$ we have

$$V_n\, \tau = \big(\tau + 2\mu_n[-1]\big)\, V_n,$$

so that

$$\begin{aligned} V_n\, (\tau - \mu_n[-1])\, (\tau + \mu_n[-1]) &= \Big((\tau + \mu_n[-1])\, V_n - 1 \Big) (\tau + \mu_n[-1]) \\ &= \big(\tau + \mu_n[-1]\big)\big(\tau + 3\mu_n[-1]\big)\, V_n, \end{aligned}$$

and (12.64) with $i = n$ also follows.

By (12.49) and (12.51), for all $m \geqslant 1$ the elements

$$\mathcal{H}_m = h_m\big(\tau + \mu_1[-1], \dots, \tau + \mu_n[-1], \tau - \mu_n[-1], \dots, \tau - \mu_1[-1]\big)1 \tag{12.65}$$

belong to $\widetilde{\mathcal{W}}(\mathfrak{sp}_{2n})$. As with Corollary 12.4.8, the following is implied by the type C part of the proof of Theorem 12.3.1.

COROLLARY 12.4.12. *Each of the families*

$$T^r \mathcal{E}_m \qquad \textit{and} \qquad T^r \mathcal{H}_m$$

with $m = 2, 4, \dots, 2n$ *and* $r = 0, 1, \dots$ *is algebraically independent and generates the classical* $\mathcal{W}$*-algebra* $\widetilde{\mathcal{W}}(\mathfrak{sp}_{2n}) \subset \mathrm{S}\big(t^{-1}\mathfrak{h}[t^{-1}]\big)$. □

12.5. Bibliographical notes

Classical $\mathcal{W}$-algebras $\mathcal{W}(\mathfrak{g})$ were defined by Drinfeld and Sokolov [**33**] and were used to introduce equations of the KdV type for arbitrary simple Lie algebras $\mathfrak{g}$. For their connections to the affine $\mathcal{W}$-algebras see the review by Arakawa [**4**]. In the case $\mathfrak{g} = \mathfrak{gl}_N$ the algebra $\mathcal{W}(\mathfrak{g})$ is isomorphic to the Adler–Gelfand–Dickey algebra [**1**], [**56**]. For the definition of the classical $\mathcal{W}$-algebras via screening operators see [**46**, Chapter 8]. In Section 12.2 we followed [**115**], where generators of $\mathcal{W}(\mathfrak{g})$ in type G_2 were also given. More general results describing generators of the classical $\mathcal{W}$-algebras $\mathcal{W}(\mathfrak{g}, f)$ for arbitrary nilpotent elements f and the Poisson brackets are given by De Sole, Kac and Valeri [**27**], [**28**]. The functions (12.47) and (12.48) can be regarded as specializations of noncommutative symmetric functions in the sense of the general theory developed by Gelfand *et al.* [**57**].

CHAPTER 13

Affine Harish-Chandra isomorphism

We will now discuss an affine version of the Harish-Chandra isomorphism (4.7) where the center $\mathrm{Z}(\mathfrak{g})$ of the universal enveloping algebra $\mathrm{U}(\mathfrak{g})$ is replaced with the center $\mathrm{Z}(\widehat{\mathfrak{g}})$ of the completed universal enveloping algebra $\widetilde{\mathrm{U}}_{-h^\vee}(\widehat{\mathfrak{g}})$ at the critical level. The role of the algebra of W-invariants in $\mathrm{U}(\mathfrak{h})$ will now be played by a completion of the classical $\mathcal{W}$-algebra associated with the Langlands dual Lie algebra ${}^L\mathfrak{g}$ (corresponding to the transposed Cartan matrix of $\mathfrak{g}$). This affine Harish-Chandra isomorphism turns out to be largely determined by an isomorphism between the Feigin–Frenkel center $\mathfrak{z}(\widehat{\mathfrak{g}})$ and the classical $\mathcal{W}$-algebra $\mathcal{W}({}^L\mathfrak{g})$. Theorem 13.1.1 stated below originates in [**39**] and its detailed proof is given in Frenkel's book [**46**, Theorem 4.3.2]. Together with [**46**, Theorem 4.3.6] these are principal results of the book. In fact, these theorems state much more than just algebra isomorphisms for the centers. The classical $\mathcal{W}$-algebras can be understood as algebras of functions on geometric objects known as *opers* and the affine Harish-Chandra isomorphisms in both versions are coordinate-independent; they are equivariant with respect to changes of variables. The proofs in [**46**] rely on the properties of the *Wakimoto modules* over $\widehat{\mathfrak{g}}$ and the fact that elements of $\mathrm{Z}(\widehat{\mathfrak{g}})$ act by scalar multiplication in these modules.

We will not go beyond the algebraic statement of the affine Harish-Chandra isomorphism and avoid using Wakimoto modules (although we will rely on Theorem 13.1.1 in type C to calculate the images of the Segal–Sugawara vectors). As before, we will only work with the Lie algebras $\mathfrak{g}$ of classical types and adopt a more direct approach based on the Yangian version of the Harish-Chandra isomorphism, as considered in Chapters 10 and 11; cf. [**51**, Section 8]. Nonetheless, we will return to the Wakimoto modules in Chapter 15 and calculate the eigenvalues of the Sugawara operators constructed in Chapters 7 and 8 which act in these modules.

13.1. Feigin–Frenkel centers and classical $\mathcal{W}$-algebras

As in Section 4.2, suppose that $\mathfrak{g} = \mathfrak{n}_- \oplus \mathfrak{h} \oplus \mathfrak{n}_+$ is a triangular decomposition of a simple Lie algebra $\mathfrak{g}$. Regard $\mathfrak{h}$ as a subalgebra of $\widehat{\mathfrak{g}}$ via the embedding taking $H \in \mathfrak{h}$ to $H[0]$. The adjoint action of $\mathfrak{h}$ on $t^{-1}\mathfrak{g}[t^{-1}]$ extends to the universal enveloping algebra and we have a natural analogue of the homomorphism (4.6) for the $\mathfrak{h}$-centralizer,

$$\mathfrak{f} : \mathrm{U}\left(t^{-1}\mathfrak{g}[t^{-1}]\right)^{\mathfrak{h}} \to \mathrm{U}\left(t^{-1}\mathfrak{h}[t^{-1}]\right) \tag{13.1}$$

which is the projection to the first summand in the direct sum decomposition

$$\mathrm{U}\left(t^{-1}\mathfrak{g}[t^{-1}]\right)^{\mathfrak{h}} = \mathrm{U}\left(t^{-1}\mathfrak{h}[t^{-1}]\right) \oplus \Big(\mathrm{U}\left(t^{-1}\mathfrak{g}[t^{-1}]\right)^{\mathfrak{h}} \cap \mathrm{U}\left(t^{-1}\mathfrak{g}[t^{-1}]\right) t^{-1}\mathfrak{n}_-[t^{-1}]\Big).$$

The second summand is the kernel of the projection which coincides with

$$\mathrm{U}\left(t^{-1}\mathfrak{g}[t^{-1}]\right)^{\mathfrak{h}} \cap t^{-1}\mathfrak{n}_+[t^{-1}]\,\mathrm{U}\left(t^{-1}\mathfrak{g}[t^{-1}]\right).$$

Note that the roles of $\mathfrak{n}_-$ and $\mathfrak{n}_+$ are interchanged as compared to the projection (4.6). This is necessary for the definition of the homomorphism to be consistent with the traditional formulas for the Wakimoto modules. It is clear that the derivation T on the algebra $\mathrm{U}\left(t^{-1}\mathfrak{g}[t^{-1}]\right)$ defined in (6.13) preserves the $\mathfrak{h}$-centralizer and that the homomorphism $\mathfrak{f}$ commutes with T.

Recall that the Feigin–Frenkel center $\mathfrak{z}(\widehat{\mathfrak{g}})$ is the center of the vertex algebra $V_{-h^\vee}(\mathfrak{g})$ and we can regard $\mathfrak{z}(\widehat{\mathfrak{g}})$ as a commutative subalgebra of $\mathrm{U}\left(t^{-1}\mathfrak{g}[t^{-1}]\right)^{\mathfrak{h}}$; see Section 6.2.

Using Theorem 12.3.1 and the isomorphism (12.38), we will regard the classical $\mathcal{W}$-algebra as a subalgebra of $\mathrm{U}\left(t^{-1}\mathfrak{h}[t^{-1}]\right) \cong \mathcal{V}(\mathfrak{h})$ and simply write $\mathcal{W}(\mathfrak{g})$ instead of $\widetilde{\mathcal{W}}(\mathfrak{g})$. We will identify the Cartan subalgebras of $\mathfrak{g}$ and its Langlands dual ${}^L\mathfrak{g}$ via a natural isomorphism so that $\mathcal{W}({}^L\mathfrak{g})$ will be viewed as a subalgebra of $\mathrm{U}\left(t^{-1}\mathfrak{h}[t^{-1}]\right)$.

The next theorem provides an affine version of the Harish-Chandra isomorphism for $\mathfrak{z}(\widehat{\mathfrak{g}})$.

THEOREM 13.1.1. *The restriction of the homomorphism* (13.1) *to the subalgebra* $\mathfrak{z}(\widehat{\mathfrak{g}})$ *yields an isomorphism*

$$\mathfrak{z}(\widehat{\mathfrak{g}}) \to \mathcal{W}({}^L\mathfrak{g}). \tag{13.2}$$

For the classical types (with the exception of type C), the theorem will follow from the calculation of the Harish-Chandra images of the families of algebraically independent generators of $\mathfrak{z}(\widehat{\mathfrak{g}})$ constructed in Chapters 7 and 8. We will show separately for types A, B and D that these images coincide with the algebraically independent generators of $\mathcal{W}({}^L\mathfrak{g})$ described in Chapter 12. In type C we will rely on Theorem 13.1.1 for the calculation of the images.

As before, we will use the element τ as defined in (7.1) and consider the natural extension of the homomorphism (13.1) to the homomorphism

$$\mathfrak{f} : \mathrm{U}\left(t^{-1}\mathfrak{g}[t^{-1}]\right)^{\mathfrak{h}} \otimes \mathbb{C}[\tau] \to \mathrm{U}\left(t^{-1}\mathfrak{h}[t^{-1}]\right) \otimes \mathbb{C}[\tau], \tag{13.3}$$

which is identical on $\mathbb{C}[\tau]$.

Type A_{N-1}. Take $\mathfrak{g} = \mathfrak{gl}_N$ and use the notation of Section 7.1. Recall also the noncommutative complete and elementary symmetric polynomials defined in (12.47) and (12.48). Set

$$\mu_i[-r] = E_{ii}[-r], \qquad i = 1, \dots, N \quad \text{and} \quad r = 1, 2, \dots.$$

PROPOSITION 13.1.2. *Under the homomorphism* (13.3) *we have*

$$\operatorname{tr}_{1,\dots,m} A^{(m)}\big(\tau + E[-1]_1\big) \dots \big(\tau + E[-1]_m\big) \mapsto e_m\big(\tau + \mu_1[-1], \dots, \tau + \mu_N[-1]\big), \tag{13.4}$$

and

$$\operatorname{tr}_{1,\dots,m} H^{(m)}\big(\tau + E[-1]_1\big) \dots \big(\tau + E[-1]_m\big) \mapsto h_m\big(\tau + \mu_1[-1], \dots, \tau + \mu_N[-1]\big). \tag{13.5}$$

PROOF. We argue as in the proof of Proposition 4.6.1. As we pointed out in Lemma 7.1.2, $\tau + E[-1]$ is a Manin matrix. By Remark 3.2.3, for a Manin matrix M we have

$$\operatorname{tr}_{1,\dots,m} A^{(m)} M_1 \dots M_m = \sum_{N \geqslant i_1 > \dots > i_m \geqslant 1} \sum_{\sigma \in \mathfrak{S}_m} \operatorname{sgn} \sigma \cdot M_{i_{\sigma(1)}\, i_1} \dots M_{i_{\sigma(m)}\, i_m}.$$

Taking $M = \tau + E[-1]$ we find that the image of the product

$$M_{i_{\sigma(1)}\, i_1} \dots M_{i_{\sigma(m)}\, i_m}$$

under the homomorphism $\mathfrak{f}$ is zero unless σ is the identity permutation. In that case the image is

$$M_{i_1 i_1} \dots M_{i_m i_m} = \big(\tau + \mu_{i_1}[-1]\big) \dots \big(\tau + \mu_{i_m}[-1]\big)$$

which proves (13.4). To verify (13.5), we use the following counterpart of (3.18) provided by Remark 3.2.3:

$$\operatorname{tr}_{1,\dots,m} H^{(m)} M_1 \dots M_m$$
$$= \sum_{N \geqslant i_1 \geqslant \dots \geqslant i_m \geqslant 1} \frac{1}{\alpha_1! \dots \alpha_N!} \sum_{\sigma \in \mathfrak{S}_m} M_{i_m\, i_{\sigma(m)}} \dots M_{i_1\, i_{\sigma(1)}},$$

where α_i is the multiplicity of index $i \in \{1, \dots, N\}$ in the multiset $\{i_1, \dots, i_m\}$. The image of the product $M_{i_m\, i_{\sigma(m)}} \dots M_{i_1\, i_{\sigma(1)}}$ under the homomorphism $\mathfrak{f}$ is zero unless σ stabilizes the multiset $\{i_1, \dots, i_m\}$. The desired formula follows by taking into account the number of permutations which stabilize the multiset; cf. the proof of Proposition 4.7.1. □

Note that (13.5) also follows from (13.4) due to the MacMahon Master Theorem (Theorem 3.2.1).

COROLLARY 13.1.3. *Under the homomorphism* (13.3) *we have*

$$\mathfrak{f} : \operatorname{cdet}\big(\tau + E[-1]\big) \mapsto \big(\tau + \mu_N[-1]\big) \dots \big(\tau + \mu_1[-1]\big) \tag{13.6}$$

and

$$\mathfrak{f} : \sum_{m=0}^{\infty} q^m \operatorname{tr}\big(\tau + E[-1]\big)^m \tag{13.7}$$
$$\mapsto \sum_{i=1}^{N} \Big(1 - q\big(\tau + \mu_1[-1]\big)\Big)^{-1} \cdots \Big(1 - q\big(\tau + \mu_i[-1]\big)\Big)^{-1}$$
$$\times \Big(1 - q\big(\tau + \mu_{i-1}[-1]\big)\Big) \cdots \Big(1 - q\big(\tau + \mu_1[-1]\big)\Big),$$

where q is an independent variable.

PROOF. Relation (13.6) is immediate from (7.10) and (13.4), while (13.7) is a consequence of the Newton identity (3.39) for the Manin matrix $M = \tau + E[-1]$. □

COROLLARY 13.1.4. *The restriction of the homomorphism* (13.1) *to the subalgebra $\mathfrak{z}(\widehat{\mathfrak{gl}}_N)$ yields an isomorphism*

$$\mathfrak{z}(\widehat{\mathfrak{gl}}_N) \to \mathcal{W}(\mathfrak{gl}_N). \tag{13.8}$$

Hence, Theorem 13.1.1 *holds for $\mathfrak{g} = \mathfrak{gl}_N$.*

PROOF. By Theorem 7.1.4, the coefficients $\phi_1, \dots, \phi_N$ of $\mathrm{cdet}\big(\tau + E[-1]\big)$ form a complete set of Segal–Sugawara vectors for $\mathfrak{gl}_N$. This means that the elements $T^r\phi_m$ with $m = 1, \dots, N$ and $r = 0, 1, \dots$ are algebraically independent generators of $\mathfrak{z}(\widehat{\mathfrak{gl}}_N)$. By Corollary 13.1.3 their respective images under the homomorphism $\mathfrak{f}$ are the elements $T^r\mathcal{E}_m$ as defined in (12.44). However, by Corollary 12.4.8 those images are algebraically independent generators of the classical $\mathcal{W}$-algebra $\mathcal{W}(\mathfrak{gl}_N)$. Hence, the homomorphism $\mathfrak{f}$ induces the isomorphism (13.8). □

We will now give an alternative proof of a slightly different form of (13.4) (and hence of Proposition 13.1.2). This argument relies on the calculations of Harish-Chandra images for the dual Yangian $\mathrm{Y}^+(\mathfrak{gl}_N)$ and it is this approach which we will be able to generalize to the other classical types below. In the following we use the notation of Section 10.5.

PROPOSITION 13.1.5. *Under the homomorphism* (13.3) *we have*

$$\mathfrak{f} : \mathrm{tr}_{1,\dots,m} A^{(m)} \big(\partial_u + E(u)_{+1}\big) \dots \big(\partial_u + E(u)_{+m}\big) \\ \mapsto e_m\big(\partial_u + \mu_1(u), \dots, \partial_u + \mu_N(u)\big),$$

where

$$\mu_i(u) = \sum_{r=0}^{\infty} \mu_i[-r-1]\, u^r.$$

PROOF. Due to the consistency of the definitions of the homomorphisms (10.24) and (13.1) and the arguments of Section 10.5, it is sufficient to calculate the classical limit of the Harish-Chandra image of the element (10.40). This element equals (10.42) and so by Corollary 10.2.2, the image is given by

$$\sum_{k=0}^{m} (-1)^k \binom{N-k}{m-k} \sum_{N \geqslant i_1 > \dots > i_k \geqslant 1} \lambda^+_{i_1}(u) \cdots \lambda^+_{i_k}(u-k+1)\, e^{-k\,\partial_u}$$

which coincides with

$$\sum_{N \geqslant i_1 > \dots > i_m \geqslant 1} \big(1 - \lambda^+_{i_1}(u) e^{-\partial_u}\big) \dots \big(1 - \lambda^+_{i_m}(u) e^{-\partial_u}\big).$$

By the definition (10.25), the classical limit of $1 - \lambda^+_i(u)e^{-\partial_u}$ coincides with the expression $\partial_u + \mu_i(u)$ thus completing the proof. □

REMARK 13.1.6. Note that relation (13.4) is equivalent to Proposition 13.1.5. Furthermore, the corresponding counterparts of relations (13.5), (13.6) and (13.7) are obtained by replacing τ with ∂_u, the matrix $E[-1]$ with $E(u)_+$, and $\mu_i[-1]$ with $\mu_i(u)$. The equivalence is a consequence of the vertex algebra structure on the vacuum module $V_{-N}(\mathfrak{gl}_N)$ as outlined in Section 6.6. Applying the state-field correspondence map to elements of $\mathfrak{z}(\widehat{\mathfrak{gl}}_N) \subset V_{-N}(\mathfrak{gl}_N)$, we get formal Laurent series in z whose coefficients are endomorphisms of $V_{-N}(\mathfrak{gl}_N)$. For the elements under discussion, these series are given by formulas which coincide with those given in (7.23), (7.24), (7.25) and (7.26) for the completed universal enveloping algebra. In accordance with the vacuum axioms, the application, say, of (7.23) to the vacuum vector $1 \in V_{-N}(\mathfrak{gl}_N)$ yields formal power series in z which appear in Proposition 13.1.5 (with the variable changed from z to u). Moreover, the evaluations of the series at $u = 0$ recover the original elements of $\mathfrak{z}(\widehat{\mathfrak{gl}}_N)$ in (13.4) and such evaluations are consistent with the Harish-Chandra homomorphism. □

Type B_n. Now take $\mathfrak{g} = \mathfrak{o}_N$ with odd $N = 2n+1$ and use the notation of Section 8.1. Consider the Segal–Sugawara vectors provided by Theorem 8.1.6. Set

$$\mu_i[-r] = F_{ii}[-r], \qquad i = 1, \dots, n \quad \text{and} \quad r = 1, 2, \dots$$

and recall the notation (12.47).

PROPOSITION 13.1.7. *Under the homomorphism* (13.3) *the image of the polynomial*

$$\gamma_m(N) \operatorname{tr}_{1,\dots,m} S^{(m)} (\tau + F[-1]_1) \dots (\tau + F[-1]_m)$$

equals

$$h_m(\tau + \mu_1[-1], \dots, \tau + \mu_n[-1], \tau - \mu_n[-1], \dots \tau - \mu_1[-1]).$$

PROOF. We will apply the argument similar to the one used in the proof of Proposition 13.1.5. As in Section 10.5, extend the ascending filtration on the dual Yangian $\widehat{\mathrm{Y}}^+(\mathfrak{o}_N)$ defined by $\deg' t_{ij}^{(-r)} = -r$ to the algebra of formal series $\widehat{\mathrm{Y}}^+(\mathfrak{o}_N)[[u, \partial_u]]$ by setting $\deg u = 1$ and $\deg \partial_u = -1$ so that the associated graded algebra is isomorphic to $\mathrm{U}\,(t^{-1}\mathfrak{o}_N[t^{-1}])[[u, \partial_u]]$. The element

$$\gamma_m(N) \operatorname{tr}_{1,\dots,m} S^{(m)} (1 - T_1^+(u) e^{-\partial_u}) \dots (1 - T_m^+(u) e^{-\partial_u}) \tag{13.9}$$

has degree $-m$ and its image in the graded algebra coincides with

$$\gamma_m(N) \operatorname{tr}_{1,\dots,m} S^{(m)} (\partial_u + F(u)_{+1}) \dots (\partial_u + F(u)_{+m}), \tag{13.10}$$

where

$$F(u)_+ = \sum_{r=1}^{\infty} F[-r]\, u^{r-1}.$$

Write (13.9) in the form

$$\gamma_m(N) \operatorname{tr}_{1,\dots,m} S^{(m)} \sum_{k=0}^{m} \sum_{1 \leqslant i_1 < \dots < i_k \leqslant m} (-1)^k\, T_{i_1}^+(u) \dots T_{i_k}^+(u-k+1)\, e^{-k\partial_u}.$$

Each product $T_{i_1}^+(u) \dots T_{i_k}^+(u-k+1)$ equals $P\, T_1^+(u) \dots T_k^+(u-k+1)\, P^{-1}$, where P is the image in (11.39) (with the identity component in the last tensor factor) of a permutation $p \in \mathfrak{S}_m$ such that $p(r) = i_r$ for $r = 1, \dots, k$. By (1.31) and the cyclic property of trace, we can bring the above expression to the form

$$\gamma_m(N) \sum_{k=0}^{m} (-1)^k \binom{m}{k} \operatorname{tr}_{1,\dots,m} S^{(m)}\, T_1^+(u) \dots T_k^+(u-k+1)\, e^{-k\partial_u}. \tag{13.11}$$

Now apply (5.32) to calculate the partial traces of the symmetrizer $S^{(m)}$ over the copies $k+1, \dots, m$ of the algebra $\operatorname{End} \mathbb{C}^N$ to get

$$\operatorname{tr}_{k+1,\dots,m} S^{(m)} = \frac{\gamma_k(N)}{\gamma_m(N)} \binom{N+m-2}{m-k} \binom{m}{k}^{-1} S^{(k)}.$$

Then (13.11) equals

$$\sum_{k=0}^{m} (-1)^k\, \gamma_k(N) \binom{N+m-2}{m-k} \operatorname{tr}_{1,\dots,k} S^{(k)}\, T_1^+(u) \dots T_k^+(u-k+1)\, e^{-k\partial_u}.$$

Applying the conjugation by the longest permutation in $\mathfrak{S}_k$ and using the cyclic property of trace we get

$$\mathrm{tr}_{1,\dots,k}\, S^{(k)}\, T_1^+(u)\dots T_k^+(u-k+1) = \mathrm{tr}_{1,\dots,k}\, T_k^+(u)\dots T_1^+(u-k+1)\, S^{(k)}$$

which equals

$$\mathrm{tr}_{1,\dots,k}\, S^{(k)}\, T_1^+(u-k+1)\dots T_k^+(u)$$

by (11.11) and the defining relations (11.37) for the dual Yangian. Hence by Theorem 11.2.1, the Harish-Chandra image of the expression (13.9) is found by

$$\sum_{k=0}^{m}(-1)^k\,\gamma_k(N)\binom{N+m-2}{m-k}\sum_{N\geqslant i_1\geqslant\dots\geqslant i_k\geqslant 1}\lambda_{i_1}^+(u-k+1)\dots\lambda_{i_k}^+(u)\,e^{-k\,\partial_u}$$

with the condition that $n+1$ occurs among the summation indices $i_1,\dots,i_k$ at most once. We can rewrite the image in the form

$$\sum_{k=0}^{m}(-1)^k\,\gamma_k(N)\binom{N+m-2}{m-k}\sum_{1\leqslant j_1\leqslant\dots\leqslant j_k\leqslant N}\lambda_{j_1}^+(u)\,e^{-\partial_u}\dots\lambda_{j_k}^+(u)\,e^{-\partial_u} \tag{13.12}$$

with the condition that $n+1$ occurs among the summation indices $j_1,\dots,j_k$ at most once.

The next step is to express (13.12) in terms of the new variables

$$\nu_i(u) = 1-\lambda_i^+(u)\,e^{-\partial_u},\qquad i=1,\dots,N. \tag{13.13}$$

This is done by a combinatorial argument as shown in the following lemma.

LEMMA 13.1.8. *The expression* (13.12) *multiplied by* $2\,(-1)^{m+1}\binom{N/2-2}{N+m-2}$ *equals*

$$\sum_{r=0}^{m}(-1)^r\binom{N/2-2}{N+r-3}\sum_{a_1+\dots+a_{1'}=r}\nu_1(u)^{a_1}\dots\nu_n(u)^{a_n}\nu_{n'}(u)^{a_{n'}}\dots\nu_{1'}(u)^{a_{1'}}+$$

$$\sum_{r=1}^{m}(-1)^r\binom{N/2-2}{N+r-3}\sum_{a_1+\dots+a_{1'}=r-1}\nu_1(u)^{a_1}\dots\nu_n(u)^{a_n}\,(\nu_{n+1}(u)-2)$$
$$\times\,\nu_{n'}(u)^{a_{n'}}\dots\nu_{1'}(u)^{a_{1'}},$$

where $a_1,\dots,a_{1'}$ *run over nonnegative integers.*

PROOF. The statement is verified by substituting (13.13) into both terms and calculating the coefficients of the sum

$$\sum_{1\leqslant i_1\leqslant\dots\leqslant i_k\leqslant N}\lambda_{i_1}^+(u)\,e^{-\partial_u}\dots\lambda_{i_k}^+(u)\,e^{-\partial_u} \tag{13.14}$$

for all $0\leqslant k\leqslant m$, where $n+1$ occurs among the summation indices $i_1,\dots,i_k$ at most once. Note the following expansion formula for the noncommutative complete symmetric functions (12.47), which is a consequence of (12.53),

$$h_r(1-x_1,\dots,1-x_p)=\sum_{k=0}^{r}(-1)^k\binom{p+r-1}{r-k}\,h_k(x_1,\dots,x_p). \tag{13.15}$$

Take $x_i=\lambda_i^+(u)\,e^{-\partial_u}$ with $i=1,\dots,n,n'\dots,1'$ and apply (13.15) with $p=2n$ to the first term in the expression of the lemma. Using a similar expansion for the second term we find that the coefficient of the sum (13.14) in the entire expression will be found as

$$\sum_{r=k}^{m}(-1)^{r-k}\binom{N/2-2}{N+r-3}\binom{N+r-3}{r-k}=\binom{N/2-2}{N+k-3}\binom{N/2+m-1}{m-k},$$

which coincides with

$$2\,(-1)^{m-k+1}\,\gamma_k(N)\binom{N/2-2}{N+m-2}\binom{N+m-2}{m-k},$$

as claimed. □

Denote the expression in Lemma 13.1.8 by A_m. Since the degree of the element (13.9) is $-m$, its Harish-Chandra image (13.12) and the expression A_m also have degree $-m$. Observe that the terms in the both sums of A_m are independent of m so that $A_{m+1}=A_m+B_{m+1}$, where

$$\begin{aligned}B_{m+1}=(-1)^{m+1}&\binom{N/2-2}{N+m-2}\\ \times&\sum_{a_1+\dots+a_{1'}=m+1}\nu_1(u)^{a_1}\dots\nu_n(u)^{a_n}\nu_{n'}(u)^{a_{n'}}\dots\nu_{1'}(u)^{a_{1'}}\\ +(-1)^{m+1}&\binom{N/2-2}{N+m-2}\sum_{a_1+\dots+a_{1'}=m}\nu_1(u)^{a_1}\dots\nu_n(u)^{a_n}\big(\nu_{n+1}(u)-2\big)\\ &\times\nu_{n'}(u)^{a_{n'}}\dots\nu_{1'}(u)^{a_{1'}}.\end{aligned}$$

Since A_{m+1} has degree $-m-1$, its component of degree $-m$ is zero, and so the sum of the homogeneous components of degree $-m$ of A_m and B_{m+1} is zero. However, each element $\nu_i(u)$ has degree -1 with the top degree component equal to $\partial_u+\mu_i(u)$, where

$$\mu_i(u)=\sum_{r=0}^{\infty}\mu_i[-r-1]\,u^r. \tag{13.16}$$

This implies that the component of A_m of degree $-m$ equals the component of degree $-m$ of the term

$$2\,(-1)^{m+1}\binom{N/2-2}{N+m-2}\sum_{a_1+\dots+a_{1'}=m}\nu_1(u)^{a_1}\dots\nu_n(u)^{a_n}\nu_{n'}(u)^{a_{n'}}\dots\nu_{1'}(u)^{a_{1'}}.$$

Taking into account the constant factor used in Lemma 13.1.8, we can conclude that the Harish-Chandra image of (13.10) is given by the noncommutative complete symmetric function

$$h_m\big(\partial_u+\mu_1(u),\dots,\partial_u+\mu_n(u),\partial_u+\mu_{n'}(u),\dots,\partial_u+\mu_{1'}(u)\big). \tag{13.17}$$

The proof is completed by setting $u=0$ in the coefficients of the polynomials in ∂_u which appear in (13.10) and (13.17); cf. Remark 13.1.6. □

COROLLARY 13.1.9. *The restriction of the homomorphism* (13.1) *to the subalgebra* $\mathfrak{z}(\widehat{\mathfrak{o}}_{2n+1})$ *yields an isomorphism*

$$\mathfrak{z}(\widehat{\mathfrak{o}}_{2n+1})\to\mathcal{W}(\mathfrak{sp}_{2n}). \tag{13.18}$$

Hence, Theorem 13.1.1 *holds for* $\mathfrak{g}=\mathfrak{o}_{2n+1}$.

PROOF. By Theorem 8.1.9, the elements $\phi_{22}, \phi_{44}, \dots, \phi_{2n\,2n}$ form a complete set of Segal–Sugawara vectors for $\mathfrak{o}_{2n+1}$. Therefore, the elements $T^r \phi_{m\,m}$ with $m = 2, 4, \dots, 2n$ and $r = 0, 1, \dots$ are algebraically independent generators of $\mathfrak{z}(\widehat{\mathfrak{o}}_{2n+1})$. By Proposition 13.1.7, their respective images under the homomorphism $\mathfrak{f}$ are the elements $T^r \mathcal{H}_m$ as defined in (12.65). By Corollary 12.4.12 these images are algebraically independent generators of the classical $\mathcal{W}$-algebra $\mathcal{W}(\mathfrak{sp}_{2n})$. Hence, the homomorphism $\mathfrak{f}$ induces the isomorphism (13.18). □

Type D_n. Take $\mathfrak{g} = \mathfrak{o}_N$ with $N = 2n$ and use the notation of Section 8.1. Consider the Segal–Sugawara vectors provided by Theorem 8.1.6. Set

$$\mu_i[-r] = F_{ii}[-r], \qquad i = 1, \dots, n \quad \text{and} \quad r = 1, 2, \dots.$$

PROPOSITION 13.1.10. *Under the homomorphism* (13.3) *the image of the polynomial*

$$\gamma_m(N)\, \mathrm{tr}_{1,\dots,m}\, S^{(m)} \big(\tau + F[-1]_1\big) \dots \big(\tau + F[-1]_m\big)$$

equals

$$\tfrac{1}{2}\, h_m\big(\tau + \mu_1[-1], \dots, \tau + \mu_{n-1}[-1], \tau - \mu_n[-1], \dots \tau - \mu_1[-1]\big)$$
$$+ \tfrac{1}{2}\, h_m\big(\tau + \mu_1[-1], \dots, \tau + \mu_n[-1], \tau - \mu_{n-1}[-1], \dots \tau - \mu_1[-1]\big).$$

PROOF. We repeat the beginning of the proof of Proposition 13.1.7 with N now taking the even value $2n$, up to the application of the Yangian character formula. This time we apply Theorem 11.2.2 to conclude that the Harish-Chandra image of the expression (13.9) is found by

$$\sum_{k=0}^{m} (-1)^k\, \gamma_k(2n) \binom{2n+m-2}{m-k} \sum_{2n \geqslant i_1 \geqslant \dots \geqslant i_k \geqslant 1} \lambda^+_{i_1}(u-k+1) \dots \lambda^+_{i_k}(u)\, e^{-k\,\partial_u}$$

with the condition that n and n' do not occur simultaneously among the summation indices $i_1, \dots, i_k$. Rewrite the image in the form

$$\sum_{k=0}^{m} (-1)^k\, \gamma_k(2n) \binom{2n+m-2}{m-k} \sum_{1 \leqslant j_1 \leqslant \dots \leqslant j_k \leqslant 2n} \lambda^+_{j_1}(u)\, e^{-\partial_u} \dots \lambda^+_{j_k}(u)\, e^{-\partial_u} \tag{13.19}$$

with the condition that n and n' do not occur simultaneously among the summation indices $j_1, \dots, j_k$. Introducing new variables by the same formulas (13.13) we come to the D_n series counterpart of Lemma 13.1.8, where we use the notation

$$c_r = -\binom{2n+r-2}{n-1}^{-1}.$$

LEMMA 13.1.11. *The expression* (13.19) *multiplied by* $2c_m$ *equals*

$$
\begin{aligned}
2c_m & \sum_{\substack{a_1+\dots+a_{1'}=m\\ a_n=a_{n'}=0}} \nu_1(u)^{a_1}\dots\nu_{1'}(u)^{a_{1'}} \\
& + c_m \sum_{\substack{a_1+\dots+a_{1'}=m\\ \text{only one of } a_n \text{ and } a_{n'} \text{ is zero}}} \nu_1(u)^{a_1}\dots\nu_{1'}(u)^{a_{1'}} \\
& - \sum_{r=1}^{m} \frac{r\,c_r}{n+r-1} \sum_{\substack{a_1+\dots+a_{1'}=r\\ a_n=a_{n'}=0}} \nu_1(u)^{a_1}\dots\nu_{1'}(u)^{a_{1'}} \\
& + \sum_{r=1}^{m} \frac{(n-1)\,c_r}{n+r-1} \sum_{\substack{a_1+\dots+a_{1'}=r\\ \text{only one of } a_n \text{ and } a_{n'} \text{ is zero}}} \nu_1(u)^{a_1}\dots\nu_{1'}(u)^{a_{1'}},
\end{aligned}
$$

where $a_1, \dots, a_{1'}$ *run over nonnegative integers.*

PROOF. Substitute (13.13) into the expression given in the lemma and calculate the coefficients of the sum

$$
\sum_{1\leqslant i_1\leqslant\dots\leqslant i_k\leqslant 2n} \lambda^+_{i_1}(u)\,e^{-\partial_u}\dots\lambda^+_{i_k}(u)\,e^{-\partial_u}. \tag{13.20}
$$

The argument splits into two cases, depending on whether neither of n and n' occurs among the summation indices $i_1, \dots, i_k$ in (13.20) or only one of them occurs. The application of the expansion formula (13.15) brings this to a somewhat lengthy but straightforward calculation with binomial coefficients which we will omit. □

Let A_m denote the four-term expression in Lemma 13.1.11. This expression equals $2c_m$ times the Harish-Chandra image of (13.9) and so A_m has degree $-m$. Hence, the component of degree $-m$ of the expression A_{m+1} is zero. On the other hand, each element $\nu_i(u)$ has degree -1 with the top degree component equal to $\partial_u + \mu_i(u)$, where $\mu_i(u)$ is defined in (13.16). This implies that the component of degree $-m$ in the sum of the third and fourth terms in A_m is zero. Therefore, the component of A_m of degree $-m$ equals the component of degree $-m$ in the sum of the first and the second terms. Taking into account the constant factor $2c_m$, we conclude that the Harish-Chandra image of (13.10) equals the component of degree $-m$ of the sum

$$
\sum_{\substack{a_1+\dots+a_{1'}=m\\ a_n=a_{n'}=0}} \nu_1(u)^{a_1}\dots\nu_{1'}(u)^{a_{1'}} + \frac{1}{2} \sum_{\substack{a_1+\dots+a_{1'}=m\\ \text{only one of } a_n \text{ and } a_{n'} \text{ is zero}}} \nu_1(u)^{a_1}\dots\nu_{1'}(u)^{a_{1'}}
$$

and hence coincides with

$$
\begin{aligned}
&\tfrac{1}{2}\,h_m\big(\partial_u+\mu_1(u),\dots,\partial_u+\mu_{n-1}(u),\partial_u-\mu_n(u),\dots\partial_u-\mu_1(u)\big)\\
&\qquad+\tfrac{1}{2}\,h_m\big(\partial_u+\mu_1(u),\dots,\partial_u+\mu_n(u),\partial_u-\mu_{n-1}(u),\dots\partial_u-\mu_1(u)\big).
\end{aligned} \tag{13.21}
$$

To complete the proof set $u=0$ in the coefficients of the polynomials in ∂_u which appear in (13.10) and (13.21); cf. Remark 13.1.6. □

Recall the additional generator Pfaffian Pf $F[-1]$ of the Feigin–Frenkel center $\mathfrak{z}(\widehat{\mathfrak{o}}_{2n})$ defined by (8.6).

PROPOSITION 13.1.12. *The image of the element* Pf $F[-1]$ *under the homomorphism* (13.1) *equals*

$$\mathfrak{f} : \text{Pf } F[-1] \mapsto \big(\mu_1[-1] - \tau\big) \dots \big(\mu_n[-1] - \tau\big)\, 1.$$

PROOF. By using the notation (12.59) we can state the proposition equivalently in the form $\mathfrak{f}$: Pf $F[-1] \mapsto (-1)^n \mathcal{E}_n^\circ$. We will use the automorphism of the Lie algebra $t^{-1}\mathfrak{o}_{2n}[t^{-1}]$ defined on the generators by

$$F_{k\,l}[r] \mapsto F_{\tilde{k}\,\tilde{l}}[r], \tag{13.22}$$

where $k \mapsto \tilde{k}$ is the involution on the set $\{1, \dots, 2n\}$ such that $n \mapsto n'$, $n' \mapsto n$ and $k \mapsto k$ for all $k \neq n, n'$. Note that this automorphism takes Pf $F[-1]$ to $-$Pf $F[-1]$. Similarly, $\mathcal{E}_n^\circ \mapsto -\mathcal{E}_n^\circ$ with respect to the automorphism of $t^{-1}\mathfrak{h}[t^{-1}]$ induced by (13.22).

Now apply Corollary 12.4.11 and observe that the elements $T^r\mathcal{H}_{2k}$ with $r \geqslant 0$ and $k = 1, \dots, n-1$ are stable under the automorphism (13.22). The Harish-Chandra image $\mathfrak{f}\big(\text{Pf } F[-1]\big)$ is a polynomial in the generators of $\mathcal{W}(\mathfrak{o}_{2n})$ and its degree with respect to the variables $\mu_1[-1], \dots, \mu_n[-1]$ does not exceed n. Therefore, $\mathfrak{f}\big(\text{Pf } F[-1]\big)$ must be proportional to $\mathcal{E}_n^\circ$. Comparing the coefficients of the product $\mu_1[-1] \dots \mu_n[-1]$ in each of these two polynomials we can conclude that $\mathfrak{f}$: Pf $F[-1] \mapsto (-1)^n \mathcal{E}_n^\circ$ as required. □

As with types A and B we have the following.

COROLLARY 13.1.13. *The restriction of the homomorphism* (13.1) *to the subalgebra* $\mathfrak{z}(\widehat{\mathfrak{o}}_{2n})$ *yields an isomorphism*

$$\mathfrak{z}(\widehat{\mathfrak{o}}_{2n}) \to \mathcal{W}(\mathfrak{o}_{2n}). \tag{13.23}$$

Hence, Theorem 13.1.1 *holds for* $\mathfrak{g} = \mathfrak{o}_{2n}$.

PROOF. By Theorem 8.1.9, the elements $T^r\phi_{m\,m}$ with $m = 2, 4, \dots, 2n-2$ together with T^r Pf $F[-1]$ for $r = 0, 1, \dots$ are algebraically independent generators of $\mathfrak{z}(\widehat{\mathfrak{o}}_{2n})$. By Propositions 13.1.10 and 13.1.12 their images under the homomorphism $\mathfrak{f}$ are the elements $T^r\mathcal{H}_m$ as defined in (12.62) and $(-1)^n T^r \mathcal{E}_n^\circ$, respectively. By Corollary 12.4.11 these images are algebraically independent generators of the classical $\mathcal{W}$-algebra $\mathcal{W}(\mathfrak{o}_{2n})$. Hence, the homomorphism $\mathfrak{f}$ induces the isomorphism (13.23). □

Type C_n. Now let $\mathfrak{g} = \mathfrak{sp}_N$ with $N = 2n$ and use the notation of Section 8.3. Consider the Segal–Sugawara vectors provided by Theorem 8.3.2. Set

$$\mu_i[-r] = F_{ii}[-r], \qquad i = 1, \dots, n \quad \text{and} \quad r = 1, 2, \dots$$

and recall the noncommutative elementary symmetric polynomials given in (12.48). By Proposition 8.3.7, expression (8.36) admits an equivalent form which is well-defined for all values $1 \leqslant m \leqslant 2n+1$. We will keep the same notation (8.36) for all these values with the understanding that the equivalent form is used for $m > n$.

PROPOSITION 13.1.14. *For all $1 \leqslant m \leqslant 2n+1$ the image of the polynomial*

$$\gamma_m(-2n)\,\mathrm{tr}_{1,\dots,m}\,S^{(m)}\big(\tau+F[-1]_1\big)\dots\big(\tau+F[-1]_m\big) \tag{13.24}$$

under the homomorphism (13.3) *equals*

$$e_m\big(\tau+\mu_1[-1],\dots,\tau+\mu_n[-1],\tau,\tau-\mu_n[-1],\dots\tau-\mu_1[-1]\big).$$

PROOF. First we will prove the statement under the assumption $m \leqslant n$. Then it will be extended to all values $m \leqslant 2n+1$ by using an argument similar to that of Section 5.5.

As for type B, extend the ascending filtration on the dual Yangian $\widehat{\mathrm{Y}}^{+}(\mathfrak{sp}_{2n})$ defined by $\deg' t_{ij}^{(-r)} = -r$ to the algebra of formal series $\widehat{\mathrm{Y}}^{+}(\mathfrak{sp}_{2n})[[u,\partial_u]]$ by setting $\deg u = 1$ and $\deg \partial_u = -1$ so that the associated graded algebra is isomorphic to $\mathrm{U}\big(t^{-1}\mathfrak{sp}_{2n}[t^{-1}]\big)[[u,\partial_u]]$. The element

$$\gamma_m(-2n)\,\mathrm{tr}_{1,\dots,m}\,S^{(m)}\big(1-T_1^{+}(u)\,e^{-\partial_u}\big)\dots\big(1-T_m^{+}(u)\,e^{-\partial_u}\big) \tag{13.25}$$

has degree $-m$ and its image in the graded algebra coincides with

$$\gamma_m(-2n)\,\mathrm{tr}_{1,\dots,m}\,S^{(m)}\big(\partial_u+F(u)_{+1}\big)\dots\big(\partial_u+F(u)_{+m}\big), \tag{13.26}$$

where

$$F(u)_{+} = \sum_{r=1}^{\infty} F[-r]\,u^{r-1}.$$

Expand (13.25) as

$$\gamma_m(-2n)\,\mathrm{tr}_{1,\dots,m}\,S^{(m)}\sum_{k=0}^{m}\sum_{1\leqslant i_1<\dots<i_k\leqslant m}(-1)^k\,T_{i_1}^{+}(u)\dots T_{i_k}^{+}(u-k+1)\,e^{-k\,\partial_u}.$$

Conjugating by appropriate permutations as in the proof of Proposition 13.1.7, we can bring the expression to the form

$$\gamma_m(-2n)\sum_{k=0}^{m}(-1)^k\binom{m}{k}\,\mathrm{tr}_{1,\dots,m}\,S^{(m)}\,T_1^{+}(u)\dots T_k^{+}(u-k+1)\,e^{-k\,\partial_u}. \tag{13.27}$$

Furthermore, taking the partial traces of the symmetrizer $S^{(m)}$ over the copies $k+1,\dots,m$ of the algebra $\mathrm{End}\,\mathbb{C}^N$ by (5.43) we get

$$\mathrm{tr}_{k+1,\dots,m}\,S^{(m)} = \frac{\gamma_k(-2n)}{\gamma_m(-2n)}\binom{2n-k+1}{m-k}\binom{m}{k}^{-1}S^{(k)}.$$

Hence (13.27) equals

$$\sum_{k=0}^{m}(-1)^k\,\gamma_k(-2n)\binom{2n-k+1}{m-k}\,\mathrm{tr}_{1,\dots,k}\,S^{(k)}\,T_1^{+}(u)\dots T_k^{+}(u-k+1)\,e^{-k\,\partial_u}.$$

By Corollary 11.2.4, the Harish-Chandra image of the expression (13.25) is found by

$$\sum_{k=0}^{m}(-1)^k\,\gamma_k(-2n)\binom{2n-k+1}{m-k} \times \sum_{2n+2\geqslant i_1>\dots>i_k\geqslant 1}\varkappa_{i_1}^{+}(u)\,e^{-\partial_u}\dots\varkappa_{i_k}^{+}(u)\,e^{-\partial_u}. \tag{13.28}$$

LEMMA 13.1.15. *Suppose that $m \leqslant n$. The expression* (13.28) *multiplied by* $2(-1)^m\binom{2n-m+1}{n+1}$ *equals*

$$\sum_{r=0}^{m}(-1)^r\binom{2n-r+2}{n+1} \times \sum_{2n+2\geqslant i_1>\dots>i_r\geqslant 1}\big(1-\varkappa^+_{i_1}(u)e^{-\partial_u}\big)\dots\big(1-\varkappa^+_{i_r}(u)e^{-\partial_u}\big). \tag{13.29}$$

PROOF. Use the expansion formula for the noncommutative elementary symmetric functions (12.48),

$$e_r(1-x_1,\dots,1-x_p)=\sum_{k=0}^{r}(-1)^k\binom{p-k}{r-k}\,e_k(x_1,\dots,x_p).$$

Taking $x_i=\varkappa^+_i(u)e^{-\partial_u}$ with $i=1,\dots,2n+2$, and simplifying the expression involving binomial coefficients, we find that the coefficient of the sum

$$\sum_{2n+2\geqslant i_1>\dots>i_k\geqslant 1}\varkappa^+_{i_1}(u)e^{-\partial_u}\dots\varkappa^+_{i_k}(u)e^{-\partial_u} \tag{13.30}$$

in (13.29) equals

$$(-1)^{m-k}\binom{n-k}{m-k}\binom{2n-k+2}{n+1}$$

which coincides with

$$2(-1)^{m-k}\gamma_k(-2n)\binom{2n-m+1}{n+1}\binom{2n-k+1}{m-k}$$

as claimed. □

The expression (13.29) is well-defined for $m\leqslant n+1$ and we denote it by A_m. The proof of Lemma 13.1.15 shows that if $m=n+1$ then the coefficient of the sum (13.30) in A_{n+1} is zero for all $0\leqslant k\leqslant n$. Furthermore, by the second part of Corollary 11.2.4, the sum (13.30) is zero for $k=n+1$. Therefore, $A_{n+1}=0$.

The degree of the element (13.25) is $-m$ and so, for $m\leqslant n$ the expression A_m also has degree $-m$. Hence, the component of degree $-m$ of the expression A_{m+1} is zero; this holds for $m=n$ as well, because $A_{n+1}=0$. We have $A_{m+1}=A_m+B_{m+1}$, where

$$B_{m+1}=(-1)^{m+1}\binom{2n-m+1}{n+1} \times \sum_{2n+2\geqslant i_1>\dots>i_{m+1}\geqslant 1}\big(1-\varkappa^+_{i_1}(u)e^{-\partial_u}\big)\dots\big(1-\varkappa^+_{i_{m+1}}(u)e^{-\partial_u}\big).$$

The sum of the homogeneous components of degree $-m$ of A_m and B_{m+1} is zero. However, each factor $1-\varkappa^+_i(u)e^{-\partial_u}$ with $i\neq n+1,n+2$ has degree -1 with the top degree component equal to $\partial_u+\mu_i(u)$, where $\mu_i(u)$ is defined in (13.16). Therefore, the components of degree $-m$ in B_{m+1} can only come from the terms corresponding to the sets $i_1>\dots>i_{m+1}$ containing indices $n+1$ or $n+2$. In the cases where only one of these indices occurs, the sum simplifies due to the relations (11.42), since

$$1-\varkappa^+_{n+1}(u)e^{-\partial_u}+1-\varkappa^+_{n+2}(u)e^{-\partial_u}=2.$$

In the cases where both indices occur, by (11.42) we have

$$(1 - \varkappa^{+}_{n+2}(u)\, e^{-\partial_u})(1 - \varkappa^{+}_{n+1}(u)\, e^{-\partial_u}) = 1 - \lambda^{+}_{n'}(u)\, \lambda^{+}_{n}(u-1)\, e^{-2\partial_u}. \tag{13.31}$$

Using (11.50), we can express $\lambda^{+}_{n'}(u) = \lambda^{+}_{n+1}(u)$ as the fraction

$$\lambda^{+}_{n+1}(u) = \frac{\lambda^{+}_{1}(u-n)\, \lambda^{+}_{2}(u-n+1) \dots \lambda^{+}_{n-1}(u-2)}{\lambda^{+}_{1}(u-n-1)\, \lambda^{+}_{2}(u-n) \dots \lambda^{+}_{n}(u-2)}. \tag{13.32}$$

Hence the product (13.31) has degree -1 with the top degree component equal to $2\partial_u$. Thus, taking Lemma 13.1.15 into account, we can conclude that the Harish-Chandra image of (13.26) equals

$$e_m\big(\partial_u + \mu_1(u), \dots, \partial_u + \mu_n(u), \partial_u, \partial_u - \mu_n(u), \dots \partial_u - \mu_1(u)\big). \tag{13.33}$$

The argument is completed by setting $u = 0$ in the coefficients of the polynomials in ∂_u which appear in (13.26) and (13.33); cf. Remark 13.1.6.

To complete the proof of the proposition, we will rely on Theorem 13.1.1 in type C [**46**, Theorem 8.1.5], which states that the restriction of the homomorphism (13.1) to the subalgebra $\mathfrak{z}(\widehat{\mathfrak{sp}}_{2n})$ yields an isomorphism

$$\mathfrak{z}(\widehat{\mathfrak{sp}}_{2n}) \to \mathcal{W}(\mathfrak{o}_{2n+1}). \tag{13.34}$$

For $a \in \{0, 1, \dots, m\}$ the coefficient ϕ_{ma} in the expansion (8.36) is a homogeneous element of $\mathrm{U}\big(t^{-1}\mathfrak{sp}_{2n}[t^{-1}]\big)$ of degree a with respect to the grading defined by $\deg F_{ij}[-r] = r$. Since the homomorphism (13.1) preserves the grading, by Corollary 12.4.10 the image of ϕ_{ma} under the isomorphism (13.34) is a polynomial in the generators $T^r \mathcal{E}_m$ with $m = 2, 4, \dots, 2n$ and $r = 0, 1, \dots$ with $r + m \leqslant a$. For a fixed value of m and varying values of n the coefficients of the polynomial are rational functions in n. Therefore, they are uniquely determined by infinitely many values of $n \geqslant m$. This allows us to extend the range of n for the Harish-Chandra image of (13.24) to all values $n \geqslant (m-1)/2$ for which it is defined. □

13.2. Yangian characters and classical $\mathcal{W}$-algebras

Here we will discuss an alternative way to see that the image of the restriction of the homomorphism (13.1) to the subalgebra $\mathfrak{z}(\widehat{\mathfrak{g}})$ is contained in the classical $\mathcal{W}$-algebra $\mathcal{W}({}^L\mathfrak{g})$. The idea is suggested by the alternative proof of Theorem 7.1.3 given in Section 10.5. Namely, for $\mathfrak{g} = \mathfrak{gl}_N$ the generators of $\mathfrak{z}(\widehat{\mathfrak{gl}}_N)$ are recovered via a classical limit procedure from some elements of the subalgebra of invariants $\mathfrak{z}(\widehat{\mathcal{V}}_{\mathrm{cri}})$. The Harish-Chandra images of those elements are polynomials in $\lambda^{+}_{i}(u + a)$ calculated in Corollary 10.2.2, while Proposition 10.6.3 states that the images belong to the intersection of the kernels of the screening operators. Therefore, to see that the Harish-Chandra images of elements of $\mathfrak{z}(\widehat{\mathfrak{gl}}_N)$ belong to $\mathcal{W}(\mathfrak{gl}_N)$, we can just verify that the classical limits of the polynomials in $\lambda^{+}_{i}(u+a)$ are annihilated by all screening operators V_i introduced in Section 12.2. In this sense, the operators V_i are thus recovered as classical limits of the screening operators S^{+}_{i}.

Type A. Consider the algebra $\mathcal{L}^{+}$ introduced in (10.49) and embed it into the algebra of formal power series in the variables $\mu_i[-r-1]$ with $i = 1, \dots, N$ and $r = 0, 1, \dots$ by setting

$$\lambda^{+}_{i}(a) \mapsto 1 - \sum_{r=0}^{\infty} \mu_i[-r-1]\, a^r. \tag{13.35}$$

Define the degrees of the variables by $\deg \mu_i[-r-1] = -r-1$. Given $A \in \mathcal{L}^+$, consider the corresponding formal power series in the variables $\mu_i[-r-1]$ and take its homogeneous component $\overline{A}$ of the maximum degree. This component is a polynomial and so we have a map

$$\text{(13.36)} \qquad \mathrm{gr} : \mathcal{L}^+ \to \mathbb{C}\big[\mu_i[-r-1] \mid i = 1, \ldots, N, \quad r = 0, 1, \ldots\big], \qquad A \mapsto \overline{A}.$$

Note its property which is immediate from the definition:

$$\text{(13.37)} \qquad \mathrm{gr}\,(AB) = \mathrm{gr}\,(A)\,\mathrm{gr}\,(B).$$

Recall the subalgebra Rep $\mathrm{Y}^+(\mathfrak{gl}_N)$ defined in (10.53).

PROPOSITION 13.2.1. *The image of the restriction of the map* (13.36) *to the subalgebra* Rep $\mathrm{Y}^+(\mathfrak{gl}_N)$ *is contained in the classical $\mathcal{W}$-algebra* $\mathcal{W}(\mathfrak{gl}_N)$ *and so it defines a map*

$$\mathrm{gr} : \mathrm{Rep}\ \mathrm{Y}^+(\mathfrak{gl}_N) \to \mathcal{W}(\mathfrak{gl}_N).$$

PROOF. Similar to (13.35), introduce variables $\sigma_i[-r-1]$ by the expansion

$$\text{(13.38)} \qquad \sigma_i^+(a) \mapsto \sum_{r=0}^{\infty} \sigma_i[-r-1]\, a^r$$

and set $\deg \sigma_i[-r-1] = -r-1$. Regarding a as a formal variable in (10.50) and (10.51), write the screening operators in terms of the new variables. Explicitly, for $i = 1, \ldots, N-1$ set

$$\text{(13.39)} \qquad S_i^\circ : \mu_j[-1] \mapsto \begin{cases} \big(1 - \mu_i[-1]\big) \sum_{k \geqslant 0} \sigma_i[-k-1] & \text{for} \quad j = i \\ -\big(1 - \mu_{i+1}[-1]\big)\, \sigma_i[-1] & \text{for} \quad j = i+1 \\ 0 & \text{for} \quad j \neq i, i+1 \end{cases}$$

and

$$\text{(13.40)} \qquad S_i^\circ : \mu_j[-r-1] \mapsto \frac{T^r}{r!} \Big(S_i^\circ\big(\mu_j[-1]\big)\Big), \qquad r \geqslant 1,$$

where the derivation T is defined in (12.46). The action of S_i° then extends to the entire algebra of formal power series in the variables $\mu_i[-r-1]$ via the Leibniz rule as in (10.52).

Now suppose that $A \in$ Rep $\mathrm{Y}^+(\mathfrak{gl}_N)$ so that $S_i^+ A = 0$ for all $i = 1, \ldots, N-1$. Denote by A° the corresponding formal power series in the variables $\mu_i[-r-1]$. By the definition of the operators S_i°, their restriction to the subalgebra $\mathcal{L}^+$ coincides with the action of the respective operators S_i^+. Therefore, $S_i^\circ A^\circ = 0$. Taking the top degree component $\overline{A}$ of A° we can write

$$S_i^\circ A^\circ = \overline{S}_i \overline{A} + \text{lower degree terms},$$

where the operator $\overline{S}_i$ is given by

$$\text{(13.41)} \qquad \overline{S}_i : \mu_j[-r-1] \mapsto \begin{cases} \sigma_i[-r-1] & \text{for} \quad j = i \\ -\sigma_i[-r-1] & \text{for} \quad j = i+1 \\ 0 & \text{for} \quad j \neq i, i+1. \end{cases}$$

On the other hand, write relations (10.50) in the form

$$\text{(13.42)} \qquad \big(1 - \mu_i(z)\big)\, e^{\partial_z} \sigma_i^+(z) - \big(1 - \mu_{i+1}(z)\big)\, \sigma_i^+(z) = 0,$$

where

$$\mu_j(z) = \sum_{r=0}^{\infty} \mu_j[-r-1]\, z^r$$

and we replaced a by a formal variable z. Setting $\deg z = 1$ and $\deg \partial_z = -1$ take the top degree components in these relations to get

$$\sigma_i^{+\,\prime}(z) = \big(\mu_i(z) - \mu_{i+1}(z)\big)\, \sigma_i^+(z).$$

Hence for the images under $\overline{S}_i$ we have

$$\overline{S}_i : \mu_i(z) \mapsto \exp \int \big(\mu_i(z) - \mu_{i+1}(z)\big)\, dz, \qquad \mu_{i+1}(z) \mapsto -\exp \int \big(\mu_i(z) - \mu_{i+1}(z)\big)\, dz,$$

and $\overline{S}_i : \mu_j(z) \mapsto 0$ for $j \neq i, i+1$. However, due to (12.42) and (12.43) this coincides with the action of the operator V_i on the series $\mu_k(z)$.

Thus, we may conclude that if an element $A \in \mathcal{L}^+$ is annihilated by all operators S_i^+, then its image $\overline{A}$ under the map (13.36) is annihilated by all operators V_i. Together with Theorem 12.4.2 this completes the proof. □

Types B, C **and** D. Now we let $\mathfrak{g} = \mathfrak{g}_N$ and use the notation of Section 11.3. We will construct a map $\mathrm{gr} : \mathrm{Rep}\, \mathrm{Y}^+(\mathfrak{g}_N) \to \mathcal{W}({}^L\mathfrak{g}_N)$ and describe its properties. First, embed $\mathcal{L}^+ = \mathcal{L}^+(\mathfrak{g}_N)$ into the algebra of formal power series in variables $\mu_i[-r-1]$ with $i = 1, \dots, N$ and $r = 0, 1, \dots$ by using (13.35) and taking the quotient by the corresponding relations (11.50). Define the degrees of the variables by $\deg \mu_i[-r-1] = -r-1$. Given $A \in \mathcal{L}^+$, consider the corresponding formal power series in the variables $\mu_i[-r-1]$ and take its homogeneous component $\overline{A}$ of the maximum degree. This component is a polynomial and so we have a map

$$\mathrm{gr} : \mathcal{L}^+ \to \mathbb{C}\big[\mu_i[-r-1] \mid i = 1, \dots, N, \quad r = 0, 1, \dots\big], \qquad A \mapsto \overline{A}. \tag{13.43}$$

Note its property (13.37). Recall the subalgebra $\mathrm{Rep}\, \widehat{\mathrm{Y}}^+(\mathfrak{g}_N)$ defined in (11.53).

Proposition 13.2.2. *The image of the restriction of the map* (13.43) *to the subalgebra* $\mathrm{Rep}\, \widehat{\mathrm{Y}}^+(\mathfrak{g}_N)$ *is contained in* $\mathcal{W}({}^L\mathfrak{g}_N)$ *and so it defines a map*

$$\mathrm{gr} : \mathrm{Rep}\, \widehat{\mathrm{Y}}^+(\mathfrak{g}_N) \to \mathcal{W}({}^L\mathfrak{g}_N).$$

Proof. The proof is quite similar to that of Proposition 13.2.1 so we only point out the changes to be made. Introduce variables $\sigma_i[-r-1]$ by (13.38) and set $\deg \sigma_i[-r-1] = -r-1$. Define operators S_i° for $i = 1, \dots, n$ as in (13.39) and (13.40), where the quotient is now taken by the respective relations (11.51) and (11.52) written in terms of the $\mu_i[-r-1]$ with a understood as a variable. Since relations (11.51) are identical to (10.50), the argument for the operators S_i° with $i = 1, \dots, n-1$ follows the same steps as for type A. To complete the proof for the operator S_n°, consider the three cases separately.

Case $\mathfrak{g} = \mathfrak{o}_{2n+1}$. As with (13.41), the corresponding operator $\overline{S}_n$ is now given by

$$\overline{S}_n : \mu_j[-r-1] \mapsto \begin{cases} 2\sigma_n[-r-1] & \text{for } \ j = n \\ 0 & \text{for } \ j \neq n. \end{cases} \tag{13.44}$$

Relations (11.50) imply

$$\lambda_{n+1}^+(a) = \frac{\lambda_1^+(a-n+1)\, \lambda_2^+(a-n+2) \dots \lambda_n^+(a)}{\lambda_1^+(a-n+1/2)\, \lambda_2^+(a-n+3/2) \dots \lambda_n^+(a-1/2)}.$$

Now use (11.52) and write $\lambda_i^+(a) = 1-\mu_i(a)$ for $i = 1,\dots,n$ to get the corresponding analogue of (13.42). As a result, we get the equation

$$\sigma_n^{+\prime}(z) = 2\mu_n(z)\,\sigma_n^+(z)$$

so that for the images under $\overline{S}_n$ we have

$$\overline{S}_n : \mu_n(z) \mapsto 2\exp 2\int \mu_n(z)\,dz,$$

and $\overline{S}_n : \mu_j(z) \mapsto 0$ for $j \neq n$. This coincides with the action of the operator $2V_n$ associated with $\mathfrak{sp}_{2n}$ on the series $\mu_n(z)$, as defined in (12.63). Hence, if an element $A \in \mathcal{L}^+$ is annihilated by all operators S_i^+, then its image $\overline{A}$ under the map (13.43) is annihilated by all operators V_i associated with $\mathfrak{sp}_{2n}$ which is the Langlands dual Lie algebra of $\mathfrak{o}_{2n+1}$.

Case $\mathfrak{g} = \mathfrak{sp}_{2n}$. Similar to (13.44), we have

$$\overline{S}_n : \mu_j[-r-1] \mapsto \begin{cases} \sigma_n[-r-1] & \text{for} \quad j = n \\ 0 & \text{for} \quad j \neq n. \end{cases}$$

By (13.32),

$$\lambda_{n+1}^+(a) = \frac{\lambda_1^+(a-n)\,\lambda_2^+(a-n+1)\dots\lambda_{n-1}^+(a-2)}{\lambda_1^+(a-n-1)\,\lambda_2^+(a-n)\dots\lambda_n^+(a-2)}.$$

Write $\lambda_i^+(a) = 1 - \mu_i(a)$ for $i = 1,\dots,n$ and use (11.52) to get the equation

$$\sigma_n^{+\prime}(z) = \mu_n(z)\,\sigma_n^+(z).$$

Hence, for the images under $\overline{S}_n$ we have

$$\overline{S}_n : \mu_n(z) \mapsto \exp\int \mu_n(z)\,dz,$$

and $\overline{S}_n : \mu_j(z) \mapsto 0$ for $j \neq n$. This coincides with the action of the operator V_n associated with $\mathfrak{o}_{2n+1}$ on the series $\mu_n(z)$; see (12.55). Therefore, if an element $A \in \mathcal{L}^+$ is annihilated by all operators S_i^+, then its image $\overline{A}$ under the map (13.43) is annihilated by all operators V_i associated with $\mathfrak{o}_{2n+1}$ which is the Langlands dual Lie algebra of $\mathfrak{sp}_{2n}$.

Case $\mathfrak{g} = \mathfrak{o}_{2n}$. Similar to (13.41), we have

$$\overline{S}_n : \mu_j[-r-1] \mapsto \begin{cases} \sigma_n[-r-1] & \text{for} \quad j = n-1, n \\ 0 & \text{for} \quad j \neq n-1, n. \end{cases}$$

We derive from (11.50) that

$$\lambda_{n+1}^+(a) = \frac{\lambda_1^+(a-n+2)\,\lambda_2^+(a-n+3)\dots\lambda_{n-1}^+(a)}{\lambda_1^+(a-n+1)\,\lambda_2^+(a-n+2)\dots\lambda_n^+(a)}.$$

Write $\lambda_i^+(a) = 1 - \mu_i(a)$ for $i = 1,\dots,n$ and use (11.52) to get the equation

$$\sigma_n^{+\prime}(z) = \big(\mu_{n-1}(z) + \mu_n(z)\big)\,\sigma_n^+(z).$$

Hence, for the images under $\overline{S}_n$ we have

$$\overline{S}_n : \mu_{n-1}(z) \mapsto \exp\int \big(\mu_{n-1}(z)+\mu_n(z)\big)\,dz, \quad \mu_n(z) \mapsto \exp\int \big(\mu_{n-1}(z)+\mu_n(z)\big)\,dz,$$

and $\overline{S}_n : \mu_j(z) \mapsto 0$ for $j \neq n-1, n$. This coincides with the action of the operator V_n associated with $\mathfrak{o}_{2n}$ on the series $\mu_{n-1}(z)$ and $\mu_n(z)$, as given in (12.58). Thus, if an element $A \in \mathcal{L}^+$ is annihilated by all operators S_i^+, then its image $\overline{A}$ under the map (13.43) is annihilated by all operators V_i associated with $\mathfrak{o}_{2n}$ which is Langlands self-dual. □

13.3. Harish-Chandra images of Sugawara operators

By Proposition 6.6.1, generators of the center $\mathrm{Z}(\widehat{\mathfrak{g}})$ of the completed universal enveloping algebra $\widetilde{\mathrm{U}}_{-h^\vee}(\widehat{\mathfrak{g}})$ at the critical level can be obtained by the application of the state-field correspondence map (6.35) to a complete set of Segal–Sugawara vectors. We will use this fact to calculate the images of the generators of $\mathrm{Z}(\widehat{\mathfrak{g}})$ in the classical types which were produced in Sections 7.2 and 8.5, under an affine version of the Harish-Chandra isomorphism involving the center $\mathrm{Z}(\widehat{\mathfrak{g}})$.

Given a triangular decomposition $\mathfrak{g} = \mathfrak{n}_- \oplus \mathfrak{h} \oplus \mathfrak{n}_+$, choose an arbitrary basis of $\mathfrak{h}$ and root bases of $\mathfrak{n}_-$ and $\mathfrak{n}_+$. Consider the corresponding basis of the affine Kac–Moody algebra $\widehat{\mathfrak{g}}$ (see (6.5)) comprised of K and the elements of the form $X[r]$ with $r \in \mathbb{Z}$, where X runs over the basis elements of $\mathfrak{g}$. Introduce any linear ordering $\prec$ on the basis elements to satisfy the following conditions. First, each basis element of $t^{-1}\mathfrak{g}[t^{-1}]$ should precede each basis element of $\mathfrak{g}[t]$ and the ordering on the corresponding basis elements of $\widehat{\mathfrak{g}}$ should be consistent with the conditions

$$\mathfrak{n}_-[t] \prec \mathfrak{h}[t] \prec \mathfrak{n}_+[t] \qquad \text{and} \qquad t^{-1}\mathfrak{n}_+[t^{-1}] \prec t^{-1}\mathfrak{h}[t^{-1}] \prec t^{-1}\mathfrak{n}_-[t^{-1}]$$

indicating the ordering between the basis elements belonging to the subspaces of $\widehat{\mathfrak{g}}$.

By the Poincaré–Birkhoff–Witt theorem, any element $x \in \mathrm{U}(\widehat{\mathfrak{g}})$ can be written as a unique linear combination of ordered monomials in the basis elements of $\widehat{\mathfrak{g}}$. Set

$$\widehat{\mathfrak{h}} = \mathfrak{h}\,[t, t^{-1}] \oplus \mathbb{C}K$$

and denote by $x_0 \in \mathrm{U}(\widehat{\mathfrak{h}})$ the component of the linear combination representing the element x, where each monomial does not contain any basis elements $X[r]$ with $X \in \mathfrak{n}_- \oplus \mathfrak{n}_+$. The linear map $\theta : x \mapsto x_0$ defines the projection $\theta : \mathrm{U}(\widehat{\mathfrak{g}}) \to \mathrm{U}(\widehat{\mathfrak{h}})$.

Consider now the universal enveloping algebra at the critical level (i.e., take the quotient of $\mathrm{U}(\widehat{\mathfrak{g}})$ by the ideal generated by $K + h^\vee$) and extend θ by continuity to get the projection

$$\theta : \widetilde{\mathrm{U}}_{-h^\vee}(\widehat{\mathfrak{g}}) \to \widetilde{\mathrm{U}}_{-h^\vee}(\widehat{\mathfrak{h}}),$$

where $\widetilde{\mathrm{U}}_{-h^\vee}(\widehat{\mathfrak{h}})$ denotes the completion of $\mathrm{U}_{-h^\vee}(\widehat{\mathfrak{h}})$ at the critical level defined as in (6.33).

Set $\Pi = \mathrm{S}\left(\mathfrak{h}[t, t^{-1}]\right)$ and define the completion $\widetilde{\Pi}$ of the commutative algebra Π as the inverse limit

$$\widetilde{\Pi} = \lim_{\longleftarrow} \Pi / I_p, \qquad p > 0, \tag{13.45}$$

where I_p denotes the ideal of Π generated by all elements $H[r]$ with $H \in \mathfrak{h}$ and $r \geqslant p$. The linear map $\eta : \mathrm{U}_{-h^\vee}(\widehat{\mathfrak{h}}) \to \Pi$ which takes each ordered monomial to the same monomial regarded as an element of the commutative algebra Π, defines a vector space isomorphism. It extends to an isomorphism of the respective completed vector spaces

$$\eta : \widetilde{\mathrm{U}}_{-h^\vee}(\widehat{\mathfrak{h}}) \to \widetilde{\Pi}.$$

Thus we get a linear map

$$\mathfrak{f} : \widetilde{\mathrm{U}}_{-h^\vee}(\widehat{\mathfrak{g}}) \to \widetilde{\Pi} \tag{13.46}$$

defined as the composition $\mathfrak{f} = \eta \circ \theta$. The next proposition provides an analogue of the Harish-Chandra homomorphism for the completed universal enveloping algebra.

PROPOSITION 13.3.1. *The restriction of the map* (13.46) *to the center* $\mathrm{Z}(\widehat{\mathfrak{g}})$ *of the algebra* $\widetilde{\mathrm{U}}_{-h^\vee}(\widehat{\mathfrak{g}})$ *is a homomorphism of commutative algebras*

$$\mathfrak{f} : \mathrm{Z}(\widehat{\mathfrak{g}}) \to \widetilde{\Pi}. \tag{13.47}$$

PROOF. For $x, y \in \mathrm{Z}(\widehat{\mathfrak{g}})$ set $x_0 = \mathfrak{f}(x)$ and $y_0 = \mathfrak{f}(y)$. Write y as a (possibly infinite) linear combination of ordered monomials in the basis elements of $\widehat{\mathfrak{g}}$. Suppose that

$$m = \prod_a X_{i_a}[-r_a - 1] \prod_b Y_{j_b}[r_b], \qquad r_a, r_b \geqslant 0,$$

is an ordered monomial which occurs in the linear combination. Note its weight property

$$\sum_a \operatorname{wt} X_{i_a} + \sum_b \operatorname{wt} Y_{j_b} = 0 \tag{13.48}$$

implied by the centrality of y, where the weight $\alpha = \operatorname{wt} X \in \mathfrak{h}^*$ of a basis vector $X \in \mathfrak{g}$ is defined by the usual condition $[H, X] = \alpha(H)\, X$ for all $H \in \mathfrak{h}$. Suppose that $m \in \ker \mathfrak{f}$. Since x is in the center, we have

$$xm = \prod_a X_{i_a}[-r_a - 1]\, x \prod_b Y_{j_b}[r_b].$$

To write xm as a linear combination of ordered monomials we will only need to use the commutation relations in $\mathfrak{g}[t]$ and those in $t^{-1}\mathfrak{g}[t^{-1}]$. Since they are weight-preserving, we derive that $xm \in \ker \mathfrak{f}$. Hence a nonzero contribution to the image $\mathfrak{f}(xy)$ can only come from $\mathfrak{f}(xy_0)$, that is, from expressions of the form

$$\prod_a H_{i_a}[-r_a - 1]\, x \prod_b H_{j_b}[r_b], \qquad r_a, r_b \geqslant 0,$$

where $H_{i_a}, H_{j_b} \in \mathfrak{h}$. If p is an ordered monomial which occurs in the linear combination representing x and $\mathfrak{f}(p) = 0$, then applying property (13.48) to the monomial p we conclude that

$$\mathfrak{f} : \prod_a H_{i_a}[-r_a - 1]\, p \prod_b H_{j_b}[r_b] \mapsto 0.$$

Finally, observe that the Lie subalgebras $\mathfrak{h}[t]$ and $t^{-1}\mathfrak{h}[t^{-1}]$ of $\widehat{\mathfrak{h}}$ are abelian which proves that $\mathfrak{f}(xy) = x_0 y_0$. □

The homomorphism (13.47) gives rise to a an analogue of the Harish-Chandra isomorphism as follows. Consider the homomorphism of commutative algebras

$$\mathrm{S}\left(t^{-1}\mathfrak{h}[t^{-1}]\right) \to \widetilde{\Pi}[[z, z^{-1}]],$$

defined on the generators by the rule

$$H[-r-1] \mapsto \frac{\partial_z^r}{r!}\, H(z), \qquad H \in \mathfrak{h}, \quad r \geqslant 0,$$

where

$$H(z) = \sum_{p \in \mathbb{Z}} H[p]\, z^{-p-1}.$$

In particular, given any element $w \in \mathcal{W}(\mathfrak{g}) \subset \mathrm{S}\left(t^{-1}\mathfrak{h}[t^{-1}]\right)$ of the classical $\mathcal{W}$-algebra $\mathcal{W}(\mathfrak{g})$ we obtain a formal Laurent series

$$w \mapsto \sum_{n\in\mathbb{Z}} w_{[n]}\, z^{-n-1}, \qquad w_{[n]} \in \widetilde{\Pi}. \tag{13.49}$$

Suppose that $\{T^r w_i\}$ is a family of algebraically independent generators of $\mathcal{W}(\mathfrak{g})$; see Chapter 12. We define the subalgebra $\widehat{\mathcal{W}}(\mathfrak{g}) \subset \widetilde{\Pi}$ as the completion of the algebra of polynomials in all variables $w_{i\,[n]}$ with respect to the topology which defines $\widetilde{\Pi}$ in (13.45); that is, the topology is determined by the system of neighborhoods of 0 formed by the ideals I_p. With this notation we have the following version of the Harish-Chandra isomorphism; see [**46**, Theorem 4.3.6].

THEOREM 13.3.2. *The homomorphism* (13.47) *is injective. Its image coincides with* $\widehat{\mathcal{W}}({}^L\mathfrak{g})$ *so that we have an isomorphism*

$$\mathfrak{f} : \mathrm{Z}(\widehat{\mathfrak{g}}) \to \widehat{\mathcal{W}}({}^L\mathfrak{g}) \tag{13.50}$$

of commutative algebras. □

It is clear from the construction that the isomorphisms described in Theorems 13.1.1 and 13.3.2 are consistent in the sense that the following diagram commutes:

$$\begin{array}{ccc} \mathfrak{z}(\widehat{\mathfrak{g}}) & \xrightarrow{\ \cong\ } & \mathcal{W}({}^L\mathfrak{g}) \\ \downarrow & & \downarrow \\ \mathrm{Z}(\widehat{\mathfrak{g}}) & \xrightarrow{\ \cong\ } & \widehat{\mathcal{W}}({}^L\mathfrak{g}), \end{array}$$

where the vertical arrows indicate the maps constructed in Section 6.6 and in (13.49). This allows us to find the Harish-Chandra images of the generators of $\mathrm{Z}(\widehat{\mathfrak{g}})$ in the classical types. The formulas stated below for type A_{N-1} follow from Proposition 13.1.2 and Corollary 13.1.3.

Type A_{N-1}. Consider the generators of the center $\mathrm{Z}(\widehat{\mathfrak{gl}}_N)$ produced in Section 7.2. We will use the noncommutative complete and elementary symmetric polynomials defined in (12.47) and (12.48). Set

$$\mu_i[r] = E_{ii}[r] \qquad \text{and} \qquad \mu_i(z) = \sum_{r\in\mathbb{Z}} \mu_i[r]\, z^{-r-1}$$

for $i = 1, \dots, N$.

PROPOSITION 13.3.3. *Under the isomorphism* (13.50) *we have*

$$: \mathrm{tr}_{1,\dots,m}\, A^{(m)}\big(\partial_z + E(z)_1\big)\dots\big(\partial_z + E(z)_m\big) : \\ \mapsto e_m\big(\partial_z + \mu_1(z), \dots, \partial_z + \mu_N(z)\big),$$

$$: \mathrm{tr}_{1,\dots,m}\, H^{(m)}\big(\partial_z + E(z)_1\big)\dots\big(\partial_z + E(z)_m\big) : \\ \mapsto h_m\big(\partial_z + \mu_1(z), \dots, \partial_z + \mu_N(z)\big),$$

$$: \mathrm{cdet}\,\big(\partial_z + E(z)\big) : \mapsto \big(\partial_z + \mu_N(z)\big)\dots\big(\partial_z + \mu_1(z)\big),$$

and

$$\sum_{m=0}^{\infty} q^m \; : \operatorname{tr}\big(\partial_z + E(z)\big)^m : \\ \mapsto \sum_{i=1}^{N} \Big(1 - q\,(\partial_z + \mu_1(z))\Big)^{-1} \cdots \Big(1 - q\,(\partial_z + \mu_i(z))\Big)^{-1} \\ \times \Big(1 - q\,(\partial_z + \mu_{i-1}(z))\Big) \cdots \Big(1 - q\,(\partial_z + \mu_1(z))\Big).$$

□

Now let $\mathfrak{g}_N$ be the classical Lie algebra of type B_n, C_n or D_n and recall the generators of the center $Z(\widehat{\mathfrak{g}}_N)$ produced in Section 8.5. The following Harish-Chandra images of the Sugawara operators are found from the respective formulas of Propositions 13.1.7, 13.1.10, 13.1.12 and 13.1.14.

Type B_n. Set

$$\mu_i[r] = F_{ii}[r] \qquad \text{and} \qquad \mu_i(z) = \sum_{r \in \mathbb{Z}} \mu_i[r]\, z^{-r-1} \tag{13.51}$$

for $i = 1, \dots, N$. Take $N = 2n + 1$.

PROPOSITION 13.3.4. *Under the isomorphism* (13.50) *the image of the series*

$$: \gamma_m(N) \operatorname{tr}_{1,\dots,m} S^{(m)} \big(\partial_z + F(z)_1\big) \dots \big(\partial_z + F(z)_m\big) :$$

equals

$$h_m\big(\partial_z + \mu_1(z), \dots, \partial_z + \mu_n(z), \partial_z - \mu_n(z), \dots \partial_z - \mu_1(z)\big).$$

□

Type D_n. Now take $N = 2n$ and use the notation (13.51).

PROPOSITION 13.3.5. *Under the isomorphism* (13.50) *the image of the series*

$$: \gamma_m(N) \operatorname{tr}_{1,\dots,m} S^{(m)} \big(\partial_z + F(z)_1\big) \dots \big(\partial_z + F(z)_m\big) :$$

equals

$$\tfrac{1}{2}\, h_m\big(\partial_z + \mu_1(z), \dots, \partial_z + \mu_{n-1}(z), \partial_z - \mu_n(z), \dots \partial_z - \mu_1(z)\big) \\ + \tfrac{1}{2}\, h_m\big(\partial_z + \mu_1(z), \dots, \partial_z + \mu_n(z), \partial_z - \mu_{n-1}(z), \dots \partial_z - \mu_1(z)\big).$$

□

Recall the noncommutative Pfaffian defined by (8.51).

PROPOSITION 13.3.6. *Under the isomorphism* (13.50) *we have*

$$\operatorname{Pf} F(z) \mapsto \big(\mu_1(z) - \partial_z\big) \dots \big(\mu_n(z) - \partial_z\big)\, 1,$$

where we assume $\partial_z\, 1 = 0$. □

Type C_n. Take $N = 2n$ and use the notation (13.51).

PROPOSITION 13.3.7. *For all $1 \leqslant m \leqslant 2n+1$ the image of the series*

$$: \gamma_m(-2n) \operatorname{tr}_{1,\dots,m} S^{(m)} \big(\partial_z + F(z)_1\big) \dots \big(\partial_z + F(z)_m\big) :$$

under the homomorphism (13.3) *equals*

$$e_m\big(\partial_z + \mu_1(z), \dots, \partial_z + \mu_n(z), \partial_z, \partial_z - \mu_n(z), \dots \partial_z - \mu_1(z)\big).$$

□

13.4. Harish-Chandra images of Casimir elements

As we pointed out in Section 6.5, generators of the center of the universal enveloping algebra $\mathrm{U}(\mathfrak{g})$ can be obtained from complete sets of Segal–Sugawara vectors via evaluation homomorphisms; see Proposition 6.5.2. Furthermore, the homomorphism (13.1) induces the Harish-Chandra homomorphism

$$\mathrm{U}(\mathfrak{g})^{\mathfrak{h}} \to \mathrm{U}(\mathfrak{h}) \tag{13.52}$$

which is the projection whose kernel is the two-sided ideal $\mathrm{U}(\mathfrak{g})^{\mathfrak{h}} \cap \mathrm{U}(\mathfrak{g})\mathfrak{n}_-$. More precisely, for any nonzero $z \in \mathbb{C}$ we have the commutative diagram

$$\begin{array}{ccc} \mathrm{U}\left(t^{-1}\mathfrak{g}[t^{-1}]\right)^{\mathfrak{h}} & \xrightarrow{\ \mathsf{f}\ } & \mathrm{U}\left(t^{-1}\mathfrak{h}[t^{-1}]\right) \\ {\scriptstyle \varrho_z}\downarrow & & \downarrow{\scriptstyle \varrho_z} \\ \mathrm{U}(\mathfrak{g})^{\mathfrak{h}} & \longrightarrow & \mathrm{U}(\mathfrak{h}) \end{array} .$$

Note that for a fixed triangular decomposition $\mathfrak{g} = \mathfrak{n}_- \oplus \mathfrak{h} \oplus \mathfrak{n}_+$, the roles of the subalgebras $\mathfrak{n}_-$ and $\mathfrak{n}_+$ in (4.6) and (13.52) are swapped.

One way to apply the results of Section 13.1 to the calculation of the Harish-Chandra images of Casimir elements with respect to the isomorphism (4.7) is to twist the homomorphism (13.52) by taking its composition with the *Chevalley involution* which is the involutive automorphism of $\mathfrak{g}$ defined in terms of the Chevalley presentation by

$$e_i \mapsto -f_i, \qquad f_i \mapsto -e_i, \qquad h_i \mapsto -h_i;$$

see Section 12.1 for the defining relations. An alternative way is to interpret the images of Casimir elements under the homomorphism (13.52) as their eigenvalues in the *lowest weight representations* of $\mathfrak{g}$ similar to (4.15) and (5.9). We will take the latter approach and consider the classical types separately.

Type A_{N-1}. Given an N-tuple of complex numbers $\mu = (\mu_1, \dots, \mu_N)$, the corresponding *irreducible lowest weight representation* $L^\circ(\mu)$ of the Lie algebra $\mathfrak{gl}_N$ is generated by a nonzero vector $\zeta \in L^\circ(\mu)$ (the *lowest vector*) such that

$$\begin{aligned} E_{ij}\,\zeta &= 0 \qquad \text{for} \quad N \geqslant i > j \geqslant 1, \qquad \text{and} \\ E_{ii}\,\zeta &= \mu_i\,\zeta \qquad \text{for} \quad N \geqslant i \geqslant 1. \end{aligned}$$

Any element $z \in \mathrm{Z}(\mathfrak{gl}_N)$ acts in $L^\circ(\mu)$ by multiplying each vector by a scalar $\mathsf{f}(z)$ which coincides with the image of z under the homomorphism (13.52) with $\mathfrak{g} = \mathfrak{gl}_N$, where μ_i is identified with the element $E_{ii} \in \mathfrak{h}$.

The representation $L^\circ(\mu)$ of $\mathfrak{gl}_N$ is finite-dimensional if and only if

$$\mu_{i+1} - \mu_i \in \mathbb{Z}_+ \qquad \text{for} \quad i = 1, \dots, N-1.$$

If these conditions hold, then $L^\circ(\mu)$ is isomorphic to a highest weight representation,

$$L^\circ(\mu) \cong L(\lambda), \tag{13.53}$$

as defined in Section 4.2, where $\lambda = (\lambda_1, \dots, \lambda_N)$ with $\lambda_i = \mu_{N-i+1}$ for $i = 1, \dots, N$.

We can now derive the following versions of Propositions 4.6.1 and 4.7.1, where we regard z as a variable and embed the universal enveloping algebra into the tensor product $\mathrm{U}(\mathfrak{gl}_N) \otimes \mathbb{C}[z^{-1}, \partial_z]$.

COROLLARY 13.4.1. *For the images under the Harish-Chandra isomorphism* (4.15) *we have*

$$\chi : \mathrm{tr}_{1,\dots,m} A^{(m)} \left(-\partial_z + E_1 z^{-1}\right) \dots \left(-\partial_z + E_m z^{-1}\right) \\ \mapsto \sum_{1 \leqslant i_1 < \dots < i_m \leqslant N} \left(-\partial_z + \lambda_{i_1} z^{-1}\right) \cdots \left(-\partial_z + \lambda_{i_m} z^{-1}\right)$$

and

$$\chi : \mathrm{tr}_{1,\dots,m} H^{(m)} \left(-\partial_z + E_1 z^{-1}\right) \dots \left(-\partial_z + E_m z^{-1}\right) \\ \mapsto \sum_{N \geqslant i_1 \geqslant \dots \geqslant i_m \geqslant 1} \left(-\partial_z + \lambda_{i_1} z^{-1}\right) \cdots \left(-\partial_z + \lambda_{i_m} z^{-1}\right).$$

PROOF. Using the homomorphism (6.30) and taking into account the property (6.31), we obtain from Proposition 13.1.2 that under the homomorphism (13.52),

$$\mathrm{tr}_{1,\dots,m} A^{(m)} \left(-\partial_z + E_1 z^{-1}\right) \dots \left(-\partial_z + E_m z^{-1}\right) \\ \mapsto e_m\left(-\partial_z + \mu_1 z^{-1}, \dots, -\partial_z + \mu_N z^{-1}\right),$$

and

$$\mathrm{tr}_{1,\dots,m} H^{(m)} \left(-\partial_z + E_1 z^{-1}\right) \dots \left(-\partial_z + E_m z^{-1}\right) \\ \mapsto h_m\left(-\partial_z + \mu_1 z^{-1}, \dots, -\partial_z + \mu_N z^{-1}\right).$$

We will regard these images as the eigenvalues of the Casimir elements in finite-dimensional lowest weight representations $L^\circ(\mu)$. Hence by (13.53), their eigenvalues in $L(\lambda)$ are found by replacing μ_i with λ_{N-i+1} for all $i = 1, \dots, N$. Since the Harish-Chandra image of any Casimir element is determined by its eigenvalues in the finite-dimensional representations, the claim follows. □

Note that the formulas of Corollary 13.4.1 are respectively equivalent to those of Propositions 4.6.1 and 4.7.1. This is implied by the relations

$$z^m \left(-\partial_z + E_1 z^{-1}\right) \dots \left(-\partial_z + E_m z^{-1}\right) = (u + E_1 + m - 1) \dots (u + E_m) \tag{13.54}$$

and

$$z^m \left(-\partial_z + \lambda_{i_1} z^{-1}\right) \cdots \left(-\partial_z + \lambda_{i_m} z^{-1}\right) = (u + \lambda_{i_1} + m - 1) \cdots (u + \lambda_{i_m}), \tag{13.55}$$

where $u = -z\partial_z$ so that $zu = (u+1)z$. For the second formula we also use Lemma 4.5.4.

We derive the following from the first formula of Corollary 13.1.3 by the same argument as in the proof of Corollary 13.4.1.

COROLLARY 13.4.2. *For the image under the Harish-Chandra isomorphism* (4.15) *we have*

$$\chi : \operatorname{cdet}\big(-\partial_z + E z^{-1}\big) \mapsto \big(-\partial_z + \lambda_1 z^{-1}\big) \dots \big(-\partial_z + \lambda_N z^{-1}\big). \qquad \square$$

This was also established in Corollary 4.6.2 in an equivalent form. The following formulas are alternatives to Corollary 4.8.2 as they deal with a modified family of Gelfand invariants.

COROLLARY 13.4.3. *Suppose that m is a positive integer. For the image under the Harish-Chandra isomorphism* (4.15) *we have*

$$\begin{aligned}\chi : \operatorname{tr}\big(-\partial_z + E z^{-1}\big)^m \\ \mapsto \sum_{i=1}^{N}\sum_{k=0}^{m} \sum_{i_1 \geqslant \dots \geqslant i_k \geqslant i < j_1 < \dots < j_{m-k}} &\big(-\partial_z + \lambda_{i_1} z^{-1}\big) \cdots \big(-\partial_z + \lambda_{i_k} z^{-1}\big) \\ &\times \big(\partial_z - \lambda_{j_1} z^{-1}\big) \cdots \big(\partial_z - \lambda_{j_{m-k}} z^{-1}\big)\end{aligned}$$

and

$$\begin{aligned}\chi : \operatorname{tr} E(E-1)\dots(E-m+1) \\ \mapsto \sum_{i=1}^{N}\sum_{k=0}^{m} \sum_{i_1 \geqslant \dots \geqslant i_k \geqslant i < j_1 < \dots < j_{m-k}} &\lambda_{i_1}\big(\lambda_{i_2} - 1\big) \cdots \big(\lambda_{i_k} - k + 1\big) \\ &\times \big(-\lambda_{j_1} + k\big) \cdots \big(-\lambda_{j_{m-k}} + m - 1\big).\end{aligned}$$

PROOF. Take the coefficients of q^m in (13.7) and argue in the same way as in the proof of Corollary 13.4.1 to get the first formula. Multiply it on the left by z^m, use (13.54) and (13.55), then set $u = -m + 1$. $\square$

Now let $\mathfrak{g}_N$ be the orthogonal Lie algebra $\mathfrak{o}_N$ or symplectic Lie algebra $\mathfrak{sp}_N$. In the notation of Section 5.1, for any n-tuple of complex numbers $\mu = (\mu_1, \dots, \mu_n)$, the corresponding *irreducible lowest weight representation* $L^\circ(\mu)$ of the Lie algebra $\mathfrak{g}_N$ is generated by a nonzero vector $\zeta \in L^\circ(\mu)$ (the *lowest vector*) such that

$$F_{ij}\,\zeta = 0 \qquad \text{for} \quad N \geqslant i > j \geqslant 1, \qquad \text{and} \tag{13.56}$$

$$F_{ii}\,\zeta = \lambda_i\,\zeta \qquad \text{for} \quad n \geqslant i \geqslant 1. \tag{13.57}$$

Type B_n. If the representation $L^\circ(\mu)$ is finite-dimensional then it is isomorphic to a highest weight representation,

$$L^\circ(\mu) \cong L(\lambda),$$

as defined in Section 5.1, where $\lambda = (\lambda_1, \dots, \lambda_n)$ with $\lambda_i = -\mu_i$ for $i = 1, \dots, n$. We will extend the range of the subscripts by setting $\lambda_{i'} = -\lambda_i$ for $i = 1, \dots, n$. We will now give an alternative way to find the Harish-Chandra images of the Casimir elements constructed in Section 5.4.

COROLLARY 13.4.4. *For the image under the Harish-Chandra isomorphism* (5.9) *we have*

$$\begin{aligned}\chi : \gamma_m(N) \operatorname{tr}_{1,\dots,m} S^{(m)}\big(-\partial_z + F_1 z^{-1}\big) \dots \big(-\partial_z + F_m z^{-1}\big) \\ \mapsto h_m\big(-\partial_z - \lambda_1 z^{-1}, \dots, -\partial_z - \lambda_n z^{-1}, -\partial_z + \lambda_n z^{-1}, \dots -\partial_z + \lambda_1 z^{-1}\big).\end{aligned}$$

Moreover, for $k \geqslant 1$ the image of the Casimir element

$$\gamma_{2k}(N)\,\mathrm{tr}_{1,\dots,2k}\,S^{(2k)}\,(F_1+k-1)\dots(F_{2k}-k)$$

under the Harish-Chandra isomorphism is given by

$$\sum_{1\leqslant i_1\leqslant\dots\leqslant i_{2k}\leqslant 1'}(\lambda_{i_1}-k)(\lambda_{i_2}-k+1)\dots(\lambda_{i_{2k}}+k-1), \tag{13.58}$$

summed over the multisets $\{i_1,\dots,i_{2k}\}$ with entries from $\{1,\dots,n,n',\dots,1'\}$.

Proof. As in the proof of Corollary 13.4.1, the first part follows from Proposition 13.1.7 by the application of the homomorphism (6.30) and replacing μ_i with $-\lambda_i$. Furthermore, using the respective counterparts of (13.54) and (13.55), we find that the Harish-Chandra image of the polynomial

$$\gamma_m(N)\,\mathrm{tr}_{1,\dots,m}\,S^{(m)}\big(u+F_1+m-1\big)\dots\big(u+F_m\big)$$

equals

$$\sum_{1\leqslant i_1\leqslant\dots\leqslant i_m\leqslant 1'}(u-\lambda_{i_1}+m-1)\dots(u-\lambda_{i_m})$$

which can also be written as

$$\sum_{1\leqslant i_1\leqslant\dots\leqslant i_m\leqslant 1'}(u+\lambda_{i_1})\dots(u+\lambda_{i_m}+m-1).$$

For $m=2k$ set $u=-k$ to get the second part of the corollary. □

Comparing this result with Proposition 5.4.1, we find that the polynomial (13.58) coincides with the factorial complete symmetric polynomial $h_k(l_1^2,\dots,l_n^2\,|\,a)$. This equality can be verified directly by using the characterization theorem for the factorial symmetric functions (Proposition 4.3.2). Namely, both elements are symmetric polynomials in $l_1^2,\dots,l_n^2$ of degree k, and their top degree components are both equal to the complete symmetric polynomial $h_k(l_1^2,\dots,l_n^2)$. It remains to verify that each of the elements (13.58) and $h_k(l_1^2,\dots,l_n^2\,|\,a)$ vanishes when $(\lambda_1,\dots,\lambda_n)$ is specialized to a partition with $\lambda_1+\dots+\lambda_n<k$ which is straightforward.

More general identities follow by comparing the formulas of Corollaries 5.4.2 and 13.4.4.

Type D_n. The finite-dimensional lowest weight representation $L^\circ(\mu)$ is isomorphic to a highest weight representation,

$$L^\circ(\mu)\cong L(\lambda),$$

as defined in Section 5.1, where $\lambda=(\lambda_1,\dots,\lambda_n)$ with $\lambda_i=-\mu_i$ for $i=1,\dots,n-1$ and $\lambda_n=(-1)^{n-1}\mu_n$. Set $\lambda_{i'}=-\lambda_i$ for $i=1,\dots,n$.

Corollary 13.4.5. *The image of the differential operator*

$$\gamma_m(N)\,\mathrm{tr}_{1,\dots,m}\,S^{(m)}\big(-\partial_z+F_1\,z^{-1}\big)\dots\big(-\partial_z+F_m\,z^{-1}\big)$$

under the Harish-Chandra isomorphism (5.9) equals

$$\tfrac{1}{2}\,h_m\big(-\partial_z-\lambda_1z^{-1},\dots,-\partial_z-\lambda_{n-1}z^{-1},-\partial_z+\lambda_nz^{-1},\dots-\partial_z+\lambda_1z^{-1}\big)$$

$$+\tfrac{1}{2}\,h_m\big(-\partial_z-\lambda_1z^{-1},\dots,-\partial_z-\lambda_nz^{-1},-\partial_z+\lambda_{n-1}z^{-1},\dots-\partial_z+\lambda_1z^{-1}\big).$$

Moreover, for $k\geqslant 1$ the image of the Casimir element

$$\gamma_{2k}(N)\,\mathrm{tr}_{1,\dots,2k}\,S^{(2k)}\,(F_1+k-1)\dots(F_{2k}-k)$$

under the Harish-Chandra isomorphism is given by

$$\frac{1}{2}\sum_{\substack{1\leqslant i_1\leqslant\dots\leqslant i_{2k}\leqslant 2n\\ i_s\neq n}}(\lambda_{i_1}-k)\dots(\lambda_{i_{2k}}+k-1)+\frac{1}{2}\sum_{\substack{1\leqslant i_1\leqslant\dots\leqslant i_{2k}\leqslant 2n\\ i_s\neq n'}}(\lambda_{i_1}-k)\dots(\lambda_{i_{2k}}+k-1), \tag{13.59}$$

summed over the multisets $\{i_1,\dots,i_{2k}\}$ with entries from $\{1,\dots,n,n',\dots,1'\}$.

PROOF. The formulas are derived from Proposition 13.1.10 in the same way as in the proof of Corollary 13.4.4. □

The polynomial (13.59) coincides with the factorial complete symmetric polynomial $h_k(l_1^2,\dots,l_n^2\,|\,a)$ which occurs as the Harish-Chandra image in Proposition 5.4.1. This equality can also be verified directly with the use of Proposition 4.3.2. More general identities follow by comparing the formulas of Corollaries 5.4.2 and 13.4.5.

Type C_n. Use the definition of the lowest weight representation $L^\circ(\mu)$ of $\mathfrak{sp}_{2n}$ given in (13.56) and (13.57). If this representation is finite-dimensional then it is isomorphic to a highest weight representation,

$$L^\circ(\mu)\cong L(\lambda),$$

as defined in Section 5.1, where $\lambda=(\lambda_1,\dots,\lambda_n)$ with $\lambda_i=-\mu_i$ for $i=1,\dots,n$. We will extend the range of the subscripts by setting $\lambda_0=0$ and $\lambda_{i'}=-\lambda_i$ for $i=1,\dots,n$.

By the results of Section 5.5, the expression (5.37) admits an equivalent form which is well-defined for all values $1\leqslant m\leqslant 2n+1$. We will keep using the same notation for this expression for all these values of m.

COROLLARY 13.4.6. *For $1\leqslant m\leqslant 2n+1$ the image of the differential operator*

$$\gamma_m(-2n)\,\mathrm{tr}_{1,\dots,m}\,S^{(m)}\big(-\partial_z+F_1\,z^{-1}\big)\dots\big(-\partial_z+F_m\,z^{-1}\big)$$

under the Harish-Chandra isomorphism (5.9) equals

$$e_m\big(-\partial_z-\lambda_1z^{-1},\dots,-\partial_z-\lambda_nz^{-1},-\partial_z,-\partial_z+\lambda_nz^{-1},\dots-\partial_z+\lambda_1z^{-1}\big).$$

Moreover, if $m=2k$ is even, then the image of the Casimir element

$$\gamma_{2k}(-2n)\,\mathrm{tr}_{1,\dots,2k}\,S^{(2k)}\,(F_1+k)\dots(F_{2k}-k+1)$$

under the Harish-Chandra isomorphism is given by

$$\sum_{1\leqslant i_1<\dots<i_{2k}\leqslant 1'}(\lambda_{i_1}+k)\dots(\lambda_{i_{2k}}-k+1), \tag{13.60}$$

summed over the subsets $\{i_1,\dots,i_{2k}\}\subset\{1,\dots,n,0,n',\dots,1'\}$, where the indices are ordered by $1<\dots<n<0<n'<\dots<1'$.

PROOF. As in the proof of Corollary 13.4.4, the formulas are implied by Proposition 13.1.14 by the application of the homomorphism (6.30). □

By Proposition 5.5.4, the polynomial (13.60) coincides with the factorial elementary symmetric polynomial $(-1)^k\, e_k(l_1^2, \dots, l_n^2 \,|\, a)$. This also follows from the characterization theorem for the factorial symmetric functions (Proposition 4.3.2). Corollary 5.5.5 implies some more general identities which follow by comparing the Harish-Chandra images of the same expressions.

13.5. Bibliographical notes

The Harish-Chandra images of generators of $\mathfrak{z}(\widehat{\mathfrak{g}})$ were calculated in [**23**] for type A and in [**111**] for types B, C and D. A direct proof of Proposition 13.1.12 can be found in Rozhkovskaya [**137**]. The calculations with the screening operators follow [**113**], but re-written in the context of the dual Yangians.

CHAPTER 14

Higher Hamiltonians in the Gaudin model

In their seminal work [**40**], Feigin, Frenkel and Reshetikhin established a connection between the center $\mathfrak{z}(\widehat{\mathfrak{g}})$ of the affine vertex algebra at the critical level and higher order Hamiltonians in the Gaudin model. They showed that any element of $S \in \mathfrak{z}(\widehat{\mathfrak{g}})$ gives rise to a Hamiltonian and used the Wakimoto modules over the affine Kac–Moody algebra $\widehat{\mathfrak{g}}$ to calculate the eigenvalues of such Hamiltonians on the Bethe vectors. In this section we apply these results and their generalizations [**42**] to get formulas for the action of the higher Gaudin Hamiltonians on tensor products of representations of the Lie algebra $\mathfrak{g}$ of a classical type in an explicit form and calculate the corresponding eigenvalues of the Bethe vectors. We will rely on the constructions of generators of $\mathfrak{z}(\widehat{\mathfrak{g}})$ obtained in Sections 7.1, 8.1 and 8.3, and their Harish-Chandra images found in Section 13.1.

14.1. Bethe ansatz equations

Recall the standard Chevalley generators e_i, h_i, f_i with $i = 1, \dots, n$ of a simple Lie algebra $\mathfrak{g}$ of rank n. The generators h_i form a basis of the Cartan subalgebra $\mathfrak{h}$ of $\mathfrak{g}$, while the e_i and f_i generate the respective nilpotent subalgebras $\mathfrak{n}_+$ and $\mathfrak{n}_-$. Let $A = [a_{ij}]$ be the Cartan matrix of $\mathfrak{g}$ so that the defining relations of $\mathfrak{g}$ take the form

$$[e_i, f_j] = \delta_{ij} h_i, \qquad [h_i, h_j] = 0,$$
$$[h_i, e_j] = a_{ij}\, e_j, \qquad [h_i, f_j] = -a_{ij}\, f_j,$$

together with the Serre relations

$$(\operatorname{ad} e_i)^{1-a_{ij}}\, e_j = 0, \qquad (\operatorname{ad} f_i)^{1-a_{ij}}\, f_j = 0, \qquad i \neq j.$$

Given any element $\chi \in \mathfrak{g}^*$ and a nonzero $z \in \mathbb{C}$, the mapping

$$\mathrm{U}\left(t^{-1}\mathfrak{g}[t^{-1}]\right) \to \mathrm{U}(\mathfrak{g}), \qquad X[r] \mapsto X\, z^r + \delta_{r,-1}\, \chi(X), \tag{14.1}$$

for $X \in \mathfrak{g}$ and $r < 0$, defines an algebra homomorphism; cf. (9.29). Using the coassociativity of the standard coproduct on $\mathrm{U}\left(t^{-1}\mathfrak{g}[t^{-1}]\right)$ defined by

$$\Delta : Y \mapsto Y \otimes 1 + 1 \otimes Y, \qquad Y \in t^{-1}\mathfrak{g}[t^{-1}],$$

for any $\ell \geqslant 1$ we get the homomorphism

$$\mathrm{U}\left(t^{-1}\mathfrak{g}[t^{-1}]\right) \to \mathrm{U}\left(t^{-1}\mathfrak{g}[t^{-1}]\right)^{\otimes \ell} \tag{14.2}$$

as an iterated coproduct map. Now fix distinct complex numbers $z_1, \dots, z_\ell$ and let u be a complex parameter. Applying homomorphisms of the form (14.1) to the tensor factors in (14.2), we get another homomorphism

$$\Psi : \mathrm{U}\left(t^{-1}\mathfrak{g}[t^{-1}]\right) \to \mathrm{U}(\mathfrak{g})^{\otimes \ell}, \tag{14.3}$$

given by

$$\Psi : X[r] \mapsto \sum_{a=1}^{\ell} X_a (z_a - u)^r + \delta_{r,-1}\, \chi(X) \in \mathrm{U}(\mathfrak{g})^{\otimes \ell},$$

where $X_a = 1^{\otimes(a-1)} \otimes X \otimes 1^{\otimes(\ell-a)}$. We will twist this homomorphism by the involutive anti-automorphism

$$\varsigma : \mathrm{U}\left(t^{-1}\mathfrak{g}[t^{-1}]\right) \to \mathrm{U}\left(t^{-1}\mathfrak{g}[t^{-1}]\right), \qquad X[r] \mapsto -X[r], \quad X \in \mathfrak{g}, \tag{14.4}$$

to get the anti-homomorphism

$$\Phi : \mathrm{U}\left(t^{-1}\mathfrak{g}[t^{-1}]\right) \to \mathrm{U}(\mathfrak{g})^{\otimes \ell}, \tag{14.5}$$

defined as the composition $\Phi = \Psi \circ \varsigma$. Since $\mathfrak{z}(\widehat{\mathfrak{g}})$ is a commutative subalgebra of $\mathrm{U}\left(t^{-1}\mathfrak{g}[t^{-1}]\right)$, the image of $\mathfrak{z}(\widehat{\mathfrak{g}})$ under Φ is a commutative subalgebra $\mathcal{A}(\mathfrak{g})_\chi$ of $\mathrm{U}(\mathfrak{g})^{\otimes \ell}$, depending on the chosen parameters $z_1, \ldots, z_\ell$, but it does not depend on u; cf. Section 9.2.

Given $\lambda \in \mathfrak{h}^*$, the Verma module M_λ is the quotient of $\mathrm{U}(\mathfrak{g})$ by the left ideal generated by $\mathfrak{n}_+$ and the elements $h_i - \lambda(h_i)$ with $i = 1, \ldots, n$. We denote the image of 1 in M_λ by 1_λ. For any weights $\lambda_1, \ldots, \lambda_\ell \in \mathfrak{h}^*$ consider the tensor product of the Verma modules $M_{\lambda_1} \otimes \ldots \otimes M_{\lambda_\ell}$. We will now describe common eigenvectors for the commutative subalgebra $\mathcal{A}(\mathfrak{g})_\chi$ in this tensor product. For a set of distinct complex numbers $w_1, \ldots, w_m$ with $w_i \neq z_j$ and a collection (multiset) of labels $i_1, \ldots, i_m \in \{1, \ldots, n\}$ introduce the *Bethe vector*

$$\phi(w_1^{i_1}, \ldots, w_m^{i_m}) \in M_{\lambda_1} \otimes \ldots \otimes M_{\lambda_\ell}$$

by the formula

$$\phi(w_1^{i_1}, \ldots, w_m^{i_m}) = \sum_{(I^1, \ldots, I^\ell)} \bigotimes_{k=1}^{\ell} \prod_{s=1}^{a_k} \frac{1}{w_{j_s^k} - w_{j_{s+1}^k}} \prod_{r \in I^k} f_{i_r}\, 1_{\lambda_k}, \tag{14.6}$$

summed over all ordered partitions $I^1 \cup I^2 \cup \cdots \cup I^\ell$ of the set $\{1, \ldots, m\}$ into ordered subsets $I^k = \{j_1^k, j_2^k, \ldots, j_{a_k}^k\}$ with the products taken from left to right, where $w_{j_{s+1}^k} := z_k$ for $s = a_k$.

Now suppose that $\chi \in \mathfrak{h}^*$. We regard χ as a functional on $\mathfrak{g}$ which vanishes on $\mathfrak{n}_+$ and $\mathfrak{n}_-$. The system of the *Bethe ansatz equations* takes the form

$$\sum_{i=1}^{\ell} \frac{\lambda_i(\check{\alpha}_{i_j})}{w_j - z_i} - \sum_{s \neq j} \frac{\alpha_{i_s}(\check{\alpha}_{i_j})}{w_j - w_s} = \chi(\check{\alpha}_{i_j}), \qquad j = 1, \ldots, m, \tag{14.7}$$

where $\alpha_l \in \mathfrak{h}^*$ and $\check{\alpha}_l \in \mathfrak{h}$ with $l = 1, \ldots, n$ denote the simple roots and coroots, respectively.

Given the above parameters, introduce the homomorphism from $\mathrm{U}\left(t^{-1}\mathfrak{h}[t^{-1}]\right)$ to the algebra of rational functions in u by the rule:

$$\varrho : H[-r-1] \mapsto \frac{\partial_u^r}{r!}\, \mathcal{H}(u), \qquad H \in \mathfrak{h}, \quad r \geqslant 0,$$

where

$$\mathcal{H}(u) = \sum_{a=1}^{\ell} \frac{\lambda_a(H)}{u - z_a} - \sum_{j=1}^{m} \frac{\alpha_{i_j}(H)}{u - w_j} - \chi(H). \tag{14.8}$$

Let $S \in \mathfrak{z}(\widehat{\mathfrak{g}})$ be a Segal–Sugawara vector. The composition $\varrho \circ \mathfrak{f}$ of this homomorphism with the isomorphism (13.2) takes S to a rational function $\varrho\big(\mathfrak{f}(S)\big)$ in u.

Furthermore, we regard the image $\Phi(S)$ of S under the anti-homomorphism (14.5) as an operator in the tensor product of Verma modules $M_{\lambda_1} \otimes \dots \otimes M_{\lambda_\ell}$. The following is essentially a reformulation of Theorems 6.5 and 6.7 from [**42**]; in the case $\chi = 0$ the result goes back to [**40**, Theorem 3].

THEOREM 14.1.1. *Suppose that the Bethe ansatz equations (14.7) are satisfied. If the Bethe vector $\phi(w_1^{i_1}, \dots, w_m^{i_m})$ is nonzero, then it is an eigenvector for the operator $\Phi(S)$ with the eigenvalue $\varrho\big(\mathfrak{f}(S)\big)$.* □

In what follows we will rely on the results of Chapter 13 to give explicit formulas for the operators $\Phi(S_i)$ and their eigenvalues $\varrho\big(\mathfrak{f}(S_i)\big)$ on the Bethe vectors for complete sets of Segal–Sugawara vectors $S_1, \dots, S_n$ in all classical types.

14.2. Gaudin Hamiltonians and eigenvalues

Type A_{N-1}. We need to calculate the images of the Segal–Sugawara vectors constructed in Section 7.1 under the involution (14.4). We extend it to the algebra $\mathrm{U}\big(t^{-1}\mathfrak{gl}_N[t^{-1}]\big) \otimes \mathbb{C}[\tau]$ with the action on $\mathbb{C}[\tau]$ as the identity map.

LEMMA 14.2.1. *For the images with respect to the involution ς we have*

$$\operatorname{tr}_{1,\dots,m} A^{(m)}\big(\tau - E[-1]_1\big)\dots\big(\tau + E[-1]_m\big) \\ \mapsto \operatorname{tr}_{1,\dots,m} A^{(m)}\big(\tau - E[-1]_1\big)\dots\big(\tau - E[-1]_m\big),$$

$$\operatorname{tr}_{1,\dots,m} H^{(m)}\big(\tau - E[-1]_1\big)\dots\big(\tau + E[-1]_m\big) \\ \mapsto \operatorname{tr}_{1,\dots,m} H^{(m)}\big(\tau - E[-1]_1\big)\dots\big(\tau - E[-1]_m\big),$$

$$\operatorname{tr}\big(\tau + E[-1]\big)^m \mapsto \operatorname{tr}\big(\tau - E^t[-1]\big)^m, \tag{14.9}$$

and

$$\operatorname{cdet}\big(\tau + E[-1]\big) \mapsto \operatorname{cdet}\big(\tau - E^t[-1]\big), \tag{14.10}$$

where t denotes the standard matrix transposition.

PROOF. The left hand side of the first relation equals a linear combination of expressions of the form

$$\operatorname{tr}_{1,\dots,m} A^{(m)} E[r_1]_{a_1} \dots E[r_p]_{a_p}\, \tau^k \tag{14.11}$$

with $1 \leqslant a_1 < \dots < a_p \leqslant m$ and $r_i < 0$. However, such an expression remains unchanged under any permutation of the factors $E[r_i]_{a_i}$. Indeed, this follows from the defining relations (7.6) implying

$$E[r]_a\, E[s]_b - E[s]_b\, E[r]_a = P_{a\,b}\, E[r+s]_b - E[r+s]_b\, P_{a\,b}$$

for $a < b$, by taking into account the identities $A^{(m)} P_{a\,b} = P_{a\,b} A^{(m)} = -A^{(m)}$ and the cyclic property of trace. Hence the image of (14.11) under ς equals

$$(-1)^p \operatorname{tr}_{1,\dots,m} A^{(m)} E[r_1]_{a_1} \dots E[r_p]_{a_p}\, \tau^k$$

which verifies the claim. The same argument proves the second relation. To prove (14.10) we use (7.10) and the fact that $\tau + E[-1]$ is a Manin matrix; see Lemma 7.1.2.

For the image of (7.10) under ς we get

$$\operatorname{tr}_{1,\dots,N} A^{(N)}\big(\tau - E[-1]_1\big)\dots\big(\tau - E[-1]_N\big) \\ = \operatorname{tr}_{1,\dots,N} A^{(N)}\big(\tau - E^t[-1]_1\big)\dots\big(\tau - E^t[-1]_N\big),$$

where we have applied the transposition $t_1 \dots t_N$ with respect to all copies of $\operatorname{End}\mathbb{C}^N$ and used the invariance of $A^{(N)}$ under this transposition. Since $\tau - E^t[-1]$ is also a Manin matrix, the resulting expression coincides with $\operatorname{cdet}\big(\tau - E^t[-1]\big)$. Finally, (14.9) follows from Theorem 3.2.10 which connects the coefficients of the polynomial in (7.9) with those of (7.11). □

With the parameters chosen as in Section 14.1, suppose that χ vanishes on the subspace $\mathfrak{n}_- \oplus \mathfrak{n}_+$ of $\mathfrak{gl}_N$ so that we can regard χ as an element of $\mathfrak{h}^*$. Set

$$E_{ij}(u) = \sum_{a=1}^{\ell} \frac{(E_{ij})_a}{u - z_a} - \chi(E_{ij}) \in \mathrm{U}(\mathfrak{gl}_N)^{\otimes \ell}.$$

Consider the row-determinant $\operatorname{rdet}\big(\partial_u + E(u)\big)$ of the matrix

$$\partial_u + E(u) = \big[\delta_{ij}\,\partial_u + E_{ij}(u)\big]$$

as a differential operator in ∂_u with coefficients in $\mathrm{U}(\mathfrak{gl}_N)^{\otimes \ell}$; see (3.25). Furthermore, in accordance with (14.8), set

$$\mathcal{E}_{ii}(u) = \sum_{a=1}^{\ell} \frac{\lambda_a(E_{ii})}{u - z_a} - \sum_{j=1}^{m} \frac{\alpha_{i_j}(E_{ii})}{u - w_j} - \chi(E_{ii}).$$

In all the following eigenvalue formulas for the Gaudin Hamiltonians we will assume that the Bethe ansatz equations (14.7) hold.

THEOREM 14.2.2. *The eigenvalue of the operator* $\operatorname{rdet}\big(\partial_u + E(u)\big)$ *on the Bethe vector* (14.6) *is found by*

$$\operatorname{rdet}\big(\partial_u + E(u)\big)\,\phi(w_1^{i_1},\dots,w_m^{i_m}) = \big(\partial_u + \mathcal{E}_{NN}(u)\big)\dots\big(\partial_u + \mathcal{E}_{11}(u)\big)\,\phi(w_1^{i_1},\dots,w_m^{i_m}).$$

PROOF. To apply Theorem 14.1.1, we will find the image of the polynomial $\operatorname{cdet}\big(\tau + E[-1]\big)$ under the anti-homomorphism Φ. Extend Φ to the map

$$\Phi : \mathrm{U}\big(t^{-1}\mathfrak{gl}_N[t^{-1}]\big) \otimes \mathbb{C}[\tau] \to \mathrm{U}(\mathfrak{gl}_N)^{\otimes \ell} \otimes \mathbb{C}[\partial_u]$$

by setting $\tau \mapsto \partial_u$. Note that by definition of the homomorphism (14.3) we have

$$\Psi : E[-1] \mapsto -E(u)$$

and so, by (14.10),

$$\Phi : \operatorname{cdet}\big(\tau + E[-1]\big) \mapsto \operatorname{cdet}\big(\partial_u + E^t(u)\big) = \operatorname{rdet}\big(\partial_u + E(u)\big).$$

On the other hand, due to (13.6) we have

$$\varrho \circ \mathfrak{f} : \operatorname{cdet}\big(\tau + E[-1]\big) \mapsto \big(\partial_u + \mathcal{E}_{NN}(u)\big)\dots\big(\partial_u + \mathcal{E}_{11}(u)\big)$$

thus completing the proof. □

The following corollaries can be derived from Theorem 14.2.2 or verified in a similar way as above with the use of Lemma 14.2.1 and the Harish-Chandra images calculated in Section 13.1.

COROLLARY 14.2.3. *The eigenvalues of the operators*

$$\mathrm{tr}_{1,\dots,m}\, A^{(m)}\big(\partial_u + E(u)_1\big)\dots\big(\partial_u + E(u)_m\big)$$

and

$$\mathrm{tr}_{1,\dots,m}\, H^{(m)}\big(\partial_u + E(u)_1\big)\dots\big(\partial_u + E(u)_m\big)$$

on the Bethe vector (14.6) *are found by the respective formulas*

$$e_m\big(\partial_u + \mathcal{E}_{11}(u),\dots,\partial_u + \mathcal{E}_{NN}(u)\big)$$

and

$$h_m\big(\partial_u + \mathcal{E}_{11}(u),\dots,\partial_u + \mathcal{E}_{NN}(u)\big).$$

COROLLARY 14.2.4. *The eigenvalue of the series*

$$\sum_{k=0}^{\infty} q^k\, \mathrm{tr}\big(\partial_u + E^t(u)\big)^k$$

on the Bethe vector (14.6) *is found by the formula*

$$\sum_{i=1}^{N} \Big(1 - q\,(\partial_u - \mathcal{E}_{11}(u))\Big)^{-1} \cdots \Big(1 - q\,(\partial_u + \mathcal{E}_{ii}(u))\Big)^{-1} \\ \times \Big(1 - q\,(\partial_u + \mathcal{E}_{i-1\,i-1}(u))\Big) \cdots \Big(1 - q\,(\partial_u + \mathcal{E}_{11}(u))\Big).$$

Types B_n and D_n. Consider the Segal–Sugawara vectors for $\mathfrak{o}_N$ constructed in Section 8.1. We extend the involution (14.4) to the algebra

$$\mathrm{U}\big(t^{-1}\mathfrak{o}_N[t^{-1}]\big) \otimes \mathbb{C}[\tau]$$

with the action on $\mathbb{C}[\tau]$ as the identity map.

LEMMA 14.2.5. *The element* (8.15) *is stable under* ς. *Moreover, in type D_n we have*

$$\varsigma : \mathrm{Pf}\ F[-1] \mapsto (-1)^n\ \mathrm{Pf}\ F[-1]. \tag{14.12}$$

PROOF. The same argument as in the proof of Lemma 14.2.1 shows that the image of (8.15) under the involution ς equals

$$\gamma_m(N)\,\mathrm{tr}_{1,\dots,m}\, S^{(m)}\big(\tau - F[-1]_1\big)\dots\big(\tau - F[-1]_m\big). \tag{14.13}$$

This is implied by the defining relations

$$F[r]_a\, F[s]_b - F[s]_b\, F[r]_a = (P_{a\,b} - Q_{a\,b})\, F[r+s]_b - F[r+s]_b\, (P_{a\,b} - Q_{a\,b})$$

for $a < b$; see (8.5). By applying the simultaneous transpositions $e_{ij} \mapsto e_{j'i'}$ to all m copies of $\mathrm{End}\,\mathbb{C}^N$ we conclude that (14.13) coincides with (8.15) because this transformation takes each factor $\tau - F[-1]_a$ to $\tau + F[-1]_a$, whereas the operator $S^{(m)}$ stays unchanged. This holds since the application of these transpositions to the right hand side of (11.11) will give the same product written in the opposite order which coincides with $S^{(m)}$; cf. an analogous property of the standard transposition used in Section 11.2. Relation (14.12) is immediate from (8.6). □

Choose parameters as in Section 14.1 and suppose that χ vanishes on the subspace $\mathfrak{n}_- \oplus \mathfrak{n}_+$ of $\mathfrak{o}_N$ so that we can regard χ as an element of $\mathfrak{h}^*$. Set

$$F_{ij}(u) = \sum_{a=1}^{\ell} \frac{(F_{ij})_a}{u - z_a} - \chi(F_{ij}) \in \mathrm{U}(\mathfrak{o}_N)^{\otimes \ell}.$$

Following (14.8) define the rational functions in u by

$$\mathcal{F}_{ii}(u)=\sum_{a=1}^{\ell}\frac{\lambda_a(F_{ii})}{u-z_a}-\sum_{j=1}^{m}\frac{\alpha_{i_j}(F_{ii})}{u-w_j}-\chi(F_{ii}).$$

In the case $\mathfrak{g}=\mathfrak{o}_{2n}$ introduce the operator

$$\mathrm{Pf}\;F(u)=\frac{1}{2^n n!}\sum_{\sigma\in\mathfrak{S}_{2n}}\operatorname{sgn}\sigma\cdot F_{\sigma(1)\,\sigma(2)'}(u)\dots F_{\sigma(2n-1)\,\sigma(2n)'}(u).$$

As before, we will assume that the Bethe ansatz equations (14.7) hold.

THEOREM 14.2.6. *The eigenvalue of the operator*

$$\gamma_m(N)\,\mathrm{tr}\,S^{(m)}\big(\partial_u+F(u)_1\big)\dots\big(\partial_u+F(u)_m\big) \tag{14.14}$$

on the Bethe vector (14.6) *is found by*

$$h_m\big(\partial_u+\mathcal{F}_{11}(u),\dots,\partial_u+\mathcal{F}_{n\,n}(u),\partial_u-\mathcal{F}_{n\,n}(u),\dots\partial_u-\mathcal{F}_{11}(u)\big)$$

for type B_n, and by

$$\tfrac{1}{2}\,h_m\big(\partial_u+\mathcal{F}_{11}(u),\dots,\partial_u+\mathcal{F}_{n-1\,n-1}(u),\partial_u-\mathcal{F}_{n\,n}(u),\dots\partial_u-\mathcal{F}_{11}(u)\big)$$

$$+\tfrac{1}{2}\,h_m\big(\partial_u+\mathcal{F}_{11}(u),\dots,\partial_u+\mathcal{F}_{n\,n}(u),\partial_u-\mathcal{F}_{n-1\,n-1}(u),\dots\partial_u-\mathcal{F}_{11}(u)\big)$$

for type D_n. Moreover, the eigenvalue of the operator Pf $F(u)$ *in type D_n is given by*

$$\big(\mathcal{F}_{11}(u)-\partial_u\big)\dots\big(\mathcal{F}_{n\,n}(u)-\partial_u\big)\,1.$$

PROOF. We apply Theorem 14.1.1 again and regard Φ as the map

$$\Phi:\mathrm{U}\big(t^{-1}\mathfrak{o}_N[t^{-1}]\big)\otimes\mathbb{C}[\tau]\to\mathrm{U}(\mathfrak{o}_N)^{\otimes\ell}\otimes\mathbb{C}[\partial_u]$$

such that $\tau\mapsto\partial_u$. By the definition of the homomorphism (14.3) we have

$$\Psi:F[-1]\mapsto -F(u).$$

Hence, using the equivalent formula (14.13) for the polynomial (8.15) we find that its image under Φ coincides with the operator (14.14). The proof of the first part of the theorem is completed by using the formulas for the Harish-Chandra images of (8.15) provided by Propositions 13.1.7 and 13.1.10. Finally, by Lemma 14.2.5, in type D_n,

$$\Phi:\mathrm{Pf}\;F[-1]\mapsto\mathrm{Pf}\;F(u)$$

so that the last claim follows from Proposition 13.1.12. □

Type C_n. Now we use the Segal–Sugawara vectors for $\mathfrak{sp}_{2n}$ constructed in Section 8.3. Extend the involution (14.4) to the algebra $\mathrm{U}\big(t^{-1}\mathfrak{sp}_{2n}[t^{-1}]\big)\otimes\mathbb{C}[\tau]$ with the action on $\mathbb{C}[\tau]$ as the identity map.

LEMMA 14.2.7. *The element* (8.36) *is stable under* ς.

PROOF. The proof is the same as for Lemma 14.2.5, which also provides an equivalent formula

$$\gamma_m(-2n)\,\mathrm{tr}\,S^{(m)}\big(\tau-F[-1]_1\big)\dots\big(\tau-F[-1]_m\big)$$

for the polynomial (8.36). □

With the parameters chosen as in Section 14.1 suppose that χ vanishes on the subspace $\mathfrak{n}_- \oplus \mathfrak{n}_-$ of $\mathfrak{sp}_{2n}$ so that we can regard χ as an element of $\mathfrak{h}^*$. Set

$$F_{ij}(u) = \sum_{a=1}^{\ell} \frac{(F_{ij})_a}{u - z_a} - \chi(F_{ij}) \in \mathrm{U}(\mathfrak{sp}_{2n})^{\otimes \ell}.$$

In accordance with (14.8) define the rational functions in u by

$$\mathcal{F}_{ii}(u) = \sum_{a=1}^{\ell} \frac{\lambda_a(F_{ii})}{u - z_a} - \sum_{j=1}^{m} \frac{\alpha_{i_j}(F_{ii})}{u - w_j} - \chi(F_{ii}).$$

As before, we will assume that the Bethe ansatz equations (14.7) hold.

THEOREM 14.2.8. *For any $1 \leqslant m \leqslant 2n+1$ the eigenvalue of the operator*

$$\gamma_m(-2n) \operatorname{tr} S^{(m)} \big(\partial_u + F(u)_1\big) \dots \big(\partial_u + F(u)_m\big)$$

on the Bethe vector (14.6) *is found by*

$$e_m\big(\partial_u + \mathcal{F}_{11}(u), \dots, \partial_u + \mathcal{F}_{nn}(u), \partial_u, \partial_u - \mathcal{F}_{nn}(u), \dots \partial_u - \mathcal{F}_{11}(u)\big).$$

PROOF. The argument relies on Proposition 13.1.14 so that the claim is derived from Theorem 14.1.1 and Lemma 14.2.7 as in the proof of Theorem 14.2.6. □

14.3. Bibliographical notes

For the origins of the formula (14.6) for the Bethe vectors see [**9**], [**40**], [**142**] and references therein. Theorem 14.2.2 was first proved in [**118**] is a different way. Our exposition follows [**113**]. The Bethe algebras of classical types were studied in a recent work by Lu, Mukhin and Varchenko [**102**] in connection with stratifications of self-dual Grassmannians.

CHAPTER 15

Wakimoto modules

The *Wakimoto modules* are representations of the affine Kac–Moody algebras $\widehat{\mathfrak{g}}$ in the Fock spaces $M(\mathfrak{g})$ associated with the Heisenberg algebras. Their construction relies on the existence of the vertex algebra homomorphism

$$V_{-h^\vee}(\mathfrak{g}) \to M(\mathfrak{g})$$

from the vacuum module at the critical level to the Fock space $M(\mathfrak{g})$. This homomorphism can be 'deformed' to the vacuum module $V_\kappa(\mathfrak{g})$ to define the Wakimoto module of an arbitrary level $\kappa \in \mathbb{C}$, as explained in the book by Frenkel [**46**, Chapter 6]. However, we will only consider such modules of the critical level $\kappa = -h^\vee$; see Section 6.2 for the definition of $V_\kappa(\mathfrak{g})$.

The Wakimoto modules play a significant role in the proof of a principal theorem of the book [**46**] describing the center of the affine vertex algebra at the critical level. Their importance is explained by the fact that elements of the center $\mathrm{Z}(\widehat{\mathfrak{g}})$ of the completed universal enveloping algebra $\widetilde{\mathrm{U}}_{-h^\vee}(\widehat{\mathfrak{g}})$ (i.e., the Sugawara operators constructed in Section 6.6) act as multiplication by scalars in the Wakimoto modules of the critical level. Our goal in this chapter is to connect the eigenvalues of the Sugawara operators with their Harish-Chandra images calculated in Section 13.3. We make this connection explicit by constructing the Wakimoto modules in all classical types and calculating the scalars.

Explicit formulas for the action of generators of $\widehat{\mathfrak{g}}$ in the Wakimoto modules depend on the *free field realization* of the Lie algebra $\mathfrak{g}$. We will start by reproducing these realizations in the classical types by following the general construction of [**46**, Chapter 5]. Closely related realizations of irreducible finite-dimensional representations of the classical groups were studied in the book by Zhelobenko [**156**].

15.1. Free field realization of $\mathfrak{gl}_N$

Consider the group GL_N of all $N\times N$ matrices over $\mathbb{C}$ with nonzero determinant. Denote by $\mathcal{B}_-$ the subgroup of all lower-triangular matrices and by $\mathcal{N}_+$ the subgroup of upper-triangular matrices with 1's on the diagonal. Elements of the Lie algebra $\mathfrak{gl}_N$ will act on the space of polynomial functions $\mathrm{Fun}\,\mathcal{N}_+$ by first order differential operators. We will identify $\mathrm{Fun}\,\mathcal{N}_+$ with the space of polynomials

$$\mathrm{Fun}\,\mathcal{N}_+ \cong \mathbb{C}[y_{ij} \mid 1 \leqslant i < j \leqslant N]$$

in the variables y_{ij} which are understood as coordinate functions on $\mathcal{N}_+$. Fix an N-tuple $\chi = (\chi_1, \dots, \chi_N) \in \mathbb{C}^N$ which is regarded as a functional on the Cartan subalgebra $\mathfrak{h}$; cf. Section 4.2.

In the following we write ∂_{ij} for the differentiation operator $\partial/\partial y_{ij}$.

PROPOSITION 15.1.1. *The formulas*

$$E_{ii} \mapsto \sum_{r=1}^{i-1} y_{ri}\,\partial_{ri} - \sum_{s=i+1}^{N} y_{is}\,\partial_{is} + \chi_i,$$

for $i = 1, \ldots, N$, *and*

$$E_{i\,i+1} \mapsto \partial_{i\,i+1} + \sum_{s=i+2}^{N} y_{i+1\,s}\,\partial_{is},$$

$$E_{i+1\,i} \mapsto \sum_{r=1}^{i-1} \bigl(y_{ri}\,y_{i\,i+1} - y_{r\,i+1}\bigr)\,\partial_{ri} \; - \sum_{r=1}^{i} y_{r\,i+1}\,y_{i\,i+1}\,\partial_{r\,i+1} + \sum_{s=i+2}^{N} y_{is}\,\partial_{i+1\,s} + \bigl(\chi_i - \chi_{i+1}\bigr)\,y_{i\,i+1}$$

for $i = 1, \ldots, N-1$, *define a representation of* $\mathfrak{gl}_N$ *in the space* $\mathbb{C}[y_{ij}]$.

PROOF. By definition, the action of an element $A \in \mathfrak{gl}_N$ on $\operatorname{Fun}\mathcal{N}_+$ is given by

$$(15.1) \qquad (Af)(Y) = \frac{d}{dt}\, f\bigl(X(t)\bigr)\Big|_{t=0}, \qquad Y \in \mathcal{N}_+, \quad f \in \operatorname{Fun}\mathcal{N}_+,$$

where $X(t) \in \mathcal{N}_+$ is uniquely determined from the decomposition

$$(15.2) \qquad e^{-tA}\,Y = X(t)\,Z(t), \qquad Z(t) \in \mathcal{B}_-,$$

which is valid for all values of t in a neighborhood of 0. The entries of the matrix $X(t)$ can be found from the well-known formulas for the Gauss decomposition.

LEMMA 15.1.2. *Given a decomposition* $C = X\,Z$ *of a* $N \times N$ *matrix* $C = [c_{ij}]$, *where* $X \in \mathcal{N}_+$ *and* $Z \in \mathcal{B}_-$, *the entries of* $X = [x_{ij}]$ *are found by*

$$x_{ij} = \begin{vmatrix} c_{i\,j} & c_{i\,j+1} & \cdots & c_{i\,N} \\ c_{j+1\,j} & c_{j+1\,j+1} & \cdots & c_{j+1\,N} \\ \cdots & \cdots & \cdots & \cdots \\ c_{N\,j} & c_{N\,j+1} & \cdots & c_{N\,N} \end{vmatrix} \begin{vmatrix} c_{j\,j} & c_{j\,j+1} & \cdots & c_{j\,N} \\ c_{j+1\,j} & c_{j+1\,j+1} & \cdots & c_{j+1\,N} \\ \cdots & \cdots & \cdots & \cdots \\ c_{N\,j} & c_{N\,j+1} & \cdots & c_{N\,N} \end{vmatrix}^{-1}$$

for all $1 \leqslant i < j \leqslant N$. □

Now take $A = E_{ii}$ and apply Lemma 15.1.2 to the matrix $C = e^{-tA}\,Y$, where $Y \in \mathcal{N}_+$ and the (i,j) entry of Y is the coordinate y_{ij} for all $i < j$. The entries of the matrix X will coincide with the respective entries of Y, except for $x_{ri} = e^t\,y_{ri}$ for $r = 1, \ldots, i-1$ and $x_{is} = e^{-t}\,y_{is}$ for $s = i+1, \ldots, N$. Hence, E_{ii} will be represented by the operator

$$E_{ii} \mapsto \sum_{r=1}^{i-1} y_{ri}\,\partial_{ri} - \sum_{s=i+1}^{N} y_{is}\,\partial_{is}.$$

Furthermore, taking $A = E_{i\,i+1}$, observe that $e^{-tA} = 1 - tE_{i\,i+1}$. Therefore, applying Lemma 15.1.2, we find that the entries of the corresponding matrix X which differ from the respective entries of Y, lie in row i and are given by $x_{i\,i+1} = y_{i\,i+1} - t$ and $x_{i\,s} = y_{i\,s} - ty_{i+1\,s}$ for $s = i+2, \dots, N$. Hence, $E_{i\,i+1}$ is represented by the operator

$$E_{i\,i+1} \mapsto -\partial_{i\,i+1} - \sum_{s=i+2}^{N} y_{i+1\,s}\,\partial_{is}.$$

Similarly, taking $A = E_{i+1\,i}$ and applying Lemma 15.1.2, we find that the entries of the matrix X which differ from the respective entries of Y, belong to row $i+1$ or columns i and $i+1$. They are given by the expressions

$$x_{r\,i} = y_{ri}\,(1 - ty_{i\,i+1}) + ty_{r\,i+1} \qquad \text{for} \quad r = 1, \dots, i-1,$$

$$x_{r\,i+1} = \frac{y_{r\,i+1}}{1 - ty_{i\,i+1}} \qquad \text{for} \quad r = 1, \dots, i,$$

and

$$x_{i+1\,s} = y_{i+1\,s} - ty_{i\,s} \qquad \text{for} \quad s = i+2, \dots, N.$$

This implies that $E_{i+1\,i}$ acts by the formula

$$E_{i+1\,i} \mapsto \sum_{r=1}^{i-1} \big(-y_{ri}\,y_{i\,i+1} + y_{r\,i+1}\big)\,\partial_{ri} + \sum_{r=1}^{i} y_{r\,i+1}\,y_{i\,i+1}\,\partial_{r\,i+1} - \sum_{s=i+2}^{N} y_{is}\,\partial_{i+1\,s}.$$

The above formulas define a representation of $\mathfrak{gl}_N$ in the space $\mathbb{C}[y_{ij}]$. Twist this representation with the automorphism of $\mathfrak{gl}_N$ which changes the signs of the generators $E_{i\,i+1}$ and $E_{i+1\,i}$ for $i = 1, \dots, N-1$ and leaves all E_{ii} unchanged. We will then get the formulas given in the proposition with $\chi = 0$. As demonstrated in [**46**, Section 5.2], given any $\chi \in \mathfrak{h}^*$, this representation can be extended in a way that each elements $H \in \mathfrak{h}$ acts on the element 1 as multiplication by the scalar $\chi(H)$, and so the formulas for the extended action follow. □

As shown in [**46**, Section 5.2], the representation of $\mathfrak{gl}_N$ provided by Proposition 15.1.1 is isomorphic to the contragredient Verma module M^*_χ.

EXAMPLE 15.1.3. For $N = 2$ the formulas of Proposition 15.1.1 define a representation of $\mathfrak{gl}_2$ in the space of polynomials in one variable $\mathbb{C}[y_{12}]$:

$$E_{11} \mapsto -y_{12}\,\partial_{12} + \chi_1, \qquad E_{22} \mapsto y_{12}\,\partial_{12} + \chi_2,$$

$$E_{12} \mapsto \partial_{12}, \qquad E_{21} \mapsto -y_{12}^2\,\partial_{12} + (\chi_1 - \chi_2)\,y_{12}.$$

If $\chi_1 - \chi_2 \notin \mathbb{Z}_+$ then this representation is irreducible. If $\chi_1 - \chi_2 = k \in \mathbb{Z}_+$ then the polynomials in y_{12} of degree not exceeding k form a $(k+1)$-dimensional irreducible submodule. □

EXAMPLE 15.1.4. The representation of the Lie algebra $\mathfrak{gl}_3$ in the space of polynomials $\mathbb{C}[y_{12}, y_{13}, y_{23}]$ is given by the formulas

$$E_{11} \mapsto -y_{12}\,\partial_{12} - y_{13}\,\partial_{13} + \chi_1,$$

$$E_{22} \mapsto y_{12}\,\partial_{12} - y_{23}\,\partial_{23} + \chi_2,$$

$$E_{33} \mapsto y_{13}\,\partial_{13} + y_{23}\,\partial_{23} + \chi_3,$$

together with

$$E_{12} \mapsto \partial_{12} + y_{23}\,\partial_{13}, \qquad E_{21} \mapsto -y_{12}^2\,\partial_{12} + y_{13}\,\partial_{23} + (\chi_1 - \chi_2)\,y_{12},$$

and

$$E_{23} \mapsto \partial_{23}, \qquad E_{32} \mapsto (y_{12}\,y_{23} - y_{13})\,\partial_{12} - y_{13}\,y_{23}\,\partial_{13} - y_{23}^2\,\partial_{23} + (\chi_2 - \chi_3)\,y_{23}.$$

15.2. Free field realization of $\mathfrak{o}_N$

Recall the classical group G_N as defined in (2.29). We point out an important property of the presentation of G_N associated with the form (2.26).

LEMMA 15.2.1. *Suppose that a matrix $C \in \mathrm{G}_N$ is written as a product $C = X\,Z$ of an upper-triangular matrix X with 1's on the diagonal and a lower-triangular matrix Z. Then both X and Z belong to G_N.*

PROOF. We have $1 = C'C = (X\,Z)'X\,Z = Z'X'X\,Z$. Therefore, $X'X = (Z')^{-1}Z^{-1}$. The matrix X' is upper-triangular with 1's on the diagonal, while Z' is lower-triangular. Therefore, $X'X = 1$ and $Z'Z = 1$, so the claim follows. □

Consider the special orthogonal group SO_N which is the subgroup of O_N whose elements are matrices C with $\det C = 1$. Now $\mathcal{B}_-$ will denote the subgroup of all lower-triangular matrices in SO_N, and $\mathcal{N}_+$ the subgroup of SO_N which consists of upper-triangular matrices with 1's on the diagonal. Due to the relation $X'X = 1$ satisfied by $X \in \mathcal{N}_+$, we can identify $\mathrm{Fun}\,\mathcal{N}_+$ with the space of polynomials

$$\mathrm{Fun}\,\mathcal{N}_+ \cong \mathbb{C}[y_{ij} \mid 1 \leqslant i < j \leqslant N \quad \text{and} \quad i + j \leqslant N]$$

regarding the y_{ij} as coordinates on $\mathcal{N}_+$.

We use the presentation of the orthogonal Lie algebra $\mathfrak{o}_N$ as in Section 5.1 and keep the notation $\mathfrak{h}$ for the Cartan subalgebra of $\mathfrak{o}_N$ spanned by the basis elements $F_{11}, \dots, F_{nn}$.

Suppose now that N is even and fix an element $\chi = (\chi_1, \dots, \chi_n) \in \mathbb{C}^n$ which is understood as a functional on $\mathfrak{h}$. As before, we use the notation $i' = N - i + 1$.

PROPOSITION 15.2.2. *The formulas*

$$F_{ii} \mapsto \sum_{r=1}^{i-1} \big(y_{ri}\,\partial_{ri} - y_{ri'}\,\partial_{ri'}\big) - \sum_{s=i+1}^{i'-1} y_{is}\,\partial_{is} + \chi_i$$

for $i = 1, \dots, n$,

$$F_{i\,i+1} \mapsto \partial_{i\,i+1} + \sum_{s=i+2}^{i'-2} y_{i+1\,s}\,\partial_{is} - \sum_{s=i+2}^{n} y_{i+1\,s}\,y_{i+1\,s'}\,\partial_{i\,i'-1}$$

for $i = 1, \dots, n-1$, and $F_{n-1\,n'} \mapsto \partial_{n-1\,n'}$, together with

$$F_{i+1\,i} \mapsto \sum_{r=1}^{i-1} \Big(\big(y_{ri}\,y_{i\,i+1} - y_{r\,i+1}\big)\,\partial_{ri} + \big(y_{r\,i'-1}\,y_{i\,i+1} + y_{r\,i'}\big)\,\partial_{r\,i'-1} - y_{ri'}\,y_{i\,i+1}\,\partial_{ri'}\Big)$$

$$- \sum_{r=1}^{i} y_{r\,i+1}\,y_{i\,i+1}\,\partial_{r\,i+1} + \sum_{s=i+2}^{i'-2} y_{is}\,\partial_{i+1\,s} - \sum_{s=i+2}^{n} y_{i\,s}\,y_{i\,s'}\,\partial_{i\,i'-1} + (\chi_i - \chi_{i+1})\,y_{i\,i+1}$$

for $i = 1, \dots, n-1$, *and*

$$F_{n'\,n-1} \mapsto \sum_{r=1}^{n-2} \Big(\big(y_{r\,n-1}\, y_{n-1\,n'} - y_{r\,n'} \big)\, \partial_{r\,n-1} + \big(y_{r\,n}\, y_{n-1\,n'} + y_{r\,n+2} \big)\, \partial_{r\,n} - y_{r\,n+2}\, y_{n-1\,n'}\, \partial_{r\,n+2} \Big) - \sum_{r=1}^{n-1} y_{r\,n'}\, y_{n-1\,n'}\, \partial_{r\,n'} + \big(\chi_{n-1} + \chi_n \big)\, y_{n-1\,n'}$$

define a representation of $\mathfrak{o}_{2n}$ *in the space* $\mathbb{C}\big[y_{ij} \mid 1 \leqslant i < j \leqslant 2n, \ \ i+j \leqslant 2n\big]$.

PROOF. The action of an element $A \in \mathfrak{o}_{2n}$ on $\operatorname{Fun} \mathcal{N}_+$ is given by the general formulas (15.1) and (15.2). We will apply Lemma 15.1.2 to the matrix $C = e^{-tA}\, Y$. Here $Y \in \mathcal{N}_+$ and we let y_{ij} be the (i,j) entry of Y for all $1 \leqslant i < j \leqslant 2n$. We will be concerned only with the (i,j) entries of the matrix X with $i < j'$. First take $A = F_{ii}$ for $i \in \{1, \dots, n\}$. Such entries will coincide with the respective entries of Y, except for $x_{ri} = e^t y_{ri}$ and $x_{ri'} = e^{-t} y_{ri'}$ for $r = 1, \dots, i-1$, as well as $x_{is} = e^{-t} y_{is}$ for $s = i+1, \dots, i'-1$. Hence, F_{ii} will act by

$$F_{ii} \mapsto \sum_{r=1}^{i-1} \big(y_{ri}\, \partial_{ri} - y_{ri'}\, \partial_{ri'} \big) - \sum_{s=i+1}^{i'-1} y_{is}\, \partial_{is}.$$

Now let $A = F_{i\,i+1}$ for $i \in \{1, \dots, n-1\}$ and apply Lemma 15.1.2 again. The entries of the corresponding matrix X which differ from the respective entries of Y are given by $x_{i\,i+1} = y_{i\,i+1} - t$ and $x_{i\,s} = y_{i\,s} - t y_{i+1\,s}$ for $s = i+2, \dots, i'-1$. Hence, $F_{i\,i+1}$ is represented by the operator

$$F_{i\,i+1} \mapsto -\partial_{i\,i+1} - \sum_{s=i+2}^{i'-1} y_{i+1\,s}\, \partial_{is}.$$

We need to express $y_{i+1\,i'-1}$ in terms of the admissible coordinates on $\mathcal{N}_+$. By taking the (i, i') entries on both sides of $YY' = 1$ we get

$$2\, y_{i\,i'} + \sum_{s=i+1}^{i'-1} y_{i\,s}\, y_{i\,s'} = 0, \qquad i = 1, \dots, n. \tag{15.3}$$

Hence,

$$F_{i\,i+1} \mapsto -\partial_{i\,i+1} - \sum_{s=i+2}^{i'-2} y_{i+1\,s}\, \partial_{is} + \sum_{s=i+2}^{n} y_{i+1\,s}\, y_{i+1\,s'}\, \partial_{i\,i'-1}.$$

Similarly, for $A = F_{n-1\,n'}$ we have $e^{-tA} = 1 - t F_{n-1\,n'}$ and so the only entry of the corresponding matrix X which differs from the respective entry of Y is $x_{n-1\,n'} = y_{n-1\,n'} - t$. Hence, $F_{n-1\,n'} \mapsto -\partial_{n-1\,n'}$. Now turn to the generators $A = F_{i+1\,i}$ for $i \in \{1, \dots, n-1\}$. By Lemma 15.1.2, we find that the entries of the matrix X which differ from the respective entries of Y are given by the expressions

$$x_{r\,i} = y_{ri}\, (1 - t y_{i\,i+1}) + t y_{r\,i+1}, \qquad x_{r\,i'} = \frac{y_{r\,i'}}{1 + t y_{i'-1\,i'}}$$

for $r = 1, \dots, i-1$,

$$x_{r\,i'-1} = y_{r\,i'-1}\, (1 + t y_{i'-1\,i'}) - t y_{r\,i'}, \qquad x_{r\,i+1} = \frac{y_{r\,i+1}}{1 - t y_{i\,i+1}}$$

for $r = 1, \dots, i$, and

$$x_{i+1\,s} = y_{i+1\,s} - t\,y_{i\,s} \qquad \text{for} \quad s = i+2, \dots, i'-2.$$

Taking into account (15.3) to express $y_{i\,i'}$ in terms of the admissible coordinates, we conclude that $F_{i+1\,i}$ acts by the formula

$$F_{i+1\,i} \mapsto -\sum_{r=1}^{i-1} \Big(\big(y_{ri}\, y_{i\,i+1} - y_{r\,i+1}\big)\, \partial_{ri} + \big(y_{r\,i'-1}\, y_{i\,i+1} + y_{r\,i'}\big)\, \partial_{r\,i'-1} - y_{ri'}\, y_{i\,i+1}\, \partial_{ri'} \Big)$$

$$+ \sum_{r=1}^{i} y_{r\,i+1}\, y_{i\,i+1}\, \partial_{r\,i+1} - \sum_{s=i+2}^{i'-2} y_{is}\, \partial_{i+1\,s} + \sum_{s=i+2}^{n} y_{i\,s}\, y_{i\,s'}\, \partial_{i\,i'-1}.$$

The calculation of the action of $F_{n'\,n-1}$ is not essentially different from the previous one; the resulting formula is obtained from the action of $F_{n\,n-1}$ by swapping the indices n and n'.

Finally, twist this representation of $\mathfrak{o}_{2n}$ with the automorphism which changes the signs of the generators $F_{i\,i+1}$ and $F_{i+1\,i}$ for $i = 1, \dots, n-1$, as well as $F_{n'\,n-1}$ and $F_{n-1\,n'}$, but leaves all F_{ii} unchanged. The action is then extended to an arbitrary $\chi \in \mathfrak{h}^*$ as in [**46**, Section 5.2] which yields the required formulas. □

EXAMPLE 15.2.3. The action of $\mathfrak{o}_4$ in the space of polynomials $\mathbb{C}[y_{12}, y_{13}]$ given by Proposition 15.2.2 takes the form

$$F_{11} \mapsto -y_{12}\,\partial_{12} - y_{13}\,\partial_{13} + \chi_1, \qquad F_{22} \mapsto y_{12}\,\partial_{12} - y_{13}\,\partial_{13} + \chi_2,$$

$$F_{12} \mapsto \partial_{12}, \qquad F_{13} \mapsto \partial_{13},$$

$$F_{21} \mapsto -y_{12}^2\,\partial_{12} + (\chi_1 - \chi_2)\, y_{12}, \qquad F_{31} \mapsto -y_{13}^2\,\partial_{13} + (\chi_1 + \chi_2)\, y_{13}.$$

This representation is isomorphic to the tensor product of two representations of $\mathfrak{sl}_2$ obtained by the restriction from the $\mathfrak{gl}_2$-action provided by Example 15.1.3. □

EXAMPLE 15.2.4. The action of the Lie algebra $\mathfrak{o}_6$ in the space of polynomials $\mathbb{C}[y_{12}, y_{13}, y_{14}, y_{15}, y_{23}, y_{24}]$ is given by

$$F_{11} \mapsto -y_{12}\,\partial_{12} - y_{13}\,\partial_{13} - y_{14}\,\partial_{14} - y_{15}\,\partial_{15} + \chi_1,$$

$$F_{22} \mapsto y_{12}\,\partial_{12} - y_{15}\,\partial_{15} - y_{23}\,\partial_{23} - y_{24}\,\partial_{24} + \chi_2,$$

$$F_{33} \mapsto y_{13}\,\partial_{13} - y_{14}\,\partial_{14} + y_{23}\,\partial_{23} - y_{24}\,\partial_{24} + \chi_3,$$

for the elements of the Cartan subalgebra, together with

$$F_{23} \mapsto \partial_{23}, \qquad F_{24} \mapsto \partial_{24}, \qquad F_{12} \mapsto \partial_{12} + y_{23}\,\partial_{13} + y_{24}\,\partial_{14} - y_{23}\, y_{24}\,\partial_{15},$$

and

$$F_{21} \mapsto -y_{12}^2\,\partial_{12} - y_{13}\, y_{14}\,\partial_{15} + y_{13}\,\partial_{23} + y_{14}\,\partial_{24} + (\chi_1 - \chi_2)\, y_{12},$$

$$F_{32} \mapsto \big(y_{12}\, y_{23} - y_{13}\big)\,\partial_{12} - y_{13}\, y_{23}\,\partial_{13} + \big(y_{14}\, y_{23} + y_{15}\big)\,\partial_{14}$$
$$- y_{15}\, y_{23}\,\partial_{15} - y_{23}^2\,\partial_{23} + (\chi_2 - \chi_3)\, y_{23},$$

$$F_{42} \mapsto \big(y_{12}\, y_{24} - y_{14}\big)\,\partial_{12} - y_{14}\, y_{24}\,\partial_{14} + \big(y_{13}\, y_{24} + y_{15}\big)\,\partial_{13}$$
$$- y_{15}\, y_{24}\,\partial_{15} - y_{24}^2\,\partial_{24} + (\chi_2 + \chi_3)\, y_{24}.$$

Now suppose that N is odd, $N = 2n+1$, and fix an element $\chi = (\chi_1, \dots, \chi_n)$ of $\mathbb{C}^n$ which is understood as a functional on the Cartan subalgebra $\mathfrak{h}$ of $\mathfrak{o}_{2n+1}$ spanned by the basis elements $F_{11}, \dots, F_{nn}$.

PROPOSITION 15.2.5. *The formulas*

$$F_{ii} \mapsto \sum_{r=1}^{i-1} \big(y_{ri}\,\partial_{ri} - y_{ri'}\,\partial_{ri'}\big) - \sum_{s=i+1}^{i'-1} y_{is}\,\partial_{is} + \chi_i,$$

$$F_{i\,i+1} \mapsto \partial_{i\,i+1} + \sum_{s=i+2}^{i'-2} y_{i+1\,s}\,\partial_{is} - \frac{1}{2}\sum_{s=i+2}^{i'-2} y_{i+1\,s}\,y_{i+1\,s'}\,\partial_{i\,i'-1}$$

for $i = 1, \dots, n$, together with

$$F_{i+1\,i} \mapsto \sum_{r=1}^{i-1} \Big(\big(y_{ri}\,y_{i\,i+1} - y_{r\,i+1}\big)\,\partial_{ri} + \big(y_{r\,i'-1}\,y_{i\,i+1} + y_{r\,i'}\big)\,\partial_{r\,i'-1} - y_{ri'}\,y_{i\,i+1}\,\partial_{ri'}\Big)$$

$$-\sum_{r=1}^{i} y_{r\,i+1}\,y_{i\,i+1}\,\partial_{r\,i+1} + \sum_{s=i+2}^{i'-2} y_{is}\,\partial_{i+1\,s} - \frac{1}{2}\sum_{s=i+2}^{i'-2} y_{i\,s}\,y_{i\,s'}\,\partial_{i\,i'-1} + \big(\chi_i - \chi_{i+1}\big)\,y_{i\,i+1}$$

for $i = 1, \dots, n-1$, and

$$F_{n+1\,n} \mapsto \sum_{r=1}^{n-1} \Big(\big(y_{r\,n}\,y_{n\,n+1} - y_{r\,n+1}\big)\,\partial_{r\,n} - y_{r\,n+2}\,y_{n\,n+1}\,\partial_{r\,n+2} + y_{r\,n+2}\,\partial_{r\,n+1}\Big)$$

$$- \frac{1}{2}\,y_{n\,n+1}^2\,\partial_{n\,n+1} + \chi_n\,y_{n\,n+1}$$

define a representation of $\mathfrak{o}_{2n+1}$ in the space

$$\mathbb{C}\big[y_{ij} \mid 1 \leqslant i < j \leqslant 2n+1, \quad i+j \leqslant 2n+1\big].$$

PROOF. Almost all the formulas are derived by the same calculation as in the proof of Proposition 15.2.2, with the use of (15.3) which holds in the same form. Additional arguments are required only for the action of the simple root generators $F_{n\,n+1}$ and $F_{n+1\,n}$. For $A = F_{n\,n+1}$ we have $e^{-tA} = 1 - t\,F_{n\,n+1} - t^2/2\,E_{n\,n+2}$, regarding the E_{ij} as the unit matrices. Hence the only entry of the corresponding matrix X which differs from the respective entry of Y is $x_{n\,n+1} = y_{n\,n+1} - t$. This gives $F_{n\,n+1} \mapsto -\partial_{n\,n+1}$ in the initial calculation. For $A = F_{n+1\,n}$ we have $e^{-tA} = 1 - t\,F_{n+1\,n} - t^2/2\,E_{n+2\,n}$. Hence, for the entries of the matrix X provided by Lemma 15.1.2 we find

$$x_{r\,n} = y_{r\,n}\,\big(1 - t\,y_{n\,n+1} + \frac{t^2}{4}\,y_{n\,n+1}^2\big) + y_{r\,n+1}\,\big(t - \frac{t^2}{2}\,y_{n\,n+1}\big) - \frac{t^2}{2}\,y_{r\,n+2}$$

for $r = 1, \dots, n-1$, where we also used the relations $y_{n+1\,n+2} = -y_{n\,n+1}$ and $y_{n\,n+2} = -1/2\,y_{n\,n+1}^2$ implied by taking the $(n, n+1)$ entry in $YY' = 1$ and by (15.3). Similarly,

$$x_{r\,n+1} = y_{r\,n+1} - \frac{t\,y_{r\,n+2}}{1 - t\,y_{n\,n+1}/2} \qquad \text{for} \quad r = 1, \dots, n,$$

and

$$x_{r\,n+2} = \frac{y_{r\,n+2}}{\big(1 - t\,y_{n\,n+1}/2\big)^2} \qquad \text{for} \quad r = 1, \dots, n-1.$$

Therefore, for the action of $F_{n+1\,n}$ we get

$$F_{n+1\,n} \mapsto \sum_{r=1}^{n-1} \Big(\big(y_{r\,n+1} - y_{r\,n}\, y_{n\,n+1}\big)\, \partial_{r\,n} + y_{r\,n+2}\, y_{n\,n+1}\, \partial_{r\,n+2} - y_{r\,n+2}\, \partial_{r\,n+1} \Big) - y_{n\,n+2}\, \partial_{n\,n+1}.$$

The argument is completed in the same way as for Proposition 15.2.2. $\square$

EXAMPLE 15.2.6. The action of $\mathfrak{o}_3$ in the space of polynomials $\mathbb{C}[y_{12}]$ takes the form

$$F_{11} \mapsto -y_{12}\, \partial_{12} + \chi_1, \qquad F_{12} \mapsto \partial_{12}, \qquad F_{21} \mapsto -\frac{1}{2}\, y_{12}^2\, \partial_{12} + \chi_1\, y_{12}.$$

This representation is isomorphic to the representation of $\mathfrak{sl}_2$ obtained by the restriction from the $\mathfrak{gl}_2$-action provided by Example 15.1.3 so that χ_1 corresponds to $(\chi_1 - \chi_2)/2$. $\square$

EXAMPLE 15.2.7. The action of the Lie algebra $\mathfrak{o}_5$ in the space of polynomials $\mathbb{C}[y_{12}, y_{13}, y_{14}, y_{23}]$ is given by

$$F_{11} \mapsto -y_{12}\, \partial_{12} - y_{13}\, \partial_{13} - y_{14}\, \partial_{14} + \chi_1, \quad F_{22} \mapsto y_{12}\, \partial_{12} - y_{14}\, \partial_{14} - y_{23}\, \partial_{23} + \chi_2,$$

together with

$$F_{12} \mapsto \partial_{12} + y_{23}\, \partial_{13} + \frac{1}{2}\, y_{23}^2\, \partial_{14}, \quad F_{21} \mapsto -y_{12}^2\, \partial_{12} + y_{13}\, \partial_{23} - \frac{1}{2}\, y_{13}^2\, \partial_{14} + (\chi_1 - \chi_2)\, y_{12},$$

and

$$F_{23} \mapsto \partial_{23}, \quad F_{32} \mapsto (y_{12}\, y_{23} - y_{13})\, \partial_{12} - y_{14}\, y_{23}\, \partial_{14} + y_{14}\, \partial_{13} - \frac{1}{2}\, y_{23}^2\, \partial_{23} + \chi_2\, y_{23}.$$

15.3. Free field realization of $\mathfrak{sp}_{2n}$

Consider the symplectic group Sp_{2n} defined in (2.29). Now $\mathcal{B}_-$ will denote the subgroup of all lower-triangular matrices in Sp_{2n}, and $\mathcal{N}_+$ the subgroup of Sp_{2n} which consists of upper-triangular matrices with 1's on the diagonal. The relation $X'X = 1$ for $X \in \mathcal{N}_+$ allows us to identify $\operatorname{Fun}\mathcal{N}_+$ with the space of polynomials

$$\operatorname{Fun}\mathcal{N}_+ \cong \mathbb{C}[y_{ij} \mid 1 \leqslant i < j \leqslant 2n \quad \text{and} \quad i + j \leqslant 2n + 1]$$

regarding the y_{ij} as coordinates on $\mathcal{N}_+$. As with the orthogonal case, we will use Lemmas 15.1.2 and 15.2.1 together with the definitions (15.1) and (15.2) of the action of the symplectic Lie algebra $\mathfrak{sp}_{2n}$ on $\operatorname{Fun}\mathcal{N}_+$. We will use the presentation of $\mathfrak{sp}_{2n}$ as in Section 5.1. Fix an element $\chi = (\chi_1, \dots, \chi_n) \in \mathbb{C}^n$ which is understood as a functional on the Cartan subalgebra $\mathfrak{h}$ of $\mathfrak{sp}_{2n}$ spanned by the basis elements $F_{11}, \dots, F_{nn}$.

PROPOSITION 15.3.1. *The formulas*

$$F_{ii} \mapsto \sum_{r=1}^{i-1} \big(y_{ri}\, \partial_{ri} - y_{ri'}\, \partial_{ri'}\big) - \sum_{s=i+1}^{i'-1} y_{is}\, \partial_{is} - 2\, y_{i\,i'}\, \partial_{i\,i'} + \chi_i$$

for $i = 1, \dots, n$,

$$F_{i\,i+1} \mapsto \partial_{i\,i+1} + \sum_{s=i+2}^{i'-1} y_{i+1\,s}\, \partial_{is} + \Big(y_{i\,i'-1} - \sum_{s=i+1}^{i'-2} \varepsilon_s\, y_{i\,s}\, y_{i+1\,s'}\Big)\partial_{i\,i'}$$

for $i = 1, \dots, n-1$, *and* $F_{n\, n+1} \mapsto 2\partial_{n\, n+1}$, *together with*

$$F_{i+1\, i} \mapsto \sum_{r=1}^{i} \Big(\big(y_{r\, i'-1}\, y_{i\, i+1} + y_{r\, i'} \big)\, \partial_{r\, i'-1} - y_{ri'}\, y_{i\, i+1}\, \partial_{ri'} - y_{r\, i+1}\, y_{i\, i+1}\, \partial_{r\, i+1} \Big)$$
$$+ \sum_{s=i+2}^{i'-2} y_{is}\, \partial_{i+1\, s} + \sum_{r=1}^{i-1} \big(y_{ri}\, y_{i\, i+1} - y_{r\, i+1} \big)\, \partial_{ri}$$
$$+ \Big(2 y_{i\, i'-1} - \sum_{s=i+2}^{i'-2} \varepsilon_s\, y_{is}\, y_{i+1\, s'} \Big) \partial_{i+1\, i'-1} + \big(\chi_i - \chi_{i+1} \big)\, y_{i\, i+1}$$

for $i = 1, \dots, n-1$, *and*

$$F_{n+1\, n} \mapsto 2 \sum_{r=1}^{n-1} \big(y_{r\, n}\, y_{n\, n+1} - y_{r\, n+1} \big)\, \partial_{r\, n} - 2 \sum_{r=1}^{n} y_{r\, n+1}\, y_{n\, n+1}\, \partial_{r\, n+1} + 2 \chi_n\, y_{n\, n+1}$$

define a representation of $\mathfrak{sp}_{2n}$ *in the space* $\mathbb{C}\big[y_{ij} \mid 1 \leqslant i < j \leqslant 2n, \;\; i+j \leqslant 2n+1 \big]$.

Proof. We argue as in the proof of Proposition 15.2.2 and for $A \in \mathfrak{sp}_{2n}$ apply Lemma 15.1.2 to the matrix $C = e^{-tA}\, Y$. Here $Y \in \mathcal{N}_+$ and we let y_{ij} be the (i, j) entry of Y for all $1 \leqslant i < j \leqslant 2n$. We choose the entries y_{ij} with the conditions $i \leqslant j'$ as coordinates on $\mathcal{N}_+$. First take $A = F_{ii}$ for $i \in \{1, \dots, n\}$. The entries of the corresponding matrix X with those conditions will coincide with the respective entries of Y, except for $x_{ri} = e^t\, y_{ri}$ and $x_{ri'} = e^{-t}\, y_{ri'}$ for $r = 1, \dots, i-1$, as well as $x_{is} = e^{-t}\, y_{is}$ for $s = i+1, \dots, i'-1$, and $x_{i\, i'} = e^{-2t}\, y_{i\, i'}$. Hence, F_{ii} will act by

$$F_{ii} \mapsto \sum_{r=1}^{i-1} \big(y_{ri}\, \partial_{ri} - y_{ri'}\, \partial_{ri'} \big) - \sum_{s=i+1}^{i'-1} y_{is}\, \partial_{is} - 2 y_{i\, i'}\, \partial_{i\, i'}.$$

Now apply Lemma 15.1.2 for $A = F_{i\, i+1}$ with $i \in \{1, \dots, n-1\}$. The entries of the corresponding matrix X which differ from the respective entries of Y are given by $x_{i\, i+1} = y_{i\, i+1} - t$ and $x_{i\, s} = y_{i\, s} - t\, y_{i+1\, s}$ for $s = i+2, \dots, i'$. For the action of $F_{i\, i+1}$ we have

$$F_{i\, i+1} \mapsto -\partial_{i\, i+1} - \sum_{s=i+2}^{i'} y_{i+1\, s}\, \partial_{is}.$$

To express $y_{i+1\, i'}$ in terms of the admissible coordinates on $\mathcal{N}_+$, take the $(i+1, i')$ entries on both sides of $YY' = 1$ to get

$$y_{i+1\, i'} - y_{i\, i'-1} + \sum_{s=i+1}^{i'-2} \varepsilon_s\, y_{i\, s}\, y_{i+1\, s'} = 0, \qquad i = 1, \dots, n-1. \tag{15.4}$$

Therefore,

$$F_{i\, i+1} \mapsto -\partial_{i\, i+1} - \sum_{s=i+2}^{i'-1} y_{i+1\, s}\, \partial_{is} + \Big(y_{i\, i'-1} - \sum_{s=i+1}^{i'-2} \varepsilon_s\, y_{i\, s}\, y_{i+1\, s'} \Big) \partial_{i\, i'}.$$

For $A = F_{n\, n+1}$ we have $e^{-tA} = 1 - 2t\, E_{n\, n+1}$ which implies that the only entry of the corresponding matrix X which differs from the respective entry of Y is $x_{n\, n+1} = y_{n\, n+1} - 2t$ and so $F_{n\, n+1} \mapsto -2\partial_{n\, n+1}$.

The calculation of the entries of the matrix X for $A = F_{i+1\,i}$ with the values $i \in \{1,\dots,n-1\}$ brings the same expressions as for $\mathfrak{o}_{2n}$, complemented by the formulas

$$x_{i\,i'} = \frac{y_{i\,i'}}{1+t\,y_{i'-1\,i'}} \qquad \text{and}$$

$$x_{i+1\,i'-1} = y_{i+1\,i'-1} + t\,\big(y_{i+1\,i'-1}\,y_{i'-1\,i'} - y_{i\,i'-1} - y_{i+1\,i'}\big) + t^2\,\big(y_{i\,i'} - y_{i\,i'-1}y_{i'-1\,i'}\big).$$

Hence using (15.4) and the relation $y_{i\,i+1} + y_{i'-1\,i'} = 0$ we get

$$F_{i+1\,i} \mapsto \sum_{r=1}^{i}\Big(-\big(y_{r\,i'-1}\,y_{i\,i+1} + y_{r\,i'}\big)\,\partial_{r\,i'-1} + y_{ri'}\,y_{i\,i+1}\,\partial_{ri'} + y_{r\,i+1}\,y_{i\,i+1}\,\partial_{r\,i+1}\Big)$$
$$- \sum_{s=i+2}^{i'-2} y_{is}\,\partial_{i+1\,s} + \sum_{r=1}^{i-1}\big(y_{r\,i+1} - y_{ri}\,y_{i\,i+1}\big)\,\partial_{ri}$$
$$- \Big(2\,y_{i\,i'-1} - \sum_{s=i+2}^{i'-2} \varepsilon_s\,y_{is}\,y_{i+1\,s'}\Big)\partial_{i+1\,i'-1}.$$

Finally, taking $A = F_{n+1\,n}$ we find $e^{-tA} = 1 - 2t\,E_{n+1\,n}$. Therefore, for the entries of the matrix X we get from Lemma 15.1.2:

$$x_{r\,n} = y_{r\,n}\,\big(1 - 2t\,y_{n\,n+1}\big) + 2t\,y_{r\,n+1} \qquad \text{for} \quad r = 1,\dots,n-1,$$

and

$$x_{r\,n+1} = \frac{y_{r\,n+1}}{1 - 2t\,y_{n\,n+1}} \qquad \text{for} \quad r = 1,\dots,n.$$

The action of $F_{n+1\,n}$ is then found by

$$F_{n+1\,n} \mapsto 2\sum_{r=1}^{n-1}\big(y_{r\,n+1} - y_{r\,n}\,y_{n\,n+1}\big)\,\partial_{r\,n} + 2\sum_{r=1}^{n} y_{r\,n+1}\,y_{n\,n+1}\,\partial_{r\,n+1}.$$

To get the required formulas we twist the action by the automorphism changing sings of the simple root vectors and extend it to an arbitrary weight χ; cf. Proposition 15.2.2. □

EXAMPLE 15.3.2. The representation of $\mathfrak{sp}_2$ in the space of polynomials $\mathbb{C}[y_{12}]$ provided by Proposition 15.3.1 takes the form

$$F_{11} \mapsto -2\,y_{12}\,\partial_{12} + \chi_1, \qquad F_{12} \mapsto 2\,\partial_{12}, \qquad F_{21} \mapsto -2\,y_{12}^2\,\partial_{12} + 2\,\chi_1\,y_{12}.$$

It is isomorphic to the representation of $\mathfrak{sl}_2$ obtained by the restriction from the $\mathfrak{gl}_2$-action of Example 15.1.3, and χ_1 corresponds to $\chi_1 - \chi_2$. □

EXAMPLE 15.3.3. The action of the Lie algebra $\mathfrak{sp}_4$ in the space of polynomials $\mathbb{C}[y_{12}, y_{13}, y_{14}, y_{23}]$ is given by

$$F_{11} \mapsto -y_{12}\,\partial_{12} - y_{13}\,\partial_{13} - 2\,y_{14}\,\partial_{14} + \chi_1, \quad F_{22} \mapsto y_{12}\,\partial_{12} - y_{13}\,\partial_{13} - 2\,y_{23}\,\partial_{23} + \chi_2,$$

together with

$$F_{12} \mapsto \partial_{12} + y_{23}\,\partial_{13} + (y_{13} - y_{12}\,y_{23})\,\partial_{14}, \qquad F_{23} \mapsto 2\,\partial_{23},$$

and

$$F_{21} \mapsto -y_{12}^2 \partial_{12} + (y_{14} + y_{12} y_{13}) \partial_{13} - y_{12} y_{14} \partial_{14} + 2 y_{13} \partial_{23} + (\chi_1 - \chi_2) y_{12},$$

$$F_{32} \mapsto 2 \left(y_{12} y_{23} - y_{13}\right) \partial_{12} - 2 y_{13} y_{23} \partial_{13} - 2 y_{23}^2 \partial_{23} + 2 \chi_2 y_{23}.$$

15.4. Wakimoto modules in type A

We will follow [**46**, Section 6.1] to reproduce the formulas defining the action of the affine Kac–Moody algebra $\widehat{\mathfrak{gl}}_N$ in the Wakimoto modules of the critical level.

The *Weyl algebra* $\mathcal{A}(\mathfrak{gl}_N)$ is generated by elements $a_{ij}[r]$ and $a^*_{ij}[r]$, where $1 \leqslant i < j \leqslant N$ and $r \in \mathbb{Z}$ with the defining relations

$$\big[a_{ij}[r], a^*_{kl}[s]\big] = \delta_{ik}\, \delta_{jl}\, \delta_{r,-s}, \qquad \big[a_{ij}[r], a_{kl}[s]\big] = \big[a^*_{ij}[r], a^*_{kl}[s]\big] = 0.$$

Let $M(\mathfrak{gl}_N)$ be the *Fock representation* of $\mathcal{A}(\mathfrak{gl}_N)$ generated by a vector $|0\rangle$ such that

$$a_{ij}[r]\,|0\rangle = 0 \qquad \text{for} \quad r \geqslant 0 \qquad \text{and} \qquad a^*_{ij}[r]\,|0\rangle = 0 \qquad \text{for} \quad r > 0.$$

Accordingly, the generators $a_{ij}[r]$ with $r \geqslant 0$ and $a^*_{ij}[r]$ with $r > 0$ are called the *annihilation operators*, whereas the generators $a_{ij}[r]$ with $r < 0$ and $a^*_{ij}[r]$ with $r \leqslant 0$ are called the *creation operators*. We can regard elements of $M(\mathfrak{gl}_N)$ as polynomials in the creation operators applied to the vector $|0\rangle$.

Introduce the formal Laurent series

$$a_{ij}(z) = \sum_{r\in\mathbb{Z}} a_{ij}[r]\, z^{-r-1} \qquad \text{and} \qquad a^*_{ij}(z) = \sum_{r\in\mathbb{Z}} a^*_{ij}[r]\, z^{-r} \tag{15.5}$$

with coefficients in $\mathcal{A}(\mathfrak{gl}_N)$. Normally ordered products of such series are defined by applying the general rule (6.9). Equivalently, for any monomial P in the generators of the Weyl algebra, the corresponding normally ordered monomial $:P:$ is obtained by moving all factors which are annihilation operators to the right of the factors which are creation operators. The normal ordering extends by linearity to arbitrary linear combinations of monomials. When applied to products of generating series (15.5), it will be understood that the normal ordering is applied to all coefficients of the series in z.

Recall the formal Laurent series $E_{ij}(z)$ introduced in (7.22) and fix an N-tuple of series $\chi(z) = \big(\chi_1(z), \dots, \chi_N(z)\big)$ with all $\chi_i(z) \in \mathbb{C}((z))$. The following formulas define the *Wakimoto module* $W_{\chi(z)}$ of the critical level.

PROPOSITION 15.4.1. *The following assignments define a representation of* $\widehat{\mathfrak{gl}}_N$ *at the critical level on the vector space* $M(\mathfrak{gl}_N)$:

$$E_{ii}(z) \mapsto \sum_{r=1}^{i-1} :a^*_{ri}(z)\, a_{ri}(z): - \sum_{s=i+1}^{N} :a^*_{is}(z)\, a_{is}(z): + \chi_i(z),$$

for $i = 1, \dots, N$, *together with*

$$E_{i\,i+1}(z) \mapsto a_{i\,i+1}(z) + \sum_{s=i+2}^{N} :a^*_{i+1\,s}(z)\, a_{is}(z):,$$

and

$$\begin{aligned}E_{i+1\,i}(z) \mapsto \sum_{r=1}^{i-1} &: \big(a^*_{ri}(z)\, a^*_{i\,i+1}(z) - a^*_{r\,i+1}(z)\big)\, a_{ri}(z) : \\ &- \sum_{r=1}^{i} : a^*_{r\,i+1}(z)\, a^*_{i\,i+1}(z)\, a_{r\,i+1}(z) : + \sum_{s=i+2}^{N} : a^*_{is}(z) a_{i+1\,s}(z) : \\ &\qquad + \big(\chi_i(z) - \chi_{i+1}(z)\big)\, a^*_{i\,i+1}(z) - (i+1)\,\partial_z\, a^*_{i\,i+1}(z).\end{aligned}$$

for $i = 1, \ldots, N-1$.

PROOF. The formulas follow from Proposition 15.1.1 providing a free field realization of $\mathfrak{gl}_N$ and the general results of [**46**, Section 6.1]. The only additional calculation is needed to get the value $c_i = -i-1$ of the coefficient c_i of the term $\partial_z\, a^*_{i\,i+1}(z)$ which is known to exist by the general results. To calculate the coefficient, note that since $K = -N$, by (7.3) we have

$$\big[E_{i\,i+1}[1], E_{i+1\,i}[-1]\big] = E_{i\,i}[0] - E_{i+1\,i+1}[0] - N. \tag{15.6}$$

Write

$$\chi_i(z) = \sum_{r\in\mathbb{Z}} \chi_i[r]\, z^{-r-1}, \qquad \chi_i[r] \in \mathbb{C}, \tag{15.7}$$

so that $\chi_i[r] = 0$ for all sufficiently large positive values of r. Apply both sides of (15.6) to the vector $|0\rangle$ and note that $E_{i\,i+1}[1]\,|0\rangle = 0$, whereas $E_{i\,i}[0]\,|0\rangle = \chi_i[0]\,|0\rangle$ for all i. Furthermore, taking the constant term in the formula for the action of $E_{i+1\,i}(z)$ we get

$$\begin{aligned}E_{i+1\,i}[-1]\,|0\rangle = \sum_{r=1}^{i-1} &\big(a^*_{ri}[0]\, a^*_{i\,i+1}[0] - a^*_{r\,i+1}[0]\big)\, a_{ri}[-1]\,|0\rangle \\ &- \sum_{r=1}^{i} a^*_{r\,i+1}[0]\, a^*_{i\,i+1}[0]\, a_{r\,i+1}[-1]\,|0\rangle + \sum_{s=i+2}^{N} a^*_{is}[0]\, a_{i+1\,s}[-1]\,|0\rangle \\ &\qquad + \sum_{p\geqslant 0} \big(\chi_i[p-1] - \chi_{i+1}[p-1]\big)\, a^*_{i\,i+1}[-p]\,|0\rangle + c_i\, a^*_{i\,i+1}[-1]\,|0\rangle.\end{aligned}$$

Now act on this vector by the operator

$$E_{i\,i+1}[1] = a_{i\,i+1}[1] + \sum_{s=i+2}^{N} \sum_{k\in\mathbb{Z}} a^*_{i+1\,s}[-k+1]\, a_{is}[k]$$

to obtain

$$E_{i\,i+1}[1]\, E_{i+1\,i}[-1]\,|0\rangle = \big(-N + i + 1 + c_i + \chi_i[0] - \chi_{i+1}[0]\big)\,|0\rangle.$$

Comparing this with the action of the right hand side of (15.6) on $|0\rangle$, we conclude that $c_i = -i-1$. □

EXAMPLE 15.4.2. The formulas for the Wakimoto module $W_{\chi(z)}$ over $\widehat{\mathfrak{gl}}_2$ with $\chi(z) = \big(\chi_1(z), \chi_2(z)\big)$ have the form

$$\begin{aligned}E_{11}(z) &\mapsto -: a^*_{12}(z)\, a_{12}(z) : + \chi_1(z), \\ E_{22}(z) &\mapsto\ : a^*_{12}(z)\, a_{12}(z) : + \chi_2(z),\end{aligned}$$

together with

$$E_{12}(z) \mapsto a_{12}(z),$$

$$E_{21}(z) \mapsto -:a_{12}^*(z)^2\, a_{12}(z): + \big(\chi_1(z) - \chi_2(z)\big)\, a_{12}^*(z) - 2\,\partial_z\, a_{12}^*(z). \qquad \square$$

By the general results of [**46**, Section 8.3.3], elements of the center $\mathrm{Z}(\widehat{\mathfrak{gl}}_N)$ of the completed universal enveloping algebra $\widetilde{\mathrm{U}}_{-N}(\widehat{\mathfrak{gl}}_N)$ act on the Wakimoto modules $W_{\chi(z)}$ by scalar multiplication. Moreover, as the next proposition shows, the eigenvalues of central elements essentially coincide with their Harish-Chandra images given in (13.47) and which were calculated in Section 13.3.

PROPOSITION 15.4.3. *The eigenvalue of any element $S \in \mathrm{Z}(\widehat{\mathfrak{gl}}_N)$ acting in the Wakimoto module $W_{\chi(z)}$ coincides with the Harish-Chandra image $\mathfrak{f}(S)$, where the generators of the algebra $\Pi = \mathrm{S}\,(\mathfrak{h}[t,t^{-1}])$ are respectively replaced by $E_{ii}[r] \mapsto \chi_i[r]$.*

PROOF. We are interested in the eigenvalue of the vector $|0\rangle \in W_{\chi(z)}$ under the action of S. Write the element S as a (possibly infinite) linear combination of ordered monomials with the ordering as defined in Section 13.3. Consider a monomial occurring in this linear combination whose rightmost factor is an element of the form $E_{ij}[r] \in \mathfrak{n}_+[t]$, so that $i < j$ and $r \geqslant 0$. By the formulas of Proposition 15.4.1, $E_{i\,i+1}[r]\,|0\rangle = 0$ for $r \geqslant 0$ which implies $E_{ij}[r]\,|0\rangle = 0$. Now suppose that the leftmost factor of a monomial occurring in the linear combination is of the form $E_{ij}[r] \in t^{-1}\mathfrak{n}_+[t^{-1}]$ with $i < j$ and $r < 0$. The result of the application of this monomial to the vector $|0\rangle$ will be a linear combination of basis elements of the Fock space $M(\mathfrak{gl}_N)$ which does not contain the vector $|0\rangle$. Hence such monomials will not contribute to the eigenvalue of S. Due to the weight property (13.48) of S, the nonzero contribution can only come from ordered monomials in the generators $E_{ii}[r]$. We find from Proposition 15.4.1, that $E_{ii}[r]\,|0\rangle = \chi_i[r]\,|0\rangle$ for all $r \geqslant 0$. Furthermore, for $r < 0$ the contribution to the coefficient of $|0\rangle$ in the expansion of $E_{ii}[r]\,|0\rangle$ into a linear combination of the basis vectors of $M(\mathfrak{gl}_N)$ is $\chi_i[r]\,|0\rangle$. Thus, the eigenvalue of S is obtained by replacing $E_{ii}[r]$ with $\chi_i[r]$, respectively, in the Harish-Chandra image $\mathfrak{f}(S)$. $\square$

COROLLARY 15.4.4. *With the replacements $\mu_i(z) \mapsto \chi_i(z)$ for $i = 1, \dots, N$, the formulas of Proposition* 13.3.3 *give the eigenvalues for the action of the elements of $\mathrm{Z}(\widehat{\mathfrak{gl}}_N)$ in the Wakimoto module $W_{\chi(z)}$.* $\square$

15.5. Wakimoto modules in types B and D

Similar to type A, consider the *Weyl algebra* $\mathcal{A}(\mathfrak{o}_N)$ generated by elements $a_{ij}[r]$ and $a_{ij}^*[r]$, where $1 \leqslant i < j \leqslant N$ with $i + j \leqslant N$ and $r \in \mathbb{Z}$. The defining relations are

$$\big[a_{ij}[r], a_{kl}^*[s]\big] = \delta_{ik}\,\delta_{jl}\,\delta_{r,-s}, \qquad \big[a_{ij}[r], a_{kl}[s]\big] = \big[a_{ij}^*[r], a_{kl}^*[s]\big] = 0.$$

The *Fock representation* $M(\mathfrak{o}_N)$ of $\mathcal{A}(\mathfrak{o}_N)$ is generated by a vector $|0\rangle$ such that

$$a_{ij}[r]\,|0\rangle = 0 \qquad \text{for} \quad r \geqslant 0 \qquad \text{and} \qquad a_{ij}^*[r]\,|0\rangle = 0 \qquad \text{for} \quad r > 0.$$

The Laurent series $a_{ij}(z)$ and $a_{ij}^*(z)$ are introduced by the same formulas (15.5) as in type A. Their normally ordered products are defined by the rule (6.9) which can also be stated in terms of the normally ordered monomials as in type A.

Suppose now that $N = 2n$ and consider the formal Laurent series $F_{ij}(z)$ introduced in (8.50). Fix an n-tuple of series

$$\chi(z) = \big(\chi_1(z), \ldots, \chi_n(z)\big), \qquad \chi_i(z) \in \mathbb{C}((z)). \tag{15.8}$$

The following formulas define the *Wakimoto module* $W_{\chi(z)}$ of the critical level.

PROPOSITION 15.5.1. *The following assignments define a representation of $\widehat{\mathfrak{o}}_{2n}$ at the critical level on the vector space $M(\mathfrak{o}_{2n})$:*

$$F_{ii}(z) \mapsto \sum_{r=1}^{i-1} \Big(: a^*_{ri}(z) a_{ri}(z) : - : a^*_{ri'}(z) a_{ri'}(z) : \Big) - \sum_{s=i+1}^{i'-1} : a^*_{is}(z) a_{is}(z) : + \chi_i(z)$$

for $i = 1, \ldots, n$,

$$F_{i\,i+1}(z) \mapsto a_{i\,i+1}(z) + \sum_{s=i+2}^{i'-2} : a^*_{i+1\,s}(z) a_{is}(z) : - \sum_{s=i+2}^{n} : a^*_{i+1\,s}(z) a^*_{i+1\,s'}(z) a_{i\,i'-1}(z) :$$

for $i = 1, \ldots, n-1$, and $F_{n-1\,n'}(z) \mapsto a_{n-1\,n'}(z)$, together with

$$\begin{aligned} F_{i+1\,i}(z) \mapsto \sum_{r=1}^{i-1} \Big(&: \big(a^*_{ri}(z)\, a^*_{i\,i+1}(z) - a^*_{r\,i+1}(z)\big)\, a_{ri}(z) : \\ &+ : \big(a^*_{r\,i'-1}(z)\, a^*_{i\,i+1}(z) + a^*_{r\,i'}(z)\big)\, a_{r\,i'-1}(z) : - : a^*_{ri'}(z)\, a^*_{i\,i+1}(z)\, a_{ri'}(z) : \Big) \\ - \sum_{r=1}^{i} &: a^*_{r\,i+1}(z)\, a^*_{i\,i+1}(z)\, a_{r\,i+1}(z) : + \sum_{s=i+2}^{i'-2} : a^*_{is}(z) a_{i+1\,s}(z) : \\ - \sum_{s=i+2}^{n} &: a^*_{i\,s}(z) a^*_{i\,s'}(z) a_{i\,i'-1}(z) : + \big(\chi_i(z) - \chi_{i+1}(z)\big)\, a^*_{i\,i+1}(z) - 2i\, \partial_z a^*_{i\,i+1}(z) \end{aligned}$$

for $i = 1, \ldots, n-1$, and

$$\begin{aligned} F_{n'\,n-1}(z) \mapsto \sum_{r=1}^{n-2} \Big(&: \big(a^*_{r\,n-1}(z)\, a^*_{n-1\,n'}(z) - a^*_{r\,n'}(z)\big)\, a_{r\,n-1}(z) : \\ + : \big(a^*_{r\,n}(z)\, a^*_{n-1\,n'}(z) &+ a^*_{r\,n+2}(z)\big)\, a_{r\,n}(z) : - : a^*_{r\,n+2}(z)\, a^*_{n-1\,n'}(z)\, a_{r\,n+2}(z) : \Big) \\ &- \sum_{r=1}^{n-1} : a^*_{r\,n'}(z)\, a^*_{n-1\,n'}(z)\, a_{r\,n'}(z) : \\ &\qquad + \big(\chi_{n-1}(z) + \chi_n(z)\big)\, a^*_{n-1\,n'}(z) - (2n-2)\, \partial_z a^*_{n-1\,n'}(z). \end{aligned}$$

PROOF. The formulas follow from Proposition 15.2.2 providing a free field realization of $\mathfrak{o}_{2n}$ and the general results of [**46**, Section 6.1]. We only need to verify the values $c_i = -2i$ of the coefficients c_i of the terms $\partial_z a^*_{i\,i+1}(z)$ and $c_n = -2n+2$ of the coefficient c_n of the term $\partial_z a^*_{n-1\,n'}(z)$. These values are known to exist by the general results. Since $K = -2n+2$ at the critical level, by (8.2) we have

$$\big[F_{i\,i+1}[1], F_{i+1\,i}[-1]\big] = F_{i\,i}[0] - F_{i+1\,i+1}[0] - 2n + 2$$

for $i = 1, \dots, n-1$. Writing $\chi_i(z)$ as in (15.7), and applying the right hand side to the vector $|0\rangle$ we get $\big(\chi_i[0] - \chi_{i+1}[0] - 2n + 2\big)\,|0\rangle$. Furthermore,

$$F_{i\,i+1}[1] = a_{i\,i+1}[1] + \sum_{s=i+2}^{N} \sum_{k+l=1} :a^*_{i+1\,s}[k]\,a_{is}[l]: \\ - \sum_{s=i+2}^{n} \sum_{k+l+m=1} :a^*_{i+1\,s}[k]\,a^*_{i+1\,s'}[l]\,a_{i\,i'-1}[m]:$$

so that $F_{i\,i+1}[1]\,|0\rangle = 0$, and all components of the vector $F_{i+1\,i}[-1]\,|0\rangle$ are annihilated by the operator $F_{i\,i+1}[1]$ except for

$$\sum_{s=i+2}^{i'-2} a^*_{is}[0]\,a_{i+1\,s}[-1]\,|0\rangle + \big(\chi_i[0] - \chi_{i+1}[0]\big)\,a^*_{i\,i+1}[-1]\,|0\rangle + c_i\,a^*_{i\,i+1}[-1]\,|0\rangle.$$

By applying $F_{i\,i+1}[1]$ to this component we get

$$F_{i\,i+1}[1]\,F_{i+1\,i}[-1]\,|0\rangle = \big(-2n + 2 + 2i + c_i + \chi_i[0] - \chi_{i+1}[0]\big)\,|0\rangle$$

and so $c_i = -2i$. The coefficient c_n is calculated in the same way by applying the operators on both sides of the relation

$$\big[F_{n-1\,n'}[1], F_{n'\,n-1}[-1]\big] = F_{n-1\,n-1}[0] + F_{n\,n}[0] - 2n + 2$$

to the vector $|0\rangle$ to get $c_n = -2n + 2$. □

Now let $N = 2n + 1$. Fix an n-tuple of series $\chi(z)$ as in (15.8) and use the series $F_{ij}(z)$ introduced in (8.50). The *Wakimoto module* $W_{\chi(z)}$ of the critical level is defined in the following proposition.

PROPOSITION 15.5.2. *The following assignments define a representation of $\widehat{\mathfrak{o}}_{2n+1}$ at the critical level on the vector space $M(\mathfrak{o}_{2n+1})$:*

$$F_{ii}(z) \mapsto \sum_{r=1}^{i-1} \Big(:a^*_{ri}(z)\,a_{ri}(z): - :a^*_{ri'}(z)\,a_{ri'}(z):\Big) - \sum_{s=i+1}^{i'-1} :a^*_{is}(z)\,a_{is}(z): + \chi_i(z)$$

$$F_{i\,i+1}(z) \mapsto a_{i\,i+1}(z) + \sum_{s=i+2}^{i'-2} :a^*_{i+1\,s}(z)\,a_{is}(z): \\ - \frac{1}{2} \sum_{s=i+2}^{i'-2} :a^*_{i+1\,s}(z)\,a^*_{i+1\,s'}(z)\,a_{i\,i'-1}(z):$$

for $i = 1, \dots, n$, together with

$$F_{i+1\,i}(z) \mapsto \sum_{r=1}^{i-1} \Big(:\big(a^*_{ri}(z)\,a^*_{i\,i+1}(z) - a^*_{r\,i+1}(z)\big)\,a_{ri}(z): \\ + :\big(a^*_{r\,i'-1}(z)\,a^*_{i\,i+1}(z) + a^*_{r\,i'}(z)\big)\,a_{r\,i'-1}(z): - :a^*_{ri'}(z)\,a^*_{i\,i+1}(z)\,a_{ri'}(z):\Big) \\ - \sum_{r=1}^{i} :a^*_{r\,i+1}(z)\,a^*_{i\,i+1}(z)\,a_{r\,i+1}(z): + \sum_{s=i+2}^{i'-2} :a^*_{is}(z)\,a_{i+1\,s}(z): \\ - \frac{1}{2} \sum_{s=i+2}^{i'-2} :a^*_{i\,s}(z)\,a^*_{i\,s'}(z)\,a_{i\,i'-1}(z): + \big(\chi_i(z) - \chi_{i+1}(z)\big)\,a^*_{i\,i+1}(z) - 2i\,\partial_z\,a^*_{i\,i+1}(z)$$

for $i = 1, \dots, n-1$, *and*

$$F_{n+1\,n}(z) \mapsto \sum_{r=1}^{n-1} \Big(: \big(a^*_{r\,n}(z)\, a^*_{n\,n+1}(z) - a^*_{r\,n+1}(z)\big)\, a_{r\,n}(z) :$$

$$- : a^*_{r\,n+2}(z)\, a^*_{n\,n+1}(z)\, a_{r\,n+2}(z) : + : a^*_{r\,n+2}(z)\, a_{r\,n}(z) : \Big)$$

$$- \frac{1}{2} : a^*_{n\,n+1}(z)^2\, a_{n\,n+1}(z) : + \chi_n(z)\, a^*_{n\,n+1}(z) - (2n-1)\, \partial_z\, a^*_{n\,n+1}(z).$$

PROOF. As with Proposition 15.5.1, we only need to calculate the coefficients c_i of the terms $\partial_z\, a^*_{i\,i+1}(z)$ and c_n of the term $\partial_z\, a^*_{n\,n+1}(z)$. This time the critical level is $K = -2n + 1$ so that (8.2) gives

$$\big[F_{i\,i+1}[1], F_{i+1\,i}[-1]\big] = F_{i\,i}[0] - F_{i+1\,i+1}[0] - 2n + 1$$

for $i = 1, \dots, n$. The same argument as for $\mathfrak{o}_{2n}$ yields the values $c_i = -2i$ for $i = 1, \dots, n-1$ and $c_n = -2n + 1$. □

The corresponding analogue of Proposition 15.4.3 holds in the same form for the orthogonal Lie algebras $\mathfrak{o}_N$. It is proved by the same argument relying on the formulas of Propositions 15.5.1 and 15.5.2. We use the homomorphism (13.47).

PROPOSITION 15.5.3. *The eigenvalue of any element* $S \in \mathrm{Z}(\widehat{\mathfrak{o}}_N)$ *acting in the Wakimoto module* $W_{\chi(z)}$ *coincides with the Harish-Chandra image* $\mathfrak{f}(S)$, *where the generators of the algebra* $\Pi = \mathrm{S}\big(\mathfrak{h}[t, t^{-1}]\big)$ *are respectively replaced by the rule* $F_{ii}[r] \mapsto \chi_i[r]$ *for* $i = 1, \dots, n$ *and* $r \in \mathbb{Z}$. □

COROLLARY 15.5.4. *With the replacements* $\mu_i(z) \mapsto \chi_i(z)$ *for* $i = 1, \dots, n$, *the formulas of Propositions* 13.3.4, 13.3.5 *and* 13.3.6 *give the eigenvalues for the action of the elements of* $\mathrm{Z}(\widehat{\mathfrak{o}}_N)$ *in the Wakimoto module* $W_{\chi(z)}$. □

15.6. Wakimoto modules in type C

The *Weyl algebra* $\mathcal{A}(\mathfrak{sp}_{2n})$ is generated by elements $a_{ij}[r]$ and $a^*_{ij}[r]$, where $1 \leqslant i < j \leqslant 2n$ with $i + j \leqslant 2n + 1$ and $r \in \mathbb{Z}$. The defining relations are

$$\big[a_{ij}[r], a^*_{kl}[s]\big] = \delta_{ik}\, \delta_{jl}\, \delta_{r,-s}, \qquad \big[a_{ij}[r], a_{kl}[s]\big] = \big[a^*_{ij}[r], a^*_{kl}[s]\big] = 0.$$

The *Fock representation* $M(\mathfrak{sp}_{2n})$ of $\mathcal{A}(\mathfrak{sp}_{2n})$ is generated by a vector $|0\rangle$ such that

$$a_{ij}[r]\, |0\rangle = 0 \quad \text{for} \quad r \geqslant 0 \quad \text{and} \quad a^*_{ij}[r]\, |0\rangle = 0 \quad \text{for} \quad r > 0.$$

We use the same formulas (15.5) as in type A to introduce the Laurent series $a_{ij}(z)$ and $a^*_{ij}(z)$, and apply the rule (6.9) to define their normally ordered products.

Fix an n-tuple of series $\chi(z)$ as in (15.8) and use the series $F_{ij}(z)$ introduced in (8.50). The following proposition defines the corresponding *Wakimoto module* $W_{\chi(z)}$ of the critical level.

PROPOSITION 15.6.1. *The following assignments define a representation of* $\widehat{\mathfrak{sp}}_{2n}$ *at the critical level on the vector space* $M(\mathfrak{sp}_{2n})$:

$$F_{ii}(z) \mapsto \sum_{r=1}^{i-1} \Big(: a^*_{ri}(z)\, a_{ri}(z) : - : a^*_{ri'}(z)\, a_{ri'}(z) : \Big) \\ - \sum_{s=i+1}^{i'-1} : a^*_{is}(z)\, a_{is}(z) : - 2 : a^*_{i\,i'}(z)\, a_{i\,i'}(z) : + \chi_i(z)$$

for $i = 1, \ldots, n$,

$$F_{i\,i+1}(z) \mapsto a_{i\,i+1}(z) + \sum_{s=i+2}^{i'-1} : a^*_{i+1\,s}(z)\, a_{is}(z) : \\ + : \Big(a^*_{i\,i'-1}(z) - \sum_{s=i+1}^{i'-2} \varepsilon_s\, a^*_{i\,s}(z)\, a^*_{i+1\,s'}(z) \Big) a_{i\,i'}(z) :$$

for $i = 1, \ldots, n-1$, *and* $F_{n\,n+1}(z) \mapsto 2 a_{n\,n+1}(z)$, *together with*

$$F_{i+1\,i}(z) \mapsto \sum_{r=1}^{i} \Big(: \big(a^*_{r\,i'-1}(z)\, a^*_{i\,i+1}(z) + a^*_{r\,i'}(z) \big)\, a_{r\,i'-1}(z) : \\ - : a^*_{ri'}(z)\, a^*_{i\,i+1}(z)\, a_{ri'}(z) : - : a^*_{r\,i+1}(z)\, a^*_{i\,i+1}(z)\, a_{r\,i+1}(z) : \Big) \\ + \sum_{s=i+2}^{i'-2} : a^*_{i\,s}(z)\, a_{i+1\,s}(z) : + \sum_{r=1}^{i-1} : \big(a^*_{ri}(z)\, a^*_{i\,i+1}(z) - a^*_{r\,i+1}(z) \big)\, a_{ri}(z) : \\ + : \Big(2 a^*_{i\,i'-1}(z) - \sum_{s=i+2}^{i'-2} \varepsilon_s\, a^*_{i\,s}(z)\, a^*_{i+1\,s'}(z) \Big)\, a_{i+1\,i'-1}(z) : \\ - (2i+1)\, \partial_z a^*_{i\,i+1}(z) + \big(\chi_i(z) - \chi_{i+1}(z) \big)\, a^*_{i\,i+1}(z)$$

for $i = 1, \ldots, n-1$, *and*

$$F_{n+1\,n}(z) \mapsto 2 \sum_{r=1}^{n-1} : \big(a^*_{rn}(z)\, a^*_{n\,n+1}(z) - a^*_{r\,n+1}(z) \big)\, a_{rn}(z) : \\ - 2 \sum_{r=1}^{n} : a^*_{r\,n+1}(z)\, a^*_{n\,n+1}(z)\, a_{r\,n+1}(z) : + 2\, \chi_n(z)\, a^*_{n\,n+1}(z) - 2(n+1)\, \partial_z a^*_{n\,n+1}(z).$$

PROOF. We use again the general construction of the Wakimoto modules [**46**, Section 6.1], so that the formulas are implied by Proposition 15.3.1. We will only verify the values of the coefficients c_i of the terms $\partial_z a^*_{i\,i+1}(z)$ and the coefficient c_n of the term $\partial_z a^*_{n\,n+1}(z)$. At the critical level we have $K = -n-1$ and so (8.32) implies

$$\big[F_{i\,i+1}[1], F_{i+1\,i}[-1] \big] = F_{i\,i}[0] - F_{i+1\,i+1}[0] - 2n - 2$$

for $i = 1, \ldots, n-1$. Write $\chi_i(z)$ as in (15.7) and apply the right hand side to the vector $|0\rangle$ to get $\big(\chi_i[0] - \chi_{i+1}[0] - 2n - 2 \big)\, |0\rangle$. Now observe that all components of the vector $F_{i+1\,i}[-1]\,|0\rangle$ are annihilated by the operator $F_{i\,i+1}[1]$ except for

$$a^*_{i\,i'}[0]\,a_{i\,i'-1}[-1]\,|0\rangle + \sum_{s=i+2}^{i'-2} a^*_{is}[0]\,a_{i+1\,s}[-1]\,|0\rangle + 2\,a^*_{i\,i'-1}[0]\,a_{i+1\,i'-1}[-1]\,|0\rangle$$
$$+ \big(\chi_i[0] - \chi_{i+1}[0]\big)\,a^*_{i\,i+1}[-1]\,|0\rangle + c_i\,a^*_{i\,i+1}[-1]\,|0\rangle.$$

By applying $F_{i\,i+1}[1]$ to this component we get

$$F_{i\,i+1}[1]\,F_{i+1\,i}[-1]\,|0\rangle = \big(-2n+1+2i+c_i+\chi_i[0]-\chi_{i+1}[0]\big)\,|0\rangle$$

and so $c_i = -2i-1$. The coefficient c_n is calculated in the same way by applying the operators on both sides of the relation

$$\big[F_{n\,n+1}[1], F_{n+1\,n}[-1]\big] = 4\,F_{n\,n}[0] - 4n - 4$$

to the vector $|0\rangle$ to get $c_n = -2n-2$. □

The following is an analogue of Proposition 15.4.3 which is verified by the same argument relying on the formulas of Proposition 15.6.1. We use the homomorphism $\mathfrak{f}$ defined in (13.47).

PROPOSITION 15.6.2. *The eigenvalue of any element $S \in \mathrm{Z}(\widehat{\mathfrak{sp}}_{2n})$ acting in the Wakimoto module $W_{\chi(z)}$ coincides with the Harish-Chandra image $\mathfrak{f}(S)$, where the generators of the algebra $\Pi = \mathrm{S}\big(\mathfrak{h}[t,t^{-1}]\big)$ are respectively replaced by the rule $F_{ii}[r] \mapsto \chi_i[r]$ for $i = 1,\dots,n$ and $r \in \mathbb{Z}$.* □

COROLLARY 15.6.3. *With the replacements $\mu_i(z) \mapsto \chi_i(z)$ for $i = 1,\dots,n$, the formulas of Proposition 13.3.7 give the eigenvalues for the action of the elements of $\mathrm{Z}(\widehat{\mathfrak{sp}}_{2n})$ in the Wakimoto module $W_{\chi(z)}$.* □

15.7. Bibliographical notes

The Wakimoto modules were originally introduced by Wakimoto [**151**] for $\widehat{\mathfrak{sl}}_2$ and by Feigin and Frenkel [**38**] for arbitrary affine Kac–Moody algebras $\widehat{\mathfrak{g}}$. The construction and its generalizations are explained in detail in the book by Frenkel [**46**, Chapter 6]. Propositions 15.4.3, 15.5.3 and 15.6.2 are particular cases of the general results which hold for arbitrary simple Lie algebras $\mathfrak{g}$ as shown in [**46**, Section 8.3.3].

Bibliography

[1] M. Adler, *On a trace functional for formal pseudo differential operators and the symplectic structure of the Korteweg-de Vries type equations*, Invent. Math. **50** (1978/79), no. 3, 219–248, DOI 10.1007/BF01410079. MR520927

[2] Henning Haahr Andersen, Catharina Stroppel, and Daniel Tubbenhauer, *Semisimplicity of Hecke and (walled) Brauer algebras*, J. Aust. Math. Soc. **103** (2017), no. 1, 1–44, DOI 10.1017/S1446788716000392. MR3679014

[3] Tomoyuki Arakawa, *Drinfeld functor and finite-dimensional representations of Yangian*, Comm. Math. Phys. **205** (1999), no. 1, 1–18, DOI 10.1007/s002200050664. MR1706920

[4] T. Arakawa, *Introduction to W-algebras and their representation theory*, `arXiv:1605.00138`.

[5] ———, *W-algebras at the critical level*, Algebraic groups and quantum groups, Contemp. Math., vol. 565, Amer. Math. Soc., Providence, RI, 2012, pp. 1–13, DOI 10.1090/conm/565/11184. MR2932418

[6] D. Arnaudon, J. Avan, N. Crampé, L. Frappat, and E. Ragoucy, *R-matrix presentation for super-Yangians* $Y(\mathrm{osp}(m|2n))$, J. Math. Phys. **44** (2003), no. 1, 302–308, DOI 10.1063/1.1525406. MR1946704

[7] Daniel Arnaudon, Alexander Molev, and Eric Ragoucy, *On the R-matrix realization of Yangians and their representations*, Ann. Henri Poincaré **7** (2006), no. 7-8, 1269–1325, DOI 10.1007/s00023-006-0281-9. MR2283732

[8] R. M. Asherova, Yu. F. Smirnov and V. N. Tolstoy, *Projection operators for simple Lie groups*, Theor. Math. Phys. **8** (1971), 813–825. MR0475353

[9] H. M. Babujian and R. Flume, *Off-shell Bethe ansatz equation for Gaudin magnets and solutions of Knizhnik-Zamolodchikov equations*, Modern Phys. Lett. A **9** (1994), no. 22, 2029–2039, DOI 10.1142/S0217732394001891. MR1290286

[10] Joan S. Birman and Hans Wenzl, *Braids, link polynomials and a new algebra*, Trans. Amer. Math. Soc. **313** (1989), no. 1, 249–273, DOI 10.2307/2001074. MR992598

[11] A. V. Bolsinov, *Commutative families of functions related to consistent Poisson brackets*, Acta Appl. Math. **24** (1991), no. 3, 253–274, DOI 10.1007/BF00047046. MR1134669

[12] ———, *Compatible Poisson brackets on Lie algebras and the completeness of families of functions in involution* (Russian), Izv. Akad. Nauk SSSR Ser. Mat. **55** (1991), no. 1, 68–92; English transl., Math. USSR-Izv. **38** (1992), no. 1, 69–90. MR1130028

[13] A. V. Bolsinov and P. Zhang, *Jordan-Kronecker invariants of finite-dimensional Lie algebras*, Transform. Groups **21** (2016), no. 1, 51–86, DOI 10.1007/s00031-015-9353-6. MR3459704

[14] Richard Brauer, *On algebras which are connected with the semisimple continuous groups*, Ann. of Math. (2) **38** (1937), no. 4, 857–872, DOI 10.2307/1968843. MR1503378

[15] Jonathan Brown and Jonathan Brundan, *Elementary invariants for centralizers of nilpotent matrices*, J. Aust. Math. Soc. **86** (2009), no. 1, 1–15, DOI 10.1017/S1446788708000608. MR2495993

[16] Jonathan Brundan and Alexander Kleshchev, *Representations of shifted Yangians and finite W-algebras*, Mem. Amer. Math. Soc. **196** (2008), no. 918, viii+107, DOI 10.1090/memo/0918. MR2456464

[17] Jean-Yves Charbonnel and Anne Moreau, *The index of centralizers of elements of reductive Lie algebras*, Doc. Math. **15** (2010), 387–421. MR2645445

[18] Vyjayanthi Chari and Andrew Pressley, *A guide to quantum groups*, Cambridge University Press, Cambridge, 1994. MR1300632

[19] I. V. Cherednik, *Special bases of irreducible representations of a degenerate affine Hecke algebra*, Funct. Analysis Appl. **20** (1986), 76–78. MR831062

[20] A. Chervov and G. Falqui, *Manin matrices and Talalaev's formula*, J. Phys. A **41** (2008), no. 19, 194006, 28, DOI 10.1088/1751-8113/41/19/194006. MR2452180

[21] A. Chervov, G. Falqui, and V. Rubtsov, *Algebraic properties of Manin matrices. I*, Adv. in Appl. Math. **43** (2009), no. 3, 239–315, DOI 10.1016/j.aam.2009.02.003. MR2549578

[22] A. Chervov, G. Falqui, V. Rubtsov, and A. Silantyev, *Algebraic properties of Manin matrices II: q-analogues and integrable systems*, Adv. in Appl. Math. **60** (2014), 25–89, DOI 10.1016/j.aam.2014.06.001. MR3256747

[23] A. V. Chervov and A. I. Molev, *On higher-order Sugawara operators*, Int. Math. Res. Not. IMRN **9** (2009), 1612–1635, DOI 10.1093/imrn/rnn168. MR2500972

[24] A. Chervov and D. Talalaev, *Quantum spectral curves, quantum integrable systems and the geometric Langlands correspondence*, arXiv:hep-th/0604128.

[25] Zajj Daugherty, Arun Ram, and Rahbar Virk, *Affine and degenerate affine BMW algebras: actions on tensor space*, Selecta Math. (N.S.) **19** (2013), no. 2, 611–653, DOI 10.1007/s00029-012-0105-3. MR3090238

[26] Alberto De Sole, Victor G. Kac, and Daniele Valeri, *Classical $\mathcal{W}$-algebras and generalized Drinfeld-Sokolov bi-Hamiltonian systems within the theory of Poisson vertex algebras*, Comm. Math. Phys. **323** (2013), no. 2, 663–711, DOI 10.1007/s00220-013-1785-z. MR3096535

[27] ———, *Structure of classical (finite and affine) $\mathcal{W}$-algebras*, J. Eur. Math. Soc. (JEMS) **18** (2016), no. 9, 1873–1908, DOI 10.4171/JEMS/632. MR3531664

[28] ———, *Classical affine $\mathcal{W}$-algebras for $\mathfrak{gl}_N$ and associated integrable Hamiltonian hierarchies*, Comm. Math. Phys. **348** (2016), no. 1, 265–319, DOI 10.1007/s00220-016-2632-9. MR3551267

[29] Jacques Dixmier, *Algèbres enveloppantes* (French), Gauthier-Villars Éditeur, Paris-Brussels-Montreal, Que., 1974. Cahiers Scientifiques, Fasc. XXXVII. MR0498737

[30] V. G. Drinfeld, *Hopf algebras and the quantum Yang–Baxter equation*, Soviet Math. Dokl. **32** (1985), 254–258. MR802128

[31] V. G. Drinfeld, *Degenerate affine Hecke algebras and Yangians*, Funct. Anal. Appl. **20** (1986), 56–58. MR831053

[32] V. G. Drinfel′d, *Quantum groups*, Proceedings of the International Congress of Mathematicians, Vol. 1, 2 (Berkeley, Calif., 1986), Amer. Math. Soc., Providence, RI, 1987, pp. 798–820. MR934283

[33] V. G. Drinfeld and V. V. Sokolov, *Lie algebras and equations of Korteweg-de Vries type*, J. Sov. Math. **30** (1985), 1975–2036.

[34] N. El Samra and R. C. King, *Dimensions of irreducible representations of the classical Lie groups*, J. Phys. A **12** (1979), no. 12, 2317–2328. MR552445

[35] Pavel Etingof and David Kazhdan, *Quantization of Lie bialgebras. II, III*, Selecta Math. (N.S.) **4** (1998), no. 2, 213–231, 233–269, DOI 10.1007/s000290050030. MR1669953

[36] ———, *Quantization of Lie bialgebras. IV. The coinvariant construction and the quantum KZ equations*, Selecta Math. (N.S.) **6** (2000), no. 1, 79–104, DOI 10.1007/s000290050003. MR1771217

[37] ———, *Quantization of Lie bialgebras. V. Quantum vertex operator algebras*, Selecta Math. (N.S.) **6** (2000), no. 1, 105–130, DOI 10.1007/s000290050004. MR1771218

[38] B. L. Feĭgin and È. V. Frenkel′, *A family of representations of affine Lie algebras* (Russian), Uspekhi Mat. Nauk **43** (1988), no. 5(263), 227–228, DOI 10.1070/RM1988v043n05ABEH001935; English transl., Russian Math. Surveys **43** (1988), no. 5, 221–222. MR971497

[39] Boris Feigin and Edward Frenkel, *Affine Kac-Moody algebras at the critical level and Gel′fand-Dikiĭ algebras*, Infinite analysis, Part A, B (Kyoto, 1991), Adv. Ser. Math. Phys., vol. 16, World Sci. Publ., River Edge, NJ, 1992, pp. 197–215. MR1187549

[40] Boris Feigin, Edward Frenkel, and Nikolai Reshetikhin, *Gaudin model, Bethe ansatz and critical level*, Comm. Math. Phys. **166** (1994), no. 1, 27–62. MR1309540

[41] Boris Feigin, Edward Frenkel, and Leonid Rybnikov, *Opers with irregular singularity and spectra of the shift of argument subalgebra*, Duke Math. J. **155** (2010), no. 2, 337–363, DOI 10.1215/00127094-2010-057. MR2736168

[42] B. Feigin, E. Frenkel, and V. Toledano Laredo, *Gaudin models with irregular singularities*, Adv. Math. **223** (2010), no. 3, 873–948, DOI 10.1016/j.aim.2009.09.007. MR2565552

[43] Dominique Foata and Guo-Niu Han, *A new proof of the Garoufalidis-Lê-Zeilberger quantum MacMahon master theorem*, J. Algebra **307** (2007), no. 1, 424–431, DOI 10.1016/j.jalgebra.2006.04.032. MR2278064

[44] ———, *A basis for the right quantum algebra and the "1 = q" principle*, J. Algebraic Combin. **27** (2008), no. 2, 163–172, DOI 10.1007/s10801-007-0080-5. MR2375489

[45] Luc Frappat, Naihuan Jing, Alexander Molev, and Eric Ragoucy, *Higher Sugawara operators for the quantum affine algebras of type A*, Comm. Math. Phys. **345** (2016), no. 2, 631–657, DOI 10.1007/s00220-015-2566-7. MR3514954

[46] Edward Frenkel, *Langlands correspondence for loop groups*, Cambridge Studies in Advanced Mathematics, vol. 103, Cambridge University Press, Cambridge, 2007. MR2332156

[47] Edward Frenkel and David Ben-Zvi, *Vertex algebras and algebraic curves*, 2nd ed., Mathematical Surveys and Monographs, vol. 88, American Mathematical Society, Providence, RI, 2004. MR2082709

[48] Edward Frenkel and Evgeny Mukhin, *Combinatorics of q-characters of finite-dimensional representations of quantum affine algebras*, Comm. Math. Phys. **216** (2001), no. 1, 23–57, DOI 10.1007/s002200000323. MR1810773

[49] ———, *The Hopf algebra* $\mathrm{Rep}\, U_q\widehat{\mathfrak{gl}}_\infty$, Selecta Math. (N.S.) **8** (2002), no. 4, 537–635, DOI 10.1007/PL00012603. MR1951206

[50] Edward Frenkel and Nikolai Reshetikhin, *Quantum affine algebras and deformations of the Virasoro and $\mathcal{W}$-algebras*, Comm. Math. Phys. **178** (1996), no. 1, 237–264. MR1387950

[51] Edward Frenkel and Nicolai Reshetikhin, *The q-characters of representations of quantum affine algebras and deformations of $\mathcal{W}$-algebras*, Recent developments in quantum affine algebras and related topics (Raleigh, NC, 1998), Contemp. Math., vol. 248, Amer. Math. Soc., Providence, RI, 1999, pp. 163–205, DOI 10.1090/conm/248/03823. MR1745260

[52] I. B. Frenkel, *Two constructions of affine Lie algebra representations and boson-fermion correspondence in quantum field theory*, J. Funct. Anal. **44** (1981), no. 3, 259–327, DOI 10.1016/0022-1236(81)90012-4. MR643037

[53] Vyacheslav Futorny and Alexander Molev, *Quantization of the shift of argument subalgebras in type A*, Adv. Math. **285** (2015), 1358–1375, DOI 10.1016/j.aim.2015.07.038. MR3406529

[54] Stavros Garoufalidis, Thang T. Q. Lê, and Doron Zeilberger, *The quantum MacMahon master theorem*, Proc. Natl. Acad. Sci. USA **103** (2006), no. 38, 13928–13931, DOI 10.1073/pnas.0606003103. MR2270337

[55] I. M. Gelfand, *Center of the infinitesimal group ring*, Mat. Sbornik **26** (1950), 103–112 (Russian). English transl. in: I. M. Gelfand, "Collected Papers", vol. II, Berlin: Springer-Verlag, 1988, pp. 22–30. MR0033831

[56] I. M. Gelfand and L. A. Dickey, *Family of Hamiltonian structures connected with integrable non-linear equations*, Preprint, IPM, Moscow (in Russian), 1978. English version in: "Collected papers of I. M. Gelfand", vol. 1, Springer-Verlag (1987), 625–646. MR542603

[57] Israel M. Gelfand, Daniel Krob, Alain Lascoux, Bernard Leclerc, Vladimir S. Retakh, and Jean-Yves Thibon, *Noncommutative symmetric functions*, Adv. Math. **112** (1995), no. 2, 218–348, DOI 10.1006/aima.1995.1032. MR1327096

[58] Roe Goodman and Nolan R. Wallach, *Higher-order Sugawara operators for affine Lie algebras*, Trans. Amer. Math. Soc. **315** (1989), no. 1, 1–55, DOI 10.2307/2001371. MR958893

[59] ———, *Representations and invariants of the classical groups*, Encyclopedia of Mathematics and its Applications, vol. 68, Cambridge University Press, Cambridge, 1998. MR1606831

[60] M. D. Gould, *Characteristic identities for semisimple Lie algebras*, J. Austral. Math. Soc. Ser. B **26** (1985), no. 3, 257–283, DOI 10.1017/S0334270000004501. MR776316

[61] Nicolas Guay and Vidas Regelskis, *Twisted Yangians for symmetric pairs of types B, C, D*, Math. Z. **284** (2016), no. 1-2, 131–166, DOI 10.1007/s00209-016-1649-2. MR3545488

[62] Nicolas Guay, Vidas Regelskis, and Curtis Wendlandt, *Representations of twisted Yangians of types B, C, D: I*, Selecta Math. (N.S.) **23** (2017), no. 3, 2071–2156, DOI 10.1007/s00029-017-0306-x. MR3663603

[63] ———, *Representations of twisted Yangians of types B, C, D: I*, Selecta Math. (N.S.) **23** (2017), no. 3, 2071–2156, DOI 10.1007/s00029-017-0306-x. MR3663603

[64] Phùng Hô Hai and Martin Lorenz, *Koszul algebras and the quantum MacMahon master theorem*, Bull. Lond. Math. Soc. **39** (2007), no. 4, 667–676, DOI 10.1112/blms/bdm037. MR2346948

[65] Takahiro Hayashi, *Sugawara operators and Kac-Kazhdan conjecture*, Invent. Math. **94** (1988), no. 1, 13–52, DOI 10.1007/BF01394343. MR958588

[66] David Hernandez, *The Kirillov-Reshetikhin conjecture and solutions of T-systems*, J. Reine Angew. Math. **596** (2006), 63–87, DOI 10.1515/CRELLE.2006.052. MR2254805

[67] Roger Howe, *Perspectives on invariant theory: Schur duality, multiplicity-free actions and beyond*, The Schur lectures (1992) (Tel Aviv), Israel Math. Conf. Proc., vol. 8, Bar-Ilan Univ., Ramat Gan, 1995, pp. 1–182, DOI 10.1007/BF02771542. MR1321638

[68] Roger Howe and Tōru Umeda, *The Capelli identity, the double commutant theorem, and multiplicity-free actions*, Math. Ann. **290** (1991), no. 3, 565–619, DOI 10.1007/BF01459261. MR1116239

[69] Jun Hu and Zhankui Xiao, *On tensor spaces for Birman-Murakami-Wenzl algebras*, J. Algebra **324** (2010), no. 10, 2893–2922, DOI 10.1016/j.jalgebra.2010.08.017. MR2725207

[70] James E. Humphreys, *Introduction to Lie algebras and representation theory*, Springer-Verlag, New York-Berlin, 1972. Graduate Texts in Mathematics, Vol. 9. MR0323842

[71] ———, *Reflection groups and Coxeter groups*, Cambridge Studies in Advanced Mathematics, vol. 29, Cambridge University Press, Cambridge, 1990. MR1066460

[72] Kenji Iohara, *Bosonic representations of Yangian double $\mathcal{D}Y_\hbar(\mathfrak{g})$ with $\mathfrak{g} = \mathfrak{gl}_N, \mathfrak{sl}_N$*, J. Phys. A **29** (1996), no. 15, 4593–4621, DOI 10.1088/0305-4470/29/15/027. MR1413221

[73] N. Iorgov, A. I. Molev, and E. Ragoucy, *Casimir elements from the Brauer-Schur-Weyl duality*, J. Algebra **387** (2013), 144–159, DOI 10.1016/j.jalgebra.2013.02.041. MR3056690

[74] A. P. Isaev and A. I. Molev, *Fusion procedure for the Brauer algebra*, Algebra i Analiz **22** (2010), no. 3, 142–154, DOI 10.1090/S1061-0022-2011-01150-1; English transl., St. Petersburg Math. J. **22** (2011), no. 3, 437–446. MR2729943

[75] A. P. Isaev, A. I. Molev, and O. V. Ogievetsky, *A new fusion procedure for the Brauer algebra and evaluation homomorphisms*, Int. Math. Res. Not. IMRN **11** (2012), 2571–2606, DOI 10.1093/imrn/rnr126. MR2926990

[76] Minoru Itoh, *Explicit Newton's formulas for $\mathfrak{gl}_n$*, J. Algebra **208** (1998), no. 2, 687–697, DOI 10.1006/jabr.1998.7510. MR1655473

[77] ———, *Two determinants in the universal enveloping algebras of the orthogonal Lie algebras*, J. Algebra **314** (2007), no. 1, 479–506, DOI 10.1016/j.jalgebra.2006.11.033. MR2331772

[78] ———, *Two permanents in the universal enveloping algebras of the symplectic Lie algebras*, Internat. J. Math. **20** (2009), no. 3, 339–368, DOI 10.1142/S0129167X09005327. MR2500074

[79] Gordon James and Adalbert Kerber, *The representation theory of the symmetric group*, Encyclopedia of Mathematics and its Applications, vol. 16, Addison-Wesley Publishing Co., Reading, Mass., 1981. With a foreword by P. M. Cohn; With an introduction by Gilbert de B. Robinson. MR644144

[80] Jens Carsten Jantzen, *Nilpotent orbits in representation theory*, Lie theory, Progr. Math., vol. 228, Birkhäuser Boston, Boston, MA, 2004, pp. 1–211. MR2042689

[81] Michio Jimbo, *A q-difference analogue of $U(\mathfrak{g})$ and the Yang-Baxter equation*, Lett. Math. Phys. **10** (1985), no. 1, 63–69, DOI 10.1007/BF00704588. MR797001

[82] N. Jing, S. Kožić, A. Molev and F. Yang, *Center of the quantum affine vertex algebra in type A*, J. Algebra **496** (2018), 138–186.

[83] N. Jing, M. Liu and A. Molev, *Isomorphism between the R-matrix and Drinfeld presentations of Yangian in types B, C and D*, `arXiv:1705.08155`.

[84] A. A. Jucis, *On the Young operators of symmetric groups* (Russian, with Lithuanian and English summaries), Litovsk. Fiz. Sb. **6** (1966), 163–180. MR0202866

[85] A. Jucis, *Factorization of Young's projection operators for symmetric groups.* (Russian, with Lithuanian and English summaries), Litovsk. Fiz. Sb. **11** (1971), 5–10. MR0290671

[86] Victor G. Kac, *Infinite-dimensional Lie algebras*, 3rd ed., Cambridge University Press, Cambridge, 1990. MR1104219

[87] Victor Kac, *Vertex algebras for beginners*, University Lecture Series, vol. 10, American Mathematical Society, Providence, RI, 1997. MR1417941

[88] Haisheng Li, *Vertex algebras and vertex Poisson algebras*, Commun. Contemp. Math. **6** (2004), no. 1, 61–110, DOI 10.1142/S0219199704001264. MR2048777

[89] V. G. Kac and D. A. Kazhdan, *Structure of representations with highest weight of infinite-dimensional Lie algebras*, Adv. in Math. **34** (1979), no. 1, 97–108, DOI 10.1016/0001-8708(79)90066-5. MR547842

[90] Masoud Kamgarpour, *On the notion of conductor in the local geometric Langlands correspondence*, Canad. J. Math. **69** (2017), no. 1, 107–129, DOI 10.4153/CJM-2016-016-1. MR3589855

[91] Harold Knight, *Spectra of tensor products of finite-dimensional representations of Yangians*, J. Algebra **174** (1995), no. 1, 187–196, DOI 10.1006/jabr.1995.1123. MR1332866

[92] Matjaž Konvalinka and Igor Pak, *Non-commutative extensions of the MacMahon Master Theorem*, Adv. Math. **216** (2007), no. 1, 29–61, DOI 10.1016/j.aim.2007.05.020. MR2353248

[93] Bertram Kostant, *The principal three-dimensional subgroup and the Betti numbers of a complex simple Lie group*, Amer. J. Math. **81** (1959), 973–1032, DOI 10.2307/2372999. MR0114875

[94] ______, *Lie group representations on polynomial rings*, Amer. J. Math. **85** (1963), 327–404, DOI 10.2307/2373130. MR0158024

[95] ______, *Fomenko-Mischenko theory, Hessenberg varieties, and polarizations*, Lett. Math. Phys. **90** (2009), no. 1-3, 253–285, DOI 10.1007/s11005-009-0344-1. MR2565041

[96] P. P. Kulish and E. K. Sklyanin, *Quantum spectral transform method. Recent developments*, Lecture Notes in Phys., vol. 151, Springer, Berlin-New York, 1982, pp. 61–119. MR671263

[97] S. Kumar, *A complete set of intertwiners for arbitrary tensor product representations via current algebras*, arXiv:1607.06115.

[98] Atsuo Kuniba and Junji Suzuki, *Analytic Bethe ansatz for fundamental representations of Yangians*, Comm. Math. Phys. **173** (1995), no. 2, 225–264. MR1355625

[99] Atsuo Kuniba, Tomoki Nakanishi, and Junji Suzuki, *T-systems and Y-systems in integrable systems*, J. Phys. A **44** (2011), no. 10, 103001, 146, DOI 10.1088/1751-8113/44/10/103001. MR2773889

[100] A. Kuniba, M. Okado, J. Suzuki, and Y. Yamada, *Difference L operators related to q-characters*, J. Phys. A **35** (2002), no. 6, 1415–1435, DOI 10.1088/0305-4470/35/6/307. MR1890924

[101] Robert Leduc and Arun Ram, *A ribbon Hopf algebra approach to the irreducible representations of centralizer algebras: the Brauer, Birman-Wenzl, and type A Iwahori-Hecke algebras*, Adv. Math. **125** (1997), no. 1, 1–94, DOI 10.1006/aima.1997.1602. MR1427801

[102] K. Lu, E. Mukhin and A. Varchenko, *Self-dual Grassmannian, Wronski map, and representations of $\mathfrak{gl}_N$, $\mathfrak{sp}_{2r}$, $\mathfrak{so}_{2r+1}$*, arXiv:1705.02048.

[103] I. G. Macdonald, *Schur functions: theme and variations*, Séminaire Lotharingien de Combinatoire (Saint-Nabor, 1992), Publ. Inst. Rech. Math. Av., vol. 498, Univ. Louis Pasteur, Strasbourg, 1992, pp. 5–39, DOI 10.1108/EUM0000000002757. MR1308728

[104] ______, *Symmetric functions and Hall polynomials*, 2nd ed., Oxford Mathematical Monographs, The Clarendon Press, Oxford University Press, New York, 1995. With contributions by A. Zelevinsky; Oxford Science Publications. MR1354144

[105] S. V. Manakov, *A remark on the integration of the Eulerian equations of the dynamics of an n-dimensional rigid body*, Functional Anal. Appl. **10** (1976), no. 4, 328–329 (1977). MR0455031

[106] Yu. I. Manin, *Some remarks on Koszul algebras and quantum groups* (English, with French summary), Ann. Inst. Fourier (Grenoble) **37** (1987), no. 4, 191–205. MR927397

[107] ______, *Quantum groups and noncommutative geometry*, Université de Montréal, Centre de Recherches Mathématiques, Montreal, QC, 1988. MR1016381

[108] A. S. Mishchenko and A. T. Fomenko, *Euler equation on finite-dimensional Lie groups*, Math. USSR-Izv. **12** (1978), 371–389. MR0482832

[109] Alexander Molev, *Yangians and classical Lie algebras*, Mathematical Surveys and Monographs, vol. 143, American Mathematical Society, Providence, RI, 2007. MR2355506

[110] A. I. Molev, *Feigin-Frenkel center in types B, C and D*, Invent. Math. **191** (2013), no. 1, 1–34, DOI 10.1007/s00222-012-0390-7. MR3004777

[111] A. I. Molev and E. E. Mukhin, *Yangian characters and classical $\mathcal{W}$-algebras*, Conformal field theory, automorphic forms and related topics, Contrib. Math. Comput. Sci., vol. 8, Springer, Heidelberg, 2014, pp. 287–334. MR3559208

[112] ______, *Invariants of the vacuum module associated with the Lie superalgebra $\mathfrak{gl}(1|1)$*, J. Phys. A **48** (2015), no. 31, 314001, 20, DOI 10.1088/1751-8113/48/31/314001. MR3371201

[113] A. I. Molev and E. E. Mukhin, *Eigenvalues of Bethe vectors in the Gaudin model*, Theor. Math. Phys., to appear; arXiv:1506.01884.

[114] A. I. Molev and E. Ragoucy, *The MacMahon master theorem for right quantum superalgebras and higher Sugawara operators for* $\widehat{\mathfrak{gl}}_{m|n}$ (English, with English and Russian summaries), Mosc. Math. J. **14** (2014), no. 1, 83–119, 171. MR3221948

[115] ———, *Classical* $\mathcal{W}$*-algebras in types* A, B, C, D *and* G, Comm. Math. Phys. **336** (2015), no. 2, 1053–1084, DOI 10.1007/s00220-014-2239-y. MR3322394

[116] A. I. Molev, E. Ragoucy, and N. Rozhkovskaya, *Segal-Sugawara vectors for the Lie algebra of type* G_2, J. Algebra **455** (2016), 386–401, DOI 10.1016/j.jalgebra.2016.03.009. MR3478867

[117] A. I. Molev and N. Rozhkovskaya, *Characteristic maps for the Brauer algebra*, J. Algebraic Combin. **38** (2013), no. 1, 15–35, DOI 10.1007/s10801-012-0388-7. MR3070117

[118] E. Mukhin, V. Tarasov, and A. Varchenko, *Bethe eigenvectors of higher transfer matrices*, J. Stat. Mech. Theory Exp. **8** (2006), P08002, 44. MR2249767

[119] Mircea Mustaţă, *Jet schemes of locally complete intersection canonical singularities*, Invent. Math. **145** (2001), no. 3, 397–424, DOI 10.1007/s002220100152. With an appendix by David Eisenbud and Edward Frenkel. MR1856396

[120] G. E. Murphy, *A new construction of Young's seminormal representation of the symmetric groups*, J. Algebra **69** (1981), no. 2, 287–297, DOI 10.1016/0021-8693(81)90205-2. MR617079

[121] Wakako Nakai and Tomoki Nakanishi, *Paths, tableaux and q-characters of quantum affine algebras: The* C_n *case*, J. Phys. A **39** (2006), no. 9, 2083–2115, DOI 10.1088/0305-4470/39/9/007. MR2211978

[122] Hiraku Nakajima, *t-analogs of q-characters of Kirillov-Reshetikhin modules of quantum affine algebras*, Represent. Theory **7** (2003), 259–274, DOI 10.1090/S1088-4165-03-00164-X. MR1993360

[123] Maxim Nazarov, *Young's orthogonal form for Brauer's centralizer algebra*, J. Algebra **182** (1996), no. 3, 664–693, DOI 10.1006/jabr.1996.0195. MR1398116

[124] ———, *Yangians and Capelli identities*, Kirillov's seminar on representation theory, Amer. Math. Soc. Transl. Ser. 2, vol. 181, Amer. Math. Soc., Providence, RI, 1998, pp. 139–163, DOI 10.1090/trans2/181/05. MR1618751

[125] ———, *Yangian of the queer Lie superalgebra*, Comm. Math. Phys. **208** (1999), no. 1, 195–223, DOI 10.1007/s002200050754. MR1729884

[126] Maxim Nazarov and Grigori Olshanski, *Bethe subalgebras in twisted Yangians*, Comm. Math. Phys. **178** (1996), no. 2, 483–506. MR1389915

[127] Maxim Nazarov and Vitaly Tarasov, *Representations of Yangians with Gelfand-Zetlin bases*, J. Reine Angew. Math. **496** (1998), 181–212, DOI 10.1515/crll.1998.029. MR1605817

[128] Andrei Okounkov, *Quantum immanants and higher Capelli identities*, Transform. Groups **1** (1996), no. 1-2, 99–126, DOI 10.1007/BF02587738. MR1390752

[129] A. Okun′kov and G. Ol′shanskiĭ, *Shifted Schur functions* (Russian, with Russian summary), Algebra i Analiz **9** (1997), no. 2, 73–146; English transl., St. Petersburg Math. J. **9** (1998), no. 2, 239–300. MR1468548

[130] Dmitri I. Panyushev, *The index of a Lie algebra, the centralizer of a nilpotent element, and the normalizer of the centralizer*, Math. Proc. Cambridge Philos. Soc. **134** (2003), no. 1, 41–59, DOI 10.1017/S0305004102006230. MR1937791

[131] D. Panyushev, A. Premet, and O. Yakimova, *On symmetric invariants of centralisers in reductive Lie algebras*, J. Algebra **313** (2007), no. 1, 343–391, DOI 10.1016/j.jalgebra.2006.12.026. MR2326150

[132] Dmitri I. Panyushev and Oksana S. Yakimova, *The argument shift method and maximal commutative subalgebras of Poisson algebras*, Math. Res. Lett. **15** (2008), no. 2, 239–249, DOI 10.4310/MRL.2008.v15.n2.a3. MR2385637

[133] A. M. Perelomov and V. S. Popov, *Casimir operators for* $\mathrm{U}(n)$ *and* $\mathrm{SU}(n)$, Soviet J. Nuclear Phys. **3** (1966), 676–680. MR0205620

[134] Mustapha Raïs and Patrice Tauvel, *Indice et polynômes invariants pour certaines algèbres de Lie* (French), J. Reine Angew. Math. **425** (1992), 123–140. MR1151316

[135] Sam Raskin, *A geometric proof of the Feigin-Frenkel theorem*, Represent. Theory **16** (2012), 489–512, DOI 10.1090/S1088-4165-2012-00417-4. MR2972556

[136] N. Yu. Reshetikhin, L. A. Takhtadzhyan, and L. D. Faddeev, *Quantization of Lie groups and Lie algebras* (Russian), Algebra i Analiz **1** (1989), no. 1, 178–206; English transl., Leningrad Math. J. **1** (1990), no. 1, 193–225. MR1015339

[137] Natasha Rozhkovskaya, *A new formula for the Pfaffian-type Segal-Sugawara vector*, J. Lie Theory **24** (2014), no. 2, 529–543. MR3235902

[138] Hebing Rui, *A criterion on the semisimple Brauer algebras*, J. Combin. Theory Ser. A **111** (2005), no. 1, 78–88, DOI 10.1016/j.jcta.2004.11.009. MR2144855

[139] L. G. Rybnikov, *The shift of invariants method and the Gaudin model* (Russian, with Russian summary), Funktsional. Anal. i Prilozhen. **40** (2006), no. 3, 30–43, 96, DOI 10.1007/s10688-006-0030-3; English transl., Funct. Anal. Appl. **40** (2006), no. 3, 188–199. MR2265683

[140] ———, *Uniqueness of higher Gaudin Hamiltonians*, Rep. Math. Phys. **61** (2008), no. 2, 247–252, DOI 10.1016/S0034-4877(08)80013-4. MR2424092

[141] Bruce E. Sagan, *The symmetric group*, 2nd ed., Graduate Texts in Mathematics, vol. 203, Springer-Verlag, New York, 2001. Representations, combinatorial algorithms, and symmetric functions. MR1824028

[142] Vadim V. Schechtman and Alexander N. Varchenko, *Arrangements of hyperplanes and Lie algebra homology*, Invent. Math. **106** (1991), no. 1, 139–194, DOI 10.1007/BF01243909. MR1123378

[143] T. A. Springer, *Linear algebraic groups*, 2nd ed., Progress in Mathematics, vol. 9, Birkhäuser Boston, Inc., Boston, MA, 1998. MR1642713

[144] H. Sugawara, *A field theory of currents*, Phys. Review **170** (1968), 1659–1662.

[145] D. V. Talalaev, *The quantum Gaudin system* (Russian), Funktsional. Anal. i Prilozhen. **40** (2006), no. 1, 86–91, DOI 10.1007/s10688-006-0012-5; English transl., Funct. Anal. Appl. **40** (2006), no. 1, 73–77. MR2223256

[146] A. A. Tarasov, *On some commutative subalgebras in the universal enveloping algebra of the Lie algebra* $\mathfrak{gl}(n, \mathbf{C})$ (Russian, with Russian summary), Mat. Sb. **191** (2000), no. 9, 115–122, DOI 10.1070/SM2000v191n09ABEH000509; English transl., Sb. Math. **191** (2000), no. 9-10, 1375–1382. MR1805600

[147] ———, *The maximality of some commutative subalgebras in Poisson algebras of semisimple Lie algebras* (Russian), Uspekhi Mat. Nauk **57** (2002), no. 5(347), 165–166, DOI 10.1070/RM2002v057n05ABEH000567; English transl., Russian Math. Surveys **57** (2002), no. 5. 1013–1014. MR1992097

[148] Tôru Umeda, *Newton's formula for* $\mathfrak{gl}_n$, Proc. Amer. Math. Soc. **126** (1998), no. 11, 3169–3175. DOI 10.1090/S0002-9939-98-04557-2. MR1468206

[149] È. B. Vinberg, *Some commutative subalgebras of a universal enveloping algebra* (Russian), Izv. Akad. Nauk SSSR Ser. Mat. **54** (1990), no. 1, 3–25, 221; English transl., Math. USSR-Izv. **36** (1991), no. 1, 1–22. MR1044045

[150] Akihito Wachi, *Central elements in the universal enveloping algebras for the split realization of the orthogonal Lie algebras*, Lett. Math. Phys. **77** (2006), no. 2, 155–168, DOI 10.1007/s11005-006-0082-6. MR2251303

[151] Minoru Wakimoto, *Fock representations of the affine Lie algebra* $A_1^{(1)}$, Comm. Math. Phys. **104** (1986), no. 4, 605–609. MR841673

[152] Hans Wenzl, *On the structure of Brauer's centralizer algebras*, Ann. of Math. (2) **128** (1988), no. 1, 173–193, DOI 10.2307/1971466. MR951511

[153] Hermann Weyl, *The Classical Groups. Their Invariants and Representations*, Princeton University Press, Princeton, N.J., 1939. MR0000255

[154] O. S. Yakimova, *The index of centralizers of elements in classical Lie algebras* (Russian, with Russian summary), Funktsional. Anal. i Prilozhen. **40** (2006), no. 1, 52–64, 96, DOI 10.1007/s10688-006-0005-4; English transl., Funct. Anal. Appl. **40** (2006), no. 1, 42–51. MR2223249

[155] Alexander B. Zamolodchikov and Alexey B. Zamolodchikov, *Factorized S-matrices in two dimensions as the exact solutions of certain relativistic quantum field theory models*, Ann. Physics **120** (1979), no. 2, 253–291, DOI 10.1016/0003-4916(79)90391-9. MR546461

[156] D. P. Želobenko, *Compact Lie groups and their representations*, American Mathematical Society, Providence, R.I., 1973. Translated from the Russian by Israel Program for Scientific Translations; Translations of Mathematical Monographs, Vol. 40. MR0473098

[157] D. P. Zhelobenko, *An introduction to the theory of S-algebras over reductive Lie algebras*, Representation of Lie groups and related topics, Adv. Stud. Contemp. Math., vol. 7, Gordon and Breach, New York, 1990, pp. 155–221. MR1104276

[158] ———, *Constructive modules and extremal projectors over Chevalley algebras* (Russian, with Russian summary), Funktsional. Anal. i Prilozhen. **27** (1993), no. 3, 5–14, 95, DOI 10.1007/BF01087533; English transl., Funct. Anal. Appl. **27** (1993), no. 3, 158–165. MR1250977

[159] ———, *Predstavleniya reduktivnykh algebr Li* (Russian, with English and Russian summaries), VO "Nauka", Moscow, 1994. MR1346309

Index

Selected Published Titles in This Series

229 **Alexander Molev,** Sugawara Operators for Classical Lie Algebras, 2018

224 **Ruy Exel,** Partial Dynamical Systems, Fell Bundles and Applications, 2017

223 **Guillaume Aubrun and Stanisław J. Szarek,** Alice and Bob Meet Banach, 2017

222 **Alexandru Buium,** Foundations of Arithmetic Differential Geometry, 2017

221 **Dennis Gaitsgory and Nick Rozenblyum,** A Study in Derived Algebraic Geometry, 2017

220 **A. Shen, V. A. Uspensky, and N. Vereshchagin,** Kolmogorov Complexity and Algorithmic Randomness, 2017

219 **Richard Evan Schwartz,** The Projective Heat Map, 2017

218 **Tushar Das, David Simmons, and Mariusz Urbański,** Geometry and Dynamics in Gromov Hyperbolic Metric Spaces, 2017

217 **Benoit Fresse,** Homotopy of Operads and Grothendieck–Teichmüller Groups, 2017

216 **Frederick W. Gehring, Gaven J. Martin, and Bruce P. Palka,** An Introduction to the Theory of Higher-Dimensional Quasiconformal Mappings, 2017

215 **Robert Bieri and Ralph Strebel,** On Groups of PL-homeomorphisms of the Real Line, 2016

214 **Jared Speck,** Shock Formation in Small-Data Solutions to 3D Quasilinear Wave Equations, 2016

213 **Harold G. Diamond and Wen-Bin Zhang (Cheung Man Ping),** Beurling Generalized Numbers, 2016

212 **Pandelis Dodos and Vassilis Kanellopoulos,** Ramsey Theory for Product Spaces, 2016

211 **Charlotte Hardouin, Jacques Sauloy, and Michael F. Singer,** Galois Theories of Linear Difference Equations: An Introduction, 2016

210 **Jason P. Bell, Dragos Ghioca, and Thomas J. Tucker,** The Dynamical Mordell–Lang Conjecture, 2016

209 **Steve Y. Oudot,** Persistence Theory: From Quiver Representations to Data Analysis, 2015

208 **Peter S. Ozsváth, András I. Stipsicz, and Zoltán Szabó,** Grid Homology for Knots and Links, 2015

207 **Vladimir I. Bogachev, Nicolai V. Krylov, Michael Röckner, and Stanislav V. Shaposhnikov,** Fokker–Planck–Kolmogorov Equations, 2015

206 **Bennett Chow, Sun-Chin Chu, David Glickenstein, Christine Guenther, James Isenberg, Tom Ivey, Dan Knopf, Peng Lu, Feng Luo, and Lei Ni,** The Ricci Flow: Techniques and Applications: Part IV: Long-Time Solutions and Related Topics, 2015

205 **Pavel Etingof, Shlomo Gelaki, Dmitri Nikshych, and Victor Ostrik,** Tensor Categories, 2015

204 **Victor M. Buchstaber and Taras E. Panov,** Toric Topology, 2015

203 **Donald Yau and Mark W. Johnson,** A Foundation for PROPs, Algebras, and Modules, 2015

202 **Shiri Artstein-Avidan, Apostolos Giannopoulos, and Vitali D. Milman,** Asymptotic Geometric Analysis, Part I, 2015

201 **Christopher L. Douglas, John Francis, André G. Henriques, and Michael A. Hill, Editors,** Topological Modular Forms, 2014

200 **Nikolai Nadirashvili, Vladimir Tkachev, and Serge Vlăduţ,** Nonlinear Elliptic Equations and Nonassociative Algebras, 2014

199 **Dmitry S. Kaliuzhnyi-Verbovetskyi and Victor Vinnikov,** Foundations of Free Noncommutative Function Theory, 2014

For a complete list of titles in this series, visit the
AMS Bookstore at **www.ams.org/bookstore/survseries/**.